DIGITAL SPECTRAL ANALYSIS

Second Edition

S. Lawrence Marple Jr.

DOVER PUBLICATIONS
Garden City, New York

Bibliographical Note

This Dover edition, first published in 2019, is a revised and updated second edition of *Digital Spectral Analysis with Applications*, originally printed by Prentice-Hall, Inc., Englewood Cliffs, New Jersey, in 1987. All the FORTRAN-related programs and support appendices have been replaced with narrative relating to MATLAB implementations and program use, all the problem sections have been deleted, and chapter appendices have either been deleted or recharacterized as "Extra" sections. MATLAB scripts and data files listed in the MATLAB Software tables are available at www.doverpublications.com/048678052x. MATLAB functions listed in the MATLAB Software tables are provided in *Digital Spectral Analysis MATLAB Software User Guide*.

Library of Congress Cataloging-in-Publication Data

Names: Marple, S. Lawrence, Jr., author.
Title: Digital spectral analysis / S. Lawrence Marple, Jr.
Description: Second edition. | Garden City, New York : Dover Publications 2019.
Identifiers: LCCN 2018045882 | ISBN 9780486780528 | ISBN 048678052X
Subjects: LCSH: Spectral theory (Mathematics) | Signal processing—Digital techniques.
Classification: LCC QA280 .M38 2019 | DDC 515/.7222—dc23
LC record available at https://lccn.loc.gov/2018045882

Manufactured in the United States of America
78052003
www.doverpublications.com

CONTENTS

CONTENTS

NOTATIONAL CONVENTIONS

- Parentheses enclose the argument of a continuous function.
  Example: $x(t)$

- Brackets enclose the argument of a discrete function.
  Example: $y[n]$

- Scalar variables and scalar functions are denoted by lowercase Latin and Greek letters.
  Examples: $\alpha \quad r_{xx}(\tau)$

- Transforms (Fourier, Z) are denoted by uppercase Latin and Greek letters. Transforms of scalar functions will use the corresponding uppercase letter for its transform.

Time Function	Transform
$x(t)$	$X(f)$
$y[n]$	$Y(z)$

  Exceptions: F—Frequency sampling interval (scalar)
  N—Number of samples (scalar)
  T—Time sampling interval (scalar)

- Vectors are denoted by lowercase bold Latin and Greek letters.
  Examples: $\mathbf{d} \quad \boldsymbol{\gamma}$

- Block vectors are denoted by underlined lowercase bold Latin and Greek letters.
  Examples: $\underline{\mathbf{c}} \quad \underline{\boldsymbol{\delta}}$

- Matrices are denoted by uppercase bold Latin and Greek letters.
  Examples: $\mathbf{A} \quad \boldsymbol{\Phi}$

- Block matrices are denoted by underlined uppercase bold Latin and Greek letters.
  Examples: $\underline{\mathbf{B}} \quad \underline{\boldsymbol{\Gamma}}$

- Special notation:
  T Vector and matrix tranposition
  H Vector and matrix complex conjugate transposition
  $*$ Scalar, vector, and matrix complex conjugation
  $\star$ Convolution operator
  $\hat{x}$ Hat denotes an estimate (of x in this case)

LIST OF KEY SYMBOLS

$a_p[k]$	Parameter of pth-order autoregression at time index k
$a_p^b[k]$	Coefficient of pth-order backward linear prediction at time index k
$a_p^f[k]$	Coefficient of pth-order forward linear prediction at time index k
ACS	Autocorrelation sequence
AIC	Akaike information criterion
AR	Autoregressive
ARMA	Autoregressive–moving average
B_e	Equivalent bandwidth in Hertz
B_s	Effective statistical bandwidth in Hertz
$b_q[k]$	Parameter of qth-order moving average at time index k
$c_{xx}[m]$	Autocovariance of discrete process $x[n]$ at lag m
$c_{xy}[m]$	Cross covariance of discrete processes $x[n]$ and $y[n]$ at lag m
CCS	Cross correlation sequence
cov$\{xy\}$	Covariance between random variables x and y
CTFS	Continuous-time Fourier series
CTFT	Continuous-time Fourier transform
det $\mathbf{A}$	Determinant of matrix $\mathbf{A}$
DTFS	Discrete-time Fourier series
DTFT	Discrete-time Fourier transform
E	Signal energy or squared error energy
$\mathcal{E}\{x[n]\}$	Expectation operation on random process $x[n]$ (ensemble average)
$\mathbf{e}(f)$	Complex sinusoid vector
$e_p^b[k]$	Backward linear prediction error at pth order
$e_p^f[k]$	Forward linear prediction error at pth order
ESD	Energy spectral density
$\mathcal{F}\{x(t)\} = X(f)$	Continuous Fourier transform of continuous-time signal $x(t)$
$\mathcal{F}^{-1}\{X(f)\} = x(t)$	Inverse continuous Fourier transform of $X(f)$
f	Frequency in Hertz
F	Frequency sampling interval in Hertz
FFT	Fast Fourier transform
h	Entropy rate
$\mathbf{I}$	Identity matrix
Im$\{x\}$	Imaginary part of complex variable x

j	Imaginary constant, $\sqrt{-1}$
$\mathbf{J}$	Reflection (exchange) matrix
$k_p = a_p[p]$	Reflection coefficient at order p
MA	Moving average
MEM	Maximum entropy method
MLE	Maximum likelihood estimate
MV	Minimum variance (spectrum)
mse$\{x\}$	Mean square error of random variable x
n	Discrete time or space index
N	Number of data samples
p	Autoregressive model order
$p(x)$	Probability density function of random variable x
$P_{xx}(f)$	Power spectral density
$P_{xy}(f)$	Cross power spectral density
$P_{xx}[m]$	Discrete power spectral density
PDF	Probability density function
PSD	Power spectral density
q	Moving average model order
Q	Statistical quality ratio of var/mean^2
$r_{xx}[m]$	Autocorrelation sequence at lag m
$r_{xy}[m]$	Cross correlation sequence at lag m
$\mathbf{R}_p$	Autocorrelation matrix composed of $r_{xx}[0]$ to $r_{xx}[p]$
Re$\{x\}$	Real part of complex variable x
s	Continuous space variable (units of ft, m, km, etc.)
S	Spatial sampling interval
$S_{xx}(f)$	Energy spectral density
$S_{xx}[m]$	Discrete energy spectral density
SVD	Singular value decomposition
t	Continuous time variable (units of seconds)
tr $\mathbf{A}$	Trace of matrix $\mathbf{A}$
T	Temporal sampling interval in seconds
$\mathbf{T}$	Toeplitz matrix
T_e	Equivalent time width in seconds
TBP	Time-bandwidth product
$\mathbf{v}_i$	Eigenvector i
var$\{x\}$	Variance of random variable x
$x[n]$	Discrete time (or space) function
$x(t)$	Continuous time (or space) function
$u[n]$	Driving noise random process
WSS	Wide-sense stationary
z	Complex variable
$\mathcal{Z}\{h[n]\} = \mathsf{H}(z)$	z-transform of $h[n]$
$\mathcal{Z}^{-1}\{\mathsf{H}(z)\} = h[n]$	Inverse z-transform of $\mathsf{H}(z)$
$\mathbf{0}_m$	Vector of m zero elements

$\delta(t)$	Continuous delta function
$\delta[n]$	Discrete delta function
λ_i	Eigenvalue i
ρ^b	Backward linear prediction squared error (variance)
ρ^f	Forward linear prediction squared error (variance)
ρ^{fb}	Forward plus backward linear prediction squared error (variance)
ρ_w	Variance of white noise process $w[n]$
$\uparrow\uparrow\uparrow(t)$	Sampling function

MATLAB SOFTWARE

FUNCTIONS (*.m)	DESCRIPTION	SEC.
ar_psd_MC	multichannel autoregressive (AR) spectral matrix	15.10.5
ar_psd_2D	two-dimensional autoregressive (AR) spectral matrix	16.7.5
arma	autoregressive-moving average parameter estimates	10.4
arma_psd	AR, MA, or ARMA spectrum	6.3
correlogram_psd	correlation-based spectral estimate	5.6
correlogram_psd_MC	multichannel correlation-based spectral matrix	15.5
correlogram_psd_2D	two-dimensional correlation-based spectral matrix	16.5
correlation_sequence	autocorrelation estimate	5.5
correlation_sequence_MC	multichannel correlation matrix estimate	15.5
correlation_sequence_2D	two-dimensional correlation matrix estimate	16.5
covariance_lp	covariance method linear prediction (LP) estimate	8.5.1
esd	energy spectral density for Prony method	11.7
fast_rls	time-recursive least squares AR parameter estimates	9.4.2
hermtoep_lineqs	Hermitian Toeplitz linear equations solution	3.8.4
lattice	lattice methods linear prediction estimates	8.4.2
lattice_MC	multichannel lattice linear prediction estimates	15.10.3
lattice_2D	two-dimensional lattice linear prediction estimates	16.7.6
levinson_recursion	linear prediction parameters from Toeplitz matrix	3.8.1
levinson_recursion_MC	multichannel block Toeplitz matrix LP parameters	15.8
levinson_recursion_2D	two-dimensional double Toeplitz matrix LP parameters	16.7.4

FUNCTIONS (*.m)	DESCRIPTION	SEC.
lms	time-recur. least-mean-square AR parameter estimates	9.3
lstsqs_prony	least squares Prony expon. parameter estimates	11.5
lstsqsvdm	least squares Vandermonde matrix solution	11.5
ma	moving average (MA) parameter estimates	10.3
minimum_eigenvalue	minimum eigen value/vector of Toeplitz matrix	3.8.5
minimum_variance_psd	minimum variance spectral estimate	12.5
minimum_variance_psd_MC	multichannel minimum variance spectral matrix	15.13
minimum_variance_psd_2D	two-dimensional minimum variance spectral matrix	16.9
modcovar_lp	modified covariance linear prediction estimates	8.5.2
noise_subspace	noise subspace eigenanalysis freq. estimates	13.6.2
periodogram_psd	Welch periodogram spectral estimation method	5.7.3
periodogram_psd_MC	multichannel periodogram spectral estimate	15.5
periodogram_psd_2D	two-dimensional periodogram spectral estimate	16.5
pisarenko	Pisarenko eigenanalysis frequency estimates	13.6.1
symcovar	symmetric covariance linear prediction estimates	11.6
toeplitz_lineqs	Toeplitz linear equations solution	3.8.4
vandermonde_lineqs	Vandermonde linear equations solution	3.9
yule_walker	Yule-Walker autoregressive spectral estimate	8.3
yule_walker_MC	multichannel Yule-Walker spectral matrix est.	15.10.1
yule_walker_2D	two-dimensional Yule-Walker spectral matrix est.	16.7.6

SCRIPTS (*.m)	DESCRIPTION	SEC.
plot_correlation	plot correlation estimates (radar & test1987)	5.5
plot_psd	plot one-dimensional spectral estimate	4.6
plot_psd_MC	plot two-channel auto & cross spectral estimates	15.2
plot_psd_2D	plot two-dimensional spectral matrix	16.2
plot_test_data	plot complex-values test data (radar & test1987)	1.3
spectrum_demo_2D	two-dimensional spectrum demonstration script	16.2
spectrum_demo_ar	autoregressive spectrum demonstration script	8.2
spectrum_demo_arma	ARMA spectrum demonstration script	10.2
specrum_demo_corr	correlation-based spectrum demonstration script	5.2
spectrum_demo_eigen	eigenanalysis frequency estimator demo script	13.2
spectrum_demo_ma	MA spectrum demonstration script	10.2
spectrum_demo_MC	two-channel spectral estimator demo script	15.2
spectrum_demo_minvar	minimum variance spectrum demo script	12.2
spectrum_demo_per	periodogram spectrum demonstration script	5.2
spectrum_demo_prony	Prony exponential estimator demo script	11.2
spectrum_demo_rls	time-recursive AR spectrum demo script	9.2

DATA (*.dat)	DESCRIPTION	SEC.
doppler_radar	simulated 64-sample complex radar data	Preface
sunspot_numbers	300 years of monthly average sunspot numbers	1.2
temperature	150 years St Louis monthly average temperatures	15.14
test1987	original 1987 text 64-sample complex test data	1.3

PREFACE

Since the publication of *Digital Spectral Analysis With Applications* in 1987 by Prentice Hall, which included a floppy disk of FORTRAN subroutines that implemented many spectral techniques presented in the text, there has been extensive feedback received by the author regarding the software. In particular, there have been many requests for more comprehensive software that includes not only additional implementations of the spectral techniques covered in the text, but also demonstration scripts for the spectral technique that (1) read in test data cases cited in the text, (2) process the data by a selected spectral technique, and (3) plot the resulting spectral estimate to yield graphics similar to those depicted in the text. The most frequent software request has been for implementations in MATLAB (trademark of The Math Works), since MATLAB has become a de facto standard for signal processing environments.

With this new edition by Dover Publications, the author is providing a large repertoire of MATLAB functions and MATLAB demonstration scripts, summarized in the table before this preface. The files for these may be found at www.doverpublications.com/048678052x.

Implementation in MATLAB has two additional benefits. First, MATLAB automatically adjusts for real-valued or complex-valued signal data, so that separate software routines to handle the real and complex cases are not needed. Second, MATLAB graphics are operating system independent and yield identical plotted output regardless of the PC or workstation used.

In the 1987 original text, two signal sampled-data records were provided for testing the spectral estimation techniques. One was the 64-sample complex-valued data set `test1987`. The second was the real-valued `sunspot_numbers` data. A third case has been created for this new edition that simulates the sampled signal of a narrow beam Doppler radar system. Such a system tracks moving objects in a rotating 360 degree field of view around the radar. A demodulated complex-valued radar pulse signal (see Sec. 2.12) is sampled, with each sample corresponding to a particular down range location related to the round trip propagation time of a radar pulse off a target. Since the radar is rotating, it is assumed that the radar beam allows only 64 complex samples to be collected per range location before the radar beam sweeps past a target. Any moving target with a velocity component in the direction of the radar will produce a Doppler shift in the transmission frequency. A spectrum analysis of the 64 signal samples after the receive down-conversion at a particular range will estimate this Doppler shift. The velocity can be estimated from the relationship

```
velocity=[doppler shift (Hz)/transmit frequency (Hz)]x[speed of light/2].
```

If the Doppler shift is positive (increases the return frequency), then the target is moving toward the radar. If the Doppler shift is negative (decreases the return frequency), then the target is moving away from the radar. For this test signal, the transmit frequency is 10 GHz and is 0 Hz after down conversion to a complex signal so that Doppler shifts actually represent positive and negative frequencies within this signal. The pulse rate (and therefore the range sample rate) is 2500 samples per second, which is needed to determine the actual Doppler shift frequencies.

The goal of this book is to provide a broad perspective of spectral estimation techniques and their implementation, with particular focus on finite data records. The practice of spectral estimation with finite data sets is not an exact science; a great deal of experimentation and subjective trade-off analysis is usually required with little statistical guidance on which to rely. Statistical analyses found in the literature typically relies on the *theory* of spectral estimation. This theory makes very restrictive assumptions about the nature of the data (e.g., the noise component of the signal is white and Gaussian) and usually applies only to the asymptotic case (the available data is permitted to grow to an infinite record length). The *practice* of spectral estimation relies more on empirical experimental observations rather than a theoretical basis. No position has been taken in this text on the relative superiority of any of the techniques. Instead, the text provides the reader with algorithms and their implementations necessary to calculate the various spectral estimators and highlights the trade-offs that should be considered for each method. Readers may then decide for themselves which methods are most appropriate for their application.

A motivation for the interest in alternative spectral estimators to classical means of spectral estimation is the apparent high-resolution performance that can be achieved for data sequences of very limited number of samples. A major emphasis in this text, therefore, is the behavior of each spectral estimator for short data records. A short data record is one in which the spectral resolution required is of the same order as the reciprocal of the signal record length (that is, the measurement time aperture). Another emphasis area of this text is that of fast computational algorithms for the solution of the matrix equations associated with each spectral estimator. A fast algorithm is a computational procedure that exploits the structure of a matrix equation to solve it in a more computationally efficient manner than can be achieved by a direct brute force solution which does not exploit the structure. The fast Fourier transform (FFT) is a classic example of a fast algorithm with which most readers may already have some familiarity. Just as the FFT algorithm made classical spectral analysis practical for processing large data records, fast algorithms in this text have also been developed to reduce the computational burden of the alternative spectral estimation methods presented, which makes them more attractive to use.

Prerequisites for this text include course material in discrete-time linear system and transform theory, such as that found in *Digital Signal Processing* by Oppenheim and Schafer (Prentice-Hall, 1989), introductory probability and statistics, such as that found in *Probability, Random Variables, and Stochastic Processes, Third Edition* by Papoulis (McGraw-Hill, 1991), and matrix algebra, such as that found in *Applied Linear Algebra, Second Edition* by Noble and Daniel (Prentice-Hall, 1977) or *Matrix Computations* by Golub and Van Loan (Johns Hopkins University Press, 1989). The chapters have been organized to facilitate a rapid implementation of most spectral estimation techniques without an in-depth investment in the mathematics of the technique. Concise summaries are provided at the beginning of each chapter for each spectral estimation technique. The intent of the text organization is to enable minimal implementation effort for those interested only in an evaluation of a technique for their application. Theoretical details follow each summary for those interested and may be read after the reader has gained some familiarity with a selected spectral estimator by exercising the software.

Following a historical introduction in Chap. 1, Chaps. 2–4 provide the necessary theoretical background and review material in linear systems, Fourier transforms, matrix algebra, random processes, and spectral statistics. Chapter 5 summarizes classical spectral estimation as it is practiced today. Chapter 6 provides the overview for parametric modeling approaches to spectral estimation. Algorithms for the parametric techniques of autoregressive (AR), moving average (MA), and autoregressive–moving average (ARMA) spectral estimation may be found in Chaps. 7–10. Exponential modeling by Prony's method, a close relative of AR spectral estimation, is presented in Chap. 11. Nonparametric approaches to spectral estimation include the minimum variance method (Chap. 12) and the frequency estimators based on eigenanalysis (Chap. 13). Chapter 14 summarizes all of the methods in Chaps. 5–13. It may serve as a starting place for those readers interested in being guided to a spectral estimation technique for their application. Extensions of many of the methods presented in Chaps. 5–13 to multichannel and two-dimensional spectral estimation are developed in Chaps. 15 and 16, respectively.

Great care has been taken to ensure that all spectral estimator expressions have the proper scaling factors to yield power spectral density in proper units of power per Hertz. The scaling, which usually involves a dependence on the sample interval and the number of samples, is often omitted in textbooks. These books choose normalized attributes (e.g., unit variance noise and unit sampling interval) for the spectral estimators to simplify the mathematical expressions. Important scaling dependencies on the physical attributes therefore are missed.

S. Lawrence Marple Jr.
Fort Worth, Texas

1

INTRODUCTION

Spectral analysis is any signal processing method that characterizes the frequency content of a measured signal. The Fourier transform is the mathematical foundation for relating a time or space signal, or a model of this signal, to its frequency-domain representation. Statistics play a significant role in spectral analysis because most signals have a noisy or random aspect. If the underlying statistical attributes of a signal were known exactly or could be determined without error from a finite interval of the signal, then spectral analysis would be an exact science. The practical reality, however, is that only an *estimate* of the spectrum can be made from a single finite segment of the signal. As a result, the practice of spectral analysis since the 1880s has tended to be a subjective craft, applying science but also requiring a degree of empirical art.

The difficulty of the spectral estimation problem is illustrated by Fig. 1.1. Two typical spectral estimates are shown in this figure, obtained by processing the same finite sample sequence by two different spectral estimation techniques. Each spectral plot represents the distribution of signal strength with frequency. A precise meaning of signal "strength" in terms of energy or energy per unit time (power) will be provided in Chaps. 2 and 4. The units of frequency, as adopted in this text, are either cycles per second (Hertz) for temporal signals or cycles per meter (wavenumber) for spatial signals. Signal strength $P(f)$ at frequency f will be computed as $10 \log_{10}[P(f)/P_{max}]$ and plotted in units of decibels (dB) relative to the maximum spectral strength P_{max} over all frequencies. The maximum relative

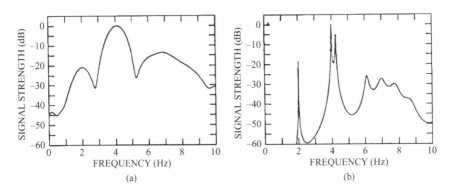

Figure 1.1. Two different spectral estimates produced from the same measured data.

1

strength plotted by this approach will, therefore, be 0 dB. The significant differences between the two spectral estimates may be attributed to differing assumptions made concerning the nature of the data and to the type of averaging used in recognition of the statistical impact of noise in the data. In a situation where no *a priori* knowledge of signal characteristics is available, one would find it difficult to select which of the two spectral estimators, if either, has represented the true underlying spectrum with better fidelity. It appears that estimate 1.1(b) has higher resolution than estimate 1.1(a), but this could be an artifact of the processing used to generate estimate 1.1(b), rather than actual detail that exists in the spectrum. This is the kind of uncertainty that arises in practice and that illustrates the subjective nature of spectral analysis.

Classical methods of spectral estimation have been well documented by several texts; the books by Blackman and Tukey [1958] and Jenkins and Watts [1968] are probably the two best known. Since the publication of these and related texts, interest has grown for devising alternative spectral estimation methods that perform better for limited data records. In particular, new methods of spectral estimation have been promoted that yield apparent improvements in frequency resolution over that achievable with classical spectral estimators. Limited data records occur frequently in practice. For example, to study intrapulse modulation characteristics within a single short radar pulse, only a few time samples may be taken from the finite duration radar pulse. In sonar, many data samples are available, but target motion necessitates that the analysis interval be short in order to assure that the target statistics are effectively unchanging within the analysis interval. The emphasis in this book is on the new, or "modern," methods of spectral estimation. In this sense, this text supplements the classical spectral estimation material covered in earlier texts. All methods described in this text assume sampled digital data, in contrast to some earlier texts that considered only continuous data.

The intent of each chapter is to provide the reader with an understanding of the assumptions that are made concerning the method or methods. The beginning of each chapter summarizes the spectral estimator technique, or techniques, and the associated software covered in that chapter, enabling scientists and engineers to implement readily each spectral estimator without being immersed in the theoretical details of the chapter. Some guidelines for applications are also provided. No attempt has been made to rank the spectral estimation methods with respect to each other. This text references a variety of MATLAB spectral estimation programs; users should probably apply several of the techniques to their experimental data. It may then be possible to extract a better understanding of the measured process from features common to all of the selected spectral estimates. For generality, complex-valued signals are assumed. The use of complex-valued signals is becoming more commonplace in digital signal processing systems. Section 2.12 describes two common sources of complex-valued signals.

An illuminating perspective of spectral analysis may be obtained by studying its historical roots. Further insight may be developed by examining some of the issues of spectral estimation. Both topics are covered in the remaining sections of this chapter. A brief description of how to use this text concludes the chapter.

1.1 HISTORICAL PERSPECTIVE

Cyclic, or recurring, processes observed in natural phenomena have instilled in humans, since the earliest times, the basic concepts that are embedded to this day in modern spectral estimation. Without performing an explicit mathematical analysis, ancient civilizations were able to devise calendars and time measures from their observations of the *periodicities* in the length of the day, the length of the year, the seasonal changes, the phases of the moon, and the motion of other heavenly bodies such as planets. In the sixth century BC, Pythagoras developed a relationship between the periodicity of pure sine vibrations of musical notes produced by a string of fixed tension and a number representing the length of the string. He believed that the essence of harmony was inherent in numbers. Pythagoras extended this empirical relationship to describe the *harmonic* motion of heavenly bodies, describing it as the "music of the spheres."

The mathematical basis for modern spectral estimation has its origins in the seventeenth-century work of the scientist Sir Isaac Newton. He observed that sunlight passing through a glass prism was expanded into a band of many colors. Thus, he discovered that each color represented a particular wavelength of light and that the white light of the sun contained all wavelengths. It was also Newton who introduced [1671] the word *spectrum* as a scientific term to describe this band of light colors. Spectrum is a variant of the Latin word "specter," meaning image or ghostly apparition. The adjective associated with spectrum is *spectral*. Thus, *spectral* estimation, rather than *spectrum* estimation, is the preferred terminology. Newton presented in his major work *Principia* [1687] the first mathematical treatment of the periodicity of wave motion that Pythagoras had empirically observed.

The solution to the wave equation for the vibrating musical string was developed by Daniel Bernoulli [1738], a mathematician who discovered the general solution for the displacement $u(x, t)$ of the string at time t and position x (the endpoints of the string are at $x = 0$ and $x = \pi$) in the wave equation to be

$$u(x, t) = A_0 + \sum_{k=1}^{\infty} \sin kx \, (A_k \cos kct + B_k \sin kct), \qquad (1.1)$$

where c, a physical quantity characteristic of the material of the string, represents the velocity of the traveling waves on the string. The term A_0 is normally zero and we will assume this here. The mathematician L. Euler [1755] demonstrated that the coefficients A_k and B_k in the series given by Eq. (1.1), which would later be called the *Fourier series*, were found as solutions to

$$A_k = \frac{2}{\pi} \int_0^{\pi} u(x, 0) \sin kx \, dx$$

$$B_k = \frac{2}{\pi} \int_0^{\pi} u(x, 0) \cos kx \, dx \qquad (1.2)$$

for which $t_0 = \pi/2kc$. The French engineer Jean Baptiste Joseph Fourier in his thesis *Analytical Theory of Heat* [1822] extended the wave equation results by asserting that any arbitrary function $u(x)$, even one with a finite number of discontinuities, could be represented as an infinite summation of sine and cosine terms,

$$u(x) = \sum_{k=1}^{\infty} (A_k \cos k\alpha x + B_k \sin k\alpha x). \qquad (1.3)$$

The mathematics of taking a function $u(x)$, or its samples, and determining its A_k and B_k coefficients has become known as *harmonic analysis*, due to the harmonic indexing of the frequencies in the sine and cosine terms.

Beginning in the mid-nineteenth century, practical scientific applications using harmonic analysis to study phenomenological data such as sound, weather, sunspot activity, magnetic deviations, river flow, and tidal variations were being made. In many of these applications, the fundamental period was either obscured by measurement error noise or was not visually evident. In addition, secondary periodic components that bore no harmonic relationship to the fundamental periodicity were often present. This yielded some problems with the estimates of the various periodicities. Manual computation of the Fourier series coefficients by direct computational techniques or by graphic-aided methods proved to be extremely tedious and were limited to very small data sets. Mechanical harmonic analyzers were developed to assist the analysis. These calculating machines were basically mechanical integrators, or planimeters, because they found the area under the curves $u(x) \sin kx$ and $u(x) \cos kx$ over the interval $0 \leq x \leq \pi$, thereby providing a calculation of the Fourier series coefficients A_k and B_k. The British physicist Sir William Thomson (aka Lord Kelvin, for whom a temperature scale was named) developed the first mechanical harmonic analyzer based on a dual-function-product planimeter, $\int u(\theta)\phi(\theta)\,d\theta$, invented by his brother James Thomson and modified to evaluate cosine and sine functions [1876, 1878]. Figure 1.2(a) and (e) illustrates versions of this device. A tracing point was guided manually along the plotted curve to be analyzed; the coefficients were then read from the integrating cylinders. A different integrating cylinder was required to evaluate each coefficient up to only the third harmonic. It was used by the British Meteorological Office to analyze graphical records of daily changes in atmospheric temperature and pressure. Observers said of this device that, due to its size and weight, it was practically a permanent fixture in the room where it was used. Improvements in harmonic analyzers were subsequently made by O. Henrici [1894] [see Fig. 1.2(b)], A. Sharp [1894], G. U. Yule [1895], and the American physicists Albert A. Michelson (who is more famous for his measurement of the speed of light) and S. W. Stratton [1898]. The Michelson-Stratton harmonic analyzer [Fig. 1.2(c)], designed using spiral springs, was particularly impressive in that it not only could perform the analysis of 80 harmonic coefficients simultaneously, but it also could work as a synthesizer (inverse Fourier transformer) to construct the superposition of the Fourier series components. Michelson used the machine in his Nobel Prize-winning optical studies. As a synthesizer, the machine

Fourier Transformer – 1 term (1876)
Thomson (Lord Kelvin) Harmonic
Analyzer
(a)

Fourier Transformer – 7 terms (1878)
Mechanical Parallel Processor
(b)

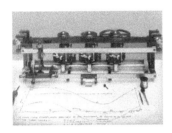

**Fourier Transformer –3 terms
(1894)**
Henrici-Coradi Harmonic
Analyzer
(c)

**Inverse Fourier
Transformer –15 terms
(1890)**
Thomson's Tide Predictor
(d)

**Fourier and Inverse
Fourier Transformer –
81 terms (1898)**
Michelson-Stratton machine
(e)

Figure 1.2. Nineteenth-century mechanical harmonic analyzers and synthesizers (forward and inverse Fourier transformers, 1876–1890). Photos courtesy The Science Museum, London.

could predict interference fringe patterns by representing them as simple harmonic curves. As an analyzer, it decomposed a visibility curve into harmonic components representing the harmonic distribution of light in a source.

The results from a harmonic analysis in this era were sometimes used to synthesize a periodic waveform from the harmonic components for purposes of prediction (a Fourier series model of a data sequence). One of the earliest uses was for tidal prediction. Using direct manual calculation, Sir William Thomson performed harmonic analyses of tidal gauge observations from British ports starting in 1866; by 1872 he had developed a tide-predicting machine that utilized the coefficients estimated from his harmonic analysis. Later versions of this machine [see Fig. 1.2(d)] could combine up to 10 harmonic tidal constituents, each a function of the port where tides were to be predicted, by initializing the machine by crank and pulley settings, as Thomson's tide predictor was a rather large machine, with a base 3 feet by 6 feet. It took approximately four hours to draw one year of tidal curves for one harbor. A tide predictor built by William Ferrel in 1882, now on display in the Smithsonian Museum in Washington, DC, was used by the U.S. Coast and Geodetic Survey to prepare tide tables from 1883 to 1910.

Although the mechanical harmonic analyzers were useful for evaluating time series with *obvious* periodicities (smoothly varying time sequences with little or no noise), numerical methods of harmonic analysis (fitting a Fourier series) were still required when evaluating very noisy data (described in the early literature as data with "irregular fluctuations") for possible *hidden* periodicities (nonobvious periodic signals), or evaluating signals with nonharmonically related periods. Of the many scientists who used harmonic analysis, Schuster [1897, 1898, 1900, 1905, 1906] made the most profound impact on what has become the classic spectral estimation technique. He suggested that a plot of the squared envelope $S_k = A_k^2 + B_k^2$ of the Fourier transform coefficients (first proposed by Stokes [1879])

$$A_k = \frac{2}{nT_0} \int_\tau^{\tau + nT_0} u(t) \cos kt \, dt$$

$$B_k = \frac{2}{nT_0} \int_\tau^{\tau + nT_0} u(t) \sin kt \, dt \qquad (1.4)$$

be computed over a range of n integral periods of T_0, where the correspondence $k = 2\pi n / T_0$ should be made. Schuster's notation has been used here. Schuster termed his method the *periodogram* [1898]. The periodogram, in concept, could be evaluated over a continuum of periods (the inverse of which are frequencies). In his papers, Schuster recognized many of the problems and peculiarities of the periodogram. Depending on the choice of starting time τ, he observed that irregular and different patterns were obtained, sometimes producing spurious peaks (he called them "accidental periodicities") where no periodicity truly existed. Schuster [1894, 1904] knew from his understanding of Fourier analysis of optical spectra that averaging of S_k, obtained by evaluation of different data segments (with the period T_0 fixed), was necessary to smooth the periodogram (obtain the "mean periodogram" in Schuster's jargon) and remove the spurious parts of the spectrum. Although he recognized the need to average, the implementation required computational means beyond the resources available in Schuster's era. Quoting Schuster [1898, p. 25],

"The periodogram as defined by the equations [1.4] will in general show an irregular outline, and also depend on the value of τ. In the optical analysis of light we are helped by the fact that the eye only receives the impression of the average of a great number of adjacent periods, and also the average, as regards time, of the intensity of radiation of any particular period .... If we were to follow the optical analogy we should have to vary the time τ ... continuously and take the average value of $r = \sqrt{A^2 + B^2}$ obtained in this way for *each* value of k ... but this would involve an almost prohibitive labor."

A thorough theoretical understanding of the statistical basis for averaging was still thirty years away in the work of Wiener, and fifty years from practical implementation of spectral averaging methods based on fast Fourier transform algorithms and digital computers to significantly reduce the "prohibitive" computational burden.

Schuster was also aware of the sidelobes (he called them "spurious periodici-ties") around mainlobe responses in the periodogram that are inherent in all Fourier analysis of finite record lengths. His cognizance of sidelobes was due to his ability to make an analogy with the diffraction patterns of the optical spectroscope caused by its limited spatial aperture ("limited resolving power"). Schuster pointed out that many researchers in his day were incorrectly asserting that all maxima in the periodogram were hidden periodicities when, in fact, many were sidelobes rather than true periodicities. In addition to the spurious periodicities in the periodogram, Schuster was also aware of the estimation bias introduced into the periodogram esti-mate when the measurement interval was not an exact integer multiple of the period under analysis. Many scientists in Schuster's day thought that white light might be a close grouping of monochromatic line components (analogous to white noise being a grouping of sinusoidal frequencies), but Schuster was able to show empirically that white light was a continuum of frequencies. Wiener was later able to extend this white light analogy to a statistical white noise process.

Schuster applied the periodogram to find hidden periodicites in meteorological, magnetic declination, and sunspot number data. The periodogram analysis of the sunspot numbers [Schuster, 1906] is of particular interest as these numbers will be used as a test case in this book. Schuster performed preprocessing averages of the monthly sunspot numbers for the years 1749 to 1894, followed by a periodogram analysis. The periodogram analysis yielded an estimate of 11.125 years for the basic sunspot cycle. This is the basis for the classic 11-year sunspot cycle often cited in astronomical literature.

Conceptually, a time series consisting of a sinusoid of frequency f_0 Hz with "superposed irregular fluctuations" (additive noise) should show a peak in the peri-odogram at a period $T_0 = 1/f_0$. However, many researchers in the early part of this century found that periodograms computed using noisy data were very erratic and did not show any dominant peaks that could confidently be considered periodicities in the data. This was true even when the data interval was increased. Examples of such periodograms are illustrated in Fig. 1.3. As more and more data samples are used, the periodogram fluctuates more and more. This behavior led to a cer-tain amount of disenchantment with the periodogram for several decades, which is unfortunate because most users were ignorant of the averaging considerations sug-gested by Schuster. Slutsky [1927], and independently Daniell [1946], observed that the fluctuations in periodograms of white noise were of the same magnitude as the mean value of the periodogram itself. The fluctuations tended to be uncorrelated from frequency to frequency, irrespective of the length of the time-series record available for analysis. Both suggested that the periodogram fluctuations could be reduced by averaging the periodogram over neighboring frequencies. This concept is the basis for the Welch method of smoothing the periodogram (see Chap. 5).

The disenchantment with the periodogram led the British statistican G. Yule to introduce in 1927 a notable alternative analysis method. Yule's idea was to model a time series with linear regression analysis in order to find one or two periodicities in the data [Yule, 1927]. His main interest was to obtain a more accurate determination

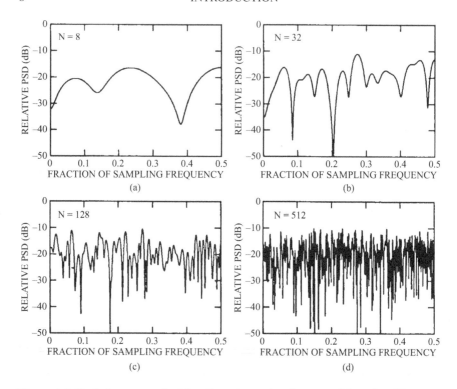

Figure 1.3. Periodograms of white Gaussian noise for record lengths. The term PSD stands for power spectral density. (a) $N = 8$. (b) $N = 32$. (c) $N = 128$. (d) $N = 512$. An increase in spectral fluctuations, rather than a tendency to a flat spectrum, results when the record length is increased.

of the primary periodicity in sunspot numbers, and to search for additional periodicities in the data. Yule felt that the "superposed irregular fluctuations" (additive noise) hypothesis of the Schuster periodogram might not hold in the case of sunspot numbers. Yule suggested that sunspot numbers might better be described by a different time-series model of the physical phenomenon, specifically a recursive harmonic process driven by a noise process, or "disturbances" as Yule called the individual noise samples. Using a simple trigonometric identity

$$\sin(kx) = 2\cos(x)\sin([k-1]x) - \sin([k-2]x)$$

with the substitution $x = 2\pi f T$, a symmetric homogeneous difference equation that describes a single discrete time harmonic variation may be written as

$$u(k) - au(k-1) + u(k-2) = 0,$$

in which $u(k) = \sin(2\pi f k T)$ is the harmonic component, T is the sample interval, f is the frequency of the harmonic variation, and $a = 2\cos(2\pi f T)$ is a coefficient characteristic of the harmonic. Yule conjectured that if the sunspot numbers had only one periodic component, then the sunspot number sequence could be generated by the process

$$u(k) = au(k-1) - u(k-2) + \epsilon(k), \qquad (1.5)$$

where $\epsilon(k)$ was some small random impulsive "disturbance" at each time index k. Yule called this the "harmonic curve equation." Thus, Yule devised the basis of what has become known as the parametric approach to spectral analysis: characterizing data measurements as the output of a recursive time-series model. Yule determined the parameter a via a least squares analysis using the 1749–1924 yearly mean sunspot numbers with the sample mean removed. He estimated that $a = 1.62374$, from which, for $T = 1$ year, he estimated a period $1/f = 10.08$ years per sunspot cycle. Not being satisfied with this number, he then filtered the sunspot numbers by taking a 3-year moving average over the data; a second least squares analysis of this filtered sunspot number sequence yielded a period of $1/f = 11.43$ years per cycle. Yule then tried to determine if there might be two periodicities in the sunspot numbers by least squares fitting a fifth-order symmetric difference equation to the data. This yielded periods of 11.95 and 1.42 years per sunspot cycle. However, he discounted a two-period model because the squared error went up, rather than down, for the two-period model.

Still not satisfied with his estimates, Yule took a cue from the structure of the "harmonic curve equation" and decided to generalize the difference equation approach and use least squares regression analysis with the model

$$u(k) = b(1)u(k-1) + b(2)u(k-2) + \epsilon(k) \qquad (1.6)$$

for arbitrary $b(1)$ and $b(2)$. This forms a process called an *autoregression* (u regressed on itself), and marks the first time that a least squares autoregression was performed for the purpose of spectral analysis (i.e., frequency determination). The least squares solution to the regression equation (1.6) is a damped sinusoid. Yule found, using Eq. (1.6), a damped sinusoidal period of 10.60 years for the unfiltered sunspot numbers, and 11.164 years for the filtered data. Although Yule could not conjecture what, if any, physical meaning to attach to the damping, he felt that the period estimates were still reasonable. Walker [1931] also used the technique of Yule to investigate a damped sinusoidal time series. The normal equations arising out of the least squares analysis have been called the *Yule-Walker equations*, in honor of these pioneers of regression modeling (these have also been called the Yule-Kendall equations by Bartlett [1948]). It is interesting to note that Walker, as well as Alter [1927], also anticipated the empirical spectral estimation techniques later espoused by Tukey. In his 1931 paper (p. 524), Walker states:

"Some light is thrown by this [regression] analysis on the utility of a series of values of the serial correlation coefficients, a *correlation-periodogram*, as a substitute for the ordinary Fourier-Schuster periodogram when there is no question of damped oscillations."

The methods used by Yule are reminiscent of another, much older data-fitting procedure by Baron de Prony [1795]. Prony discovered a method of fitting an exponential model to a set of data representing pressure and volume relationships of gases. Prony's procedure involved an exact, rather than least squares, regression fit to the data. The regression coefficients were used as coefficients for a polynomial, whose roots became the model exponents. The amplitude of each exponential component was then found by a second pass through the data.

The year 1930 marked a major turning point for spectral analysis. Norbert Wiener published his classic paper *Generalized Harmonic Analysis* [1930]. This paper placed spectral analysis on a firm statistical foundation in its treatment of random processes. Several significant contributions were made. Precise statistical definitions of *autocorrelation* and *power spectral density* (PSD) were given for stationary random processes. These two functions of a random process were shown to be related via a continuous Fourier transform, which is the basis of what is known today as the Wiener-Khintchine theorem (Khintchine [1934], who made related contributions, is jointly honored). It has been recently recognized that Einstein may also have anticipated in 1914 a similar Fourier transform relationship between the autocorrelation and the power spectral density [Einstein, 1914; Gardner and Yaglom, 1987]. The use of the Fourier transform, rather than the Fourier series of traditional harmonic analysis, enabled Wiener to define spectra in terms of a *continuum* of frequencies, rather than as *discrete* harmonic frequencies. White noise was shown by Wiener to have a uniform PSD, that is, to be composed of all frequencies equally, a result he deduced from his studies of Brownian motion and from Schuster's optical analogies. Other early contributions to the statistical characterization of spectral analysis include Bartlett [1946, 1948, 1950] and Kendall [1945].

The regression equation of Yule may alternatively be viewed as a prediction of a signal by a linear combination of past samples of that signal plus a noise contribution, thereby constituting a so-called *linear prediction* problem. Building on the probabilistic framework for time series developed by Khintchine and Slutsky, Wold [1938] developed a unified stochastic linear difference equation model for discrete-time series. Wold introduced the terms *moving average* for time-series models originally described by Slutsky (who used the term "moving summation"), and *linear autoregression* for time-series models originally described by Yule. He was also the first to describe the relationship between the autoregressive parameters and the autocorrelation sequence as the "Yule-Walker equations." In his 1938 monograph, Wold also introduced a very important decomposition theorem for stationary time series. This theorem asserted that any stationary random process could be expressed as the sum of a deterministic component and a one-sided moving average process of possibly infinite order driven by white noise. This theorem led Kolmogoroff [1939] to the formulation and solution of the linear prediction problem. Computing a spectrum from the autoregressive coefficients was suggested as early as 1948 by Bartlett [1948, p. 686], who used a second-order autoregressive power spectral density. The equations of the linear prediction problem and the Yule-Walker equations have a special structure first studied by the German mathematician O. Toeplitz [1911]. This structure was exploited by N. Levinson, a colleague of Wiener, to develop a very efficient computational procedure to solve the Yule-Walker equations [Levinson, 1947]. Extensions of the linear prediction concept to digital seismic data was promoted by the Geophysical Analysis Group (GAG) at MIT during the 1950s. This work, particularly that of Enders Robinson [1954], made significant contributions to the modern digital methods of spectral analysis and signal processing.

Whereas Wiener might be considered the father of modern *theoretical* spectral analysis, John Tukey might be considered the father of modern *empirical* spectral analysis. In a paper at Woods Hole, Massachusetts, in June 1949, Tukey laid out an empirical foundation for spectral estimation using correlation estimates produced from finite time sequences. His procedure also included proper data collection techniques and estimates of the approximate statistical distribution of the spectral estimator. A classic treatise describing this procedure in detail was published with Blackman in 1958. Many of the terms of modern spectral estimation (such as "aliasing," "windowing," "prewhitening," "tapering," "smoothing," "decimation") can be attributed to Tukey.

The next key contribution to digital spectral analysis came in the form of a very efficient computational algorithm to solve the discrete Fourier transform, a version of the Fourier transform applicable to digital data processing. Although there is evidence of contributions by many researchers to the development of a fast Fourier transform (FFT) [Heideman et al., 1984], the concise paper by Jim Cooley and John Tukey [1965] focused the attention of the digital signal processing community on the practicality of efficient Fourier transform computation. The FFT, perhaps more than any other contribution, has led to the increased application of digital spectral analysis as a signal processing tool.

The prime catalyst for the current interest in high-resolution spectral estimation from limited time or space records may be attributed to the work of John Burg [1967]. His high-resolution spectral estimator, which he described in the context of a maximum entropy mathematical formalism, has been instrumental in the development of parametric, or modeling, approaches to high-resolution spectral estimation. The maximum entropy spectrum is very closely related to autoregressive spectral analysis (see Chap. 7). Both Bartlett [1948] and Parzen [1957] had suggested the use of autoregressive spectral estimation, but not until Burg's work did interest in this method receive significant attention.

The development of fast computational algorithms continues to have an impact on newer methods of spectral estimation, much as the FFT has impacted classical spectral estimators. Fast algorithms make real-time implementation feasible for many of the newer spectral estimation methods. Fast algorithms for fitting least squares linear parametric models to data samples have been a topic of research since the 1970s. Additional historical material on spectral estimation may be found in a survey paper by Robinson [1982].

1.2 SUNSPOT NUMBERS

Due to their historic interest for spectral estimation, the sunspot number series will also be used as a test example for the spectral estimation methods presented in this text. Some discussion of their origin is therefore appropriate. Sunspots have been observed with the naked eye since at least 300 AD, as they are mentioned in old Chinese records of this era. Telescopic observations of sunspots have been made in Europe since 1610 AD. Modern systematic measurements of sunspots began in 1835. By 1843, Hofrat Heinrich Schwabe had discovered the basic cyclic nature between minimum and maximum numbers of spots on the sun's surface.

In order to quantify the results of the observations, the *relative* sunspot number R was introduced by Rudolf Wolf [1868] in 1848 as a measure of sunspot activity. The value of R attempts to account for the fact that sunspots tend to appear in groups. The subjective determination of R is strictly by manual observation on a daily basis. Two separate measurements are actually taken. The first is a count s of the total number of individual spots, regardless of their size. The second is a count g of the numbers of sunspot groups. The original Wolf relative sunspot number was simply a weighted combination $R = 10g + s$. This arbitrary weighting of the two counts reflected the relative importance of a new sunspot. A greater increase in activity is observed when a new spot appears in a region which had been spotless before (thus forming a new group) than in a region which already had an existing group. Note that no measurement of sunspot *area* is made, which would seem to be a more sensitive measure. This is a reflection of the limitations in instrumentation in Wolf's day. However, later analysis has shown that the *yearly* means of R are almost linearly proportional to the *yearly* spot area means.

Because 10–50 observers are typically used to obtain a single daily *average* relative sunspot number, some adjustment for the type of observation equipment (for example, magnification power), geographic location, locale (for example, typical atmospheric and weather conditions), and observer (his counting method and personal judgment) must be made. Each observer's R value is therefore weighted by $k(10g + s)$, where k is an observer scaling factor specific to the observatory's locale and equipment. The Wolf baseline instrument and location provides the reference value $k = 1$. It may be seen that the relative sunspot numbers are quite subjective. The factor k is a variable that must be determined empirically for each observer. The daily value of R is a function of the number of observers actually reporting. Weather conditions and days off taken by the observers may cause fluctuations in the daily sunspot numbers.

Although initiated in 1848, Wolf researched archives to find estimates for the relative sunspot numbers for years prior to 1848. Daily records are now available back to the year 1818, estimates of monthly means have been made back to 1749, estimates of annual means have been made back to 1700, and the epochs of maximum and minimum points in sunspot activity have been noted in recorded observations since 1610. Wolf assumed an 11.1-year cycle to guide his back-dated estimates of sunspot numbers. Current research has produced estimates of annual sunspot numbers back to 1500 AD based on correlation of sunspot activity with terrestrial meterological data, climate variations (droughts and famines), radiocarbon chronology, dendrochronology (tree-ring studies), and paleomagnetic investigations. Variations in mean annual temperatures and mean annual magnetic declinations have been found, for example, to have an approximate 19- to 22-year cycle, roughly twice that of the basic sunspot cycle.

The monthly means of R from 1700 to 1984 are in file sunspot_numbers.dat at the website cited in the preface. These are plotted in Fig. 1.4. The Sunspot Index Data Center in Brussels, Belgium, has continued Wolf's numbers to the present and is the international source for R.

The key to reasonable and meaningful sunspot numbers is averaging. A large number of observers from a variety of geographical locations are used to obtain a daily average. The daily average is probably not sufficient averaging, as Fig. 1.5 illustrates. This figure plots the daily numbers for a typical month, comparing the international relative sunspot numbers (R_I) prepared in Brussels with the American

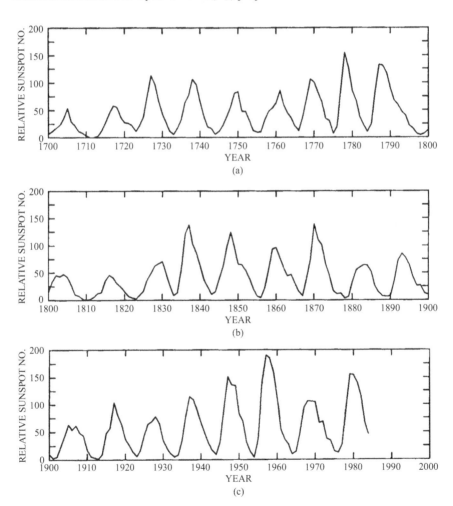

Figure 1.4. Annual mean relative sunspot numbers since 1700.

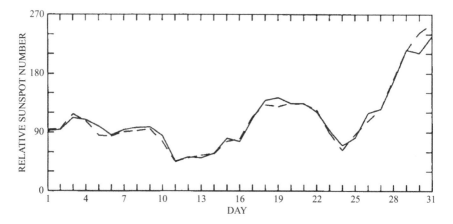

Figure 1.5. Daily relative sunspot numbers for January 1982 (solid line for international numbers and dashed line for AAVSO numbers).

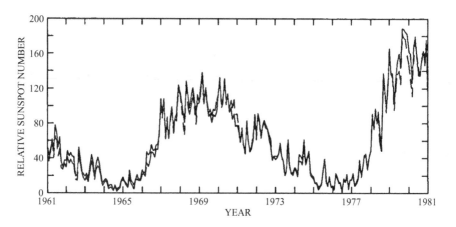

Figure 1.6. Monthly mean relative sunspot numbers for 1961–1983 (solid line for international numbers and dashed line for AAVSO numbers).

Association of Variable Star Observers (AAVSO) relative sunspot numbers (R_A). One can see that the sunspot activity may vary greatly over a 1-month interval. Although the two curves generally track each other, there are slight differences between the two curves. This is partly explained by the rotation period of the sun. The sun does not have a single rotation period; it rotates faster at the equator than at other solar latitudes. The mean rotation period at the equator is estimated to be in a range from 25 to 27 days, while at 60° latitude the mean rotation period is about 31 days. Sunspots themselves vary in duration from a few hours to several months. Figure 1.6 illustrates the differences between the two sunspot number sources using monthly averaging. The monthly mean is probably the minimum averaging interval

in which sunspot numbers start becoming stable, with yearly means proving to be
the most stable numbers.

1.3   A TEST CASE

Although the sunspot number data are a classical time series, there is no way to ascer-
tain the true spectrum from such a relatively short data sequence. A synthetic test
data sequence has been devised in order to provide evidence of the short data record
characterization of each spectral estimator developed in this text with a scenario
of known deterministic signals and additive noise. The data record consists of 64
samples (assume one sample per second) of a complex-valued process consisting of
four complex sinusoids and a complex additive colored noise process. Figure 1.7(a)
and (b) plots the real and imaginary components of the data record respectively.
The data record is contained in file test1987.dat and at the website cited in the
preface. MATLAB plotting script plot_test_data.m will reproduce Fig. 1.7.

  The true spectrum, calculated by analytical means, is illustrated in Fig. 1.8. The
frequency axis represents the frequency as a fraction of the sampling frequency, that
is, the actual frequency f in Hz normalized by dividing by the sampling frequency

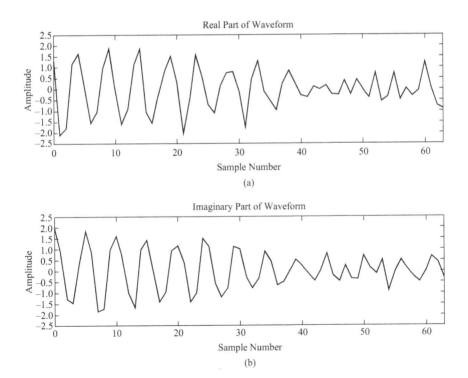

Figure 1.7. Real and imaginary components of complex data record test1987.

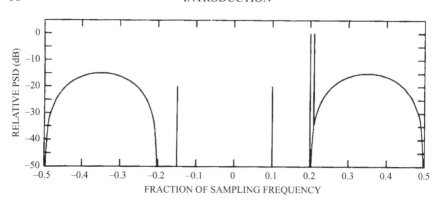

Figure 1.8. Analytically determined spectrum of 1987 test data.

$f_s = 1/T$, where T is the data sample interval. Due to the sampling theorem for signals (see Chap. 2), the fractional frequencies must lie between -0.5 and 0.5. In both test cases, two of the sinusoids in the spectrum have been placed at very close fractional frequencies in order to test the resolution capability of a spectral estimator. Weaker complex sinusoidal signals of 20 dB less power than the stronger sinusoids were placed at other frequency locations. These sinusoidal signals were selected to test a spectral estimator's capability to pick out weaker signal components among stronger signals. The colored noise process was generated by passing two independently generated zero-mean real white noise processes through identical moving average filters (see Chaps. 6 and 10) to separately generate the real and imaginary components of the test data noise process. The height of the lines representing the sinusoidal components of the spectrum are adjusted to reflect the power of each sinusoid relative to the *total* power in the colored noise process. Thus, the two stronger sinusoids have more power than the noise, while the weaker sinusoids have less power than the noise.

1.4   ISSUES IN SPECTRAL ESTIMATION

The interest in alternative spectral analysis methods has been promoted by the promise of improved performance, such as better frequency resolution, increased signal detectability, or maintenance of spectral shape "fidelity" with fewer parameters. It has been difficult to characterize analytically the performance of the alternative methods in the limited data (short data record) case; only very limited empirical results may be found in the literature. A "short" data record is one in which the resolution required is of the same order as the reciprocal of the record length. This has led to a number of issues in modern spectral estimation. Some of these key issues are now highlighted.

1.4.1 Resolution

Spectral resolution is high on the list of key issues in contemporary spectral estimation, especially in regard to limited data sequences. Just what constitutes resolution is a subjective area. One of the earliest definitions was provided by Lord Rayleigh [1879, p. 269], who provided this working definition for the resolution of telescopes with finite spatial aperture:

> "As the power of a telescope is measured by the closeness of the double stars which it can resolve, so the power of a spectroscope ought to be measured by the closeness of the closest double lines in the spectrum which it is competent to resolve."

Analogous definitions for resolution of finite-time aperture signals have been proposed. Formal definitions of resolution are presented in Chaps. 2 and 5. A common practice in the literature is to convey the relative resolutions of two spectral estimators based on visual impressions. Consider in Fig. 1.9 the results of two spectral estimators. The spectral estimate shown in Fig. 1.9(a) has a value of unity at its peak magnitude and barely yields two distinguishable peaks to indicate the presence of two close spectral components. If one simply forms the new, but arbitrary, spectral function

$$P_B(f) = \frac{1}{1 - P_A(f)},$$

which is plotted in Fig. 1.9(b), then the visual impression is that of a spectrum with "higher" resolution. The truth is that both spectra contain the same information. Visual comparisons alone are, therefore, not sufficient to judge resolution performance [Kay and Demeure, 1984].

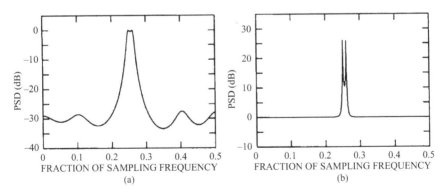

Figure 1.9. Visual enhancement of spectral resolution. (a) Original spectrum. (b) Enhanced spectrum obtained by $1/(1 - x)$ mapping.

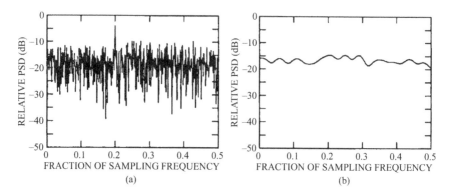

Figure 1.10. (a) Unsmoothed periodogram with one sinusoid detectable. (b) Smoothed periodogram with no sinusoids detectable.

1.4.2 Signal Detectability

The use of spectral estimation to detect the presence of signals suggests another issue. Consider the periodograms illustrated in Fig. 1.10. Both periodograms are based on the same 1024-point realization of a process consisting of a single sinusoid in additive white noise at -17-dB signal-to-noise ratio (SNR). Figure 1.10(a) is a single periodogram based on the entire 1024 samples. The sinusoid located at fractional frequency 0.2 is detectable above the noise level in the spectral estimate. Figure 1.10(b) is a smoothed periodogram obtained by segmenting the data into intervals of 32 data samples and averaging the periodograms made from each of the resulting 32 segments. The flat characteristic of white noise is more evident in this spectral estimate, but the response due to the presence of the sinusoid is not detectable. In fact, it may be shown that the unsmoothed periodogram yields an improved detectability over that of the smoothed periodogram by roughly the square root of the number of segments, or $10 \log \sqrt{32} \approx 7.5$ dB in this case. In terms of characterizing the entire spectrum, the smoothed spectral estimate of Fig. 1.10(b) is better; in terms of detectability, however, the unsmoothed estimate is better. Thus, the motivation for the type of spectral estimator selected will depend on whether one is interested in a smooth estimate over the entire range of frequencies, or better detectability over selected portions of the spectral frequency range. Spectral estimation algorithms are, therefore, not necessarily good detection algorithms.

1.4.3 Confusing Parameter Estimation with Spectral Estimation

The spectral estimation problem involves the estimation of a *function* of frequency. The performance of a spectral estimator may be judged by how well it matches the known spectrum of a test signal over a continuous range of frequencies. There are applications where only the behavior of the spectral estimator local to specific frequency regions is of interest. Estimating the frequency of a sinusoid in white noise, the example used in Fig. 1.10, is one such application. As a *spectral estimator*, the estimate of Fig. 1.10(b) is better because it more closely resembles the flat spectrum of white noise. As a *sinusoidal frequency estimator*, the estimate of Fig. 1.10(a) is better because the sinusoidal frequency may be more accurately determined. There is some confusion in the literature because many *parameter estimation* problems are presented in spectral estimation contexts, even though the spectral estimator function is not the ultimate quantity sought.

The parameter estimation problem and the spectral estimation problem require separate statistical treatments, although the two estimation problems often overlap for specific cases. The performance of a *spectral* estimator should, therefore, not be judged entirely in terms of its performance as a *parameter* estimator. The comparison made in Fig. 1.10 shows, for example, that the smoothed periodogram is a good *spectral* estimator, but a poor *parameter* estimator, the parameter in this case being the sinusoidal frequency.

1.4.4   One Estimate Is Not Enough

A spectral estimate produced from a finite data record represents a guess as to the true limit spectrum had an infinite data record been available. The behavior and performance of spectral estimators must, therefore, be characterized in statistical terms. The bias and variance are common statistical measures of estimator performance. As analytical determination of bias and variance is usually not mathematically tractable, a practical scheme is to overlap spectral estimates from several realizations and visually examine the bias and variance among the estimates as a function of frequency. Areas in the overlapped spectral plots where the experimental variance is large would indicate that any spectral features seen in this area of the spectrum may not be statistically significant. On the other hand, features in the overlapped spectra where the variance is small are more confidently attributed to actual frequency components in the signal. However, the application of spectral estimation to short data records often does not permit multiple spectral estimates to be made. Statistical analysis of single spectral estimates made from short data records is a difficult problem in general. For these reasons, the reader is cautioned not to draw any statistical conclusions when comparing the various spectral estimates computed using the demonstration data.

1.4.5   The Whole Picture

The formal definition of the spectrum, to be found in Chap. 4, will show that it is a function strictly of the second-order statistics. The second-order statistics are also assumed to remain unchanged, or *stationary*, over time. Thus, the spectrum is not a complete statistical picture of a random process that may have other information contained in the third- and higher-order statistics. Many common signals analyzed in practice are also not stationary, in which case the methods of this text could not be applied to the whole data record. In practice, though, segments of short data intervals from the longer data record may be considered as locally stationary, so that the spectral estimation methods of this text can be used to account for the nonstationary situation by allowing the spectral estimate to vary from interval to interval. The time-varying statistics may thus be tracked by examining the variations in spectral estimates from segment to segment.

1.5 HOW TO USE THIS TEXT

This text has been structured to permit easy implementation of many of the spectral estimators described. Each chapter that contains implementations of spectral estimators begins with a summary section that highlights specific estimator algorithms implemented in that chapter, including step-by-step instructions for running the spectral estimator demonstration scripts. MATLAB functions and scripts related to a chapter are in the resource cited in the preface. Readers should experiment with the demonstration scripts discussed in each chapter summary in order to obtain "hands-on" experience with one or more algorithms developed in the chapter, before proceeding deeper into a chapter to get the theoretical and mathematical details.

  Those readers desiring assistance in the selection of a spectral estimation technique may find the summary table in Chap. 14 helpful. All key spectral estimation algorithms are listed in this table, including location within the text where each algorithm is discussed.

  The chapters of the book have been organized into four major components: background material, spectral estimators for one-dimensional single-channel data sequences, summary of the one-dimensional single-channel spectral estimators, and spectral estimators for multichannel or two-dimensional data sequences. The background material (Chaps. 2–4) provides the mathematical foundation for the balance of the text, including the definition of the power spectral density. Spectral estimators are grouped according to three classes: classical spectral estimators (Chap. 5), parametric spectral estimators (Chaps. 6–11), and nonparametric spectral estimators (Chaps. 12–13). Multichannel and two-dimensional spectral estimators are presented in Chaps. 15 and 16, respectively. Multichannel refers to the spectral analysis of vector data (e.g., data from an array of sensors) that is a function of a scalar (e.g., time). Multidimensional refers to the spectral analysis of scalar data (e.g., image data) that is a function of a vector (e.g., two spatial coordinates).

References

Alter, D., A Group or Correlation Periodogram, with Application to the Rainfall of the British Isles, *Mon. Weather Rev.*, vol. 56, pp. 263–266, 1927.
Bartlett, M. S., On the Theoretical Specification of Sampling Properties of Autocorrelated Time-Series, *J. R. Stat. Soc.*, Ser. B, vol. 8, pp. 27–41, 1946.

Bartlett, M. S., Smoothing Periodograms from Time Series with Continuous Spectra, *Nature*, London, vol. 161, pp. 686–687, May 1948.

Bartlett, M. S., Periodogram Analysis and Continuous Spectra, *Biometrika*, vol. 37, pp. 1–16, June 1950.

Bernoulli, Daniel, *Hydrodynamica*, Basel, Switzerland, 1738.

Blackman, R. B., and J. W. Tukey, The Measurements of Power Spectra from the Point of View of Communications Engineering, *Bell Syst. Tech. J.*, vol. 33, pp. 185–282, 485–569, 1958; also republished by Dover Publications, New York, 1959.

Burg, John P., Maximum Entropy Spectral Analysis, in *Proceedings 37th Meeting of Society of Exploration Geophysicists*, Oklahoma City, Okla., October 1967. (Also in *Modern Spectrum Analysis*, D. G. Childers, ed., IEEE Press Selected Reprint Series, New York, 1978.)

Cooley, J. W., and J. W. Tukey, An Algorithm for the Machine Calculation of Complex Fourier Series, *Math. Comput.*, vol. 19, pp. 297–301, April 1965.

Daniell, P. J., Discussion of "On the Theoretical Specification and Sampling Properties of Autocorrelated Time-Series," *J. R. Stat. Soc.*, ser. B, vol. 8, pp. 88-90, 1946.

Einstein, Albert, Method for the Determination of the Statistical Values of Observations Concerning Quantities Subject to Irregular Fluctuations, *Archives des Sciences et Naturelles*, vol. 37, pp. 254–256, 1914 (translation in *IEEE ASSP Magaine*, vol. 4, October 1987).

Euler, L., *Institutiones Calculi Differentialis*, St. Petersburg, Russia, 1755.

Fourier, Jean Baptiste Joseph, *Théorie analytique de la chaleur* (Analytical Theory of Heat), Paris, France, 1822.

Gardner, William A., Introduction to Einstein's Contribution to Time-Series Analysis, and A. M. Yaglom, Einstein's 1914 Paper on the Theory of Irregularly Fluctuating Series of Observations, *IEEE ASSP Magazine*, vol. 4, October 1987.

Heideman, M. T., D. H. Johnson, and C. S. Burrus, Gauss and the History of the Fast Fourier Transform, *IEEE Acoust. Speech, and Signal Process. Magazine*, vol. 1, pp. 14–21, October 1984.

Henrici, O., On a New Harmonic Analyser, *Philos. Mag.*, vol. 38, pp. 110–121, 1894.

Jenkins, G. M., and D. G. Watts, *Spectral Analysis and Its Applications*, Holden-Day, Inc., San Francisco, 1968.

Kay, S. M., and Cedric Demeure, The High-Resolution Spectrum Estimator—A Subjective Entity, *Proc. IEEE*, vol. 72, pp. 1815–1816, December 1984.

Kendall, M. G., On the Analysis of Oscillatory Time Series, *J. R. Stat. Soc.*, vol. 108, p. 93, 1945.

Khintchine, Aleksandr Jakovlevich, Korrelationstheorie der Stationären Stochastischen Prozesse, *Math. Ann.*, vol. 109, pp. 604–615, 1934.

Kolmogoroff, Andrei N., Sur l'interpolation et l'extrapolation des suites stationnaires, *C. R. Acad. Sci.*, vol. 208, p. 2043, 1939.

Levinson, Norman, The Wiener (Root Mean Square) Error Criterion in Filter Design and Prediction, *J. Math. Phys.*, vol. 25, pp. 261–278, 1947.

Michelson, Albert Abraham, and S. W. Stratton, A New Harmonic Analyser, *Philos. Mag.*, vol. 45, pp. 85–91, 1898.

Newton, Isaac, *Phil. Trans.*, vol. VI, 1671. On page 3076: "Comparing the length of this coloured spectrum with its breadth . . .".

Newton, Isaac, *Principia*, London, England, 1687; *Optics*, London, England, 1704.

Parzen, Emanuel, On Consistent Estimates of the Spectrum of a Stationary Time Series, *Ann. Math. Stat.*, vol. 28, pp. 329–348, 1957.

de Prony, Baron (Gaspard Riche), Essai expérimental et analytique: sur les lois de la dilatabilité des fluides élastiques et sur celles de la force expansive de la vapeur de l'eau et de la vapeur de l'alkool, à différentes températures, *J. Ec. Polytech.*, vol. 1, cahier 2, pp. 24–76, 1795.

Rayleigh, Lord, Investigations in Optics, with Special References to the Spectroscope, *Philos. Mag. J. Sci.*, ser. 5, vol. 8, no. 49, pp. 261–274, October 1879.

Rayleigh, Lord, On the Resultant of a Large Number of Vibrations of the Same Pitch and of Arbitrary Phase, *Philos. Mag.*, vol. 10, pp. 73–78, August 1880.

Robinson, Enders A., Predictive Decomposition of Time Series with Applications to Seismic Exploration, MIT Geophysical Analysis Group (GAG), Cambridge, Mass., 1954; reprinted in *Geophysics*, vol. 32, pp. 418–484, 1967.

Robinson, Enders A., A Historical Perspective of Spectrum Estimation, *Proc. IEEE*, vol. 70, pp. 885–907, September 1982.

Schuster, Sir Arthur, On Interference Phenomenon, *Philos. Mag.*, vol. 37, pp. 509–545, June 1894.

Schuster, Sir Arthur, On Lunar and Solar Periodicities of Earthquakes, *Proc. R. Soc. London*, vol. 61, pp. 455–465, 1897.

Schuster, Sir Arthur, On the Investigation of Hidden Periodicities with Application to a Supposed Twenty-Six-Day Period of Meteorological Phenomena, *Terr. Mag.*, vol. 3, no. 1, pp. 13–41, March 1898.

Schuster, Sir Arthur, The Periodogram of Magnetic Declination as Obtained from the Records of the Greenwich Observatory during the Years 1871–1895, *Trans. Cambridge Philos. Soc.*, vol. 18, pp. 107–135, April 1900.

Schuster, Sir Arthur, *The Theory of Optics*, Cambridge University Press, London, 1904.

Schuster, Sir Arthur, The Periodogram and Its Optical Analogy, *Proc. R. Soc. London*, ser. A, vol. 77, pp. 136–140, 1905. Also, On Sun Spot Periodicities—Preliminary Notice, pp. 141–145, 1905.

Schuster, Sir Arthur, On the Periodicities of Sunspots, *Philos. Trans. R. Soc. London*, ser. A, vol. 206, pp. 69–100, 1906.

Sharp, Archibald, Harmonic Analyser, Giving Direct Readings of the Amplitude and Epoch of the Various Constituent Simple Harmonic Terms, *Philos. Mag.*, vol. 38, pp. 121–125, 1894.

Slutsky, Evgenievich, The Summation of Random Causes as the Source of Cyclic Processes (in Russian), *Probl. Econ. Conditions* (Moscow), vol. 3, no. 1, 1927; English translation in *Econometrica*, vol. 5, pp. 105, 1937.

Stokes, George Gabriel, Comments on the "Preliminary Report to the Committee on Solar Physics on a Method of Detecting the Unknown Inequalities of a Series of Observations," *Proc. R. Soc. London*, vol. 29, pp. 122–123, 29 May 1879.

Thomson, Sir William (Lord Kelvin), On an Instrument for Calculating the Integral of the Product of Two Given Functions, *Proc. R. Soc. London*, vol. 24, pp. 266–268, 1876.

Thomson, Sir William, Harmonic Analyzer, *Proc. R. Soc. London*, vol. 27, pp. 371–373, 1878.

Toeplitz, Otto, Zur Theorie der Quadratischen und Bilinearen Formen von Unendlichvielen Veränderlichen, *Math. Ann.*, vol. 70, pp. 351–376, 1911.

Tukey, John W., The Sampling Theory of Power Spectrum Estimates, *Proc. Symp. Applied Autocorrelation Analysis of Physical Problems*, U.S. Office of Naval Research (NAVEXOS-P-725), pp. 47–67, 1949. Reprinted in *J. Cycle Res.*, vol. 6, pp. 31–52, 1957.

Walker, Sir Gilbert, On Periodicity in Series of Related Terms, *Proc. R. Soc. London*, ser. A, vol. 131, pp. 518–532, 1931.

Wiener, Norbert, Generalized Harmonic Analysis, *Acta Math.*, vol. 55, pp. 117–258, 1930.

Wold, Herman O. A., A Study in the Analysis of Stationary Time Series, dissertation, Uppsala University, 1938; reprinted by Almqvist & Wiksell Forlag, Stockholm, 1954.

Wolf, R., Tafel der Relativzahlen, *Astron. Mitteil. Eidgen. Sternwarte*, Zurich, No. 24, vol. 111, 1868.

Yule, George Udny, On a Simple Form of Harmonic Analyses, *Philos. Mag.*, vol. 39, pp. 367–374, 1895.

Yule, George Udny, On a Method of Investigating Periodicities in Disturbed Series, with Special Reference to Wolfer's Sunspot Numbers, *Philos. Trans. R. Soc. London*, ser. A, vol. 226, pp. 267–298, July 1927.

2

REVIEW OF LINEAR SYSTEMS AND TRANSFORM THEORY

2.1 INTRODUCTION

As this is a review chapter, it is assumed that the reader has had some exposure to linear systems theory and Fourier transform theory from an introductory course in both continuous-time and discrete-time systems. Details of properties presented in this chapter may be found in the reference list at the end of the chapter. Another function of this chapter is to introduce many of the notational conventions used throughout the text.

Key concepts in linear systems theory are presented separately for continuous-time and discrete-time signals in Secs. 2.3 and 2.4, respectively. Section 2.5 is devoted to Fourier transform theory for continuous-time signals. Although Fourier transform theory for discrete-time signals is often presented in the literature independently of the continuous-time transform, Secs. 2.6 and 2.7 illustrate that the discrete-time transform may be viewed as a special case of the continuous-time transform. This development provides the link between the continuous-time Fourier transform (CTFT) and the discrete-time Fourier series (DTFS) of a finite sequence of samples of the continuous-time signal. This is in recognition that most realistic scenarios involve processing sequences from uniformly sampled continuous temporal or spatial waveforms. The DTFS used in this book differs slightly from the traditional discrete Fourier transform (DFT) definition found in the literature. The difference is due to the inclusion of the sampling interval as a scaling factor in the DTFS. This factor will be important for yielding spectral estimators with proper power units. Section 2.10 summarizes the fast Fourier transform (FFT), a very efficient computational algorithm to compute the DTFS. Finally, the important concept of time-bandwidth product (TBP) is discussed at length in Sec. 2.11. The TBP forms the basis of a measure for resolution.

The material in this chapter is based on the assumption that the signal is deterministic. The additional mathematical theory required to treat random signals is covered in Chap. 4.

2.2 SIGNAL NOTATION

A *continuous* signal is any real or complex time waveform $f(t)$, defined as a function of the continuous real temporal variable t, or any real or complex space waveform $g(s)$, defined as a function of the continuous real spatial variable s. Continuous signals are often termed *analog* if they can take on a continuum of values at any point t or s. Unless stated otherwise, all signals considered in this text will be assumed complex valued. The use of complex data has become prevalent with the increase in digital signal processing. Complex data is the natural output, for example, of a complex demodulation process (see Sec. 2.12). Some signals, such as found in radar and communication, have in-phase and quadrature components that lend themselves naturally to a complex representation, as the radar data set has illustrated.

A *discrete* signal is any function $f[n]$ that is a sequence of real or complex numbers defined for every integer n. A continuous temporal function $f(t)$ that is sampled at uniform intervals of T seconds, or a continuous spatial function $g(s)$ that is sampled at uniform intervals of S meters (or other appropriate spatial units), will generate the discrete sequences $f[n] = f(nT)$ or $g[n] = g(nS)$, respectively. Note that the distinguishing feature between continuous and discrete function notation is, respectively, the use of parentheses ( ) versus brackets [ ]. Parentheses are smooth, *continuous* line segments, whereas a bracket is formed by three *discrete* line segments. These symbols should avoid ambiguity between the continuous and discrete functions that use similar letter symbols.

A continuous or discrete signal whose amplitudes can take on a finite number of values, rather than a continuum, at any time t or at any index n is said to be *digital*. Discrete signals are, in practice, most commonly digital. Hardware analog-to-digital converters naturally quantize continuous-time, continuous-amplitude waveforms into discrete-time, discrete-amplitude samples. Digital computers, for example, use finite-length registers to represent data values.

2.3 CONTINUOUS LINEAR SYSTEMS

A *linear system*, be it continuous or discrete, is simply a system in which super-position of responses may be applied. Thus, the response of a linear system to the sum of two input signals is simply the sum of the separate system responses to each individual input signal. A linear system will be *time-invariant* if an input $x(t)$ produces an output response $y(t)$ and delayed input $x(t - t_0)$ produces a corresponding delayed output response $y(t - t_0)$ for any time (or space) shift t_0.

A special input signal is the *unit impulse function* $\delta(t)$, also called the *continuous delta function*. The unit impulse function may be conceptualized as a continuous signal of zero width and infinite height

$$\delta(t) = \begin{cases} \infty & \text{if } t = 0 \\ 0 & \text{if } t \neq 0 \end{cases} \tag{2.1}$$

but with unit area

$$\int_{-\infty}^{\infty} \delta(t)\, dt = 1.$$ (2.2)

The impulse function has the *sifting* property

$$\int_{-\infty}^{\infty} f(\tau)\delta(t - \tau)\, d\tau = \int_{-\infty}^{\infty} f(t - \tau)\delta(\tau)\, d\tau = f(t)$$ (2.3)

in that it can isolate, or reproduce, a particular value of a function $f(t)$, provided that the function is continuous at t. *Note that $\delta(t)$ has mathematical meaning only when used within the context of an integral.* For example, the product of a continuous function and an impulse function, $f(t)\delta(t)$, may be interpreted to mean $f(0)\delta(t)$, because

$$\int_{-\infty}^{\infty} g(t - \tau)[f(\tau)\delta(\tau)]\, d\tau = \int_{-\infty}^{\infty} [g(t - \tau)f(\tau)]\delta(\tau)\, d\tau = g(t)f(0).$$ (2.4)

Thus, the product of the function $f(t)$ with the impulse function has "sampled" the function at $t = 0$.

The output response $g(t)$ of a linear time-invariant (LTI) system to an arbitrary input signal $f(t)$ is determined by the *continuous convolution integral*

$$g(t) = \int_{-\infty}^{\infty} h(\tau)f(t - \tau)\, d\tau = \int_{-\infty}^{\infty} h(t - \tau)f(\tau)\, d\tau,$$ (2.5)

where $h(t)$ is the *continuous impulse response* of the system, that is, the output response of the system to an impulse function as input

$$h(t) = \int_{-\infty}^{\infty} h(\tau)\delta(t - \tau)\, d\tau.$$ (2.6)

A concise notation to represent the convolution operation is the $\star$ symbol. Thus, Eq. (2.5) could be expressed as

$$g(t) = h(t) \star f(t).$$ (2.7)

If the input to a continuous linear time-invariant system with impulse response $h(t)$ is a complex exponential of the form $f(t) = \exp(st)$, where s is an arbitrary complex value, then the output $g(t)$ has the form

$$g(t) = \int_{-\infty}^{\infty} h(\tau)\exp(s[t - \tau])\, d\tau = H(s)\exp(st),$$ (2.8)

where

$$H(s) = \int_{-\infty}^{\infty} h(\tau)\exp(-s\tau)\, d\tau.$$ (2.9)

The function $H(s)$ is the *continuous system function*; it is also recognizable as the *Laplace transform* of the impulse response. The term $\exp(st)$ is said to be an *eigenfunction* of the linear system because both input and output are proportional to it. $H(s)$ is then an *eigenvalue* of the system because it represents the complex scaling of the eigenfunction.

2.4   DISCRETE LINEAR SYSTEMS

To develop a framework for discrete linear systems, one may proceed in a manner analogous to that for continuous linear systems. A special input signal for discrete systems is the *unit impulse function* $\delta[n]$, also called the *discrete delta sequence*. It is simply defined as

$$\delta[n] = \begin{cases} 1 & \text{if } n = 0 \\ 0 & \text{if } n \neq 0. \end{cases} \tag{2.10}$$

Unlike its continuous-time counterpart, there are no analytical difficulties in defining $\delta[n]$. Any arbitrary sequence $f[n]$ may be written as a weighted sum of discrete-time impulse functions

$$f[n] = \sum_{k=-\infty}^{\infty} f[k]\delta[n-k] = \sum_{k=-\infty}^{\infty} f[n-k]\delta[k]. \tag{2.11}$$

Let $h[n]$ denote the *discrete impulse response* of a discrete linear system excited by a unit impulse function. The output response $g[n]$ to an arbitrary input sequence $f[n]$ is then provided by the *discrete convolution summation*

$$g[n] = \sum_{k=-\infty}^{\infty} h[n-k]f[k] = h[n] \star f[n]. \tag{2.12}$$

Note that brackets or parentheses on the functions surrounding the operator $\star$ will determine whether a discrete convolution or a continuous convolution operation is intended. A *causal discrete system* is one in which the output at time n depends only on the input $f[k]$ for $k \leq n$, which implies $h[k] = 0$ for $k < 0$.

If the input to a discrete linear system with impulse response $h[n]$ is an exponential sequence $f[n] = z^n$, where z is an arbitrary complex value, then the output response is

$$g[n] = H(z)z^n, \tag{2.13}$$

where

$$H(z) = \sum_{n=-\infty}^{\infty} h[n]z^{-n} = \mathcal{Z}\{h[n]\}. \tag{2.14}$$

Thus, z^n is an eigenfunction of a discrete linear system. $H(z)$ is the *discrete system function*; it is defined for those values of z such that the summation in Eq. (2.14) converges. $H(z)$ may be recognized as the two-sided *z-transform* of the sequence $h[n]$. The symbol $\mathcal{Z}$ will be used to designate the *z*-transform operation. Note that $H(z)$

is a continuous function of z, even though it is defined from a discrete sequence. The function $H(z)$ forms a polynomial in which the powers of z^{-1} correspond to the sequence time index (e.g., $h[n]$ is associated with the term z^{-n}). In this sense, z^{-1} may be considered a single sample delay operator. Some important properties of z-transforms used in this text are summarized in Table 2.1. Note that these properties are valid only for the common region of convergence for $F(z)$ and $G(z)$. For proof of these properties, consult Oppenheim and Schafer [1975, 1989].

An important class of discrete system functions are those with rational function z-transforms, that is, a ratio of polynomials in z. Let a causal discrete linear system be described by a constant coefficient linear difference equation, in which the input sequence $x[n]$ and output sequence $y[n]$ are related by

$$\sum_{k=0}^{P} a[k]y[n-k] = \sum_{k=0}^{Q} b[k]x[n-k] \tag{2.15}$$

for $n \geq 0$. The sequences $a[0], \ldots, a[P]$ and $b[0], \ldots, b[Q]$ characterize the linear system. Without loss of generality, one may assume $a[0] = 1$. Given a set of initial conditions of the x and y sequences for indices $n < 0$ and the input sequence $x[n]$ for indices $n \geq 0$, then the output sequence $y[n]$ for indices $n \geq 0$ may be directly computed. For zero initial conditions, the z-transform of Eq. (2.15) yields

$$Y(z)\sum_{k=0}^{P} a[k]z^{-k} = X(z)\sum_{k=0}^{Q} b[k]z^{-k}, \tag{2.16}$$

where the convolution property from Table 2.1 has been used. The system function $H(z)$ relating input and output is therefore

$$H(z) = \frac{Y(z)}{X(z)} = \frac{\displaystyle\sum_{k=0}^{Q} b[k]z^{-k}}{1 + \displaystyle\sum_{k=1}^{P} a[k]z^{-k}}, \tag{2.17}$$

Table 2.1. z-Transform Properties

Property	Discrete Time Function	Z-Transform
Definitions	$f[n], g[n]$	$F(z), G(z)$
Linearity	$af[n] + bg[n]$	$aF(z) + bG(z)$
Shift	$f[n-m]$	$z^{-m}F(z)$
Scaling	$a^{-n}f[n]$	$F(az), a > 0$
Conjugation	$f^*[n]$	$F^*(z^*)$
Time Reversal	$f[-n]$	$F(1/z)$
	$f^*[-n]$	$F^*(1/z^*)$
Convolution	$f[n] \star g[n]$	$F(z) \cdot G(z)$

which is a rational polynomial in z^{-1}. An equivalent factored form of $H(z)$ is

$$H(z) = \frac{b[0] \prod\limits_{k=1}^{Q} (1 - z_k z^{-1})}{\prod\limits_{k=1}^{P} (1 - p_k z^{-1})}, \qquad (2.18)$$

so that $b[0]$ may be regarded as a gain scaling factor. The roots $z_1, \ldots, z_Q$ of the numerator polynomial are called the *zeros* of $H(z)$; the roots $p_1, \ldots, p_P$ of the denominator polynomial are called the *poles* of $H(z)$. To have a stable system, it can be shown [Oppenheim and Schafer, 1989] that all poles must have the property $|p_k| < 1$. A *minimum-phase* linear system is a causal system in which all the poles and zeros of the system function $H(z)$ lie inside the unit z-plane circle, that is, $|z_k| < 1$ and $|p_k| < 1$.

The system function (2.17) may also be expressed as a partial fraction expansion. Assuming $Q \leq P$ and no repeated poles exist, then

$$H(z) = \sum_{k=1}^{P} \frac{r[k]}{1 - p_k z^{-1}} = \sum_{k=1}^{P} \frac{r[k]z}{z - p_k}. \qquad (2.19)$$

The residues $r[k]$ for $k = 1$ to $k = P$ may be found by using the residue theorem

$$r[k] = \left[\frac{H(z)(z - p_k)}{z} \right]_{z = p_k}. \qquad (2.20)$$

Repeated poles require a slightly more complicated treatment. The inverse z-transform of Eq. (2.19) will show [Oppenheim and Schafer, 1989] that the discrete system impulse response for zero initial conditions is a sum of discrete exponentials

$$h(n) = \begin{cases} \sum\limits_{k=1}^{P} r[k] p_k^n & \text{for } n \geq 0 \\ 0 & \text{for } n < 0. \end{cases} \qquad (2.21)$$

2.5 CONTINUOUS-TIME FOURIER TRANSFORM

Although the Laplace transform $H(s)$ permits general analysis of continuous linear systems for arbitrary exponential inputs, the analysis often of most interest is the evaluation along the imaginary axis, $s = j2\pi f$. The Laplace transform thus reduces to a function of f,

$$H(f) = \int_{-\infty}^{\infty} h(t)\exp(-j2\pi ft)\,dt = \mathcal{F}\{h(t)\}. \qquad (2.22)$$

In this form, $H(f)$ is called the *continuous-time Fourier transform* (CTFT). The symbol $\mathcal{F}$ will be used to designate the CTFT operation. The variable f in the complex sinusoid $\exp(-j2\pi ft)$ corresponds to frequency with units in Hertz (cycles per second), assuming t has units of time (seconds). For spatial signal processing, the spatial variable x replaces the time variable t, and spatial frequency k (the wavenumber in cycles per spatial unit) replaces temporal frequency f. The CTFT, in effect, identifies the frequencies and amplitudes of those complex sinusoids that combine to form an arbitrary waveform. The *inverse Fourier transform* is given by

$$h(t) = \int_{-\infty}^{\infty} H(f)\exp(j2\pi ft)\,df = \mathcal{F}^{-1}\{H(f)\}. \qquad (2.23)$$

The symbol $\mathcal{F}^{-1}$ will be used to designate the inverse CTFT operation. Expressions (2.22) and (2.23) form the continuous-time Fourier transform pair.

The existence of a CTFT and its inverse for a given function has been determined under a variety of conditions. One sufficient condition is that a signal $h(t)$ be absolute integrable in the sense

$$\int_{-\infty}^{\infty} |h(t)|\,dt < \infty. \qquad (2.24)$$

A less restrictive sufficient condition for the existence of the CTFT is that the signal have finite energy,

$$\int_{-\infty}^{\infty} |h(t)|^2\,dt < \infty.$$

Another sufficient condition for the existence of a CTFT applies to signals that can be represented in the form $h(t) = a(t)\sin(2\pi ft + \theta)$. The function $a(t)$ must be damped, i.e., $a(t+k) < a(t)$ for $k > 0$, and $h(t)$ must also satisfy

$$\int_{-\infty}^{\infty} \left|\frac{h(t)}{t}\right|\,dt < \infty. \qquad (2.25)$$

If the theory of distributions is allowed (a topic which is beyond the mathematical scope intended here), even the CTFT of undamped periodic signals and impulse functions may be defined.

Some key properties and functions of the continuous Fourier transform used frequently in the text are tabulated in Table 2.2. Derivations of these relationships, particularly those involving impulse functions, may be found in many standard

Table 2.2. Important CTFT Properties and Functions

Property or Function	Function	Transform
Linearity	$ag(t) + bh(t)$	$aG(f) + bH(f)$
Time Shift	$h(t - t_0)$	$H(f)\exp(-j2\pi f t_0)$
Frequency Shift (Modulation)	$h(t)\exp(j2\pi f_0 t)$	$H(f - f_0)$
Scaling	$(1/|\alpha|)h(t/\alpha)$	$H(\alpha f)$
Temporal Convolution Theorem	$g(t) \star h(t)$	$G(f) \cdot H(f)$
Frequency Convolution Theorem	$g(t) \cdot h(t)$	$G(f) \star H(f)$
Window Function	$A\,\mathrm{wind}(t/T_0)$	$2AT_0\,\mathrm{sinc}(2T_0 f)$
Sinc Function	$2AF_0\,\mathrm{sinc}(2F_0 t)$	$A\,\mathrm{wind}(f/F_0)$
Impulse Function	$A\delta(t)$	A
Sampling (Replicating) Function	$\uparrow\!\uparrow\!\uparrow_T(t)$	$F\uparrow\!\uparrow\!\uparrow_F(f),\ F = 1/T$

texts [Brigham, 1974; Rabiner and Gold, 1975; Oppenheim and Schafer, 1989; Bracewell, 1978]. Note that the rectangular *window* function is defined as

$$\mathrm{wind}(x) = \begin{cases} 1 & |x| < 1 \\ 1/2 & |x| = 1 \\ 0 & |x| > 1 \end{cases} \tag{2.26}$$

and the *sinc* function is defined as

$$\mathrm{sinc}(x) = \frac{\sin(\pi x)}{\pi x}. \tag{2.27}$$

Of special interest in Sec. 2.6 is the *sampling function*, defined as

$$\uparrow\!\uparrow\!\uparrow_W(x) = \sum_{n=-\infty}^{\infty} \delta(x - nW), \tag{2.28}$$

which has a transform that is also a sampling function. It has also been called a *replicating function* for reasons to be explained in the next section.

Another significant property is *Parseval's theorem* for two functions $g(t)$ and $h(t)$,

$$\int_{-\infty}^{\infty} g(t)h^*(t)\,dt = \int_{-\infty}^{\infty} G(f)H^*(f)\,df. \tag{2.29}$$

By setting $g(t) = h(t)$, Parseval's theorem becomes the *energy theorem*

$$\mathsf{E} = \int_{-\infty}^{\infty} |h(t)|^2\,dt = \int_{-\infty}^{\infty} |H(f)|^2\,df \tag{2.30}$$

because the left-hand side of Eq. (2.30) is the total energy E in the signal $h(t)$. Expression (2.30) is simply a statement of conservation of energy in the two domains. Thus, the function

$$S(f) = |H(f)|^2 \qquad (2.31)$$

represents the distribution of energy as a function of frequency, and is therefore called the *energy spectral density* (ESD) of the deterministic signal $h(t)$.

In addition to the ESD, two other spectral definitions relevant to deterministic signals have been described in the linear system literature. If $h(t)$ represents the impulse response of a filter, then its CTFT

$$H(f) = \int_{-\infty}^{\infty} h(t) \exp(-j2\pi f t)\, dt \qquad (2.32)$$

is called the *frequency response function*. $H(f)$ is, in general, a complex quantity and is therefore expressible in polar format as

$$H(f) = |H(f)| \exp(j\theta(f)), \qquad (2.33)$$

where the real functions $|H(f)|$ and $\theta(f)$ are defined as

$$|H(f)| = [\text{Re}\{H(f)\}^2 + \text{Im}\{H(f)\}^2]^{1/2}$$
$$\theta(f) = \tan^{-1}[\text{Im}\{H(f)\}/\text{Re}\{H(f)\}].$$

The function $|H(f)|$ is termed the *amplitude spectrum* of $h(t)$ and the function $\theta(f)$ is termed the *phase spectrum*.

2.6 SAMPLING AND WINDOWING OPERATIONS

In Sec. 2.7, the discrete-time Fourier series (DTFS) is developed as a special case of the CTFT. The development makes use of two basic signal processing operations: *sampling* and *windowing*. The effect of each operation on a signal and its transform is examined in this section. This will simplify the presentation in Sec. 2.7. A key result of this section is the sampling theorem, which indicates the conditions for which a sampled continuous-time signal or continuous-space signal may be recovered from its samples with no loss of information. The following discussions assume signals that are functions of time, although the presentation could analogously be developed for signals that are functions of space (see Sec. 2.13).

Table 2.3 lists the windowing and sampling functions to be used in this discussion. The uniform *sampling interval* is T seconds. The *sampling frequency*, or *sampling rate*, is $F = 1/T$ Hz. Note that the windowing and sampling functions in the time domain are designated as TW and TS, respectively, whereas the windowing and sampling functions in the frequency domain are designated as FW and FS, respectively.

Assume that a continuous real-valued signal $x(t)$ to be sampled is *band-limited* (BL) with maximum frequency F_0 Hz, that is, the CTFT of $x(t)$ is zero for $|f| > F_0$. The CTFT of a real signal $x(t)$ is always a symmetrical function with a total

Table 2.3. Windowing and Sampling Functions

Operation	Time Function	Transform Function
Time Windowing (NT-sec timewidth)	$TW = \text{wind}(2t/NT - 1)$	$\mathcal{F}\{TW\} = NT\,\text{sinc}(NTf)$ $\cdot \exp(-j\pi NTf)$
Frequency Windowing ($1/T$-Hz bandwidth)	$\mathcal{F}^{-1}\{FW\} = \frac{1}{T}\,\text{sinc}(t/T)$	$FW = \text{wind}(2Tf)$
Time Sampling (T-sec intervals)	$TS = T \uparrow\uparrow\uparrow_T (t)$	$\mathcal{F}\{TS\} = \uparrow\uparrow\uparrow_{1/T}(f)$
Frequency Sampling ($1/NT$-Hz intervals)	$\mathcal{F}^{-1}\{FS\} = \uparrow\uparrow\uparrow_{NT}(t)$	$FS = \frac{1}{NT}\uparrow\uparrow\uparrow_{1/NT}(f)$

frequency bandwidth of $2F_0$ Hz, as depicted in Fig. 2.1(a). A sampled function of $x(t)$ may be constructed by multiplication with the sampling function

$$x_S(t) = x(t) \cdot TS = T \sum_{n=-\infty}^{\infty} x(nT)\delta(t - nT). \qquad (2.34)$$

The reader is cautioned that the above product of $x(t)$ with a train of impulse functions must be interpreted in the manner of Eq. (2.4). By the frequency convolution theorem, the CTFT of $x_S(t)$ is simply the convolution of the transforms of $x(t)$ and TS,

$$X_S(f) = X(f) \star \mathcal{F}\{TS\} = \sum_{k=-\infty}^{\infty} X(f - kF). \qquad (2.35)$$

Convolving $X(f)$ with the sampling function $\mathcal{F}\{TS\} = \uparrow\uparrow\uparrow_{1/T}(f)$ simply *replicates* $X(f)$ at frequency separations of $1/T$ Hz, corresponding to the location of each impulse function. Thus, $X_S(f)$ is a periodic replication of $X(f)$. For this reason, $\uparrow\uparrow\uparrow(f)$ is often called a *replicating function*. In general, sampling in one domain forces a periodic replication in the transformed domain. If the sampling rate is selected too low such that $F < 2F_0$, then the replicated transforms will overlap with adjacent transforms, as illustrated in Fig. 2.1(c). This overlap is a distorting form of *frequency aliasing*. If the sampling rate is increased such that $F \geq 2F_0$, then there will be no overlap, as indicated in Fig. 2.1(d). The sampling frequency $F_N = 2F_0$ is known as the *Nyquist sampling rate* or *foldover frequency*.

In order to recover the original time signal from its samples, that is, to *interpolate* a continuum of values between the samples, one can filter the sampled data with an ideal rectangular low-pass filter FW, as illustrated in Fig. 2.1(e). The result, illustrated in Fig. 2.1(f),

$$X(f) = X_s(f) \cdot FW = [X(f) \star \mathcal{F}\{TS\}] \cdot FW \qquad (2.36)$$

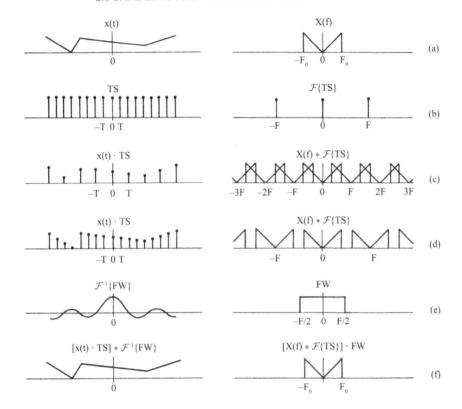

Figure 2.1. Illustration of the temporal sampling theorem for a real band-limited signal. (a) Original band-limited time function and its transform. (b) Temporal sampling function and its transform. (c) Sampled time function and replicated transform when $F_0 > 1/2T$. (d) Sampled time function and replicated transform when $F_0 < 1/2T$. (e) Frequency window (ideal low-pass filter) and its transform (a sinc function). (f) Original time function recovered by convolution with sinc function.

recovers the original CTFT. Using both the temporal and frequency convolution theorems, then

$$x(t) = x_s(t) \star \mathcal{F}^{-1}\{\text{FW}\} = [x(t) \cdot \text{TS}] \star \mathcal{F}^{-1}\{\text{FW}\}$$

$$= \sum_{n=-\infty}^{\infty} x(nT)\text{sinc}([t - nT]/T). \qquad (2.37)$$

Expressions (2.36) and (2.37) constitute a concise mathematical statement of the *temporal sampling theorem*. This theorem proves that a band-limited real signal can be reconstructed exactly, using the band-limited interpolation formula (2.37), from knowledge of an *infinite* number of samples taken at a sample rate $F \geq 2F_0$.

The same arguments can also be used with complex-valued signals with a bandwidth of $2F_0$ Hz.

A dual theorem to expression (2.37) is the *frequency sampling theorem*. Rather than a band-limited signal, consider a real function that is *time-limited* (TL), that is, a signal $x(t)$ which is zero for $|t| > T_0$. If one now samples in the frequency domain, rather than the time domain, at frequency intervals of $F \leq 1/2T_0$ (otherwise, there will be *time aliasing*), then a nonoverlapping, periodic extension of the TL signal is created. Placement of an ideal rectangular time window will recover the original TL signal; in the frequency domain, this corresponds to a filtering operation for reconstruction of the original transform. The time-domain operations that characterize the frequency sampling theorem are

$$x(t) = [x(t) \star \mathcal{F}^{-1}\{FS\}] \cdot TW, \tag{2.38}$$

while the corresponding transform operations

$$X(f) = [X(f) \cdot FS] \star \mathcal{F}\{TW\}$$

$$= \sum_{n=-\infty}^{\infty} X(n/2T_0)\operatorname{sinc}(2T_0[f - n/2T_0])$$

$$= \sum_{n=-\infty}^{\infty} X(nF)\operatorname{sinc}([f - nF]/F), \tag{2.39}$$

in which $F = 1/2T_0$, will represent a time-limited interpolation procedure. Thus, the CTFT $X(f)$ of a time-limited signal may be uniquely recovered from the uniformly sampled transform values if a frequency sampling interval $F \leq 1/2T_0$ Hz is selected.

2.7 RELATING THE CONTINUOUS AND DISCRETE TRANSFORMS

In order to generate spectral estimates from data samples that are scaled properly for energy and power units, it is necessary to use a slightly different definition and terminology for the usual definition of the discrete Fourier transform (DFT) found in most digital signal processing textbooks. The transform pair for the usual DFT definition of an N-point time sequence $x[n]$ and its N-point transform sequence $X[k]$ are

$$X[k] = \sum_{n=0}^{N-1} x[n]\exp(-j2\pi kn/N) \tag{2.40}$$

$$x[n] = \frac{1}{N}\sum_{k=0}^{N-1} X[k]\exp(j2\pi kn/N). \tag{2.41}$$

In this text, the *discrete-time Fourier series* (DTFS) pair

$$X[k] = T \sum_{n=0}^{N-1} x[n] \exp(-j2\pi kn/N)$$

$$x[n] = \frac{1}{NT} \sum_{k=0}^{N-1} X[k] \exp(j2\pi kn/N),$$

defined for $0 \le k \le N - 1$ and $0 \le n \le N - 1$, will instead be used because the dependence on the sampling interval T is incorporated explicitly into the transform definitions. The justification for the DTFS name and definition, as well as the justification for other forms of the Fourier transform, is the subject of this section. The sampling and windowing operations introduced in Sec. 2.6 will be used to directly relate the CTFT to the DTFS and other Fourier transforms. Although the presentation approach here is rather unique, more detailed explanations of some of the developments within this section may be found in books by Brigham [1974], Bloomfield [1976], Papoulis [1977], Geçkinli and Yavuz [1983], and Oppenheim and Schafer [1989].

The DTFS may be viewed as an approximation of the CTFT based on a finite number of data samples. To develop the exact nature of the relationship, a four-step sequence of linear operations is required. These operations are *windowing* in both the time and frequency domains, and *sampling* in both the time and frequency domains. As was demonstrated in Sec. 2.6, if windowing is performed in one domain, filtering (convolution) with the sinc function is performed in the other domain as a result of the convolution theorem. Similarly, if sampling is performed in one domain, periodic replication is performed in the other domain. Because windowing and sampling are linear and commutative operations, there are many possible ways in which to order the four operations, all of which will yield the identical end result, although the intermediate results will be different. Figure 2.2 suggests two possible sequences for the four operations. Consider first the sequence designated by node numbers 1, 2, 4, 6, and 8. The original continuous-time function $x(t)$ and its transform $X(f)$ are not restricted and may be, in general, nonperiodic, non-band-limited, and non-time-limited. Assuming that N time samples of $x(t)$ are to be taken at uniform intervals of T seconds, the first operation, FW, band limits (windows in frequency) the signal to $\pm 1/2T$ Hz. The second operation, TS, samples the band-limited signal in time at T-second intervals. The third operation, TW, time-limits the sampled signal to N samples. The fourth operation, FS, samples in frequency at intervals of $1/NT$ Hz, which periodically replicates the original N time samples. Using the function definitions in Table 2.3 that represent these four operations, the sequence of time-domain and frequency-domain operations that lead to the DTFS pair are given by

$$x_{\text{DTFS}} = \left(\left((x(t) \star \mathcal{F}^{-1}\{\text{FW}\}) \cdot \text{TS}\right) \cdot \text{TW}\right) \star \mathcal{F}^{-1}\{\text{FS}\} \quad (2.42)$$

$$X_{\text{DTFS}} = \left(\left((X(f) \cdot \text{FW}) \star \mathcal{F}\{\text{TS}\}\right) \star \mathcal{F}\{\text{TW}\}\right) \cdot \text{FS}. \quad (2.43)$$

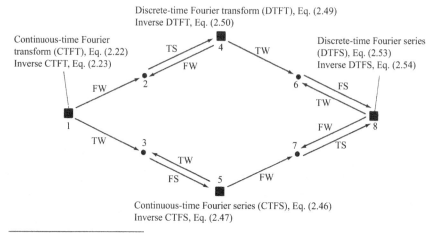

Discrete-time Fourier transform (DTFT), Eq. (2.49)
Inverse DTFT, Eq. (2.50)

Continuous-time Fourier transform (CTFT), Eq. (2.22)
Inverse CTFT, Eq. (2.23)

Discrete-time Fourier series (DTFS), Eq. (2.53)
Inverse DTFS, Eq. (2.54)

Continuous-time Fourier series (CTFS), Eq. (2.46)
Inverse CTFS, Eq. (2.47)

FW: Frequency-domain window
TW: Time-domain window
FS: Frequency-domain sampling
TS: Time-domain sampling

Figure 2.2. Two possible sequences of two windowing and two sampling operations to relate the CTFT to the DTFS.

The sampling operations between nodes 2–4 and 6–8 are reversible by application of the temporal sampling theorem or the frequency sampling theorem, respectively. The appropriate operations are designated as FW and TW on reverse arrows between nodes 4–2 and 8–6, respectively. The other suggested path illustrated in Fig. 2.2 uses the node sequence 1, 3, 5, 7, and 8. This switches the ordering of the time-window/frequency-sample pair of operations with the frequency-window/time-sample pair of operations.

As a result of the windowing and sampling operations, four different types of Fourier relationships will be encountered at nodes 1, 4, 5, and 8, as indicated on Fig. 2.2: continuous-time Fourier transform, discrete-time Fourier transform, continuous-time Fourier series, and discrete-time Fourier series. A distinction has been made between Fourier *transform* and Fourier *series*. Those nodes at which the frequency-domain function is continuous are termed Fourier transforms; those nodes at which the frequency-domain function is discrete are termed Fourier series. Note that the historical term "discrete Fourier transform" (DFT), in reference to Eqs. (2.40) and (2.41), does *not* follow this convention.

The bandlimiting and time-limiting operations required to relate the DTFS and the CTFT show that the DTFS is a distorted variant of the original continuous-time signal $x(t)$ and transform $X(f)$. The effect of the bandlimiting creates a filtered time function at node 2 of

$$x_{BL}(t) = x(t) \star \mathcal{F}^{-1}\{FW\}, \tag{2.44}$$

while the effect of time-limiting creates a filtered transform function at node 3 of

$$X_{TL}(f) = X(f) \star \mathcal{F}\{TW\}. \tag{2.45}$$

Samples of these functions are designated as $x_{BL}[n] = x_{BL}(nT)$ and $X_{TL}[k] = X_{TL}(k/NT)$.

Two other interesting observations can be made concerning the sequences of windowing and sampling operations depicted on Fig. 2.2. At node 5, the time windowing and frequency sampling operations create the *continuous-time Fourier series* (CTFS). The Fourier relationship at this node relates a continuous-time periodic function of period NT seconds to a discrete line spectrum. The following Fourier transform pairs may easily be established for this case, using the properties in Table 2.2 and the function definitions in Table 2.3, as

$$X_{TL}[k] = \int_0^{NT} x(t) \exp(-j2\pi kt/NT) \, dt \ \ k = 0, \pm1, \pm2, \dots, \pm\infty \tag{2.46}$$

$$x(t) = \frac{1}{NT} \sum_{k=-\infty}^{\infty} X_{TL}[k] \exp(j2\pi kt/NT) \ \ 0 \le t \le NT. \tag{2.47}$$

Note that Eq. (2.47) is a periodic function that matches the original time function at node 1 only over the 0-to-NT time interval. Expression (2.47) represents the traditional Fourier series representation of a periodic signal. Early literature referred to the Fourier analysis of periodic signals, Eq. (2.46), as *harmonic analysis*. The energy theorem for the CTFS is simply

$$\int_0^{NT} |x(t)|^2 \, dt = \frac{1}{NT} \sum_{k=-\infty}^{\infty} |X_{TL}[k]|^2. \tag{2.48}$$

This represents the energy over a period of NT seconds.

At node 4, the frequency windowing and time sampling operations have created the *discrete-time Fourier transform* (DTFT). It is characterized by a periodic *transform* function of period $1/T$ Hz. The following Fourier transform pairs may be established for this case, between the discrete-time sequence $x[n]$ to the continuous periodic transform $X(f)$

$$X(f) = T \sum_{n=-\infty}^{\infty} x_{BL}[n] \exp(-j2\pi fnT) \ \ -1/2T \le f \le 1/2T \tag{2.49}$$

$$x_{BL}[n] = \int_{-1/2T}^{1/2T} X(f) \exp(j2\pi fnT) \, df \ \ n = 0, \pm1, \pm2, \dots, \pm\infty. \tag{2.50}$$

Note that Eq. (2.49) is a periodic function that matches the original transform function at node 1 only over the $-1/2T$-to-$1/2T$ Hertz frequency interval. The energy theorem for the DTFT is

$$T \sum_{n=-\infty}^{\infty} \left| x_{BL}[n] \right|^2 = \int_{-1/2T}^{1/2T} \left| X(f) \right|^2 df. \tag{2.51}$$

This represents the energy over a frequency period of $1/T$ Hz. Expression (2.49) is related to the z-transform of a discrete sequence as follows:

$$X(f) = \left. T X(z) \right|_{z=\exp(j2\pi fT)}, \tag{2.52}$$

where $X(z)$ is the z-transform of the sequence $x_{BL}[n]$. It is assumed that the region of convergence of the z-transform includes the unit circle. Thus, the discrete-time Fourier transform is simply the z-transform which has been evaluated along the unit circle of the z-plane and scaled by T. The discrete-time Fourier transform will be used frequently throughout this text.

Regardless of which of the two sequences of four operations is selected, the final result at node 8 is the same. At this node, the *discrete-time Fourier series* (DTFS) has been created, with the following transform relationships

$$X_{TL}[k] = T \sum_{n=0}^{N-1} x_{BL}[n] \exp(-j2\pi kn/N) \quad k = -\frac{N}{2}, \ldots, \frac{N}{2} - 1 \tag{2.53}$$

$$x_{BL}[n] = \frac{1}{NT} \sum_{k=-N/2}^{N/2-1} X_{TL}[k] \exp(j2\pi kn/N) \quad n = 0, \ldots, N-1 \tag{2.54}$$

which follow from use of the properties in Table 2.3. The energy theorem for the DTFS is simply

$$T \sum_{n=0}^{N-1} \left| x_{BL}[n] \right|^2 = \frac{1}{NT} \sum_{k=-N/2}^{N/2-1} \left| X_{TL}[k] \right|^2, \tag{2.55}$$

representing the energy over N data samples.

Both sequences $x_{BL}[n]$ and $X_{TL}[k]$ are periodic modulo N, so the property $X_{TL}[-k] = X_{TL}[N-k]$ for $-N/2 \leq k \leq 1$ holds. Expression (2.54) may thus be equivalently written as

$$x_{BL}[n] = \frac{1}{NT} \sum_{n=0}^{N-1} X_{TL}[k] \exp(j2\pi kn/N) \tag{2.56}$$

for $0 \leq n \leq N - 1$. Equations (2.53) and (2.54) constitute the discrete-time Fourier series pair related to the original continuous-time and continuous-frequency pair $x(t)$ and $X(f)$. The distinguishing differences between the DTFS equations and the usual discrete Fourier transform equations (2.40) and (2.41) is the factor T. The factor T, which represents the time sample interval in seconds, and the factor $1/NT$, which represents the frequency sample interval in Hz, are required in order that Eqs. (2.53) and (2.54) represent, in effect, a rectangular integration approximation of the transform integral

$$\sum_{n=0}^{N-1} x[n] \exp(-j2\pi nfT)\, T \approx \int_{0}^{NT} x(t) \exp(-j\pi ft)\, dt. \qquad (2.57)$$

The T and $1/NT$ factors scale the DTFS by the sample rate, so that the correctly scaled energy and power values will be computed.

An examination of Fig. 2.2 will indicate the exact relationship required of the DTFS time sequence or the DTFS transform sequence in order for each to be the original continuous-time function or the original continuous-transform function, respectively. If $x(t)$ is inherently band-limited to $1/T$ Hz total bandwidth, then the DTFS time sequence preserves the original $x(t)$ values at the sample points, but the DTFS transform sequence is composed of samples of a smeared version of the original transform $X(f)$ [see Eq. (2.44)]. Conversely, if $x(t)$ is inherently time-limited to NT seconds in extent, then the DTFS transform sequence preserves the original $X(f)$ values at the sample points, but the DTFS time sequence is composed of samples of a smeared version of the original signal $x(t)$ [see Eq. (2.43)]. The smeared effects may be mitigated by decreasing T (so that $1/T$ is a broader bandwidth) or by increasing N (so that NT is a broader time duration), thereby making the DTFS a better approximation of the CTFT. Note that the DTFS will be identical to the CTFT only for periodic signals that can be represented as a sum of the N complex sinusoids of frequencies k/NT Hz, for $k = 0$ to $k = N - 1$.

2.8   THE ISSUE OF SCALING FOR POWER DETERMINATION

In a manner analogous to definition (2.31) for the energy spectral density of the continuous-time Fourier transform, the energy spectral density of the discrete-time Fourier transform is

$$S_{\text{DTFT}}(f) = \left|X(f)\right|^2 = \left|T \sum_{n=-\infty}^{\infty} x_{BL}[n] \exp(-j2\pi fnT)\right|^2, \qquad (2.58)$$

based on energy theorem (2.51) and defined for $-1/2T \le f \le 1/2T$. The energy spectral density of the continuous-time Fourier series is

$$S_{\text{CTFS}}[k] = \left|X_{TL}[k]\right|^2 = \left|\int_{0}^{NT} x(t) \exp(-j2\pi kt/NT)\, dt\right|^2, \qquad (2.59)$$

based on energy theorem (2.48) and defined for $-\infty < k < \infty$. The energy spectral density of the discrete-time Fourier series is

$$S_{\text{DTFS}}[k] = \left|X_{TL}[k]\right|^2 = \left|T\sum_{n=0}^{N-1} x_{BL}[n]\exp(-j2\pi kn/N)\right|^2, \qquad (2.60)$$

based on energy theorem (2.55) and defined for $0 \le k \le N - 1$.

Both the CTFS and DTFS have periodic time functions with period NT seconds. Power is energy per unit time. Therefore, dividing the ESDs represented by Eqs. (2.59) and (2.60) by NT will yield the deterministic *power spectral densities* (PSD)

$$P_{\text{CTFS}}[k] = \frac{1}{NT} S_{\text{CTFS}}[k], \qquad (2.61)$$

defined for $-\infty < k < \infty$, and

$$P_{\text{DTFS}}[k] = \frac{1}{NT} S_{\text{DTFS}}[k] = \frac{T}{N}\left|\sum_{n=0}^{N-1} x_{BL}[n]\exp(-j2\pi kn/N)\right|^2, \qquad (2.62)$$

defined for $0 \le k \le N - 1$. Both are in units of power per Hz. The total power P in one period is then obtained by rectangular integration of the *area* under the PSD values,

$$\mathsf{P}_{\text{CTFS}} = F\sum_{k=-\infty}^{\infty} P_{\text{CTFS}}[k]$$

$$\mathsf{P}_{\text{DTFS}} = F\sum_{k=0}^{N-1} P_{\text{DTFS}}[k],$$

where $F = 1/NT$ is the discrete frequency sample interval in Hz. Some texts define a DTFS "spectrum" as

$$F \cdot P_{\text{DTFS}}[k] = \left|\frac{1}{N}\sum_{n=0}^{N-1} x_{BL}[n]\exp(-j2\pi kn/N)\right|^2 \qquad (2.63)$$

because it does not depend on the sample interval T. The value of Eq. (2.63) at each index k will yield exactly the power of a sinusoid of frequency k/NT Hz. However, Eq. (2.63) does not represent the correctly scaled PSD function; Eq. (2.62) does.

2.9   THE ISSUE OF ZERO PADDING

The discrete-time Fourier series may be modified to interpolate transform values between the N original transform values by a process called *zero padding*. Consider adding N zero values $x[N], \ldots, x[2N - 1]$ to the available data samples

$x[0], \ldots, x[N-1]$. The DTFS of the zero-augmented $2N$-point data sequence is

$$X[k] = T \sum_{n=0}^{2N-1} x[n] \exp(-j2\pi nk/2N) = T \sum_{n=0}^{N-1} x[n] \exp(-j2\pi nk/2N),$$
(2.64)

where the range on the right-hand summation has been adjusted to reflect the zero samples. Let $k = 2l$, so that

$$X[l] = T \sum_{n=0}^{N-1} x[n] \exp(-j2\pi nl/N)$$
(2.65)

for $l = 0, 1, \ldots, N-1$ represents the even values of $X[k]$. The $2N$-point discrete-time Fourier series, therefore, reduces to that of the N-point discrete-time Fourier series at the even index values. The odd values of k represent the interpolated DTFS values between the original N-point DTFS values.

As more zeros are padded to the original N-point sequence, even more interpolation is obtained. Figure 2.3 illustrates the interpolative impact of zero padding. In the limit of infinite zero padding, the discrete-time Fourier series is equivalent to computing the discrete-time Fourier transform of an N-point windowed data sequence

$$X(f) = T \sum_{n=0}^{N-1} x[n] \exp(-j2\pi f nT).$$
(2.66)

The transform represented by Eq. (2.66) corresponds to node 6 in Fig. 2.2. There is a common misconception that zero padding improves the inherent resolution because it increases the data interval and decreases the interval between frequency evaluation points. As may be observed from Fig. 2.3, zero padding does *not* enhance the underlying resolution of the transform made from a given finite sample sequence. The zero padding simply provides an interpolated transform with a smoother appearance. It also removes ambiguities sometimes encountered in practice with the discrete-time Fourier series, due to narrowband signal components with center frequencies that lie between the N frequency evaluation points of a non-zero-padded discrete-time Fourier series, as illustrated in Fig. 2.3. The accuracy of estimating the frequency of spectral peaks is also enhanced with zero padding.

2.10 THE FAST FOURIER TRANSFORM

The fast Fourier transform (FFT) is not another Fourier transform. The name encompasses a number of computationally efficient *algorithms* for rapidly evaluating the discrete-time Fourier series. The concepts behind these algorithms have been discussed at length in numerous references [Otnes and Enochson, 1972; Brigham, 1974; Rabiner and Gold, 1975; McClellan and Rader, 1979; Elliott and Rao, 1983;

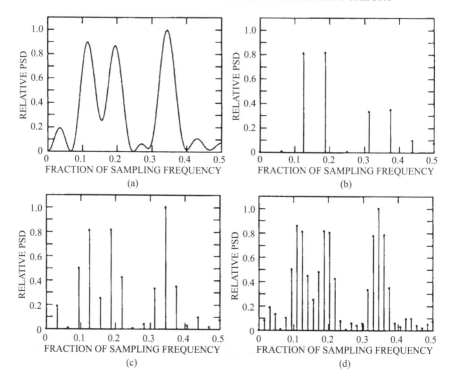

Figure 2.3. Transform interpolation by zero padding. (a) Magnitude of DTFT for a 16-point data record containing three sinusoids. (b) Magnitude of DTFS with no zero padding. Ambiguities are unresolved. (c) Magnitude of DTFS with double padding. Ambiguities are resolved because the three sinusoids can be distinguished. (d) Magnitude of DTFS with eight times zero padding.

Bergland, 1969; Cochran et al., 1967; Bracewell, 1989]. It is not the purpose of this section to present yet another tutorial on this topic, but rather to provide a brief overview of the basis for the computational savings afforded by the FFT algorithm. The FFT is the first of several fast computational algorithms to be presented in this text for implementing spectral estimation methods.

The central concept of the FFT is to divide an N-point DTFS into two or more smaller DTFSs, each of which can be individually computed and linearly combined to yield the DTFS of the original N-point sequence. These smaller DTFSs can, in turn, be divided into even smaller DTFSs of correspondingly smaller subsequences. In general, an N-point DTFS requires $\log_2 N$ stages with N adds and $N/2$ multiplies at each stage. Thus, an N-point FFT requires roughly $N \log_2 N$ adds and $N \log_2 N/2$ complex multiplies. This is a significant savings relative to the N^2 operations that would be required to individually evaluate N transform values from the N data points. If using zero padding, then even further reductions in computations may be achieved by eliminating, or *pruning*, computational paths involving only zero values. Details on pruning may be found in papers by Markel [1971], Skinner [1976], and Sreenivas and Rao [1979]. The book software uses the MATLAB built-in FFT function. More sophisticated FFT program codes may be found in Chap. 1 of the IEEE Press text *Programs for Digital Signal Processing* [1979] or in the text by McClellan and Rader [1979].

2.11   RESOLUTION AND THE TIME-BANDWIDTH PRODUCT

A common rule-of-thumb often used for spectral analysis suggests that the spectral resolution in Hertz will be approximately the reciprocal of the signal observation time in seconds. Because the frequency spacing of the DTFS is $1/NT$ Hz, where NT is the total data record time interval, each DTFS frequency "bin," or "cell," will, according to this rule, represent roughly the spectral resolution limit. This section examines the basis for this rule-of-thumb and places it in proper perspective as a measure of resolution. The results in this section apply to deterministic, finite-energy signals. The time-bandwidth product (TBP) will be revisited in Chap. 5, where it will be shown that a modification into a three-term statistical stability-time-bandwidth product is required to handle random signals of finite power.

A signal $x[n]$ cannot simultaneously be both time-limited and band-limited. Although a function may not be strictly time-limited or band-limited, it can still be characterized by an interval of T_e seconds in which most of its temporal energy is concentrated and by an interval B_e Hz in which most of its spectral energy is concentrated. In order to quantify the *temporal concentration* of a discrete-time signal sequence and the associated *frequency concentration* of its transform, several measures have been proposed. Two of these measures are presented here.

The *equivalent time width* T_e of discrete-time signal $x[n]$ is defined as

$$T_e = \frac{T \sum_{n=-\infty}^{\infty} x[n]}{x[0]}, \tag{2.67}$$

that is, the "area" of the discrete signal divided by its central value. The equivalent time width of a signal is the time width of a rectangular signal with height equal to the value of $x[n]$ at the origin and area equal to that of the original signal. This is illustrated in Fig. 2.4(a) and (b). Note that T_e will not, in general, be an integral multiple of the sample interval T. The *equivalent bandwidth* B_e of the discrete-time Fourier transform $X(f)$ of $x[n]$ is defined analogously as

$$B_e = \frac{\int_{-1/2T}^{1/2T} X(f) \, df}{X(0)}. \tag{2.68}$$

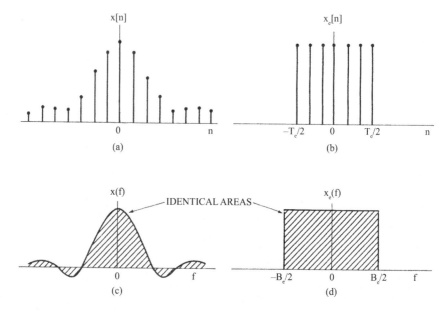

Figure 2.4. (a) Original signal sequence. (b) Equivalent time-width of rectangular signal of equal area and equal height as original signal. (c) Transform of original signal. (d) Equivalent bandwidth of rectangular transform of equal area and equal height as original transform.

This is illustrated in Fig. 2.4(c) and (d). These two measures of the temporal concentration and the frequency concentration apply only to signals that are real-valued and symmetric, that have the maximum value at the origin, and that integrate to a finite nonzero area. These conditions are appropriate for the window functions to be considered in Chap. 5 as part of the classical spectral estimators.

It is obvious from the transform relationships (2.49) and (2.50) that

$$X(0) = T \sum_{n=-\infty}^{\infty} x[n]$$

$$x[0] = \int_{-1/2T}^{1/2T} X(f)\,df, \tag{2.69}$$

from which it follows that the time-bandwidth product

$$T_e B_e = \frac{T \sum_{n=-\infty}^{\infty} x[n]}{x[0]} \cdot \frac{\int_{-1/2T}^{1/2T} X(f)\,df}{X(0)} = 1, \tag{2.70}$$

that is, the equivalent widths of a signal and its transform are reciprocals. It is this unity TBP relationship that is responsible for the rule-of-thumb cited at the beginning of this section, for which T_e is considered to be the observation interval.

An alternative TBP may be defined, based upon mean-square width. The mean-square width is a measure of the variance (mean-square deviation) of a function about its mean value. In order to handle a broader class of functions than possible with equivalent width definitions (2.67) and (2.68), the mean-square width definitions are based on the squared-magnitude functions $|x[n]|^2$ and $|X(f)|^2$. This will permit complex-valued functions, oscillatory functions, and functions that have zero integrated area to have meaningful mean-square widths. The *mean-square time width* $\tilde{T}_e$ of discrete-time signal $x[n]$ is defined as

$$(\tilde{T}_e)^2 = \frac{T \displaystyle\sum_{n=-\infty}^{\infty} (nT)^2 |x[n]|^2}{T \displaystyle\sum_{n=-\infty}^{\infty} |x[n]|^2}, \tag{2.71}$$

and the *mean-square bandwidth* $\tilde{B}_e$ of the DTFT $X(f)$ of $x[n]$ is defined analogously as

$$(\tilde{B}_e)^2 = \frac{\displaystyle\int_{-1/2T}^{1/2T} f^2 |X(f)|^2 \, df}{\displaystyle\int_{-1/2T}^{1/2T} |X(f)|^2 \, df}. \tag{2.72}$$

These definitions assume that the mean values of $x[n]$ and $X(f)$ are located at the origins $x[0]$ and $X(0)$, although the definitions can be modified to handle the case for mean values not at the origin.

Noting that the energy E of $x[n]$ and $X(f)$ is given by Eq. (2.51) as

$$\mathsf{E} = T \sum_{n=-\infty}^{\infty} |x[n]|^2 = \int_{-1/2T}^{1/2T} |X(f)|^2 \, df,$$

then the mean-square width product $(\tilde{T}_e)^2 (\tilde{B}_e)^2$ is

$$(\tilde{T}_e)^2 (\tilde{B}_e)^2 = \frac{T \displaystyle\sum_{n=-\infty}^{\infty} (nT)^2 |x[n]|^2 \int_{-1/2T}^{1/2T} f^2 |X(f)|^2 \, df}{\mathsf{E}^2}. \tag{2.73}$$

It can be shown [Bracewell, 1978; Papoulis, 1977] using a variation of the Schwartz inequality and assuming that the signal has finite energy E and satisfies

$$\lim_{n \to \infty} (nT)|x[n]|^2 = 0,$$

then $(\tilde{T}_e)^2 (\tilde{B}_e)^2 \geq 1/16\pi^2$. Therefore, the classical *root-mean-square* (rms) *time-bandwidth product* inequality satisfies

$$\tilde{T}_e \tilde{B}_e \geq \frac{1}{4\pi}. \tag{2.74}$$

If $\tilde{T}_e$ and T_e are approximately the same time width, then the TBP of (2.74) establishes a smaller concentration bandwidth than given by the TBP of (2.70).

The *half-power* (3-dB) *bandwidth* B_h is often used in practive as an approximation for $\tilde{B}_e$ because it is easy to measure. This is the bandwidth about the origin of the transform at which the squared magnitude of the transform is half its squared magnitude at the origin, i.e., $|X(\pm B_h/2)|^2 = |X(0)|^2 /2$.

The two TBP expressions (2.70) and (2.74) establish a relationship between the time concentration of a *single* signal and the spectral concentration of its transform. The TBP does *not* explicitly quantify the capability to *resolve* the spectral responses due to *two* or more signals. As a result, other proposed definitions of *resolution* have appeared in the literature. Most concern the capability to resolve the responses of *two* sinusoids that are close in frequency and equal in amplitude. These definitions rely on some measure of how close two sinusoids may be before their spectral responses are indistinguishable. The general concept behind these definitions assumes that the frequency separation of the sinusoids may be no closer than the equivalent bandwidth of the window through which a segment of the two sinusoids is observed. Because most resolution measures deal with signals as seen through windows, the time-bandwidth product $T_e B_e$ is most often used to establish the equivalent bandwidth, and therefore, the resolution. Thus, resolution in Hertz is said to be approximately equal to the reciprocal of the observation time. Some spectral estimation methods in this text can achieve resolutions better than this. This suggests that the TBP must be interpreted carefully if used to infer resolution. The high-resolution methods effectively extrapolate the measured signal beyond its observed interval. The effective time interval has a larger time concentration than the original observation interval. This means T_e is effectively larger, so that B_e will be smaller and the resolution correspondingly better.

2.12   EXTRA: SOURCE OF COMPLEX-VALUED SIGNALS

Although most discrete-time signals originate as real-valued samples of a continuous-time signal, modern signal processing systems very often convert the real-valued samples into complex-valued samples at some point in the processing chain because real samples are multiplied with a complex sinusoid $\exp(j\omega t)$. For example, processing in the frequency domain is often a system requirement. This means that the complex-valued DTFS $X[k]$

$$X[k] = T \sum_{n=0}^{N-1} x[n] \exp(-j2\pi kn/N)$$

will be used in the processing, which involves complex sinusoid products.

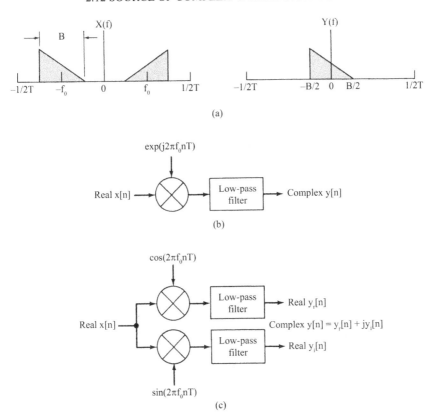

Figure 2.5. Complex demodulation process. (a) Spectra of original real signal (left) and demodulated complex signal (right). (b) Complex demodulator. (c) Scheme to implement complex demodulation.

Complex demodulation of real-valued bandpass signals to complex-valued baseband signals is another common source of complex data (originally introduced in the Weaver demodulator [1956]). Consider in Fig. 2.5(a) the symmetric spectrum of the real-valued bandpass signal and the desired complex-valued baseband signal centered at 0 Hz. This baseband signal of bandwidth B Hz is created by forming the product $x[n] \exp(j2\pi f_0 nT)$, in which f_0 is the center frequency of the bandpass signal, and then filtering the resulting complex signal by a low-pass filter with real-valued coefficients and band edges at $\pm B/2$ Hz, as shown in Fig. 2.5(b). Noting that

$$\exp(j2\pi f_0 nT) = \cos(2\pi f_0 nT) + j \sin(2\pi f_0 nT),$$

then Fig. 2.5(c) presents a practical implementation of complex demodulation. The advantages of complex data include: less aliasing in the frequency domain (and, therefore, less bias in spectral estimators); the sampling rate can be halved, which

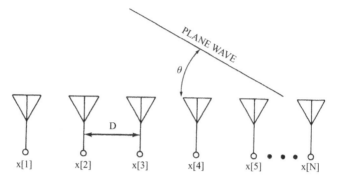

Figure 2.6. Linear array of N sensors with spacing D.

may be easier to implement in hardware due to lower clock frequencies; the model orders of the parametric spectral analysis techniques of Chaps. 8–13 can be halved, which saves computations.

2.13   EXTRA: WAVENUMBER PROCESSING WITH LINEAR SPATIAL ARRAYS

A temporal complex sinusoid has the exponential form $\exp(j2\pi f nT)$ for $1 \leq n \leq N$ when sampled at uniform increments of T seconds across a time aperture of NT seconds. Spectral analysis is used to determine the sinusoidal frequency f from the N data samples. The analogous spatial problem is encountered when processing signals from an array of uniformly spaced sensors, as illustrated in Fig. 2.6. A monochromatic plane wave of frequency f impinging on the array at an angle θ will have the exponential form $\exp(j2\pi knD)$ for $1 \leq n \leq N$ when sampled at uniform inter-sensor spacing increments of D across a spatial aperture of length ND. Spectral analysis is used to determine the wavenumber k from the N sensor samples. The wavenumber is related to the angle-of-incidence θ as follows

$$k = \frac{f}{c} \sin \theta = \frac{1}{\lambda} \sin \theta$$

in which c is the plane wave velocity of propagation and λ is the plane wave wavelength. Thus, spatial spectral analysis may be use for direction finding by relating the angle θ to the estimated wavenumber k.

References

Bergland, G. D., A Guided Tour of the Fast Fourier Transform, *IEEE Spectrum*, vol. 6, pp. 41–52, July 1969.

Bloomfield, Peter, *Fourier Analysis of Time Series: An Introduction*, John Wiley & Sons, Inc., New York, 1976.

Bracewell, R. N., *The Fourier Transform and Its Applications*, second ed., McGraw-Hill Book Company, New York, 1978.

Bracewell, R. N., The Fourier Transform, *Scientific American*, vol. 260, pp. 86–95, June 1989.

Brigham, E. Oran, *The Fast Fourier Transform*, Prentice-Hall, Inc., Englewood Cliffs, N. J., 1974.

Cochran, W. T., et al., What Is the Fast Fourier Transform?, *IEEE Trans. Audio Electroacoust.*, vol. AU-15, pp. 45–55, June 1967.

Digital Signal Processing Committee, ed., Fast Fourier Transform Subroutines, Chapter 1 in *Programs for Digital Signal Processing*, IEEE Press, New York, 1979.

Elliott, D. F., and K. R. Rao, *Fast Algorithms–Applications, Analysis, and Application*, Academic Press, Inc., New York, 1983.

Geçkinli, N. C., and Davras Yavuz, *Discrete Fourier Transformation and Its Applications to Power Spectra Estimation*, Elsevier Scientific Publishing Company, Amsterdam, 1983.

Markel, J. D., FFT Pruning, *IEEE Trans. Audio Electroacoust.*, vol. AU-19, pp. 305–311, December 1971.

McClellan, J. H., and C. M. Rader, eds., *Number Theory in Digital Signal Processing*, Prentice-Hall, Inc., Englewood Cliffs, N. J., 1979.

Oppenheim, A. V., and R. W. Schafer, *Digital Signal Processing*, Prentice-Hall, Inc., Englewood Cliffs, N. J., 1975.

Oppenheim, A. V., and R. W. Schafer, *Discrete-Time Signal Processing*, Prentice-Hall, Inc., Englewood Cliffs, N. J., 1989.

Otnes, R. K., and L. Enochson, *Digital Time Series Analysis*, John Wiley & Sons, Inc., New York, 1972.

Papoulis, Athanasios, *The Fourier Integral and Its Application*, McGraw-Hall Book Company, New York, 1962.

Papoulis, Athanasios, *Signal Analysis*, McGraw-Hill Book Company, New York, 1977.

Rabiner, L. R., and Ben Gold, *Theory and Application of Digital Signal Processing*, Prentice-Hall, Inc., Englewood Cliffs, N. J., 1975.

Skinner, D. P., Pruning the Decimation-in-Time FFT Algorithm, *IEEE Trans. Acoust. Speech Signal Process.*, vol. ASSP-24, pp. 193–194, April 1976.

Sreenivas, T. V., and P. V. S. Rao, FFT Algorithm for Both Input and Output Pruning, *IEEE Trans. Acoust. Speech Signal Process.*, vol. ASSP-27, pp. 291–292, June 1979.

Weaver, D. K. Jr., A Third Method of Generation and Detection of Single-Sideband Signals, *Proc. IRE*, vol. 44, pp. 1703–1705, December 1956.

3

REVIEW OF MATRIX ALGEBRA

3.1 INTRODUCTION

An indispensible tool for describing the mathematical relationships of most spectral estimation techniques is linear algebra, with a particular emphasis on matrix operations. Matrices provide a convenient, concise method for systematic representation of the detailed algebraic and numerical relationships that occur frequently in spectral analysis. One matrix structure, the Toeplitz matrix, will receive special treatment because it is particularly relevant for many spectral estimation methods. The goal of this chapter is to introduce basic matrix notation and matrix algebra operations that will be used extensively throughout this book. Some familiarity on the part of the reader with the subject of linear algebra is assumed and, therefore, most results provided in this chapter are provided without proofs. Proofs and other details may be found by consulting the references listed at the end of this chapter, particularly those of Noble and Daniel [1977] and Golub and Van Loan [1989].

3.2 MATRIX ALGEBRA BASICS

A *matrix* is defined as a set of real or complex numbers arranged in a rectangular array

$$\mathbf{A} = \{a[i, j]\} = \begin{pmatrix} a[1, 1] & a[1, 2] & \cdots & a[1, n] \\ a[2, 1] & a[2, 2] & \cdots & a[2, n] \\ \vdots & \vdots & \ddots & \vdots \\ a[m, 1] & a[m, 2] & \cdots & a[m, n] \end{pmatrix} \quad (3.1)$$

of m rows and n columns. The individual entries in the matrix, such as $a[i, j]$, are the *elements* of the matrix. The mn elements compose a matrix said to be of *dimension*, or size, $m \times n$. Here $a[i, j]$ designates the element in the ith row and jth column of the matrix, where $1 \leq i \leq m$ and $1 \leq j \leq n$, although other index ranges may be useful, such as $0 \leq i < m$. Matrices throughout this text will be denoted by boldface capital letters.

A *block matrix* is a matrix composed of elements that are themselves matrices:

$$\underline{\mathbf{A}} = \{\mathbf{A}[i, j]\} = \begin{pmatrix} \mathbf{A}[1, 1] & \cdots & \mathbf{A}[1, q] \\ \vdots & \ddots & \vdots \\ \mathbf{A}[p, 1] & \cdots & \mathbf{A}[p, q] \end{pmatrix}. \qquad (3.2)$$

Each matrix element $\mathbf{A}[i, j]$ has the same $m \times n$ dimension. The *block dimension* of $\underline{\mathbf{A}}$ is said to be $p \times q$. The *scalar dimension* of the matrix $\underline{\mathbf{A}}$ is then $pm \times qn$. A block matrix will be denoted by an underlined boldface capital letter. The block definition will be useful later for the presentations on multichannel and two-dimensional spectral estimation.

A *row vector* is a $1 \times n$ matrix

$$\mathbf{b} = (b[1] \ b[2] \ \cdots \ b[n])$$

and is denoted simply with a bold lowercase letter. A *column vector* is an $m \times 1$ matrix

$$\mathbf{c} = \begin{pmatrix} c[1] \\ \vdots \\ c[m] \end{pmatrix}$$

and is also denoted by a bold lowercase letter. A *block column vector* is simply a block matrix of block dimension $p \times 1$ (scalar dimension $pm \times n$)

$$\underline{\mathbf{c}} = \begin{pmatrix} \mathbf{C}[1] \\ \vdots \\ \mathbf{C}[p] \end{pmatrix}$$

and is denoted by an underlined bold lowercase letter.

Matrices $\mathbf{A}$ and $\mathbf{B}$ are said to be equal if and only if $\mathbf{A}$ and $\mathbf{B}$ have the same row dimension and the same column dimension and $a[i, j] = b[i, j]$ for all i and j. The sum or difference of two $m \times n$ matrices $\mathbf{A}$ and $\mathbf{B}$ to form a new matrix $\mathbf{C}$

$$\mathbf{C} = \mathbf{A} \pm \mathbf{B} \qquad (3.3)$$

is obtained simply by summing or differencing the corresponding elements

$$c[i, j] = a[i, j] \pm b[i, j]$$

for $1 \leq i \leq m$ and $1 \leq j \leq n$. Note that matrix addition is both associative and commutative. Multiplication of a matrix $\mathbf{A}$ by a scalar α is defined by

$$\alpha \mathbf{A} = \{\alpha a[i, j]\}. \qquad (3.4)$$

Two matrices $\mathbf{A}$ and $\mathbf{B}$ can be multiplied together in the order $\mathbf{AB}$ to yield the resultant matrix $\mathbf{C}$ if and only if the number of columns in $\mathbf{A}$ equals the number of rows in **B**. The (i, k) element of the product **AB** is obtained by multiplication of the elements of the ith row of **A** with the elements of the kth column of **B**. That is, if **A** = $\{a[i, j]\}$ is $m \times n$ and **B** = $\{b[j, k]\}$ is $n \times p$, then the matrix **C** = **AB** is an $m \times p$ matrix with elements

$$c[i, k] = \sum_{j=1}^{n} a[i, j] b[j, k]. \tag{3.5}$$

Note that matrix multiplication is associative, but is not, in general, commutative (**AB** ≠ **BA**). Note also that **AB** = **0**, where **0** is the all-zeros matrix, does not necessarily imply that either **A** = **0** or **B** = **0**. The relationship **AB** = **AC** also does not necessarily imply that **B** = **C**.

The *transpose* of an $m \times n$ matrix **A** is the $n \times m$ matrix $\mathbf{A}^T$ obtained by interchanging the rows and columns of **A**

$$\mathbf{A}^T = \begin{pmatrix} a[1, 1] & a[2, 1] & \cdots & a[m, 1] \\ a[1, 2] & a[2, 2] & \cdots & a[m, 2] \\ \vdots & \vdots & \ddots & \vdots \\ a[1, n] & a[2, n] & \cdots & a[m, n] \end{pmatrix} \tag{3.6}$$

so that the element of $\mathbf{A}^T$ indexed by (i, j) is simply the element of **A** indexed by (j, i). The superscript ^{T} will be used to denote the transpose operation. It can be shown that the transpose of the sum of two matrices is the sum of their transposes

$$(\mathbf{A} + \mathbf{B})^T = \mathbf{A}^T + \mathbf{B}^T, \tag{3.7}$$

while the transpose of the product of two matrices is the product of their transposes in reverse order

$$(\mathbf{AB})^T = \mathbf{B}^T \mathbf{A}^T. \tag{3.8}$$

Using the symbol $*$ to denote complex conjugation, the matrix $\mathbf{A}^*$ will designate a matrix whose (i, j) element is the complex conjugate of the (i, j) element of **A**. The *Hermitian transpose* of an $m \times n$ complex element matrix **A** is then the $n \times m$ matrix $\mathbf{A}^H$ obtained by complex conjugating all the elements of **A** and transposing the elements in the usual fashion, or transposing then conjugating $[\mathbf{A}^H = (\mathbf{A}^T)^* = (\mathbf{A}^*)^T]$,

$$\mathbf{A}^H = \begin{pmatrix} a^*[1, 1] & a^*[2, 1] & \cdots & a^*[m, 1] \\ a^*[1, 2] & a^*[2, 2] & \cdots & a^*[m, 2] \\ \vdots & \vdots & & \vdots \\ a^*[1, n] & a^*[2, n] & \cdots & a^*[m, n] \end{pmatrix}. \tag{3.9}$$

The superscript ^{H} will be used throughout the text to denote the Hermitian transpose operation. Analogous to the transpose, it may be shown that

$$(\mathbf{A} + \mathbf{B})^H = \mathbf{A}^H + \mathbf{B}^H \tag{3.10}$$

$$(\mathbf{AB})^H = \mathbf{B}^H \mathbf{A}^H. \tag{3.11}$$

The *block Hermitian transpose* of a $p \times q$ complex block-element matrix $\underline{\mathbf{A}}$ is the $q \times p$ block matrix $\underline{\mathbf{A}}^H$ obtained by Hermitian transposing all the elements of $\underline{\mathbf{A}}$

$$\underline{\mathbf{A}}^H = \begin{pmatrix} \mathbf{A}^H[1,1] & \cdots & \mathbf{A}^H[p,1] \\ \vdots & \ddots & \vdots \\ \mathbf{A}^H[1,q] & \cdots & \mathbf{A}^H[p,q] \end{pmatrix}. \tag{3.12}$$

The use of bold lowercase letters to designate both column and row vectors is ambiguous. To prevent confusion in the text, bold lowercase letters will be reserved only for column vectors. References to "vector" in this book will mean column vector, unless explicitly indicated to be a row vector. A row vector can always be devised by simply transposing a column vector

$$\mathbf{c}^T = \begin{pmatrix} c[1] & \cdots & c[m] \end{pmatrix}. \tag{3.13}$$

A useful quantity encountered frequently is the vector *inner product* between two $n \times 1$ vectors

$$\mathbf{w} = \begin{pmatrix} w[1] \\ \vdots \\ w[n] \end{pmatrix} \qquad \mathbf{v} = \begin{pmatrix} v[1] \\ \vdots \\ v[n] \end{pmatrix}$$

defined as the scalar

$$\langle \mathbf{w}, \mathbf{v} \rangle = \sum_{j=1}^{n} w^*[j]\,v[j]. \tag{3.14}$$

Note that this can be expressed with normal matrix multiplication rules as

$$\langle \mathbf{w}, \mathbf{v} \rangle = \mathbf{w}^H \mathbf{v}. \tag{3.15}$$

In an analogous manner, the vector *outer product*, defined as $\mathbf{w}\mathbf{v}^H$, forms an $m \times m$ square matrix from the $m \times 1$ vectors $\mathbf{w}$ and $\mathbf{v}$

$$\mathbf{w}\mathbf{v}^H = \begin{pmatrix} w[1]v^*[1] & w[1]v^*[2] & \cdots & w[1]v^*[m] \\ \vdots & \vdots & & \vdots \\ w[m]v^*[1] & w[m]v^*[2] & \cdots & w[m]v^*[m] \end{pmatrix}. \tag{3.16}$$

A sense of the "size" or "length" of vectors and matrices can be expressed by either of the *vector norms*

$$\|\mathbf{v}\|_1 = \sum_{i=1}^{m} |v[i]| \tag{3.17}$$

$$\|\mathbf{v}\|_2 = \left(\sum_{i=1}^{m} |v[i]|^2 \right)^{1/2} = \langle \mathbf{v}, \mathbf{v} \rangle^{1/2} \tag{3.18}$$

or the *Frobenius matrix norm*

$$\|\mathbf{A}\|_F = \left(\sum_{i=1}^{m} \sum_{j=1}^{n} |a[i, j]|^2 \right)^{1/2}. \tag{3.19}$$

Each of these norms yields nonnegative scalar values.

3.3 SPECIAL VECTOR AND MATRIX STRUCTURES

This section identifies special vectors and matrices to be encountered frequently in the chapters ahead. The reader should become familiar with the notations presented here, due to their frequent use throughout this text.

An $m \times 1$ *zero vector* of m zeros and an $m \times m$ *zero matrix* of mn zeros are denoted, respectively, by

$$\mathbf{0}_m = \begin{pmatrix} 0 \\ \vdots \\ 0 \end{pmatrix}, \quad \mathbf{0}_{mn} = \begin{pmatrix} 0 & \cdots & 0 \\ \vdots & \ddots & \vdots \\ 0 & \cdots & 0 \end{pmatrix} \tag{3.20}$$

in which the zero vector is distinguished from the zero matrix by the number of subscripts. The sequence $\exp(j2\pi f mT)$ is a *complex sinusoid* of frequency f, sampled at T second intervals. The *complex sinusoid vector* $\mathbf{e}_m(f)$ of frequency f is defined as the $(m + 1) \times 1$ vector

$$\mathbf{e}_m(f) = \begin{pmatrix} 1 \\ \exp(j2\pi f T) \\ \vdots \\ \exp(j2\pi f mT) \end{pmatrix} \tag{3.21}$$

in which m designates the highest time index. The discrete-time Fourier transform (DTFT) for N data samples can be expressed, for example, as the scaled vector inner product

$$X(f) = T \, \mathbf{e}_{N-1}^{H}(f)\mathbf{x}_{N-1}, \tag{3.22}$$

for which the data vector $\mathbf{x}$ is defined as

$$\mathbf{x}_{N-1} = \begin{pmatrix} x[0] \\ \vdots \\ x[N-1] \end{pmatrix}.$$

A matrix for which the number of rows, n, equals the number of columns is said to be a *square matrix* of size $n \times n$. The elements $a[i, i]$ for $i = 1$ to n are said to lie on the *principal diagonal* (or *main diagonal*) of the square matrix **A**. The elements $a[i, n + 1 - i]$ are said to be on the *cross diagonal*.

A *symmetric matrix* **S** is a square matrix such that

$$\mathbf{S}^T = \mathbf{S}, \tag{3.23}$$

which means $s[i, j] = s[j, i]$ symmetrically about the principal diagonal. An example of a 4×4 symmetric matrix is

$$\mathbf{S} = \begin{pmatrix} s[1, 1] & s[1, 2] & s[1, 3] & s[1, 4] \\ s[1, 2] & s[2, 2] & s[2, 3] & s[2, 4] \\ s[1, 3] & s[2, 3] & s[3, 3] & s[3, 4] \\ s[1, 4] & s[2, 4] & s[3, 4] & s[4, 4] \end{pmatrix}. \tag{3.24}$$

A *Hermitian matrix* is a square $n \times n$ matrix **G** of complex elements with complex conjugate symmetry

$$\mathbf{G}^H = \mathbf{G}. \tag{3.25}$$

A *diagonal matrix* **D** is a square matrix in which all the elements are zero, except for the elements along the principal diagonal (some of which may be zero),

$$\mathbf{D} = \begin{pmatrix} d[1] & 0 & \cdots & 0 \\ 0 & d[2] & \cdots & 0 \\ \vdots & \vdots & \ddots & \vdots \\ 0 & 0 & \cdots & d[n] \end{pmatrix}. \tag{3.26}$$

The abbreviation $\mathbf{D} = \mathrm{diag}\,(d[1], \ldots, d[n])$ will be used as a more compact description of a diagonal matrix.

The *identity matrix* **I** is a special case diagonal matrix in which all the diagonal elements are unity

$$\mathbf{I} = \begin{pmatrix} 1 & 0 & \cdots & 0 \\ 0 & 1 & \cdots & 0 \\ \vdots & \vdots & \ddots & \vdots \\ 0 & 0 & \cdots & 1 \end{pmatrix}. \tag{3.27}$$

If the size of the identity matrix is needed within a matrix equation, the dimension will be explicitly noted as a subscript. Thus, $\mathbf{I}_m$ denotes an $m \times m$ identity matrix. For any $m \times n$ matrix **A**,

$$\mathbf{I}_m \mathbf{A} = \mathbf{A} \mathbf{I}_n = \mathbf{A}.$$

If the dimension subscript is omitted from **I**, it will be understood from the context what dimension is required.

A close relative of the identity matrix is the *reflection matrix* $\mathbf{J}$ (sometimes called an *exchange matrix* or a *reversing matrix*)

$$\mathbf{J} = \begin{pmatrix} 0 & \cdots & 0 & 1 \\ 0 & \cdots & 1 & 0 \\ \vdots & \ddots & \vdots & \vdots \\ 1 & \cdots & 0 & 0 \end{pmatrix}, \tag{3.28}$$

which has unity elements along the cross diagonal. The $\mathbf{J}$ matrix reverses, or exchanges, the order of the rows or columns of a matrix or vector. Premultiplying an $m \times n$ matrix $\mathbf{A}$ by $\mathbf{J}_m$, where the subscript m denotes the reflection matrix size, yields a matrix with reversed ordering of rows (reflection about the central horizontal axis)

$$\mathbf{J}_m\mathbf{A} = \begin{pmatrix} a[m,1] & a[m,2] & \cdots & a[m,n] \\ \vdots & \vdots & & \vdots \\ a[2,1] & a[2,2] & \cdots & a[2,n] \\ a[1,1] & a[1,2] & \cdots & a[1,n] \end{pmatrix}, \tag{3.29}$$

whereas postmultiplying by $\mathbf{J}_n$ yields a matrix with reversed ordering of columns (reflection about the central vertical axis)

$$\mathbf{A}\mathbf{J}_n = \begin{pmatrix} a[1,n] & \cdots & a[1,2] & a[1,1] \\ a[2,n] & \cdots & a[2,2] & a[2,1] \\ \vdots & & \vdots & \vdots \\ a[m,n] & \cdots & a[m,2] & a[m,1] \end{pmatrix}. \tag{3.30}$$

As with the identity matrix, the dimension subscript of the $\mathbf{J}$ matrix will often be omitted when its size can be deduced from the context of a matrix equation. In an analogous manner for matrices, $\mathbf{Jc}$ will reverse the element order of the column vector $\mathbf{c}$, and $\mathbf{c}^T\mathbf{J}$ will reverse the element order of the row vector $\mathbf{c}^T$. It may be easily shown that $\mathbf{J}^T = \mathbf{J}$ and $\mathbf{J}^2 = \mathbf{JJ} = \mathbf{I}$.

The *shift matrix* $\mathbf{Z}$ is defined as a square matrix of dimension $m \times m$ with $m-1$ ones along the diagonal just below the main diagonal. An example of a 4×4 shift matrix is

$$\mathbf{Z} = \begin{pmatrix} 0 & 0 & 0 & 0 \\ 1 & 0 & 0 & 0 \\ 0 & 1 & 0 & 0 \\ 0 & 0 & 1 & 0 \end{pmatrix}. \tag{3.31}$$

The effect of $\mathbf{Z}$ (or $\mathbf{Z}^T$) applied by multiplication on the left to a matrix $\mathbf{A}$ is to shift the rows down (up). When applied by multiplication on the right, it shifts the columns right (left).

A *persymmetric matrix* $\mathbf{P}$ is an $n \times n$ square matrix that is symmetric about its cross diagonal, as illustrated by the 4×4 example

$$\mathbf{P} = \begin{pmatrix} p[1,1] & p[1,2] & p[1,3] & p[1,4] \\ p[2,1] & p[2,2] & p[2,3] & p[1,3] \\ p[3,1] & p[3,2] & p[2,2] & p[1,2] \\ p[4,1] & p[3,1] & p[2,1] & p[1,1] \end{pmatrix}. \tag{3.32}$$

Thus, $p[i,j] = p[n-j+1, n-i+1]$. It may easily be shown that $\mathbf{P}^T = \mathbf{JPJ}$ and $\mathbf{P} = \mathbf{JP}^T\mathbf{J}$.

A *centrosymmetric matrix* $\mathbf{R}$ is an $n \times n$ square matrix with the property $r[i,j] = r^*[n-i+1, n-j+1]$. A special class of centrosymmetric matrices are those that are doubly symmetric, that is, Hermitian about the principal diagonal and persymmetric about the cross diagonal, as illustrated by the 4×4 complex matrix

$$\mathbf{R} = \begin{pmatrix} r[1,1] & r^*[2,1] & r^*[3,1] & r^*[4,1] \\ r[2,1] & r[2,2] & r^*[3,2] & r^*[3,1] \\ r[3,1] & r[3,2] & r[2,2] & r^*[2,1] \\ r[4,1] & r[3,1] & r[2,1] & r[1,1] \end{pmatrix}. \tag{3.33}$$

In this special case, $r[i,j] = r^*[j,i] = r[n-j+1, n-i+1] = r^*[n-i+1, n-j+1]$. It may easily be shown that $\mathbf{R} = \mathbf{JRJ}$ if $\mathbf{R}$ is real, or $\mathbf{R} = \mathbf{JR}^*\mathbf{J}$ if $\mathbf{R}$ has complex elements. When the dimension of a centrosymmetric matrix is even ($n = 2r$), it may be partitioned into the form

$$\mathbf{R}_{even} = \begin{pmatrix} \mathbf{A} & \mathbf{B} \\ \mathbf{JB}^*\mathbf{J} & \mathbf{JA}^*\mathbf{J} \end{pmatrix}, \tag{3.34}$$

where $\mathbf{A}$ and $\mathbf{B}$ are general $r \times r$ matrices of no special structure. Similarly, when the dimension is of odd order, $n = 2r + 1$, then a centrosymmetric matrix may be partitioned as

$$\mathbf{R}_{odd} = \begin{pmatrix} \mathbf{A} & \mathbf{x} & \mathbf{B} \\ \mathbf{x}^H & \alpha & \mathbf{x}^H\mathbf{J} \\ \mathbf{JB}^*\mathbf{J} & \mathbf{Jx} & \mathbf{JA}^*\mathbf{J} \end{pmatrix}, \tag{3.35}$$

where $\mathbf{A}$ and $\mathbf{B}$ are general $r \times r$ matrices, $\mathbf{J}$ is an $r \times r$ reflection matrix, $\mathbf{x}$ is a column vector of dimension r, and α is a real scalar.

A very important matrix in spectral estimation is the *Toeplitz matrix* $\mathbf{T}$ (named after the German mathematician O. Toeplitz). It has the property that all the elements along any diagonal are identical, that is, $t[i,j] = t[i-j]$. A Toeplitz matrix can be, in general, nonsquare. An example of a 5×4 Toeplitz matrix is

$$\mathbf{T} = \begin{pmatrix} t[0] & t[-1] & t[-2] & t[-3] \\ t[1] & t[0] & t[-1] & t[-2] \\ t[2] & t[1] & t[0] & t[-1] \\ t[3] & t[2] & t[1] & t[0] \\ t[4] & t[3] & t[2] & t[1] \end{pmatrix}. \tag{3.36}$$

Note that a square Toeplitz matrix is a special case of a persymmetric matrix,

$$\mathbf{T} = \mathbf{J}\mathbf{T}^T\mathbf{J}. \tag{3.37}$$

If a square Toeplitz matrix has the additional property that it is Hermitian, $t^*[k] = t[-k]$, then $\mathbf{T} = \mathbf{J}\mathbf{T}^*\mathbf{J}$. Therefore, a Hermitian Toeplitz matrix is centrosymmetric.

A close relative of the Toeplitz matrix is the *Hankel matrix* $\mathbf{H}$, which has the property that all the elements along any cross diagonal are identical, that is, $h[i, j] = h[i + j - n - 1]$. A Hankel matrix can be, in general, nonsquare. An example of a 4×5 Hankel matrix is

$$\mathbf{H} = \begin{pmatrix} h[-4] & h[-3] & h[-2] & h[-1] & h[0] \\ h[-3] & h[-2] & h[-1] & h[0] & h[1] \\ h[-2] & h[-1] & h[0] & h[1] & h[2] \\ h[-1] & h[0] & h[1] & h[2] & h[3] \end{pmatrix}. \tag{3.38}$$

Note that a square $n \times n$ Hankel matrix is a special case of a Hermitian matrix, that is, $\mathbf{H}^H = \mathbf{H}$. Hankel matrices can be related to Toeplitz matrices.

A *circulant matrix* $\mathbf{C}$ is an $n \times n$ square matrix composed from n unique complex elements. If the matrix elements have the relationship

$$c[i, j] = \begin{cases} c[j - i] & \text{for } j - i \geq 0 \\ c[n - j + 1] & \text{for } j - i < 0 \end{cases} \tag{3.39}$$

over the range $1 \leq i, j \leq n$, then the matrix $\mathbf{C}$ is said to be *right-circulant*. An example of a 4×4 right-circulant matrix is

$$\mathbf{C} = \begin{pmatrix} c[0] & c[1] & c[2] & c[3] \\ c[3] & c[0] & c[1] & c[2] \\ c[2] & c[3] & c[0] & c[1] \\ c[1] & c[2] & c[3] & c[0] \end{pmatrix}. \tag{3.40}$$

Observe that each row is formed from the row above it by shifting right one element and bringing the rightmost element to the left-hand side of the row. A *left-circulant* matrix may be defined in an analogous manner for left shifts.

Another matrix form that is found in spectral estimation is the $m \times n$ *Vandermonde matrix* $\mathbf{V}$ (named after Alexandre Vandermonde [1735–1796]) with elements that are expressed in terms of powers of the n base parameters $z_1, \ldots, z_n$,

$$v[i, j] = z_j^{i-1} \quad \text{for} \quad 1 \le i \le m, 1 \le j \le n. \tag{3.41}$$

Therefore, an $m \times n$ Vandermonde matrix would have the structure

$$\mathbf{V} = \begin{pmatrix} 1 & 1 & \cdots & 1 \\ z_1 & z_2 & \cdots & z_n \\ z_1^2 & z_2^2 & \cdots & z_n^2 \\ \vdots & \vdots & & \vdots \\ z_1^{m-1} & z_2^{m-1} & \cdots & z_n^{m-1} \end{pmatrix}. \tag{3.42}$$

An *upper triangular matrix* $\mathbf{U}$ is an $n \times n$ square matrix with zero elements below its principal diagonal, i.e., $u[i, j] = 0$ for $j < i$. A 3×3 example is

$$\mathbf{U} = \begin{pmatrix} u[1, 1] & u[1, 2] & u[1, 3] \\ 0 & u[2, 2] & u[2, 3] \\ 0 & 0 & u[3, 3] \end{pmatrix}. \tag{3.43}$$

$\mathbf{U}$ is also called a *right triangular matrix* because the nonzero elements are to the right of the main diagonal. A *lower triangular matrix* $\mathbf{L}$ is an $n \times n$ square matrix with zero elements above its principal diagonal, i.e., $l[i, j] = 0$ for $j > i$. An example is

$$\mathbf{L} = \begin{pmatrix} l[1, 1] & 0 & 0 \\ l[2, 1] & l[2, 2] & 0 \\ l[3, 1] & l[3, 2] & l[3, 3] \end{pmatrix}. \tag{3.44}$$

$\mathbf{L}$ is also called a *left triangular matrix* because the nonzero elements are to the left of the main diagonal.

3.4 MATRIX INVERSE

An $n \times n$ square matrix $\mathbf{A}^{-1}$ such that $\mathbf{A}^{-1}\mathbf{A} = \mathbf{A}\mathbf{A}^{-1} = \mathbf{I}$, if such a matrix $\mathbf{A}^{-1}$ exists, is called the *inverse* of the square matrix $\mathbf{A}$. A square matrix that possesses such an inverse is said to be *nonsingular*. If it does not possess an inverse, the matrix is said to be *singular*. If $\mathbf{A}$, $\mathbf{B}$, and the product $\mathbf{AB}$ are all nonsingular, an important identity that will be used frequently is

$$(\mathbf{AB})^{-1} = \mathbf{B}^{-1}\mathbf{A}^{-1}. \tag{3.45}$$

A set of $n \times 1$ vectors $\mathbf{v}_1, \ldots, \mathbf{v}_n$ is said to be *linearly dependent* if there exist numbers $\alpha_1, \ldots, \alpha_n$, not all zero, such that

$$\alpha_1 \mathbf{v}_1 + \alpha_2 \mathbf{v}_2 + \cdots + \alpha_n \mathbf{v}_n = 0, \tag{3.46}$$

otherwise, the vector set is *linearly independent*. If the columns of a rectangular $m \times n$ matrix $\mathbf{A}$ are considered as individual $m \times 1$ vectors, then the *column rank* of $\mathbf{A}$ is the number of linearly independent column vectors. Similarly, if the rows of $\mathbf{A}$ are partitioned as individual $1 \times n$ row vectors, then the *row rank* of $\mathbf{A}$ is the number of linearly independent row vectors. It can be shown [Noble and Daniel, 1977] for any matrix that the row rank is equal to the column rank, so it is sufficient to indicate a single rank number to be associated with each matrix. A square $m \times m$ matrix $\mathbf{A}$ is nonsingular if and only if its rank is m (i.e., it is invertible). A *rank one* matrix can be expressed in the form of a vector outer product $\mathbf{v}\mathbf{w}^H$, because every column is a multiple of $\mathbf{v}$ and every row is a multiple of $\mathbf{w}^H$. If the inverse does not exist, then the rank k of the matrix $\mathbf{A}$ is less than n. The *null space* of the matrix $\mathbf{A}$ is the linear subspace of nonzero vectors $\mathbf{x}$ that satisfy $\mathbf{Ax} = \mathbf{0}$. The null space will have dimension $n - k$.

Another indicator of the singularity of a square $n \times n$ matrix $\mathbf{A}$ is a unique scalar number called the *determinant*, denoted det $\mathbf{A}$. It is defined by the expression

$$\det \mathbf{A} = \sum_{j=1}^{n} (-1)^{1+j} a[1, j] A_{1j} \tag{3.47}$$

where $a[1, j]$ is an element from the top row of matrix $\mathbf{A}$ and A_{1j} is the determinant of the $(n - 1) \times (n - 1)$ matrix formed by omitting the first row, and jth column of the $n \times n$ matrix $\mathbf{A}$. This is a recursive definition which starts with the fact that a 1×1 matrix (a single element) has a determinant equal to that element. More details on computing determinants may be found in standard linear algebra texts, such as Sec. 6.5 of Noble and Daniel [1977].

An $n \times n$ matrix $\mathbf{A}$ is singular if and only if det $\mathbf{A} = 0$. Thus, if det $\mathbf{A} \neq 0$, then the matrix is invertible. It may also be shown that the product of two $n \times n$ square matrices $\mathbf{A}$ and $\mathbf{B}$ has the determinant

$$\det \mathbf{AB} = \det \mathbf{A} \det \mathbf{B}. \tag{3.48}$$

It is also true that if $\mathbf{A}$ is Hermitian ($\mathbf{A}^H = \mathbf{A}$), then det $\mathbf{A}$ is real-valued.

Two matrix inversion lemmas are frequently used in spectral analysis studies. If $\mathbf{A}$ is a nonsingular $n \times n$ matrix, $\mathbf{C}$ a nonsingular $m \times m$ matrix, $\mathbf{B}$ an $n \times m$ matrix, and $\mathbf{D}$ an $m \times n$ matrix, then the augmented matrix $\mathbf{A} + \mathbf{BCD}$ has the inverse

$$(\mathbf{A} + \mathbf{BCD})^{-1} = \mathbf{A}^{-1} - \mathbf{A}^{-1}\mathbf{B}\left(\mathbf{DA}^{-1}\mathbf{B} + \mathbf{C}^{-1}\right)^{-1}\mathbf{DA}^{-1} \qquad (3.49)$$

assuming the inverse of $\mathbf{DA}^{-1}\mathbf{B} + \mathbf{C}^{-1}$ exists. This relationship is known as the *augmented matrix inversion lemma*. One specialized form of this lemma is of interest. If $\mathbf{C}$ is the unity scalar (a 1×1 matrix), $\mathbf{B}$ is an $n \times 1$ column vector $\mathbf{v}$, and $\mathbf{D}$ is a conjugated $1 \times n$ row vector $\mathbf{w}^H$, then

$$\left(\mathbf{A} + \mathbf{vw}^H\right)^{-1} = \frac{\mathbf{A}^{-1} - \left(\mathbf{A}^{-1}\mathbf{v}\right)\left(\mathbf{w}^H\mathbf{A}^{-1}\right)}{1 + \mathbf{w}^H\mathbf{A}^{-1}\mathbf{v}} \qquad (3.50)$$

is the inverse of a matrix formed by augmenting $\mathbf{A}$ with a rank one matrix $\mathbf{vw}^H$.

Let the $n \times n$ matrix $\mathbf{Y}$ be partitioned into submatrices $\mathbf{A}, \mathbf{B}, \mathbf{C}, \mathbf{D}$ as follows

$$\mathbf{Y} = \begin{pmatrix} \mathbf{A} & \mathbf{D} \\ \mathbf{C} & \mathbf{B} \end{pmatrix}, \qquad (3.51)$$

where $\mathbf{A}$ is an $m \times m$ matrix, $\mathbf{B}$ is an $(n-m) \times (n-m)$ matrix, $\mathbf{C}$ is an $(n-m) \times m$ matrix, and $\mathbf{D}$ is an $m \times (n-m)$ matrix. The inverse of $\mathbf{Y}$ is given by

$$\begin{aligned}
\mathbf{Y}^{-1} &= \begin{pmatrix} \mathbf{A}^{-1} + \mathbf{A}^{-1}\mathbf{D}\Delta^{-1}\mathbf{CA}^{-1} & -\mathbf{A}^{-1}\mathbf{D}\Delta^{-1} \\ -\Delta^{-1}\mathbf{CA}^{-1} & \Delta^{-1} \end{pmatrix} \\
&= \begin{pmatrix} \Lambda^{-1} & -\Lambda^{-1}\mathbf{DB}^{-1} \\ -\mathbf{B}^{-1}\mathbf{C}\Lambda^{-1} & \mathbf{B}^{-1} + \mathbf{B}^{-1}\mathbf{C}\Lambda^{-1}\mathbf{DB}^{-1} \end{pmatrix},
\end{aligned} \qquad (3.52)$$

assuming either that the inverses of $\mathbf{A}$ and $\Delta = \mathbf{B} - \mathbf{CA}^{-1}\mathbf{D}$ exist, or that the inverses of $\mathbf{B}$ and $\Lambda = \mathbf{A} - \mathbf{DB}^{-1}\mathbf{C}$ exist, whichever is appropriate. The matrices Δ and Λ have been termed the *Schur complements* of $\mathbf{A}$. Equation (3.52) is known as the *partitioned matrix inversion lemma*. In the special case where $\mathbf{Y}$ is $n \times n$, $\mathbf{A}$ is an $(n-1) \times (n-1)$ matrix, $\mathbf{D} = \mathbf{v}$ is an $(n-1) \times 1$ column matrix, $\mathbf{C} = \mathbf{w}^H$ is an $1 \times (n-1)$ row vector, and $\mathbf{B} = \alpha$ is an 1×1 matrix (i.e., a scalar), then

$$\mathbf{Y}^{-1} = \begin{pmatrix} \mathbf{A}^{-1} + \beta\mathbf{A}^{-1}\mathbf{vw}^H\mathbf{A}^{-1} & -\beta\mathbf{A}^{-1}\mathbf{v} \\ -\beta\mathbf{w}^H\mathbf{A}^{-1} & \beta \end{pmatrix}, \qquad (3.53)$$

where the scalar $\beta = (\alpha - \mathbf{w}^H\mathbf{A}^{-1}\mathbf{v})^{-1}$. Thus, the inverse of a matrix formed by bordering the original matrix with an additional row and column (or by omitting a row and column) may be computed simply. Matrices created in this way are known as *border matrices*.

Table 3.1 lists the form of the inverse of the special matrices introduced in Sec. 3.3. Many of these are simple to prove. For example, to prove that the inverse

Table 3.1. Matrix Inverse Structures

Matrix Form	Matrix Inverse Form
Hermitian (Symmetric)	Hermitian (Symmetric)
Persymmetric	Persymmetric
Centrosymmetric	Centrosymmetric
Toeplitz	Persymmetric
Hankel	Symmetric
Right Circulant	Right Circulant
Left Circulant	Left Circulant
Triangular	Triangular

of a persymmetric matrix $\mathbf{P}$ is also persymmetric, use the fact that $\mathbf{JJ} = \mathbf{I}$ to write

$$\mathbf{PP}^{-1} = \mathbf{PJJP}^{-1} = \mathbf{I}$$

and pre- and postmultiply by $\mathbf{J}$

$$\mathbf{J}(\mathbf{PJJP}^{-1})\mathbf{J} = \mathbf{P}^T(\mathbf{JP}^{-1}\mathbf{J}) = \mathbf{I}.$$

This yields

$$\mathbf{JP}^{-1}\mathbf{J} = (\mathbf{P}^{-1})^T. \tag{3.54}$$

The property $(\mathbf{P}^T)^{-1} = (\mathbf{P}^{-1})^T$ has been used.

An $n \times n$ matrix $\mathbf{Q}$ such that $\mathbf{Q}^H\mathbf{Q} = \mathbf{QQ}^H$ is called *orthogonal*. Furthermore, if $\mathbf{Q}$ has the property $\mathbf{Q}^H\mathbf{Q} = \mathbf{QQ}^H = \mathbf{I}$, then it is said to be an *orthonormal*, or a *unitary*, matrix. Because $\mathbf{Q}^{-1}\mathbf{Q} = \mathbf{I}$, then $\mathbf{Q}^{-1} = \mathbf{Q}^H$. A unitary matrix has the property that $|\det \mathbf{Q}| = 1$.

3.5 SOLUTION OF LINEAR EQUATIONS

Many spectral estimation methods require the solution of a set of simultaneous linear equations, which can be expressed in the generic matrix form

$$\mathbf{Az} = \mathbf{b}, \tag{3.55}$$

in which $\mathbf{A}$ is an arbitrary $n \times n$ square matrix, $\mathbf{b}$ is some arbitrary $n \times 1$ column vector, and $\mathbf{z}$ is an $n \times 1$ vector of unknown values whose solution is to be found. Note that it is assumed that there are as many linear equations as there are unknowns so that the solution is uniquely determined (assuming the matrix $\mathbf{A}$ is not singular, i.e., it has full rank n). There are also situations in which the matrix products $\mathbf{A}^{-1}\mathbf{B}$ or $\mathbf{BA}^{-1}$ must be computed, in which $\mathbf{A}$ and $\mathbf{B}$ are both $n \times n$ square matrices. These products are equivalent to solving the linear equations $\mathbf{AZ} = \mathbf{B}$ or $\mathbf{ZA} = \mathbf{B}$ for matrix $\mathbf{Z}$. Three methods to solve nonsingular linear equations (3.55) are presented in this section.

3.5.1 Direct Inversion

An obvious brute-force method for obtaining the solution vector $\mathbf{z}$ is to multiply both sides of (3.55) on the left by the inverse $\mathbf{A}^{-1}$ (which exists if rank $\mathbf{A} = n$) yielding

$$\mathbf{z} = \mathbf{A}^{-1}\mathbf{b}. \tag{3.56}$$

However, this is neither the most numerically efficient nor the most numerically stable method for solving linear equations.

3.5.2 Triangular Decomposition

Consider the special case of an $n \times n$ matrix $\mathbf{A}$ that is upper triangular and nonsingular (always the case if all the diagonal elements are nonzero), then the system of linear equations involving $\mathbf{A}$ can be solved by a method known as *back substitution* in a number of computations proportional to n^3. Consider the 3×3 upper triangular matrix case

$$\begin{pmatrix} a[1,1] & a[1,2] & a[1,3] \\ 0 & a[2,2] & a[2,3] \\ 0 & 0 & a[3,3] \end{pmatrix} \begin{pmatrix} z[1] \\ z[2] \\ z[3] \end{pmatrix} = \begin{pmatrix} b[1] \\ b[2] \\ b[3] \end{pmatrix}. \tag{3.57}$$

From the last row, observe that $a[3,3]z[3] = b[3]$, from which the solution $z[3] = b[3]/a[3,3]$ may be calculated. If $z[3]$ is *substituted back* into the matrix expression, then the second row will have only the one unknown $z[2]$, found by solution of $a[2,2]z[2] + a[2,3]z[3] = b[2]$. Continuing to back substitute will eventually solve for all the unknown z components.

For general square $n \times n$ nonsingular matrix $\mathbf{A}$, an alternative to (3.56) for the solution of $\mathbf{Az} = \mathbf{b}$ is a three-step procedure known as triangular decomposition (also called LU factorization). A nonsingular matrix $\mathbf{A}$ can always be uniquely decomposed into a product of an upper triangular matrix $\mathbf{U}$ and a lower triangular matrix $\mathbf{L}$ (with ones along the main diagonal),

$$\mathbf{A} = \mathbf{LU} \tag{3.58}$$

using the Gaussian elimination technique that requires computations proportional to $2n^3/3$ (this technique often interchanges the rows of $\mathbf{A}$ to improve the numerical accuracy of the solution). The triangular matrix $\mathbf{U}$ is sometimes factored as $\mathbf{U} = \mathbf{DU}'$ in which $\mathbf{D}$ is a diagonal matrix and $\mathbf{U}'$ is an upper triangular matrix with ones along the diagonal. Once the LU (or LDU′) decomposition has been accomplished, "forward" substitution starting with element $y[1]$ is used to solve for $\mathbf{y}$ in the intermediate triangular system of linear equations

$$\mathbf{Ly} = \mathbf{b}. \tag{3.59}$$

With solution vector **y** now known, "back" substitution starting with element $x[n]$ is used to find the solution vector **x** in the triangular system of linear equations

$$\mathbf{Ux} = \mathbf{y}. \tag{3.60}$$

The LU decomposition technique for solving square linear equations requires a number of computations proportional to n^3 and memory storage proportional to n^2.

   If matrix **A** is Hermitian, then the triangular factorization will have the special form

$$\mathbf{A} = \mathbf{R}^H\mathbf{R} \quad \text{or} \quad \mathbf{A} = \mathbf{R}^H\mathbf{DR} \tag{3.61}$$

in which **D** is a diagonal matrix with nonzero real elements and **R** is an upper triangular matrix with either real nonzero principal diagonal elements ($\mathbf{R}^H\mathbf{R}$ case) or ones on the principal diagonal ($\mathbf{R}^H\mathbf{DR}$ case). This special triangular decomposition for Hermitian matrices is called *Cholesky factorization*; it is also called *matrix square-root factorization*. The corresponding forward and back substitution operations to (3.59) and (3.60) for the Hermitian case are

$$\mathbf{R}^H\mathbf{y} = \mathbf{b} \tag{3.62}$$

and

$$\mathbf{Rx} = \mathbf{y}. \tag{3.63}$$

Due to the Hermitian symmetry of **A**, the number of computations required to solve a set of Hermitian equations by Cholesky factorization is about half that required for general triangular decomposition because only one matrix **R** needs to be determined, rather than the two matrices **L** and **U**.

3.5.3   Eigenvalue Decomposition

A third approach for the solution of determined square linear equations exploits eigenvalue decomposition. It was shown in Sec. 2.3 that the exponential $\exp(j2\pi ft)$ was an eigenfunction of a linear system. A linear system with an exponential at the input reproduces the exponential form at the output, unchanged except for a complex amplitude scaling. In an analogous manner, if **q** is a nonzero vector that is left unchanged, except for scaling by a factor λ, when operated on by the linear transformation represented by the $n \times n$ square matrix **A**

$$\mathbf{Aq} = \lambda\mathbf{q}, \tag{3.64}$$

then **q** is said to be an *eigenvector* of the matrix **A**. The scalar λ associated with the eigenvector is an *eigenvalue*. *Principal eigenvectors* are those eigenvectors associated with the largest magnitude eigenvalues of **A**. The order n polynomial $f(\lambda) = \det(\mathbf{A} - \lambda\mathbf{I})$ is called the *characteristic polynomial* of **A**. The n roots of

$f(\lambda) = 0$ yield the n eigenvalues λ_i,

$$f(\lambda) = \prod_{i=1}^{n} (\lambda_i - \lambda) = \sum_{k=0}^{n} \gamma_k \lambda^k = 0. \tag{3.65}$$

There is always at least one eigenvector associated with each distinct eigenvalue, and eigenvectors associated with distinct eigenvalues are linearly independent of each other. The highest-order coefficient γ_n in (3.65) is $(-1)^n$ and the lowest-order coefficient γ_0 can easily be shown to satisfy

$$\gamma_0 = \det \mathbf{A} = \prod_{i=1}^{n} \lambda_i. \tag{3.66}$$

Defining the trace of a matrix as the sum of its principal diagonal elements

$$\operatorname{tr} \mathbf{A} = \sum_{i=1}^{n} a[i, i], \tag{3.67}$$

then it may also be shown [Noble and Daniel, 1977] that the sum of eigenvalues satisfy

$$\sum_{i=1}^{n} \lambda_i = \operatorname{tr} \mathbf{A} \tag{3.68}$$

and

$$\gamma_{n-1} = (-1)^{n-1} \operatorname{tr} \mathbf{A} \tag{3.69}$$

is another term of the characteristic polynomial. The matrix $\mathbf{A} + k\mathbf{I}$ has eigenvalues $\lambda_i + k$. Note that all the eigenvalues of a unitary matrix $\mathbf{Q}$ have unit modulus.

Suppose an $n \times n$ eigenvector matrix $\mathbf{Q}$ is constructed from the n column eigenvectors $\mathbf{q}_1$ to $\mathbf{q}_n$ of the $n \times n$ square matrix $\mathbf{A}$ and an $n \times n$ diagonal matrix $\Lambda = \operatorname{diag}(\lambda_1, \ldots, \lambda_n)$ is constructed from the n distinct eigenvalues of $\mathbf{A}$. Based on (3.64), we then have the relation $\mathbf{AQ} = \mathbf{Q}\Lambda$. The matrix $\mathbf{A}$ will therefore have a linearly independent set of n eigenvectors if and only if $\mathbf{A}$ can be expressed as the eigenvalue decomposition (EVD)

$$\mathbf{A} = \mathbf{Q}\Lambda\mathbf{Q}^{-1}, \quad \Lambda = \mathbf{Q}^{-1}\mathbf{AQ}, \tag{3.70}$$

in which $\mathbf{Q}$ and Λ are nonsingular. The inverse $\mathbf{A}^{-1}$ can be expressed in terms of the EVD as

$$\mathbf{A}^{-1} = \mathbf{Q}\Lambda^{-1}\mathbf{Q}^{-1} \tag{3.71}$$

and therefore the eigenvalues of the inverse $\mathbf{A}^{-1}$ are $1/\lambda_i$. The solution to a square set of nonsingular linear equations can be expressed in terms of the EVD as

$$\mathbf{z} = \mathbf{A}^{-1}\mathbf{b} = \mathbf{Q}\Lambda^{-1}\mathbf{Q}^{-1}\mathbf{b}. \tag{3.72}$$

The *quadratic form* associated with Hermitian matrix $\mathbf{A}$ is the real scalar quantity $\langle \mathbf{x}, \mathbf{A}\mathbf{x} \rangle = \mathbf{x}^H \mathbf{A}\mathbf{x} = \sum_i \sum_j a_{ij} x_i^* x_j$. The quadratic form is said to be *positive definite* if $\langle \mathbf{x}, \mathbf{A}\mathbf{x} \rangle > 0$ for all nonnull vectors $\mathbf{x}$, and *positive semidefinite*, if $\langle \mathbf{x}, \mathbf{A}\mathbf{x} \rangle \geq 0$ for all $\mathbf{x}$. The matrix $\mathbf{A}$ is said to be positive definite or semidefinite if its corresponding quadratic form is positive definite or semidefinite. If and only if all of the eigenvalues of a Hermitian matrix are positive, then the matrix is positive definite; if and only if the eigenvalues are all nonnegative, then the matrix is positive semidefinite.

Spectral analysis concepts discussed in Chap. 13 require the computation of the minimum or maximum eigenvalue of a Hermitian matrix. The *power iteration method* is one technique for finding these particular eigenvalues. For example, picking some initial vector $\mathbf{x}_0$, then the linear equation

$$\mathbf{y}_{k+1} = \mathbf{A}\mathbf{x}_k \tag{3.73}$$

is used iteratively, normalizing at each step to yield

$$\mathbf{x}_{k+1} = \frac{\mathbf{y}_{k+1}}{\sigma_{k+1}}, \quad \sigma_{k+1} = \mathbf{y}_{k+1}^H \mathbf{y}_{k+1} \tag{3.74}$$

until $\mathbf{x}_k$ converges. The value of σ_k after convergence will be the maximum eigenvalue λ_{max} and $\mathbf{x}_k$ will be the associated eigenvector. Convergence is assured provided $\mathbf{x}_0$ is not orthogonal to the eigenvector associated with the maximum eigenvalue.

To determine the minimum eigenvalue and its associated eigenvector, the inverse matrix $\mathbf{A}^{-1}$, rather than $\mathbf{A}$, is used in the linear equation

$$\mathbf{y}_{k+1} = \mathbf{A}^{-1}\mathbf{x}_k, \tag{3.75}$$

which may more conveniently be solved for $\mathbf{y}_{k+1}$ as

$$\mathbf{A}\mathbf{y}_{k+1} = \mathbf{x}_k. \tag{3.76}$$

Note that a process of deflation may be used to remove each minimum (maximum) eigenvalue as it is computed, a consequence of Eq. (3.64),

$$\mathbf{A} - \sigma \mathbf{x}\mathbf{x}^H \tag{3.77}$$

and the process repeated to find the remaining eigenvalues and eigenvectors.

3.5.4   Eigenvalue Decomposition of Special Matrices

The specific EVD of Hermitian, triangular, circulant, and centrosymmetric matrices is summarized in this section. The Toeplitz matrix EVD is discussed in Sec. 3.8.

If a nonsingular matrix $\mathbf{A}$ has Hermitian symmetry, then the EVD of $\mathbf{A}$ will have the additional property

$$\mathbf{A} = \mathbf{Q}\Lambda\mathbf{Q}^{-1} = \mathbf{A}^H = (\mathbf{Q}^{-1})^H \Lambda^H \mathbf{Q}^H. \tag{3.78}$$

From this it may be inferred that $\mathbf{Q}^{-1} = \mathbf{Q}^H$ and $\Lambda = \Lambda^H$ when $\mathbf{A}$ is Hermitian. Thus, a Hermitian matrix has real eigenvalues and, if all the eigenvalues are distinct, the eigenvector matrix $\mathbf{Q}$ is orthonormal because $\mathbf{Q}\mathbf{Q}^H = \mathbf{I}$. Hermitian matrices are therefore diagonalized by a unitary transformation with the orthogonal EVD

$$\mathbf{A} = \mathbf{Q}\Lambda\mathbf{Q}^H = \sum_{i=0}^{n} \lambda_i \mathbf{q}_i \mathbf{q}_i^H. \tag{3.79}$$

Expression (3.79) has unfortunately been called the "spectrum" by mathematicians, but this nomenclature should not be confused with the spectral estimation concepts covered by this text. Based on the relationship

$$\mathbf{A}^{-1} = (\mathbf{Q}^H)^{-1} \Lambda^{-1} \mathbf{Q}^{-1} = \mathbf{Q}\Lambda^{-1}\mathbf{Q}^H, \tag{3.80}$$

which assumes $\mathbf{Q}$ and Λ are invertible, it then follows that

$$\mathbf{A}^{-1} = \sum_{i=1}^{n} \frac{1}{\lambda_i} \mathbf{q}_i \mathbf{q}_i^H, \tag{3.81}$$

so that the solution to (3.55) for Hermitian matrix $\mathbf{A}$ can be expressed as a weighted sum of eigenvectors

$$\mathbf{z} = \mathbf{A}^{-1}\mathbf{b} = \sum_{i=1}^{n} \left(\frac{(\mathbf{q}_i^H \mathbf{b})}{\lambda_i} \right) \mathbf{q}_i \tag{3.82}$$

in which $\mathbf{q}_i^H \mathbf{b}$ is the scalar inner product between the right-hand side vector of the linear equations and the eigenvector associated with eigenvalue λ_i.

It is easily proved that a triangular matrix $\mathbf{A}$ has the determinant, which is the product of its diagonal elements

$$\det \mathbf{A} = \prod_{i=1}^{n} a[i, i]. \tag{3.83}$$

Comparing this with Eq. (3.66), it can be seen that the product of the eigenvalues of a triangular matrix are equal to the product of the diagonal elements.

A circulant matrix has eigenvalues and eigenvectors related to the discrete-time Fourier series (DTFS) [Cornyn, 1974; McClellan and Parks, 1972]. Recall from Sec. 2.7 that the n-point DTFS $X[0], \ldots, X[n-1]$ of an n-point sequence

$x[0], \ldots, x[n-1]$ for $T = 1$ is

$$X[k] = \sum_{i=0}^{n-1} x[i] \exp(-j2\pi ik/n) \quad \text{for} \quad k = 0, \ldots, n-1, \quad (3.84)$$

which may be expressed in matrix form as

$$\begin{pmatrix} X[0] \\ \vdots \\ X[n-1] \end{pmatrix} = \mathbf{Q}_{\text{DTFS}} \begin{pmatrix} x[0] \\ \vdots \\ x[n-1] \end{pmatrix}, \quad (3.85)$$

where the elements of $\mathbf{Q}_{\text{DTFS}}$ are the exponential coefficients

$$q_{\text{DTFS}}[i, k] = \exp(-j2\pi ik/n) \quad \text{for} \quad 0 \leq i, k \leq n-1. \quad (3.86)$$

The right-circulant matrix $\mathbf{C}$ may be expressed as the eigenvalue decomposition

$$\mathbf{C} = \left(\frac{1}{\sqrt{n}} \mathbf{Q}_{\text{DTFS}} \right) \mathbf{\Lambda}_{\text{DTFS}} \left(\frac{1}{\sqrt{n}} \mathbf{Q}_{\text{DTFS}}^H \right), \quad (3.87)$$

where the diagonal matrix $\mathbf{\Lambda}_{\text{DTFS}}$ of eigenvalues is given by

$$\begin{pmatrix} \lambda_0 \\ \lambda_1 \\ \vdots \\ \lambda_{n-1} \end{pmatrix} = \mathbf{Q}_{\text{DTFS}} \begin{pmatrix} c_0 \\ c_1 \\ \vdots \\ c_{n-1} \end{pmatrix}. \quad (3.88)$$

Thus, the eigenvalues of an arbitrary circulant matrix are the DTFS of the circulant elements $c_0, \ldots, c_{n-1}$. The columns of the DTFS matrix $\mathbf{Q}_{\text{DTFS}}$ form the eigenvectors. The normalized eigenvectors $\mathbf{q}_k$ are composed of powers of the n roots of unity,

$$q_k = \frac{1}{\sqrt{n}} \begin{pmatrix} 1 \\ w_n^k \\ \vdots \\ w_n^{(n-1)k} \end{pmatrix}, \quad w_n = \exp(j2\pi/n). \quad (3.89)$$

A similar development for the left-circulant matrix can be performed using the *inverse* DTFS, rather than the *forward* DTFS. Linear equations such as $\mathbf{Cx} = \mathbf{y}$ may be solved efficiently with FFT algorithms.

The eigenvectors of a centrosymmetric matrix $\mathbf{C}$ have either conjugate symmetry or anti-conjugate symmetry [Cantoni and Butler, 1976]. A vector $\mathbf{v}$ is conjugate symmetric if $\mathbf{Jv} = \mathbf{v}^*$, where $\mathbf{J}$ is the reflection matrix. A vector $\mathbf{v}$ is anti-conjugate symmetric if $\mathbf{Jv} = -\mathbf{v}^*$. It may be inferred from this definition that the center element for a vector of odd dimension must be real if the vector is conjugate symmetric or zero if the vector is anti-conjugate symmetric. Assuming unique eigenvalues, a square centrosymmetric matrix of *even* dimension n will have $n/2$ conjugate symmetric orthonormal eigenvectors $\mathbf{v}_i$ for $i = 1, \ldots, n/2$, with corresponding eigenvalues σ_i of the form

$$\mathbf{v}_i = \frac{1}{\sqrt{2}} \begin{pmatrix} \mathbf{y}_i \\ \mathbf{J}\mathbf{y}_i^* \end{pmatrix}, \qquad (3.90)$$

where $\mathbf{y}_i$ are the orthonormal eigenvectors of the matrix $\mathbf{A} + \mathbf{J}\mathbf{B}$, that is,

$$(\mathbf{A} + \mathbf{J}\mathbf{B})\mathbf{y}_i = \sigma_i \mathbf{y}_i, \qquad (3.91)$$

and where $\mathbf{A}$ and $\mathbf{B}$ are as given in Eq. (3.35). It will also have $n/2$ anti-conjugate symmetric orthonormal eigenvectors $\mathbf{w}_i$ for $i = 1, \ldots, n/2$, with corresponding eigenvalues λ_i of the form

$$\mathbf{w}_i = \frac{1}{\sqrt{2}} \begin{pmatrix} \mathbf{z}_i \\ -\mathbf{J}\mathbf{z}_i^* \end{pmatrix}, \qquad (3.92)$$

where $\mathbf{z}_i$ are the orthonormal eigenvectors of the matrix $\mathbf{A} - \mathbf{J}\mathbf{B}$, that is,

$$(\mathbf{A} - \mathbf{J}\mathbf{B})\mathbf{z}_i = \lambda_i \mathbf{z}_i. \qquad (3.93)$$

If a square centrosymmetric matrix has *odd* dimension n and structure as given by Eq. (3.34), then it will have $(n+1)/2$ conjugate symmetric orthonormal eigenvectors $\mathbf{v}_i$ for $i = 1, \ldots, (n+1)/2$, with corresponding eigenvalues σ_i of the form

$$\mathbf{v}_i = \frac{1}{\sqrt{2}} \begin{pmatrix} \mathbf{y}_i \\ 2\beta_i \\ \mathbf{J}\mathbf{y}_i^* \end{pmatrix}, \qquad (3.94)$$

where vector $\mathbf{y}_i$ and real scalar β_i are the orthonormal eigenvectors of the augmented matrix

$$\begin{pmatrix} \mathbf{A} + \mathbf{J}\mathbf{B} & \sqrt{2}\mathbf{x} \\ \sqrt{2}\mathbf{x}^H & \alpha \end{pmatrix} \begin{pmatrix} \mathbf{y}_i \\ \beta_i \end{pmatrix} = \sigma_i \begin{pmatrix} \mathbf{y}_i \\ \beta_i \end{pmatrix}, \qquad (3.95)$$

and $(n-1)/2$ anti-conjugate symmetric orthonormal eigenvectors $\mathbf{w}_i$ for $i = 1, 2, 3, 4, \ldots, (n-1)/2$ with corresponding eigenvalues λ_i of the form

$$\mathbf{w}_i = \frac{1}{\sqrt{2}} \begin{pmatrix} \mathbf{z}_i \\ 0 \\ -\mathbf{J}\mathbf{z}_i^* \end{pmatrix}, \qquad (3.96)$$

where $\mathbf{z}_i$ are the orthonormal eigenvectors of the matrix $\mathbf{A} - \mathbf{J}\mathbf{B}$, that is,

$$(\mathbf{A} - \mathbf{J}\mathbf{B})\mathbf{z}_i = \lambda_i \mathbf{z}_i. \qquad (3.97)$$

The proof of the above relationships is based on the fact that the centrosymmetric matrix $\mathbf{R}$ is orthogonally similar to the matrices $\mathbf{A} - \mathbf{J}\mathbf{B}$ and $\mathbf{A} + \mathbf{J}\mathbf{B}$, that is, if n is even, then the orthogonal similarity matrix

$$Q = \frac{1}{\sqrt{2}} \begin{pmatrix} I & -J \\ I & J \end{pmatrix}, \quad QQ^T = I \tag{3.98}$$

transforms the centrosymmetric matrix R_n to

$$QRQ^H = \begin{pmatrix} A - JB & 0 \\ 0 & A + JB \end{pmatrix}, \tag{3.99}$$

and if n is odd, then

$$Q = \frac{1}{\sqrt{2}} \begin{pmatrix} I & 0 & -J \\ 0 & \sqrt{2} & 0 \\ I & 0 & J \end{pmatrix} \tag{3.100}$$

transforms R_n to

$$QRQ^H = \begin{pmatrix} A - JB & 0 & 0 \\ 0 & \alpha & \sqrt{2}x^H \\ 0 & \sqrt{2}x & A + JB \end{pmatrix}. \tag{3.101}$$

3.6 OVERDETERMINED AND UNDERDETERMINED LINEAR EQUATIONS

Section 3.5 summarized approaches computing the unique solution vector z of the determined linear equations $Az = b$ in which the number of unknowns m (i.e., the dimension of vector z) and the number of linear equations n (i.e., the dimension of vector b) were equal ($m = n$), which means the matrix A is square. This section will summarize methodologies developed to achieve unique solutions in two other cases: (1) *overdetermined* linear equations in which the number of unknowns is *less* than the number of linear equations ($m < n$) and (2) *underdetermined* linear equations in which the number of unknowns is *greater* than the number of equations ($m > n$).

3.6.1 Least Squares Linear Equations

Consider first the overdetermined linear equation case in which it is desired to form an approximation $\hat{y}[n]$ of the complex sequence $y[n]$ by a linear combination of the m complex sequences $x_1[n], x_2[n], \ldots, x_m[n]$

$$\hat{y}[k] = \sum_{i=1}^{m} a_i x_i[k] \tag{3.102}$$

over a sequence index range of $1 \leq k \leq n$. For the overdetermined case, $m < n$. The method used to obtain a unique solution for the unknown approximation parameters

$a_1, \ldots, a_m$ is the *least squares method*, which seeks to minimize the sum of squared errors

$$E = \sum_{k=1}^{n} |e[k]|^2, \qquad (3.103)$$

in which $e[k] = y[k] - \hat{y}[k]$ is the complex error between the actual sample value $y[k]$ and its linear approximation $\hat{y}[k]$.

A matrix representation of the n error equations is

$$\mathbf{y} = \mathbf{X}\mathbf{a} + \mathbf{e}, \qquad (3.104)$$

in which the rectangular $n \times m$ data matrix $\mathbf{X}$, the $m \times 1$ column vector $\mathbf{a}$, the $n \times 1$ column data vector $\mathbf{y}$, and the $n \times 1$ column error vector $\mathbf{e}$ are defined as

$$\mathbf{X} = \begin{pmatrix} x_1[1] & \cdots & x_m[1] \\ \vdots & & \vdots \\ x_1[n] & \cdots & x_m[n] \end{pmatrix}, \quad \mathbf{a} = \begin{pmatrix} a_1 \\ \vdots \\ a_m \end{pmatrix}$$

and

$$\mathbf{y} = \begin{pmatrix} y[1] \\ \vdots \\ y[n] \end{pmatrix}, \quad \mathbf{e} = \begin{pmatrix} e[1] \\ \vdots \\ e[n] \end{pmatrix}.$$

By defining the augmented $n \times (m+1)$ matrix $(\mathbf{y} \ \mathbf{X})$, the error vector $\mathbf{e}$ of (3.106) may alternatively be expressed as

$$(\mathbf{y} \ \mathbf{X}) \begin{pmatrix} 1 \\ -\mathbf{a} \end{pmatrix} = \mathbf{e}. \qquad (3.105)$$

The squared error E of (3.103) may also be expressed as (using 3.104)

$$E = \|\mathbf{e}\|_2 = \mathbf{e}^H \mathbf{e} = \mathbf{y}^H \mathbf{y} - \mathbf{y}^H \mathbf{X}\mathbf{a} - \mathbf{a}^H \mathbf{X}^H \mathbf{y} + \mathbf{a}^H \mathbf{X}^H \mathbf{X}, \qquad (3.106)$$

in which the vector inner product $\mathbf{y}^H \mathbf{y}$ forms a scalar, the matrix-vector product $\mathbf{X}^H \mathbf{y}$ forms an $m \times 1$ column vector, and the matrix-matrix product $\mathbf{X}^H \mathbf{X}$ forms an $m \times m$ square Hermitian matrix. Note that the least squares approach is equivalent to finding minimum $\|\mathbf{e}\|_2$, which is the *minimum norm* of the error vector.

The minimization of E may be found by inspection after adding and subtracting $\mathbf{y}^H \mathbf{X}(\mathbf{X}^H \mathbf{X})^{-1}\mathbf{X}^H \mathbf{y}$ in Eq. (3.108) (a procedure called "completing the square") so that E can be expressed as

$$E = \mathbf{y}^H \mathbf{y} - \mathbf{y}^H \mathbf{X}(\mathbf{X}^H \mathbf{X})^{-1}\mathbf{X}^H \mathbf{y}$$
$$+ (\mathbf{X}^H \mathbf{y} - \mathbf{X}^H \mathbf{X}\mathbf{a})^H (\mathbf{X}^H \mathbf{X})^{-1}(\mathbf{X}^H \mathbf{y} - \mathbf{X}^H \mathbf{X}\mathbf{a}). \qquad (3.107)$$

Only the quadratic scalar term $(\mathbf{X}^H \mathbf{y} - \mathbf{X}^H \mathbf{X}\mathbf{a})^H (\mathbf{X}^H \mathbf{X})^{-1}(\mathbf{x}^H \mathbf{y} - \mathbf{x}^H \mathbf{x}\mathbf{a})$ is a function of the unknown parameter vector $\mathbf{a}$. Due to its Hermitian symmetry, this term can only be real and positive; setting this term to zero will then minimize E. This is accomplished by selecting **a** such that

$$\mathbf{X}^H\mathbf{y} - \mathbf{X}^H\mathbf{X}\mathbf{a} = \mathbf{0}_m, \tag{3.108}$$

where $\mathbf{0}_m$ is the all-zeros vector. The unique vector **a** that minimizes the squared error is therefore found by solving the set of m linear equations expressed as

$$\mathbf{X}^H\mathbf{X}a = \mathbf{X}^H\mathbf{y} \text{ or } \mathbf{a} = \left(\mathbf{X}^H\mathbf{X}\right)^{-1}\mathbf{X}^H\mathbf{y} \tag{3.109}$$

or as

$$\mathbf{A}\mathbf{z} = \mathbf{b} \tag{3.110}$$

by making the associations $\mathbf{A} = \mathbf{X}^H\mathbf{X}$, $\mathbf{b} = \mathbf{X}^H\mathbf{y}$, and $\mathbf{z} = \mathbf{a}$. The minimum squared error is obtained by substituting Eq. (3.108) into Eq. (3.107), yielding

$$\mathsf{E}_{min} = \mathbf{y}^H\mathbf{y} - \mathbf{y}^H\mathbf{X}\mathbf{a}.$$

Equations (3.108) and above may be combined into a consolidated set of $m + 1$ linear equations as follows:

$$\begin{pmatrix} \mathbf{y}^H\mathbf{y} & \mathbf{y}^H\mathbf{X} \\ \mathbf{X}^H\mathbf{y} & \mathbf{X}^H\mathbf{X} \end{pmatrix} \begin{pmatrix} 1 \\ -\mathbf{a} \end{pmatrix} = (\mathbf{y} \quad \mathbf{X})^H (\mathbf{y} \quad \mathbf{X}) \begin{pmatrix} 1 \\ -\mathbf{a} \end{pmatrix} = \begin{pmatrix} \mathsf{E}_{min} \\ \mathbf{0}_m \end{pmatrix}. \tag{3.111}$$

Equations (3.109) and (3.111) are often identified as the *normal equations* of least squares analysis of overdetermined linear equations. The term "normal" refers to the orthogonality of the approximation error $e[k]$ with the approximation $\hat{y}[k]$. Note that $(\mathbf{y} \quad \mathbf{X})^H (\mathbf{y} \quad \mathbf{X})$ also forms a Hermitian matrix.

The squared error expression is a special case of a matrix quadratic equation of general form

$$\mathsf{E} = r - \mathbf{r}^H\mathbf{a} - \mathbf{a}^H\mathbf{r} + \mathbf{a}^H\mathbf{R}\mathbf{a}, \tag{3.112}$$

in which r is a known positive scalar, **r** is a known $m \times 1$ column vector, and **r** is a known $m \times m$ Hermitian matrix. The normal equations that provide the solution for the $m \times 1$ vector **a** that minimize the positive real-valued scalar E are given by

$$\begin{pmatrix} r & \mathbf{r}^H \\ \mathbf{r} & \mathbf{r} \end{pmatrix} \begin{pmatrix} 1 \\ -\mathbf{a} \end{pmatrix} = \begin{pmatrix} \mathsf{E}_{min} \\ \mathbf{0}_m \end{pmatrix}. \tag{3.113}$$

This follows directly from Eq. (3.111) by making the substitution of r for $\mathbf{y}^H\mathbf{y}$, **r** for $\mathbf{x}^H\mathbf{y}$, and **r** for $\mathbf{x}^H\mathbf{x}$.

3.6.2 Minimum Norm Linear Equations

In the underdetermined linear equation case ($m > n$), there are either an infinite number of possible solutions to the relationship

$$\mathbf{y} = \mathbf{xa} \tag{3.114}$$

or no solution at all. A procedure that will yield a unique solution to Eq. (3.114) is to impose the constraint that unknown vector $\mathbf{a}$ has minimum norm $\|\mathbf{a}\|_2 = \mathbf{a}^H \mathbf{a}$. The solution can be shown to satisfy [Golub and Van Loan, 1989]

$$\mathbf{a} = \mathbf{X} \left(\mathbf{XX}^H\right)^{-1} \mathbf{y} \tag{3.115}$$

in which $\mathbf{XX}^H$ forms an $n \times n$ square Hermitian matrix.

3.7 SOLUTION OF OVERDETERMINED AND UNDERDETERMINED LINEAR EQUATIONS

The solution of the least squares linear equations (3.109) for the overdetermined case and the minimum norm linear equations (3.15) for the underdetermined case could utilize the triangular decomposition or eigenvalue decomposition techniques of Sec. 3.5 by making the association $\mathbf{A} = \mathbf{X}^H \mathbf{X}$ in the least squares case and $\mathbf{A} = \mathbf{XX}^H$ in the minimum norm case. Better numerical accuracy, algorithm stability, and rank determination is obtained, however, with two alternative factorizations known as *orthogonal decomposition* (QRD) and *singular-value decomposition* (SVD). These two decompositions are distinguished from the LU and EVD decompositions of Sec. 3.5 because they factor the *data matrix* $\mathbf{X}$ rather than the Hermitian *product matrix* $\mathbf{A} = \mathbf{X}^H \mathbf{X}$ or $\mathbf{A} = \mathbf{XX}^H$. A special application of SVD known as *total least squares* is also presented in this section.

3.7.1 Orthogonal Decomposition

Any $n \times m$ rectangular matrix $\mathbf{X}$ can be decomposed into the product

$$\mathbf{X}_{n,m} = \mathbf{Q}_{n,m} \begin{pmatrix} \mathbf{R}_{m,m} \\ \mathbf{0}_{n-m,m} \end{pmatrix} \tag{3.116}$$

in which $\mathbf{Q}$ is an $n \times n$ rectangular *orthonormal* matrix ($\mathbf{Q}^H \mathbf{Q} = \mathbf{I}$), $\mathbf{R}$ is an $m \times m$ upper triangular matrix, and $\mathbf{0}_{n-m,m}$ is a rectangular $(n-m) \times m$ all zeros matrix. The structure of Eq. (3.116) assumes the overdetermined least squares case ($n > m$). The factorization is normally called *QR decomposition*. It has also been assumed that the matrix $\mathbf{X}$ has full rank (specifically, rank $\mathbf{X} = m$). Triangular matrix $\mathbf{R}$, which generally has arbitrary real elements along its main diagonal, may also be expressed as $\Phi \mathbf{R}'$ in which Φ is a diagonal matrix with real diagonal elements and $\mathbf{R}'$ is an upper triangular matrix with ones along the main diagonal. The orthogonal decomposition will have the form $\mathbf{X} = \mathbf{Q}\Phi\mathbf{R}'$ in this case. Variations of (3.116) also exist for the square, undetermined, and rank deficient cases [Golub and Van Loan, 1989]. The orthogonal matrix $\mathbf{Q}$ is typically determined by using the Gram-Schmidt orthogonalization procedure [Noble, 1977], which requires computations proportional to n^3. Note that $\mathbf{A} = \mathbf{X}^H\mathbf{X} = \mathbf{R}^H\mathbf{Q}^H\mathbf{Q}\mathbf{R} = \mathbf{R}^H\mathbf{R}$, which means the triangular factor $\mathbf{R}$ of the QR decomposition of $\mathbf{X}$ is identical to the triangular Cholesky factor $\mathbf{R}$ of the Hermitian product matrix $\mathbf{X}^H\mathbf{X}$. Substituting (3.116) into the least squares solution $\mathbf{X}^H\mathbf{X}\mathbf{a} = \mathbf{X}^H\mathbf{y}$ will yield

$$\left(\mathbf{R}^H\mathbf{Q}^H\right)(\mathbf{Q}\mathbf{R})\,\mathbf{a} = \left(\mathbf{R}^H\mathbf{Q}^H\right)\mathbf{y}. \tag{3.117}$$

Multiplication on the left by triangular matrix inverse $(\mathbf{R}^H)^{-1}$ and use of the property $\mathbf{Q}^H\mathbf{Q} = \mathbf{I}$ will produce

$$\mathbf{R}\mathbf{a} = \mathbf{Q}^H\mathbf{y} \quad \text{or} \quad \Phi\mathbf{R}'\mathbf{a} = \mathbf{Q}^H\mathbf{y} \tag{3.118}$$

depending on whether $\mathbf{R}$ or $\mathbf{R}'$ is used. Because $\mathbf{R}$ is upper triangular, the least squares solution vector $\mathbf{a}$ can easily be computed by the back substitution method, Eq. (3.57), after the QR decomposition of $\mathbf{x}$ has been performed.

3.7.2 Singular Value Decomposition

The eigenvalue decomposition concept for nonsingular square $n \times n$ matrices can be extended to an arbitrary $n \times m$ complex-valued matrix $\mathbf{X}$ of rank $k \leq \min(m, n)$ with a factorization called the *singular-value decomposition* (SVD). This is the most powerful of the decomposition techniques and requires approximately $12n^3$ operations to calculate. The SVD theorem [Depretters, 1988; Forsythe et al., 1977; Klema and Laub, 1980; Lawson and Hanson, 1974; Comon and Golub, 1990] asserts that there exist positive real numbers $\sigma_1 \geq \sigma_2 \geq \cdots \geq \sigma_k > 0$ (the *singular values* of $\mathbf{X}$), an $n \times n$ unitary matrix $\mathbf{U} = [\mathbf{u}_1 \cdots \mathbf{u}_n]$ with column vectors $\mathbf{u}_i$ (the *left singular vectors*) and the property $\mathbf{U}^H\mathbf{U} = \mathbf{I}$, and an $m \times m$ unitary matrix $\mathbf{V} = [\mathbf{v}_1 \cdots \mathbf{v}_m]$ with column vectors $\mathbf{v}_i$ (the *right singular vectors*) and the property $\mathbf{V}^H\mathbf{V} = \mathbf{I}$ such that matrix $\mathbf{X}$ can be decomposed into the product

$$\mathbf{X} = \mathbf{U}\Sigma\mathbf{V}^H = \sum_{i=1}^{k} \sigma_i\mathbf{u}_i\mathbf{v}_i^H, \tag{3.119}$$

where the $n \times m$ singular value matrix Σ has the structure

$$\Sigma = \begin{pmatrix} \mathbf{D} & \mathbf{0} \\ \mathbf{0} & \mathbf{0} \end{pmatrix} \tag{3.120}$$

and $\mathbf{D} = \text{diag}(\sigma_1, \ldots, \sigma_k)$ is a $k \times k$ diagonal matrix containing the singular values. Noting that $\mathbf{X}^H = \mathbf{V}\Sigma^H\mathbf{U}^H$, $\mathbf{U}^H\mathbf{U} = \mathbf{I}$, and $\mathbf{V}^H\mathbf{V} = \mathbf{I}$, then

$$\mathbf{X}^H\mathbf{X} = \mathbf{V}(\Sigma^H\Sigma)\mathbf{V}^H, \quad \mathbf{X}\mathbf{X}^H = \mathbf{U}(\Sigma\Sigma^H)\mathbf{U}^H \qquad (3.121)$$

and

$$\mathbf{X}^H\mathbf{X}v_i = \sigma_i^2\mathbf{v}_i, \quad \mathbf{X}\mathbf{X}^H\mathbf{u}_i = \sigma_i^2\mathbf{u}_i \qquad (3.122)$$

for $1 \leq i \leq k$. The matrix products $\Sigma^H\Sigma$ and $\Sigma\Sigma^H$ are diagonal matrices of size $m \times m$ and $n \times n$, respectively, with diagonal elements σ_i^2. The matrix products $\mathbf{X}^H\mathbf{X}$ and $\mathbf{X}\mathbf{X}^H$ are Hermitian matrices of size $m \times m$ and $n \times n$, respectively. The columns of $\mathbf{U}$ are, therefore, the orthonormal eigenvectors of $\mathbf{X}\mathbf{X}^H$ and the columns of $\mathbf{V}$ are, therefore, the orthonormal eigenvectors of $\mathbf{X}^H\mathbf{X}$. Both share the same eigenvalues σ_i^2 for $1 \leq i \leq k$. The singular values are, therefore, simply the positive square roots of the nonzero eigenvalues of $\mathbf{X}^H\mathbf{X}$ and $\mathbf{X}\mathbf{X}^H$, which relate the SVD to the EVD of the least squares and minimum norm linear equations of Sec. 3.6.

Using the unitary properties of $\mathbf{U}$ and $\mathbf{V}$ and Eq. (3.121), then

$$\mathbf{X}\mathbf{V} = \mathbf{U}\Sigma\mathbf{V}^H\mathbf{V} = \mathbf{U}\Sigma, \quad \mathbf{U}^H\mathbf{X} = \mathbf{U}^H\mathbf{U}\Sigma\mathbf{V}^H = \Sigma\mathbf{V}^H \qquad (3.123)$$

or

$$\mathbf{X}\mathbf{v}_i = \sigma_i\mathbf{u}_i, \quad \mathbf{X}^H\mathbf{u}_i = \sigma_i\mathbf{v}_i \qquad (3.124)$$

for $1 \leq i \leq k$. Equation (3.124) provides a pair of relationships between the eigenvectors $\mathbf{u}_i$ and $\mathbf{v}_i$ that share the common singular value σ_i. The eigenvectors that are associated with the zero singular values, $\mathbf{u}_i$ for $k+1 \leq i \leq n$ and $\mathbf{v}_i$ for $k+1 \leq i \leq m$, are not relatable in the manner of Eqs. (3.123) and (3.124).

The *generalized inverse* $\mathbf{X}^\#$ of the $n \times m$ matrix $\mathbf{X}$ of rank k is defined in terms of the SVD matrix components as the unique matrix

$$\mathbf{X}^\# = \mathbf{V}\Sigma^\#\mathbf{U}^H = \sum_{i=1}^{k} \sigma_i^{-1}\mathbf{v}_i\mathbf{u}_i^H, \qquad (3.125)$$

where

$$\Sigma^\# = \begin{pmatrix} \mathbf{D}^{-1} & \mathbf{0} \\ \mathbf{0} & \mathbf{0} \end{pmatrix}. \qquad (3.126)$$

The generalized inverse $\mathbf{X}^\#$ provides a unique minimum norm solution $\mathbf{a} = \mathbf{X}^\#\mathbf{y}$ to the problem of finding the $m \times 1$ vector $\mathbf{a}$ that simultaneously minimizes the squared equation error $\|\mathbf{X}\mathbf{a} - \mathbf{y}\|_2$, for some given $n \times 1$ vector $\mathbf{y}$, and minimizes the solution vector length $\|\mathbf{a}\|_2$. The minimum length constraint is required in order to ensure a unique solution for those cases in which more than one vector $\mathbf{a}$ will minimize $\|\mathbf{X}\mathbf{a} - \mathbf{y}\|_2$ (always the case for $m > n$). If $m = n$ and rank $\mathbf{X}$ is n (i.e., it is nonsingular), then the pseudoinverse reduces to the usual square matrix inverse $\mathbf{X}^\# = \mathbf{X}^{-1}$. If $n > m$ and rank $\mathbf{X} = m$, then $\mathbf{X}^\# = (\mathbf{X}^H\mathbf{X})^{-1}\mathbf{X}^H$ and $\mathbf{a} = (\mathbf{X}^H\mathbf{X})^{-1}\mathbf{X}^H\mathbf{y}$.

This is the usual least squares solution for a set of overdetermined equations (see Sec. 3.6). If $m > n$ and rank $\mathbf{X} = n$, then $\mathbf{X}^{\#} = \mathbf{X}^{H}(\mathbf{X}\mathbf{X}^{H})^{-1}$ and $\mathbf{a} = \mathbf{X}^{H}(\mathbf{X}\mathbf{X}^{H})^{-1}\mathbf{y}$. This solution is called the *minimum norm solution* for a set of underdetermined equations that will yield the vector $\mathbf{a}$ of minimum Euclidean length.

Computing the pseudoinverse $\mathbf{X}^{\#}$ directly from the SVD Eq. (3.119) has two advantages over direct computation of $\mathbf{X}^{\#}$ in the form of either $(\mathbf{X}^{H}\mathbf{X})^{-1}\mathbf{X}^{H}$, in the case $n > m$, or $\mathbf{X}^{H}(\mathbf{X}\mathbf{X}^{H})^{-1}$, in the case $m > n$. First, the SVD assists in the determination of the rank of $\mathbf{X}$ by selecting only the nonnegligible singular values (zero out the negligible singular values before computing $\mathbf{X}^{\#}$). A *low rank approximation* $\hat{\mathbf{X}}$ to matrix $\mathbf{X}$ is generated by forming the product $\hat{\mathbf{X}} = \mathbf{U}\Sigma\mathbf{V}^{H}$ using only the nonnegligible singular values in Σ. Second, the computational precision to calculate $\mathbf{X}^{\#}$ to a given accuracy with the SVD is approximately half that required to calculate $\mathbf{X}^{\#}$ by direct computation of $(\mathbf{X}^{H}\mathbf{X})^{-1}$ or $(\mathbf{X}\mathbf{X}^{H})^{-1}$.

3.7.3 Total Least Squares

The underlying presumption of the least squares linear equations approach of Sec. 3.6.1 places the source of the approximation error, or noise, $\mathbf{e}$ entirely within the $n \times 1$ data vector $\mathbf{y}$ of Eq. (3.104). A unique $m \times 1$ solution vector $\mathbf{a}$ was obtained by minimizing the squared error, represented by the norm $\|\mathbf{e}\|_2$. Least squares estimation techniques based on linear prediction, to be encountered in Chaps. 8, 11, and 13, will have noise components within the elements of the $n \times m$ data matrix $\mathbf{X}$ as well as the data vector $\mathbf{y}$. To better account simultaneously for the noise effects in both $\mathbf{X}$ and $\mathbf{y}$, the least squares problem can be reformulated by a technique known as the *total least squares* (TLS) [Golub and Van Loan, 1989]. The error model

$$(\mathbf{X} + \mathbf{E})\mathbf{a} = (\mathbf{y} + \mathbf{e}) \quad \text{or} \quad ((\mathbf{e}\mathbf{y}) + (\mathbf{e}\mathbf{E})) \begin{pmatrix} 1 \\ -\mathbf{a} \end{pmatrix} = \begin{pmatrix} 0 \\ \mathbf{0}_m \end{pmatrix} \qquad (3.127)$$

introduces an error matrix $\mathbf{E}$ in addition to the usual error vector $\mathbf{e}$. The TLS approach selects for the solution the vector $\mathbf{a}$ which minimizes the Frobenius norm $\|(\mathbf{e}\mathbf{E})\|_F$ of the total error $(\mathbf{e}\mathbf{E})$. The unique solution [Golub and Van Loan, 1980] which minimizes this norm is simply

$$\mathbf{a} = - \begin{pmatrix} v_{m+1}[2]/v_{m+1}[1] \\ \vdots \\ v_{m+1}[m+1]/v_{m+1}[1] \end{pmatrix} \qquad (3.128)$$

in which $\mathbf{v}_{m+1} = (v_{m+1}[1], \ldots, v_{m+1}[m+1])^T$ is the $(m+1) \times 1$ right singular vector associated with the minimum singular value σ_{m+1} of the augmented matrix singular value decomposition

$$\begin{pmatrix} \mathbf{y} & \mathbf{X} \end{pmatrix} = \mathbf{U}\Sigma\mathbf{V}^{H} = \sum_{i=1}^{m+1} \sigma_i \mathbf{u}_i \mathbf{v}_i^{H} \qquad (3.129)$$

which assumes the augmented matrix has full rank, i.e., rank $\begin{pmatrix} \mathbf{y} & \mathbf{X} \end{pmatrix} = m$, and the minimum singular value is not repeated (has no multiplicity), otherwise appropriate modifications with Householder transformations must be added to the procedure in order to obtain a unique solution with minimum norm. Based on the augmented least squares matrix of Eq. (3.111) and the eigenequation (3.124), the following relationships

$$\begin{pmatrix} \mathbf{y} & \mathbf{X} \end{pmatrix}^H \begin{pmatrix} \mathbf{y} & \mathbf{X} \end{pmatrix} \mathbf{v}_{m+1} = \begin{pmatrix} \mathbf{y}^H\mathbf{y} & \mathbf{y}^H\mathbf{X} \\ \mathbf{x}^H\mathbf{y} & \mathbf{X}^H\mathbf{X} \end{pmatrix} \mathbf{v}_{m+1} = \sigma_{m+1}^2 \mathbf{v}_{m+1} \qquad (3.130)$$

and

$$\mathbf{v}_{m+1} = v_{m+1}[1] \begin{pmatrix} 1 \\ -\mathbf{a} \end{pmatrix} \qquad (3.131)$$

lead to the conclusion

$$\left(\mathbf{X}^H\mathbf{X} - \sigma_{m+1}^2 \mathbf{I} \right) \mathbf{a} = \mathbf{X}^H\mathbf{y}. \qquad (3.132)$$

The TLS solution may therefore be viewed, when compared to the usual least squares solution of Eq. (3.110), as a modified least squares solution in which the minimum eigenvalue of the augmented data matrix product $(\mathbf{y}\mathbf{X})^H (\mathbf{y} \quad \mathbf{X})$ is removed from the main diagonal of the data matrix product $\mathbf{x}^H\mathbf{X}$. The TLS procedure is not only applicable to the case of an overdetermined system of linear equations, but it can also be applied to an underdetermined system of linear equations by using the minimum norm solution of the augmented matrix $(\mathbf{y}\mathbf{X})$.

3.8 THE TOEPLITZ MATRIX

The Toeplitz matrix structure comes up frequently in the development of various spectral estimation methods. Because of its importance, this entire section devotes an extended emphasis on the Toeplitz matrix, including the Toeplitz inverse, Toeplitz linear equation solution, and Toeplitz eigenanalysis. The special structure of the Toeplitz matrix will lead to very computationally efficient "fast" algorithms for these three situations.

3.8.1 Toeplitz Inverse

The inverse of a square non–Hermitian Toeplitz matrix is considered first [Trench, 1964; Zohar, 1969]. Denote $\mathbf{T}_M$ as a square $(M + 1) \times (M + 1)$ Toeplitz matrix with $t[M]$ and $t[-M]$ as its highest indexed coefficients

$$\mathbf{T}_M = \begin{pmatrix} t[0] & t[-1] & \cdots & t[-M] \\ t[1] & \ddots & \ddots & \vdots \\ \vdots & \ddots & \ddots & t[-1] \\ t[M] & \cdots & t[1] & t[0] \end{pmatrix}. \qquad (3.133)$$

In general, $t[k] \neq t[-k]$ for $k = 1, \ldots, M$. Denote the inverse of $\mathbf{T}_M$ by

$$\mathbf{U}_M = \mathbf{T}_M^{-1} = \begin{pmatrix} u[0,0] & \cdots & u[0,M] \\ \vdots & \ddots & \vdots \\ u[M,0] & \cdots & u[M,M] \end{pmatrix}. \tag{3.134}$$

Because a Toeplitz matrix is persymmetric, its inverse $\mathbf{U}_M$ must also be persymmetric. The inverse is not Toeplitz, although it will be shown to be composed of sums of products of triangular Toeplitz matrices.

To obtain the complete inverse $\mathbf{U}_M$, it is necessary only to solve for the first column of the inverse

$$\begin{pmatrix} t[0] & t[-1] & \cdots & t[-M] \\ t[1] & \ddots & \ddots & \vdots \\ \vdots & \ddots & \ddots & t[-1] \\ t[M] & \cdots & t[1] & t[0] \end{pmatrix} \begin{pmatrix} u[0,0] \\ u[1,0] \\ \vdots \\ u[M,0] \end{pmatrix} = \begin{pmatrix} 1 \\ 0 \\ \vdots \\ 0 \end{pmatrix} \tag{3.135}$$

and the last column of the inverse

$$\begin{pmatrix} t[0] & t[-1] & \cdots & t[-M] \\ t[1] & \ddots & \ddots & \vdots \\ \vdots & \ddots & \ddots & t[-1] \\ t[M] & \cdots & t[1] & t[0] \end{pmatrix} \begin{pmatrix} u[0,M] \\ \vdots \\ u[M-1,M] \\ u[M,M] \end{pmatrix} = \begin{pmatrix} 0 \\ \vdots \\ 0 \\ 1 \end{pmatrix}. \tag{3.136}$$

It will be shown that the remaining elements of $\mathbf{U}_M$ may then be obtained recursively by exploiting the persymmetry property.

An alternative representation to that of Eq. (3.135) is obtained by dividing the vectors on both sides of the equal sign in (3.137) by $u[0,0]$ (assuming it is nonzero), substituting $a_M[k] = u[k,0]/u[0,0]$ for $k = 1, \ldots, M$, and selecting $\rho_M^a = 1/u[0,0]$ to yield

$$\begin{pmatrix} t[0] & t[-1] & \cdots & t[-M] \\ t[1] & \ddots & \ddots & \vdots \\ \vdots & \ddots & \ddots & t[-1] \\ t[M] & \cdots & t[1] & t[0] \end{pmatrix} \begin{pmatrix} 1 \\ a_M[1] \\ \vdots \\ a_M[M] \end{pmatrix} = \begin{pmatrix} \rho_M^a \\ 0 \\ \vdots \\ 0 \end{pmatrix}. \tag{3.137}$$

Similarly, an alternative to Eq. (3.136) may be obtained by dividing the vectors on both sides of the equal sign in (3.138) by $u[M,M]$ (assuming it is nonzero), substituting $b_M[k] = u[M-k,M]/u[M,M]$ for $k = 1, \ldots, M$, and selecting $\rho_M^b = 1/u[M, M]$ to yield

$$
\begin{pmatrix}
t[0] & t[-1] & \cdots & t[-M] \\
t[1] & \ddots & \ddots & \vdots \\
\vdots & \ddots & \ddots & t[-1] \\
t[M] & \cdots & t[1] & t[0]
\end{pmatrix}
\begin{pmatrix}
b_M[M] \\
\vdots \\
b_M[1] \\
1
\end{pmatrix}
=
\begin{pmatrix}
0 \\
\vdots \\
0 \\
\rho_M^b
\end{pmatrix}.
\tag{3.138}
$$

Due to the persymmetry of $\mathbf{U}_M$, it must be true that $u[0, 0] = u[M, M]$ which implies equality between the complex-valued scalars $\rho_M^a = \rho_M^b = \rho_M$. Equations (3.137) and (3.139) are related to the normal equations of an autoregressive process (see Chaps. 6 and 7) where the physical meaning of these parameters will be discussed at length.

The solution for the $a_M[k]$ and $b_M[k]$ coefficients may be obtained with a recursive algorithmic procedure that successively yields the coefficients at each order m, starting at $m = 1$ and working recursively to $m = M$. The original recursive algorithm was developed by Levinson [1947] in connection with the solution of the normal equations for linear prediction of a stationary discrete-time process. It was rediscovered by Durbin [1960] in connection with the problem of fitting an autoregressive model to a given autocorrelation sequence. The key components of this recursion are the following partitions of $\mathbf{T}_m$ that depend on the special structure of a Toeplitz matrix

$$
\begin{aligned}
\mathbf{T}_m &= \begin{pmatrix} \mathbf{T}_{m-1} & \mathbf{J}_m \mathbf{s}_m \\ \mathbf{r}_m^T \mathbf{J}_m & t[0] \end{pmatrix} = \mathbf{Z}_{m+1}^T \mathbf{T}_m \mathbf{Z}_{m+1} + \begin{pmatrix} \mathbf{0}_{mm} & \mathbf{J}_m \mathbf{s}_m \\ \mathbf{r}_m^T \mathbf{J}_m & t[0] \end{pmatrix} \\
&= \begin{pmatrix} t[0] & \mathbf{s}_m^T \\ \mathbf{r}_m & \mathbf{T}_{m-1} \end{pmatrix} = \mathbf{Z}_{m+1} \mathbf{T}_m \mathbf{Z}_{m+1}^T + \begin{pmatrix} t[0] & \mathbf{s}_m^T \\ \mathbf{r}_m & \mathbf{0}_{mm} \end{pmatrix},
\end{aligned}
\tag{3.139}
$$

where the $m \times 1$-dimensional column vectors $\mathbf{r}_m$ and $\mathbf{s}_m$ are given by

$$
\mathbf{r}_m = \begin{pmatrix} t[1] \\ \vdots \\ t[m] \end{pmatrix}
\qquad
\mathbf{s}_m = \begin{pmatrix} t[-1] \\ \vdots \\ t[-m] \end{pmatrix}.
\tag{3.140}
$$

$\mathbf{J}_m$ is an $m \times m$ reflection matrix, and $\mathbf{Z}_{m+1}$ is an $(m + 1) \times (m + 1)$ shift matrix of Eq. (3.31). Using a process of induction, assume that the vector of $a_m[k]$ coefficients may be formed as a linear combination of the vectors of $a_{m-1}[k]$ and $b_{m-1}[k]$ coefficients

$$
\begin{pmatrix}
1 \\
a_m[1] \\
\vdots \\
a_m[m-1] \\
a_m[m]
\end{pmatrix}
=
\begin{pmatrix}
1 \\
a_{m-1}[1] \\
\vdots \\
a_{m-1}[m-1] \\
0
\end{pmatrix}
+ \alpha
\begin{pmatrix}
0 \\
b_{m-1}[m-1] \\
\vdots \\
b_{m-1}[1] \\
1
\end{pmatrix},
\tag{3.141}
$$

in which the right-hand-side vectors have been augmented with an additional zero element and α is a scalar to be determined. Note that $a_m[m] = \alpha$. Equation (3.141) asserts that the first column of the inverse matrix for order m is obtained as a linear combination of the first and last columns of the inverse matrix for order $m - 1$. By defining the column vectors

$$\mathbf{a}_m = \begin{pmatrix} a_m[1] \\ \vdots \\ a_m[m] \end{pmatrix} \qquad \mathbf{b}_m = \begin{pmatrix} b_m[1] \\ \vdots \\ b_m[m] \end{pmatrix}, \qquad (3.142)$$

Eq. (3.143) may be more concisely expressed as

$$\begin{pmatrix} 1 \\ \mathbf{a}_m \end{pmatrix} = \begin{pmatrix} 1 \\ \mathbf{a}_{m-1} \\ 0 \end{pmatrix} + a_m[m] \begin{pmatrix} 0 \\ \mathbf{Jb}_{m-1} \\ 1 \end{pmatrix}. \qquad (3.143)$$

In order to determine the scalar $a_m[m]$, it is necessary to multiply both sides of (3.143) on the left by $\mathbf{T}_m$ and substitute the appropriate partition from Eq. (3.139), yielding

$$\mathbf{T}_m \begin{pmatrix} 1 \\ \mathbf{a}_m \end{pmatrix} = \begin{pmatrix} \mathbf{T}_{m-1} & \mathbf{Js}_m \\ \mathbf{r}_m^T \mathbf{J} & t[0] \end{pmatrix} \begin{pmatrix} 1 \\ \mathbf{a}_{m-1} \\ 0 \end{pmatrix} \qquad (3.144)$$

$$+ \alpha_m[m] \begin{pmatrix} t[0] & \mathbf{s}_m^T \\ \mathbf{r}_m & \mathbf{T}_{m-1} \end{pmatrix} \begin{pmatrix} 0 \\ \mathbf{Jb}_{m-1} \\ 1 \end{pmatrix}. \qquad (3.145)$$

This reduces, based on (3.137) and (3.139), to the result

$$\begin{pmatrix} \rho_m \\ 0 \\ \vdots \\ 0 \\ 0 \end{pmatrix} = \begin{pmatrix} \rho_{m-1} \\ 0 \\ \vdots \\ 0 \\ \Delta_m \end{pmatrix} + a_m[m] \begin{pmatrix} \nabla_m \\ 0 \\ \vdots \\ 0 \\ \rho_{m-1} \end{pmatrix} \qquad (3.146)$$

in which

$$\Delta_m = \mathbf{r}_m^T \mathbf{J} \begin{pmatrix} 1 \\ \mathbf{a}_{m-1} \end{pmatrix} = \sum_{k=1}^{m} t[k] a_{m-1}[m-k], \quad a_{m-1}[0] = 1 \qquad (3.147)$$

$$\nabla_m = \mathbf{s}_m^T \begin{pmatrix} \mathbf{Jb}_{m-1} \\ 1 \end{pmatrix} = \sum_{k=1}^{m} t[-k] b_{m-1}[m-k], \quad b_{m-1}[0] = 1. \qquad (3.148)$$

Note that Δ_m and ∇_m require knowledge of the order $m - 1$ coefficients only. In order to have equality in Eq. (3.146), the bottom element on the right-hand side should be zero

$$\Delta_m + a_m[m]\rho_{m-1} = 0, \tag{3.149}$$

which implies

$$a_m[m] = -\Delta_m/\rho_{m-1}. \tag{3.150}$$

The top elements on both sides of Eq. (3.146) are therefore related as

$$\rho_m = \rho_{m-1} + a_m[m]\nabla_m = \rho_{m-1} - \Delta_m \nabla_m/\rho_{m-1}. \tag{3.151}$$

The analogous order update to Eq. (3.143) for the vector $\mathbf{b}_m$ can also be proved by induction to be

$$\begin{pmatrix} 1 \\ \mathbf{b}_m \end{pmatrix} = \begin{pmatrix} 1 \\ \mathbf{b}_{m-1} \\ 0 \end{pmatrix} + b_m[m] \begin{pmatrix} 0 \\ \mathbf{Ja}_m \\ 1 \end{pmatrix}. \tag{3.152}$$

When each side of (3.152) is multiplied on the left by $\mathbf{T}_m$, the result reduces to

$$\begin{pmatrix} \rho_m \\ 0 \\ \vdots \\ 0 \\ 0 \end{pmatrix} = \begin{pmatrix} \rho_{m-1} \\ 0 \\ \vdots \\ 0 \\ \nabla_m \end{pmatrix} + b_m[m] \begin{pmatrix} \Delta_m \\ 0 \\ \vdots \\ 0 \\ \rho_{m-1} \end{pmatrix}, \tag{3.153}$$

where Δ_m and ∇_m were defined previously by Eqs. (3.147) and (3.148). Again, for equality to hold in Eq. (3.153), the assignment

$$b_m[m] = -\nabla_m/\rho_{m-1} \tag{3.154}$$

must be made. Substituting for ∇_m in Eq. (3.151),

$$\rho_m = \rho_{m-1} + a_m[m]\left(-\rho_{m-1}b_m[m]\right) = \rho_{m-1}\left(1 - a_m[m]b_m[m]\right). \tag{3.155}$$

Note from (3.137) that $t[0] = \rho_0$ at order $m = 0$, which initializes the recursion. The complex scalar ρ_m is related to the determinant of the Toeplitz matrix by the expression

$$\det \mathbf{T}_m = \prod_{k=0}^{m} \rho_k = \rho_0^{m+1} \prod_{k=1}^{m} (1 - a_k[k]b_k[k])^{m+1-k}. \tag{3.156}$$

To prove this result, use Eqs. (3.53) and (3.155) and note that the following matrix product construction

$$\mathbf{T}_m \begin{pmatrix} \mathbf{I}_m & \mathbf{b}_m \\ \mathbf{0}_m^T & 1 \end{pmatrix} = \begin{pmatrix} \mathbf{T}_{m-1} & \mathbf{Js}_m \\ \mathbf{r}_m^T\mathbf{J} & t[0] \end{pmatrix} \begin{pmatrix} \mathbf{I}_m & \mathbf{b}_m \\ \mathbf{0}_m^T & 1 \end{pmatrix} = \begin{pmatrix} \mathbf{T}_{m-1} & \mathbf{0}_m \\ \mathbf{r}_m^T\mathbf{J} & \rho_m \end{pmatrix} \tag{3.157}$$

has the determinant $\det \mathbf{T}_m = \rho_m \det \mathbf{T}_{m-1}$, based on the property $\det \mathbf{AB} = \det \mathbf{A} \det \mathbf{B}$. On an element-by-element basis, the vector recursions of (3.145) and

(3.153) yield

$$a_m[k] = a_{m-1}[k] + a_m[m]b_{m-1}[m-k] \tag{3.158}$$
$$b_m[k] = b_{m-1}[k] + b_m[m]a_{m-1}[m-k] \tag{3.159}$$

for $k = 1, \ldots, m-1$. The six equations (3.146), (3.150), (3.151), (3.153), (3.155), and (3.157) together constitute a closed-form recursion ($m = 0, \ldots, M$), known as the Levinson recursion, for the vectors $\mathbf{a}_m$ and $\mathbf{b}_m$. This, in effect, provides the first and last columns of the Toeplitz inverse. Note that the submatrices $\mathbf{T}_0, \mathbf{T}_1, \ldots, \mathbf{T}_M$ must all be invertible (det $\mathbf{T}_m \neq 0$) for this recursion to work. The vector updates given by (3.145) and (3.153), as well as the vector inner products (3.148) and (3.149), require a number of computations proportional to M at each step in the recursion. Because there are M steps, the total computational requirements of this algorithm are proportional to M^2.

A MATLAB function `levinson_recursion.m` for the specialized case of a Hermitian Toeplitz matrix is available through Dover Publications text web page:

```
function [rho,a]=levinson_recursion(r)
% Levinson recursion algorithm for solving Hermitian Toeplitz linear equation
% Ra =rho.  Typically vector  a  contains the LP/AR parameters and  rho  is
% the white noise variance (see Section 7.3.1)
%
%      [rho,a] = levinson_recursion(r)
%
% r    -- left-side column vector of Toeplitz matrix elements r(1:p+1) for
%         MATLAB  [ r(0:p) in text ]
% rho -- top element of right-hand-side all-zeros vector
% a   -- solution vector a(1:p)
```

In order to compute the remaining elements of the inverse, first note that the inverse Toeplitz matrix may be expressed in the partitioned form

$$\mathbf{U}_M = \rho_M^{-1} \begin{pmatrix} 1 & \mathbf{b}_M^T \\ \mathbf{a}_M & \rho_M \mathbf{U}_{M-1} + \mathbf{a}_M \mathbf{b}_M^T \end{pmatrix}. \tag{3.160}$$

Because the inverse is persymmetric,

$$\mathbf{J}\mathbf{U}_M^T\mathbf{J} = \mathbf{U}_M, \tag{3.161}$$

an alternative to Eq. (3.161) is

$$\mathbf{U}_M = \rho_M^{-1} \begin{pmatrix} \rho_M \mathbf{U}_{M-1} + \mathbf{J}\mathbf{b}_M\mathbf{a}_M^T\mathbf{J} & \mathbf{J}\mathbf{b}_M \\ \mathbf{a}_M^T\mathbf{J} & 1 \end{pmatrix}. \tag{3.162}$$

From (3.161) and (3.163), the following relations for the elements $u_M[j,k]$ of the order M Toeplitz inverse may be found

$$u_M[0,0] = u_M[M,M] = \rho_m^{-1} \tag{3.163}$$
$$u_M[0,k] = u_M[M-k,M] = b_M[k]\rho_M^{-1} \quad \text{for } 1 \le k \le M \tag{3.164}$$

$$u_M[j, 0] = u_M[M, M - j] = a_M[j]\rho_M^{-1} \quad \text{for } 1 \leq j \leq M \qquad (3.165)$$

$$u_M[j + 1, k + 1] = u_{M-1}[j, k] + a_M[j + 1]b_M[k + 1]\rho_M^{-1} \quad \text{for } 0 \leq j, k \leq M - 1$$
$$(3.166)$$

$$u_M[j, k] = u_{M-1}[j, k] + b_M[M - j]a_M[M - k]\rho_M^{-1} \quad \text{for } 0 \leq j, k \leq M - 1. \qquad (3.167)$$

By combining the last two of the above expressions, the element $u_{M-1}[j, k]$ can be eliminated:

$$u_M[j + 1, k + 1]$$
$$= u_M[j, k] + \rho_M^{-1}\big(a_M[j + 1]b_M[k + 1] - b_M[M - j]a_M[M - k]\big), \qquad (3.168)$$

which holds for $0 \leq j, k \leq M - 1$. Equation (3.164) is a recursive relationship that permits all inverse elements to be computed from knowledge of only the $\mathbf{a}_M$ and $\mathbf{b}_M$ vectors. The following may also be developed from Eq. (3.164),

$$\mathbf{T}_M^{-1} = \rho_M^{-1} \begin{pmatrix} 1 & 0 & \cdots & 0 \\ a_M[1] & \ddots & \ddots & \vdots \\ \vdots & \ddots & \ddots & 0 \\ a_M[M] & \cdots & a_M[1] & 1 \end{pmatrix} \begin{pmatrix} 1 & b_M[1] & \cdots & b_M[M] \\ 0 & \ddots & \ddots & \vdots \\ \vdots & \ddots & \ddots & b_M[1] \\ 0 & \cdots & 0 & 1 \end{pmatrix}$$
$$- \rho_M^{-1} \begin{pmatrix} 0 & 0 & \cdots & 0 \\ b_M[M] & \ddots & \ddots & \vdots \\ \vdots & \ddots & \ddots & 0 \\ b_M[1] & \cdots & b_M[M] & 0 \end{pmatrix} \begin{pmatrix} 0 & a_M[M] & \cdots & a_M[1] \\ 0 & \ddots & \ddots & \vdots \\ \vdots & \ddots & \ddots & a_M[M] \\ 0 & \cdots & 0 & 0 \end{pmatrix}. \qquad (3.169)$$

Although the inverse of a general Toeplitz matrix is not Toeplitz, it can be factored into the sum of products of upper and lower triangular Toeplitz matrices.

A case of special interest is that of the inverse of a Hermitian Toeplitz matrix $\mathbf{H}_M$, which has the form

$$\mathbf{H}_M = \begin{pmatrix} t[0] & t^*[1] & \cdots & t^*[M] \\ t[1] & t[0] & \cdots & t^*[M - 1] \\ \vdots & \vdots & \ddots & \vdots \\ t[M] & t[M - 1] & \cdots & t[0] \end{pmatrix} \qquad (3.170)$$

for complex data. Note that $\mathbf{H}_M^H = \mathbf{H}_M = \mathbf{J}\mathbf{H}_M^*\mathbf{J}$. In this situation, the vectors of Eq. (3.141) are related by complex symmetry

$$\mathbf{s}_M = \mathbf{r}_M^*, \qquad (3.171)$$

so that the recursions simplify, reducing the number of computations for finding the inverse. In particular, it easily follows that the vectors

$$\mathbf{b}_M^* = \mathbf{a}_M \qquad (3.172)$$

and the scalars

$$\nabla_M^* = \Delta_M \qquad (3.173)$$

have complex conjugate relationships. The scalar term ρ_M, which now satisfies

$$\rho_M = \rho_{M-1}\left(1 - |a_M[M]|^2\right), \qquad (3.174)$$

is real and positive. This implies the following:

$$\det \mathbf{H}_M = \rho_0 \prod_{k=1}^{M}\left(1 - |a_k[k]|^2\right) > 0. \qquad (3.175)$$

The number of computations needed to solve for the $\mathbf{a}_M$ vector is reduced approximately by a factor of 2 in the Hermitian case over the non–Hermitian case due to relationships (3.168), (3.169), and (3.170).

An alternative solution to the classical Levinson algorithm that exploits the symmetric redundancies further to reduce the number of multiplications is based on a concept called the *split Levinson algorithm* [Delsarte and Genin, 1986]. Instead of the two-term recursions of Eqs. (3.145) and (3.153) based on two order $m - 1$ parameters to recursively obtain the order m parameters, the split Levinson uses a three-term recursion based on order $m - 1$ and order $m - 2$ parameters to obtain the order m parameters. The essence of the split Levinson is to solve for a symmetric vector $\mathbf{f}_M$ that solves a symmetric form of Hermitian Toeplitz linear equations

$$\mathbf{T}_M\mathbf{f}_M = \begin{pmatrix} \tau_M \\ \mathbf{0}_{m-1} \\ \tau_M \end{pmatrix} \qquad (3.176)$$

in which the solution can be shown to satisfy a three-term recursion

$$\mathbf{f}_M = \begin{pmatrix} \mathbf{f}_{M-1} \\ 0 \end{pmatrix} + \begin{pmatrix} 0 \\ \mathbf{f}_{M-1} \end{pmatrix} - \alpha_{M-1}\begin{pmatrix} 0 \\ \mathbf{f}_{M-2} \\ 0 \end{pmatrix}. \qquad (3.177)$$

Another relationship enables α_M to be computed from $\mathbf{f}_M$.

The matrix inverse (3.172) for a general Toeplitz matrix specializes in the Hermitian Toeplitz matrix case to

$$\mathbf{H}_M^{-1} = \rho_M^{-1}\begin{pmatrix} 1 & 0 & \cdots & 0 \\ a_M[1] & \ddots & \ddots & \vdots \\ \vdots & \ddots & \ddots & 0 \\ a_M[M] & \cdots & a_M[1] & 1 \end{pmatrix}\begin{pmatrix} 1 & a_M^*[1] & \cdots & a_M^*[M] \\ 0 & \ddots & \ddots & \vdots \\ \vdots & \ddots & \ddots & a_M^*[1] \\ 0 & \cdots & 0 & 1 \end{pmatrix}$$
$$- \rho_M^{-1}\begin{pmatrix} 0 & 0 & \cdots & 0 \\ a_M^*[M] & \ddots & \ddots & \vdots \\ \vdots & \ddots & \ddots & 0 \\ a_M^*[1] & \cdots & a_M^*[M] & 0 \end{pmatrix}\begin{pmatrix} 0 & a_M[M] & \cdots & a_M[1] \\ 0 & \ddots & \ddots & \vdots \\ \vdots & \ddots & \ddots & a_M[M] \\ 0 & \cdots & 0 & 0 \end{pmatrix}. \qquad (3.178)$$

3.8.2 Hermitian Toeplitz Triangular Decomposition

The Levinson algorithm also provides the basis for an efficient Cholesky (triangular) decomposition of the inverse of a Hermitian Toeplitz matrix [Burg, 1968; Therrien, 1983]. Because all lower-order solution vectors from $\mathbf{a}_1$ up to $\mathbf{a}_M$ are computed as part of the algorithm, it is then possible to construct the matrix product

$$\mathbf{H}_M \mathbf{A}_M = \mathbf{Q}_M, \tag{3.179}$$

where $\mathbf{H}_M$ is the Hermitian Toeplitz matrix and $\mathbf{A}_M$ is a lower-triangular matrix constructed from the solution coefficient vectors $\mathbf{a}_m$ from $m = 0$ to $m = M$,

$$\mathbf{A}_M = \begin{pmatrix} 1 & 0 & \cdots & 0 & 0 \\ a_M[1] & 1 & \cdots & 0 & 0 \\ a_M[2] & a_{M-1}[1] & \cdots & 0 & 0 \\ \vdots & \vdots & \ddots & \vdots & \vdots \\ a_M[M-1] & a_{M-1}[M-2] & \cdots & 1 & 0 \\ a_M[M] & a_{M-1}[M-1] & \cdots & a_1[1] & 1 \end{pmatrix}. \tag{3.180}$$

Using Eq. (3.139), it is easily seen that the right-hand matrix is itself a triangular matrix with the form

$$\mathbf{Q}_M = \begin{pmatrix} \rho_M & x & \cdots & x & x \\ 0 & \rho_{M-1} & \cdots & x & x \\ \vdots & \vdots & \ddots & \vdots & \vdots \\ 0 & 0 & \cdots & \rho_1 & x \\ 0 & 0 & \cdots & 0 & \rho_0 \end{pmatrix}, \tag{3.181}$$

in which the x denote nonzero off-diagonal quantities whose specific values are not of interest. If both sides of (3.183) are multiplied on the left by the conjugate transpose of $\mathbf{A}_M$, then

$$\mathbf{A}_M^H \mathbf{H}_M \mathbf{A}_M = \mathbf{A}_M^H \mathbf{Q}_M = \mathbf{P}_M, \tag{3.182}$$

where matrix $\mathbf{P}_M$ is the diagonal matrix

$$\mathbf{P}_M = \begin{pmatrix} \rho_M & 0 & \cdots & 0 & 0 \\ 0 & \rho_{M-1} & \cdots & 0 & 0 \\ \vdots & \vdots & \ddots & \vdots & \vdots \\ 0 & 0 & \cdots & \rho_1 & 0 \\ 0 & 0 & \cdots & 0 & \rho_0 \end{pmatrix} \tag{3.183}$$

formed as the result of the product of two upper triangular matrices. Thus, an alternative decomposition to Eq. (3.172) for the inverse of a Hermitian Toeplitz matrix is the triangular factorization

$$\mathbf{H}_M^{-1} = \mathbf{A}_M \mathbf{P}_M^{-1} \mathbf{A}_M^H. \tag{3.184}$$

3.8.3 Solution of Toeplitz Linear Equations

The Levinson algorithm also provides the foundation for algorithms that solve linear equations involving Toeplitz and Hermitian Toeplitz matrices. Consider the general Toeplitz linear equations of the form

$$\mathbf{T}_M \mathbf{x}_M = \mathbf{z}_M, \tag{3.185}$$

where $\mathbf{z}_M$ is a known right-hand side $(M+1)$-dimensional column vector

$$\mathbf{z}_M = \begin{pmatrix} z[0] \\ \vdots \\ z[M] \end{pmatrix} \tag{3.186}$$

and $\mathbf{x}_M$ is the unknown $(M+1)$-dimensional column vector

$$\mathbf{x}_M = \begin{pmatrix} x_M[0] \\ \vdots \\ x_M[M] \end{pmatrix} \tag{3.187}$$

whose solution is desired. The solution for $\mathbf{x}_M$ may be embedded within the recursions of the Levinson algorithm by noting that the order m solution vector $\mathbf{x}_m$ for the linear equations

$$\mathbf{T}_m \mathbf{x}_m = \mathbf{z}_m, \tag{3.188}$$

where $m \leq M$ and $\mathbf{z}_m$ is an $(m+1)$-dimensional vector formed from the first $m+1$ elements of $\mathbf{z}_M$, may be computed recursively from the order $m-1$ solution vector $\mathbf{x}_{m-1}$ as follows:

$$\mathbf{x}_m = \begin{pmatrix} \mathbf{x}_{m-1} \\ 0 \end{pmatrix} + \alpha_m \begin{pmatrix} \mathbf{J}\mathbf{b}_m \\ 1 \end{pmatrix} \tag{3.189}$$

for some scalar α_m, to be determined, and $\mathbf{b}_m$ is the vector of Eq. (3.144). To prove Eq. (3.185), simply multiply both sides of (3.185) on the left by $\mathbf{T}_m$ and substitute the appropriate partition from Eq. (3.141) to yield

$$\mathbf{z}_m = \begin{pmatrix} \mathbf{z}_{m-1} \\ \beta_m \end{pmatrix} + \alpha_m \begin{pmatrix} 0 \\ \vdots \\ 0 \\ \rho_m \end{pmatrix}, \tag{3.190}$$

where

$$\beta_m = \mathbf{r}_m^T \mathbf{J}\mathbf{x}_{m-1} = \sum_{k=0}^{m-1} x[k]t[m-k]. \tag{3.191}$$

Equality will result by selecting

$$\alpha_m = (z[m] - \beta_m)/\rho_m. \tag{3.192}$$

The recursion is initialized at order zero by setting

$$t[0]x[0] = z[0], \tag{3.193}$$

the solution vector is updated by Eq. (3.185) at each stage of the Levinson algo-rithm for $1 \le m \le M$. For the case of a Toeplitz matrix with Hermitian symmetry, the simplifying substitution $\mathbf{a}_m^* = \mathbf{b}_m$ is made. This fast algorithm requires $2M^2$ multiplication/addition operations and memory storage of $4M$ (or $3M$ if in-place memory operations are used). In comparison, solution by the Cholesky method would require a number of operations proportional to M^3 and storage proportional to M^2. Cybenko [1980] has found that Levinson-type algorithms have comparable numerical stability to the more computationally inefficient Cholesky decomposition method of linear equation solution (Sec. 3.5.2). Other implementation aspects are discussed by Carayannis et al. [1982], Farden [1976], Roebuck [1978], Yarlagadda [1980], and Zohar [1974, 1979, 1981].

3.8.4   Toeplitz Linear Equation Solutions

The Levinson recursion forms the internal basis of fast algorithmic solutions of linear equations involving Toeplitz or Hermitian Toeplitz structures. MATLAB func-tion `toeplitz_lineqs.m` will solve the nonsymmetric Toeplitz linear equation $\mathbf{T}_M \mathbf{x}_M = \mathbf{y}_M$:

```
function x=toeplitz_lineqs(T_left,T_top,y)
% Finds linear equation solution for nonsymmetric Toeplitz matrix: xT = y
%
%              x = toeplitz_lineqs(T_left,T_top,y)
%
% T_left -- left column of Toeplitz matrix (all elements generally complex)
% T_top  -- top row of Toeplitz matrix (all elements generally complex-valued)
% y      -- given y row vector (generally complex-valued)
% x      -- solution row vector (can be complex-valued)
```

and the MATLAB function `hermtoep_lineqs.m` will solve the Hermitian-symmetric Toeplitz linear equation $\mathbf{H}_M \mathbf{x}_M = \mathbf{z}_M$:

```
function x=hermtoep_lineqs(t,z)
% Finds solution of Hermitian Toeplitz set of linear equations  Hx = z.
%
%     x = hermtoep_lineqs(t,z)
%
% t -- left-side column vector of Toeplitz matrix consisting of elements
%      t(0:m) [ (1:m+1) in MATLAB ]
% z -- right-hand-side vector with elements z(0:m) [ (1:m+1) in MATLAB ]
% x -- solution vector with elements x(0:m)        [ (1:m+1) in MATLAB ]
```

Both fast linear equation solution algorithms require a number of operations proportional to M^2 and storage proportional to M.

3.8.5 Toeplitz Eigenanalysis

Despite its special structure, there are few known special properties of the eigenvalues and eigenvectors of finite-dimensional Toeplitz matrices [Datta, 1984; Makhoul, 1981]. Let

$$A(z) = \sum_{k=0}^{M} v[k]z^{-k} \qquad (3.194)$$

be the eigenpolynomial formed from the elements $v[k]$ of an eigenvector of the $(M+1) \times (M+1)$ Toeplitz matrix $\mathbf{T}_M$. If $\mathbf{T}_M$ is Hermitian, then $\mathbf{T}_M$ is a special case of a centrosymmetric matrix, in which case the eigenvectors will be either conjugate symmetric or anti-conjugate symmetric (see Sec. 3.5.4). The polynomial $A(z)$ will similarly be conjugate symmetric or anti-conjugate symmetric. Such symmetry in a polynomial can be shown to possess roots that occur in inverse pairs [Makhoul, 1981]. Thus, if z_i is a root, then so is $1/z_i$. If z_i is inside the unit circle, then $1/z_i$ will be outside the unit circle. If z_i is on the unit circle, $1/z_i = z_i^*$ will be its complex conjugate (and on the unit circle as well). If the coefficients of $A(z)$ are real, then the complex zeros will also come in complex conjugate pairs.

  If the *maximum* eigenvalue of $\mathbf{T}_M$ is distinct, then it can be shown that the polynomial formed from the eigenvector of the maximum eigenvalue has its roots *on* the unit circle. The same is true for the eigenvector corresponding to the *minimum* eigenvalue, if it also is distinct. The roots of the polynomials formed from the remaining eigenvectors may or may not have their roots on the unit circle; the same will be true if the maximum or minimum eigenvalues are not unique.

  From Eqs. (3.68) and (3.157), the eigenvalues λ_k of a Toeplitz matrix and the ρ_k elements are related by

$$\det \mathbf{T}_M = \prod_{k=0}^{M} \rho_k = \prod_{k=0}^{M} \lambda_k. \qquad (3.195)$$

The trace of an $(M+1) \times (M+1)$ Toeplitz matrix $\mathbf{T}_M$ is simply $(M+1)t[0]$. Using (3.68), the following result holds:

$$t[0] = \frac{1}{M+1} \sum_{k=0}^{M} \lambda_k. \qquad (3.196)$$

As will be shown in Chap. 6, the element $t[0]$ corresponds to the zero-lag autocorrelation term of a random process, which represents the total power in a signal. Thus, the average of the eigenvalues also represent the total power in a signal.

  It is possible to make asymptotic statements concerning the behavior of eigenvalues of Toeplitz matrices by relating them to the eigenvalues of a circulant matrix, for which the analytical structure of the eigenvalues is known. For example, the 4×4 tridiagonal Toeplitz matrix

$$\begin{pmatrix} t[0] & t[-1] & 0 & 0 \\ t[1] & t[0] & t[-1] & 0 \\ 0 & t[1] & t[0] & t[-1] \\ 0 & 0 & t[1] & t[0] \end{pmatrix} \tag{3.197}$$

may be approximated by the circulant matrix

$$\begin{pmatrix} t[0] & t[-1] & 0 & t[1] \\ t[1] & t[0] & t[-1] & 0 \\ 0 & t[1] & t[0] & t[-1] \\ t[-1] & 0 & t[1] & t[0] \end{pmatrix} \tag{3.198}$$

by simply adding the appropriate elements to the opposite corners. As the 4×4 dimension is permitted to grow to dimension $n \times n$, the two matrices become closer in the sense that the n^2 elements agree in all positions, except at the two corners. As $n \to \infty$, the matrices become asymptotically equivalent. Details of the asymptotic behavior of the eigenvalues of Toeplitz matrices approximated as circulant matrices may be found in Grenander and Szegö [1958] and Gray [1972].

There are spectral estimation techniques (see Sec. 13.6.1) that require the minimum or maximum eigenvalues of a Toeplitz matrix and the associated eigenvectors. The classical power method may be used to find these eigenvalues, with some computational reduction possible by exploiting the Toeplitz structure. The eigenvector associated with the minimum eigenvalue may be computed by the power method (Sec. 3.5.3) as the asymptotic solution to a sequence of vectors $\mathbf{v}(k)$, each found as the solution to

$$\mathbf{v}(k+1) = \mathbf{T}^{-1}\mathbf{v}(k) \quad \text{for} \quad k = 0, 1, \ldots \tag{3.199}$$

starting with some initial guess for vector $\mathbf{v}(0)$. Equation (3.195) may be rewritten as

$$\mathbf{T}\mathbf{v}(k+1) = \mathbf{v}(k), \tag{3.200}$$

forming a set of linear equations for the unknown vector $\mathbf{v}(k+1)$. Use of the MATLAB function hermtoep_lineqns, for example, can solve (3.196) in a number of computations per iteration that is proportional to M^2. After each iteration of (3.204), the interim solution vector $\mathbf{v}(k)$ is scaled by the square root of its magnitude $\mathbf{v}^H(k)\mathbf{v}(k)$ to produce a unit length vector. Typically, a few iterations are all that are needed for the vector to converge to the eigenvector of the minimum eigenvalue, with the minimum eigenvalue given by the Rayleigh quotient

$$\lambda_{min} = \frac{\mathbf{v}^H(k)\mathbf{T}\mathbf{v}(k)}{\mathbf{v}^H(k)\mathbf{v}(k)}. \tag{3.201}$$

Convergence can be slowed if $\lambda_{max}/\lambda_{min} \gg 1$. The maximum eigenvector may similarly be found by iterating the matrix equation

$$\mathbf{T}\mathbf{v}(k) = \mathbf{v}(k+1) \tag{3.202}$$

until convergence is achieved. Simple matrix-vector multiplication is all that is required here, so that computations will be proportional to M^2.

MATLAB function `minimum_eigenvalue.m` solves for the minimum eigenvalue and associated eigenvector of a Hermitian Toeplitz matrix using the power iteration approach:

```
function [eigval,eigvec]=minimum_eigenvalue(t)

% Computes the minimum eigenvalue and associated eigenvector for a hermitian
% toeplitz matrix T.
%
%    [eigval,eigvec] = minimum_eigenvalue(t)
%
% t      -- left side column vector of matrix T
% eigval -- minimum eigenvalue of the matrix T
% eigvec -- eigenvector associated with minimum eigenvalue
```

3.9 THE VANDERMONDE MATRIX

The Vandermonde matrix structure will be encountered in the Prony (Chap. 11) and Pisarenko (Chap. 13) techniques for spectral analysis. This section illustrates how the special Vandermonde structure can be exploited to develop a fast computational algorithm for solution of linear equations involving Vandermonde matrices. Consider the linear equation set

$$\begin{pmatrix} v_1^0 & v_2^0 & \cdots & v_p^0 \\ \vdots & \vdots & & \vdots \\ v_1^{p-1} & v_2^{p-1} & \cdots & v_p^{p-1} \end{pmatrix} \begin{pmatrix} x[1] \\ \vdots \\ x[p] \end{pmatrix} = \begin{pmatrix} y[1] \\ \vdots \\ y[p] \end{pmatrix} \quad \text{or} \quad \mathbf{V}\mathbf{x} = \mathbf{y}. \tag{3.203}$$

The key to a solution is the recognition of the following decomposition

$$\mathbf{V}\mathbf{D} - \mathbf{Z}\mathbf{V} = \begin{pmatrix} 1 \\ 0 \\ \vdots \\ 0 \end{pmatrix} \begin{pmatrix} v_1^{-1} & \cdots & v_p^{-1} \end{pmatrix} \tag{3.204}$$

in which

$$\mathbf{D} = \begin{pmatrix} v_1^{-1} & 0 & \cdots & 0 \\ 0 & v_2^{-1} & \ddots & \vdots \\ \vdots & \ddots & \ddots & 0 \\ 0 & \cdots & 0 & v_p^{-1} \end{pmatrix}, \quad \mathbf{Z} = \begin{pmatrix} 0 & 0 & \cdots & 0 \\ 1 & 0 & \ddots & \vdots \\ & \ddots & \ddots & 0 \\ 0 & & 1 & 0 \end{pmatrix}. \tag{3.205}$$

Solution of (3.199) requires $5p^2/2$ operations by the algorithm of Björck and Pereyra [1970], which has nicely been summarized in section 4.6 of the text by Golub and Van Loan [1989]. The key to the fast algorithmic solution is the connection to the dual linear equation problem $\mathbf{v}^T\mathbf{w} = \mathbf{z}$, which is equivalent to polynomial interpolation of degree $p - 1$ through p sample points, i.e., $\sum_{i=0}^{p-1} w_i v_j^{i-1} = z_j$. The solution is obtained by utilizing the LU factorization of the inverse of $\mathbf{v}$, $\mathbf{x} = \mathbf{v}^{-1}\mathbf{y} = \mathbf{U}\mathbf{L}\mathbf{y}$, in which

$$\mathbf{u} = \left[\mathbf{L}_1^T(1)\mathbf{D}_1^{-1} \cdots \mathbf{L}_{p-1}^T(1)\mathbf{D}_{p-1}^{-1} \right], \tag{3.206}$$

$$\mathbf{L} = \left[\mathbf{L}_{p-1}(v_{p-1}) \cdots \mathbf{L}_1(v_1) \right], \tag{3.207}$$

$$\mathbf{L}_k(\alpha) = \begin{pmatrix} \mathbf{I}_k & \mathbf{0}_{k,p-k} \\ \mathbf{0}_{p-k,k} & \mathbf{I}_{p-k} - \alpha\mathbf{Z}_{p-k} \end{pmatrix}, \tag{3.208}$$

and

$$\mathbf{D}_k = \mathrm{diag}\left[1, \ldots, 1, v_{k+1} - v_1, \ldots, v_p - v_{p-k-1} \right] \tag{3.209}$$

is a special diagonal matrix of dimension $p \times p$, $\mathbf{I}_k$ is an identity matrix of dimension $k \times k$, and $\mathbf{Z}_{p-k}$ is a shift matrix of dimension $(p - k) \times (p - k)$.

MATLAB function vandermonde_lineqs solves the Vandermonde linear equation system $\mathbf{V}\mathbf{x} = \mathbf{y}$:

```
function x=vandermonde_lineqs(v,y)
% Solves the complex linear simultaneous equations  Vx = y  in which  V is a
% Vandermonde matrix using the fast computational algorithm of Bjorck and
% Pereyra (section 3.9).
%
%    x = vandermonde_linens(v,y)
%
% v -- vector or complex Vandermonde basis elements
% y -- right-hand-side vector with complex elements
% x -- solution vector
```

References

Björck, Ake, and Victor Pereyra, Solution of Vandermonde Systems of Equations, *Mathematics of Computation*, vol. 24, pp. 893–903, October 1970.

Burg, John P., A New Analysis Technique for Time Series Analysis, NATO Advanced Study Institute on Signal Processing with Emphasis on Underwater Acoustics, Enschede, The Netherlands, August 1968. Paper reprinted in *Modern Spectrum Analysis*, Don Childers, ed., IEEE Press, New York, 1978 (see p. 48).

Cantoni, Antonio, and Paul Butler, Eigenvalues and Eigenvectors of Symmetric Centrosymmetric Matrices, *Linear Algebra and Its Applications*, vol. 13, pp. 275–288, March 1976.

Carayannis, George, Nicholas Kalouptsidis, and Dimitris G. Manolakis, Fast Recursive Algorithms for a Class of Linear Equations, *IEEE Trans. Acoust. Speech Signal Process.*, vol. ASSP-30, pp. 227–239, April 1982.

Comon, P., and G. H. Golub, Tracking a Few Extreme Singular Values and Vectors in Signal Processing, *Proc. IEEE*, vol. 78, pp. 1327–1343, August 1990.

Cornyn, John J. Jr., Direct Methods for Solving Systems of Linear Equations Involving Toeplitz or Hankel Matrices, *Naval Research Laboratory (NRL) Memorandum Report 2920*, October 1974 (available from NTIS, AD/A-002 931).

Cybenko, George G., The Numerical Stability of the Levinson-Durbin Algorithm for Toeplitz Systems of Equations, *SIAM J. Sci. Stat. Comput.*, vol. 1, pp. 303–319, September 1980.

Datta, Lokesh, and Salvatore D. Morgera, Comments and Corrections "On the Eigenvectors of Symmetric Toeplitz Matrices," *IEEE Trans. Acoust. Speech Signal Process.*, vol. ASSP-32, pp. 440–441, April 1984; reply by Makhoul, vol. ASSP-33, pp. 737–738, June 1985.

Delsarte, Philippe, and Yves Genin, The Split Levinson Algorithm, *IEEE Trans. Acoust. Speech Signal Process.*, vol. ASSP-34, pp. 470–478, June 1986.

Deprettere, Ed. F., ed., *SVD and Signal Processing*, North-Holland, 1988.

Durbin, J., The Fitting of Time Series Models, *Rev. Inst. Int. Stat.*, vol. 28, pp. 233–244, 1960.

Farden, David C., Solution of a Toeplitz Set of Linear Equations, *IEEE Trans. Antennas Propag.*, vol. AP-24, pp. 906–907, November 1976.

Forsythe, G. E., M. A. Malcolm, and C. B. Moler, *Computer Methods for Mathematical Computations*, Prentice-Hall, Inc., Englewood Cliffs, N. J., 1977.

Golub, Gene H., and Charles F. Van Loan, *Matrix Computations*, 2nd ed., The Johns Hopkins University Press, Baltimore, 1989.

Golub, Gene H., and Charles F. Van Loan, An Analysis of the Total Least Squares Problem, *SIAM J. Numer. Anal.*, vol. 17, pp. 883–893, December 1980.

Gray, R. M., On the Asymptotic Eigenvalue Distribution of Toeplitz Matrices, *IEEE Trans. Inf. Theory*, vol. IT-18, pp. 725–730, November 1972.

Grenander, O., and G. Szegö, *Toeplitz Forms and Their Applications*, University of California Press, Berkeley, Calif., 1958.

Klema, Virginia C., and Alan J. Laub, The Singular Value Decomposition: Its Computation and Some Applications, *IEEE Trans. Autom. Control*, vol. AC-25, pp. 164–176, April 1980.

Lawson, C. L., and R. J. Hanson, *Solving Least Squares Problems*, Prentice-Hall, Inc., Englewood Cliffs, N. J., 1974.

Levinson, Norman, The Wiener RMS (Root Mean Square) Error Criterion in Filter Design and Prediction, *J. Math. Phys.*, vol. 25, pp. 261–278, January 1947.

Makhoul, John, On the Eigenvectors of Symmetric Toeplitz Matrices, *IEEE Trans. Acoust. Speech Signal Process.*, vol. ASSP-29, pp. 868–872, August 1981.

McClellan, J. H., and T. W. Parks, Eigenvalue and Eigenvector Decomposition of the Discrete Fourier Transform, *IEEE Trans. Audio Electroacoust.*, vol. AU-20, pp. 66–74, March 1972. See also comments in vol. AU-21, p. 65, February 1973.

Noble, Ben, and James W. Daniel, *Applied Linear Algebra*, 2nd ed., Prentice-Hall, Inc., Englewood Cliffs, N. J., 1977.

Roebuck, Pamela A., and Stephen Barnett, A Survey of Toeplitz and Related Matrices, *Int. J. Syst. Scie.*, vol. 9, pp. 921–934, 1978.

Therrien, C. W., On the Relation between Triangular Matrix Decomposition and Linear Prediction, *Proc. IEEE*, vol. 71, pp. 1459–1460, December 1983.

Trench, William F., An Algorithm for the Inversion of Finite Toeplitz Matrices, *J. Soc. Ind. Appl. Math.*, vol. 12, pp. 515–522, September 1964.

Yarlagadda, R., and B. N. Suresh Babu, A Note on the Application of FFT to the Solution of a System of Toeplitz Normal Equations, *IEEE Trans. Circuits Syst.*, vol. CAS-27, pp. 151–154, February 1980.

Zohar, Shalhav, Toeplitz Matrix Inversion: The Algorithm of W. F. Trench, *J. Assoc. Comput. Mach.*, vol. 16, pp. 592–601, October 1969.

Zohar, Shalhav, The Solution of a Toeplitz Set of Linear Equations, *J. Assoc. Comput. Mach.*, vol. 21, pp. 272–276, April 1974.

Zohar, Shalhav, FORTRAN Subroutines for the Solution of Toeplitz Sets of Linear Equations, *IEEE Trans. Acoust. Speech Signal Process.*, vol. ASSP-27, pp. 656–658, December 1979. See also corrections in: vol. ASSP-28, p. 601, October 1980; vol. ASSP-29, p. 1212, December 1981.

4

REVIEW OF RANDOM
PROCESS THEORY

4.1 INTRODUCTION

An assumption is made that the reader already has some familiarity with probability, random variable, and random process theory. This chapter serves simply as a review. A primary topic of this chapter is the formal introduction of the *power spectral density* (PSD) concept. It will be seen that the PSD is a statistical extension of the energy spectral density of Chap. 2. Another important topic is the property of *ergodicity*, which permits the substitution of time averages for statistical averages and leads to an alternative definition of power spectral density. These two definitions of PSD provide the foundation for the classical methods of power spectral density estimation that are covered in Chap. 5. For those readers needing a more extensive review than covered in this chapter, the excellent texts by Papoulis [1977, 1984, 1990], Brillinger [1974], Gardner [1985], and Therrien [1992] are recommended.

4.2 PROBABILITY AND RANDOM VARIABLES

Consider a discrete event E, which is one of a finite number of possible discrete outcomes of a random experiment. The *probability* of this event, denoted as $\Pr(E)$, may be intuitively considered to be the limit of the ratio of the number of times E occurred to the number of times the experiment was attempted. The probability is bounded, therefore, between 0 and 1. A *random variable* x is, in simplest terms, a variable which takes on values at random from a continuum of possible values. The manner of specifying the probability with which different values are taken by the random variable is quantified by the *probability distribution function* $F(x)$, defined by $F(x) = \Pr(x \leq x)$, that is, the probability that random variable x has a value less than or equal to x is given by $F(x)$. The *probability density function* (PDF) $p(x)$ is the derivative of the probability distribution, $p(x) = \partial F(x)/\partial x$.

The *expectation* of a random variable x, denoted as $\mathcal{E}\{x\}$, is given by

$$\mathcal{E}\{x\} = \int_{-\infty}^{\infty} x\, p(x)\, dx = \bar{x}. \qquad (4.1)$$

This is also called the *mean value* of x, or the *first moment* of x. It represents the value toward which the average of a number of observations of x tends, in the probabilistic sense, as the number of observations increases. The expectation of a function of random variable x, say $g(x)$, can be directly calculated from the PDF of x by

$$\mathcal{E}\{g(x)\} = \int_{-\infty}^{\infty} g(x)p(x)\,dx. \tag{4.2}$$

The expectation of the squared magnitude of x

$$\mathcal{E}\{|x|^2\} = \int_{-\infty}^{\infty} |x|^2 p(x)\,dx \tag{4.3}$$

is called the *mean squared value*, or the *second moment* of x. The *variance* ρ of a random variable is the mean squared deviation of the random variable from its mean,

$$\text{var}\{x\} = \int_{-\infty}^{\infty} |x - \mathcal{E}\{x\}|^2 p(x)\,dx = \mathcal{E}\{|x|^2\} - |\mathcal{E}\{x\}|^2 = \rho. \tag{4.4}$$

Note that the mean squared value and the variance are equal only for a zero-mean random variable. An indication of the statistical correlation to which one random variable is related to another random variable is given by the *covariance*,

$$\begin{aligned}
\text{cov}\{xy\} &= \mathcal{E}\{(x - \mathcal{E}\{x\})(y^* - \mathcal{E}\{y\}^*)\} \\
&= \int_{-\infty}^{\infty}\int_{-\infty}^{\infty} (x - \mathcal{E}\{x\})(y^* - \mathcal{E}\{y\}^*)p(x, y)\,dx\,dy \\
&= \mathcal{E}\{xy^*\} - \mathcal{E}\{x\}\mathcal{E}\{y\}^*
\end{aligned} \tag{4.5}$$

in which $p(x, y)$ represents the *joint* probability density function of random variables x and y. In this expression, $\mathcal{E}\{xy\}$ represents the joint second moment of x and y. For convenience, the contrasting notation between a random variable x and its value x will not be used past this point in order to simplify this review.

In the presentation of various parameter estimators in later chapters, several terms are used to describe the behavior of the estimators. These terms are defined here. The *bias* $B(\hat{\alpha})$ of an estimator $\hat{\alpha}$ of the parameter α is the difference between the true value of the parameter and the expected value of the estimate,

$$B(\hat{\alpha}) = \alpha - \mathcal{E}\{\hat{\alpha}\}. \tag{4.6}$$

An *unbiased* estimator is one for which $B(\hat{\alpha}) = 0$. The *mean square error* (mse) of the estimator $\hat{\alpha}$ is

$$\text{mse}\{\hat{\alpha}\} = \mathcal{E}\{|\alpha - \hat{\alpha}|^2\} = \text{var}\{\hat{\alpha}\} + |B(\hat{\alpha})|^2. \tag{4.7}$$

An estimator is said to be *consistent* if, as the number of observations increases, the bias and variance both tend to zero. A *maximum likelihood estimator* (MLE) is one for which the value of the parameter is selected to maximize the probability of having obtained the observed value. The Cramer-Rao (CR) inequality is frequently used to lower bound the variance performance of an unbiased estimator of a scalar parameter. Assuming that the random vector $\mathbf{x}$ is used to estimate parameter α, then the CR inequality takes the form

$$\text{var}\{\hat{\alpha}\} \geq \mathcal{E} \left\{ \left(\frac{\partial \ln p(\mathbf{x}|\alpha)}{\partial \alpha} \right)^2 \right\}^{-1} \tag{4.8}$$

evaluated at the true value of the parameter α. Here $p(\mathbf{x}|\alpha)$ is the PDF of random variable $\mathbf{x}$ with dependence on α shown explicitly. An estimator that attains the CR bound is said to be *efficient*.

Two important probability density functions are used in this book. The *uniform distribution* of a real variable x is characterized by a uniform (constant) PDF

$$p(x) = \frac{1}{b-a} \tag{4.9}$$

over the finite real interval $a \leq x \leq b$. The *Gaussian distribution* of a real random variable x with mean $\bar{x}$ and variance ρ is characterized by a PDF given by

$$p(x) = [2\pi\rho]^{-1/2} \exp\left[-\frac{1}{2\rho}(x-\bar{x})^2 \right] \tag{4.10}$$

for $-\infty < x < \infty$. A *multivariate* Gaussian PDF for a vector of K real random variables $\mathbf{x} = [x_1\, x_2 \ldots x_K]^T$ has the structure

$$p(\mathbf{x}) = [(2\pi)^K \det \mathbf{C}_K]^{-1/2} \exp\left[-\frac{1}{2}(\mathbf{x}-\bar{\mathbf{x}})^T \mathbf{C}_K^{-1}(\mathbf{x}-\bar{\mathbf{x}}) \right], \tag{4.11}$$

in which the K-element mean vector $\bar{\mathbf{x}}$ is defined as

$$\bar{\mathbf{x}} = \mathcal{E}\{\mathbf{x}\} \tag{4.12}$$

and the $K \times K$ covariance matrix $\mathbf{C}_K$ is defined as

$$
\begin{aligned}
\mathbf{C}_K &= \mathcal{E}\{(\mathbf{x}-\bar{\mathbf{x}})(\mathbf{x}-\bar{\mathbf{x}})^T\} \\
&= \begin{pmatrix}
\text{var}\{x_1\} & \text{cov}\{x_1 x_2\} & \cdots & \text{cov}\{x_1 x_K\} \\
\text{cov}\{x_2 x_1\} & \text{var}\{x_2\} & \cdots & \text{cov}\{x_2 x_K\} \\
\vdots & \vdots & \ddots & \vdots \\
\text{cov}\{x_K x_1\} & \text{cov}\{x_K x_2\} & \cdots & \text{var}\{x_K\}
\end{pmatrix}.
\end{aligned}
\tag{4.13}
$$

A Gaussian PDF for complex random variables can also be defined (see Appendix E of Monzingo and Miller [1980]). Let $z = x_r + jx_i$ be the complex random variable, where x_r, the real part of z, is a real random variable with mean $\bar{x}_r$ and variance $\rho_z/2$ and x_i, the imaginary part of z, is a real random variable with mean $\bar{x}_i$ and identical variance $\rho_z/2$. The random variables x_r and x_i are assumed to be independent, i.e., $\text{cov}\{x_r x_i\} = 0$. A Gaussian PDF for the complex random variable z may then be embedded in a special case *bivariate* Gaussian PDF for two real random variables, in which

$$\mathbf{x} = \begin{pmatrix} x_r \\ x_i \end{pmatrix}, \quad \bar{\mathbf{x}} = \begin{pmatrix} \bar{x}_r \\ \bar{x}_i \end{pmatrix}, \quad \mathbf{C}_2 = \begin{pmatrix} \rho_z/2 & 0 \\ 0 & \rho_z/2 \end{pmatrix}. \tag{4.14}$$

Substituting these into Eq. (4.11) for $K = 2$ will yield, after some argument simplification,

$$p(z) = p(\mathbf{x}) = [\pi \rho_z]^{-1} \exp\left[-\frac{1}{\rho_z} |z - \bar{z}|^2 \right], \tag{4.15}$$

in which $\bar{z} = \mathcal{E}\{z\} = \bar{x}_r + j\bar{x}_i$ and $\text{var}\{z\} = \text{var}\{x_r\} + \text{var}\{x_i\} = \rho_z$. A multivariate Gaussian PDF for a vector $\mathbf{z}$ of K complex random variables can be embedded into a Gaussian PDF for $2K$ real random variables with the form

$$p(\mathbf{z}) = [\pi^K \det \mathbf{C}_K]^{-1} \exp[-(\mathbf{z} - \bar{\mathbf{z}})^H \mathbf{C}_K^{-1} (\mathbf{z} - \bar{\mathbf{z}})], \tag{4.16}$$

in which complex mean $\bar{\mathbf{z}}$ and Hermitian covariance matrix $\mathbf{C}_K$ are given by

$$\bar{\mathbf{z}} = \mathcal{E}\{\mathbf{z}\} \tag{4.17}$$

$$\mathbf{C}_K = \mathcal{E}\{(\mathbf{z} - \bar{\mathbf{z}})(\mathbf{z} - \bar{\mathbf{z}})^H\}. \tag{4.18}$$

The assumption has been made that the real and imaginary components of each complex random variable z_k in complex random vector $\mathbf{z}$ are independent and have identical variance, that is, $\text{cov}\{\text{Re}[z_k]\text{Im}[z_k]\}$ is zero for $k = l$ and nonzero for $k \neq l$, and $\text{var}\{\text{Re}[z_k]\} = \text{var}\{\text{Im}[z_k]\}$.

4.3 RANDOM PROCESSES

A discrete *random process* may be thought of as a collection, or *ensemble*, of real or complex discrete sequences of time (or space), any one of which might be observed on any trial of an experiment. The ensemble of sequences will be denoted as $x[n; i]$, in which i is the ith sequence out of the ensemble and n is the time index. For a given i representing the ensemble member under observation, the abbreviated notation $x[n]$ will be used. For a fixed time index n, the value of the observed element over all ensemble sequences will be a random variable. The range of values will, in general, be a continuum, even though $x[n;i]$ is discrete in both n and i. The probability that $x[n]$ takes values in a certain range α is given by the probability distribution function $F(\alpha;n) = \Pr(x[n] \leq \alpha)$, in which the dependence on the time of observation is shown explicitly in the notation. The corresponding PDF is $p(\alpha;n) = \partial F(\alpha;n)/\partial \alpha$.

The *mean* or *expected value* of a random process $x[n]$ at time index n is defined as

$$\bar{x}[n] = \mathcal{E}\{x[n]\}. \tag{4.19}$$

The *autocorrelation* of a random process at two different time indices n_1 and n_2 is defined as

$$r_{xx}[n_1, n_2] = \mathcal{E}\{x[n_1]x^*[n_2]\}. \tag{4.20}$$

Equation (4.20) is the engineering definition for autocorrelation, as first suggested by Wiener. Statisticians reserve the term autocorrelation for a related quantity that is normalized to have magnitude lying between zero and unity. The autocorrelation of the process $x[n]$ with the mean removed is the *autocovariance*, defined as

$$c_{xx}[n_1, n_2] = \mathcal{E}\{(x[n_1] - \bar{x}[n_1])(x^*[n_2] - \bar{x}^*[n_2])\}, \tag{4.21}$$

which can be shown to satisfy

$$c_{xx}[n_1, n_2] = r_{xx}[n_1, n_2] - \bar{x}[n_1]\bar{x}^*[n_2]. \tag{4.22}$$

If the random process has zero mean for all n, then the autocorrelation and the autocovariance are identical

$$c_{xx}[n_1, n_2] = r_{xx}[n_1, n_2].$$

The terms "correlation" and "covariance" are often used synonymously in the literature, but are formally identical only for zero-mean processes.

When two different random processes $x[n]$ and $y[n]$ are involved, then one may define the *cross correlation* as

$$r_{xy}[n_1, n_2] = \mathcal{E}\{x[n_1]y^*[n_2]\} \tag{4.23}$$

and the *cross covariance* as

$$c_{xy}[n_1, n_2] = \mathcal{E}\{(x[n_1] - \bar{x}[n_1])(y^*[n_2] - \bar{y}^*[n_2])\}$$
$$= r_{xy}[n_1, n_2] - \bar{x}[n_1]\bar{y}^*[n_2]. \tag{4.24}$$

Two zero-mean random processes are said to be *uncorrelated* if $r_{xy}[n_1, n_2] = 0$ for all n_1 and n_2.

The definitions to this point have shown an explicit dependence on the time indices. A random process is *wide-sense stationary* (WSS) if its mean is constant for all time indices (i.e., independent of time) and its autocorrelation depends only on the time index difference $m = n_2 - n_1$. A pair of random processes are jointly wide-sense stationary if their cross correlation depends only on the time index difference. Processes that are jointly stationary must also be individually stationary. Note that WSS applies *only* to the first- and second-moment statistics of random processes. Higher-order moments are not constrained. In summary, WSS discrete random process $x[n]$ is statistically characterized by a constant mean

$$\bar{x}[n] = \bar{x}, \tag{4.25}$$

an *autocorrelation sequence* (ACS)[1]

$$r_{xx}[m] = \mathcal{E}\{x[n+m]x^*[n]\} \tag{4.26}$$

that is a function of the time-difference index m, and an *autocovariance sequence*

$$c_{xx}[m] = \mathcal{E}\{(x[n+m] - \bar{x})(x^*[n] - \bar{x}^*)\} = r_{xx}[m] - |\bar{x}|^2. \tag{4.27}$$

The jointly WSS discrete random processes $x[n]$ and $y[n]$ are statistically characterized by a *cross correlation sequence* (CCS)

$$r_{xy}[m] = \mathcal{E}\{x[n+m]y^*[n]\} \tag{4.28}$$

that is a function of the time-difference index m, and a *cross covariance sequence*

$$c_{xy}[m] = \mathcal{E}\{(x[n+m] - \bar{x})(y^*[n] - \bar{y}^*)\} = r_{xy}[m] - \bar{x}\bar{y}^*. \tag{4.29}$$

Useful properties are

$$r_{xx}[0] \geq |r_{xx}[m]|$$
$$r_{xx}[-m] = r_{xx}^*[m]$$
$$r_{xx}[0]r_{yy}[0] \geq |r_{xy}[m]|^2$$
$$r_{xy}[-m] = r_{yx}^*[m], \tag{4.30}$$

which apply for all integers m. It is easy to see from these properties that the ACS of a WSS random process must be a maximum at the origin ($m = 0$). If the Hermitian Toeplitz autocorrelation matrix

$$\mathbf{R}_M = \begin{pmatrix} r_{xx}[0] & r_{xx}[-1] & \cdots & r_{xx}[-M] \\ r_{xx}[1] & r_{xx}[0] & \cdots & r_{xx}[-M+1] \\ \vdots & \vdots & \ddots & \vdots \\ r_{xx}[M] & r_{xx}[M-1] & \cdots & r_{xx}[0] \end{pmatrix} \tag{4.31}$$

is formed from $(M+1)$ ACS components, then the quadratic form

$$\mathcal{E}\left\{ \left| \sum_{m=0}^{M} a[m]x[n-m] \right|^2 \right\} = \mathbf{a}^H \mathbf{R}_M \mathbf{a}$$

$$= \sum_{m=0}^{M} \sum_{n=0}^{M} a[m]a^*[n] r_{xx}[m-n] \geq 0 \tag{4.32}$$

---

[1] Eqs. (4.26) and (4.28) are the predominant definitions for the ACS and CCS found in the literature. Alternative, but less common, definitions are $r_{xx}[m] = \mathcal{E}\{x[n]x^*[n+m]\}$ and $r_{xy}[m] = \mathcal{E}\{x[n]y^*[n+m]\}$, which are complex conjugates of this text's definitions.

must be positive semidefinite for any arbitrary $M \times 1$ vector $\mathbf{a} = \left(a[0] \ldots a[m]\right)^T$ if $r_{xx}[m]$ is a valid autocorrelation sequence. The ACS is said to have a positive-semidefinite property.

The *power spectral density* (PSD) is defined as the discrete-time Fourier transform (DTFT) of the autocorrelation sequence

$$P_{xx}(f) = T \sum_{m-\infty}^{\infty} r_{xx}[m] \exp(-j2\pi f mT). \tag{4.33}$$

The PSD, which is assumed to be band-limited to $\pm 1/2T$ Hz, is periodic in frequency with period $1/T$ Hz. The power spectral density function describes how the variance of a random process is distributed with frequency. To justify the name, consider the inverse DTFT

$$r_{xx}[m] = \int_{-1/2T}^{1/2T} P_{xx}(f) \exp(j2\pi f mT) \, df \tag{4.34}$$

evaluated for $m = 0$,

$$r_{xx}[0] = \int_{-1/2T}^{1/2T} P_{xx}(f) \, df. \tag{4.35}$$

It will be shown in Sec. 4.4 that the autocorrelation at zero lag ($m = 0$) represents the *average power* in the random process. The area under $P_{xx}(f)$ must also represent the average power, as indicated by Eq. (4.35). Therefore, $P_{xx}(f)$ is a density function (power per unit of frequency) that represents the distribution of power with frequency. The Fourier transform pairs (4.33) and (4.34) are often termed the discrete-time *Wiener-Khintchine theorem*. Due to the property $r_{xx}[-m] = r_{xx}^*[m]$, the PSD must be strictly a real, positive function. If the ACS is strictly real-valued, then $r_{xx}[-m] = r_{xx}[m]$ and the PSD can be expressed as the one-sided cosine transform

$$P_{xx}(f) = 2T \sum_{m=0}^{\infty} r_{xx}[m] \cos(2\pi f mT), \tag{4.36}$$

which also makes $P_{xx}(f) = P_{xx}(-f)$, a symmetric function.

The *cross power spectral density* (CPSD) of two jointly stationary processes $x[n]$ and $y[n]$ is defined as the DTFT of the cross correlation sequence

$$P_{xy}(f) = T \sum_{m=-\infty}^{\infty} r_{xy}[m] \exp(-j2\pi f mT). \tag{4.37}$$

Because $r_{xy}[-m] \neq r_{xy}^*[m]$, the CPSD is, in general, a complex-valued function. However, the property $P_{xy}(f) = P_{yx}^*(f)$ does hold.

A zero-mean random process $w[n]$ of special interest is that of a discrete-time white noise. A white noise process is uncorrelated with itself for all lags, except at $m = 0$, for which its variance is ρ_w. The ACS has the form

$$r_{ww}[m] = \rho_w \delta[m], \tag{4.38}$$

where $\delta[m]$ is the discrete delta sequence of Eq. (2.10). The PSD of the white noise autocorrelation sequence is, therefore,

$$P_{ww}(f) = T\rho_w, \tag{4.39}$$

which is a constant for all frequencies, justifying the name *white* noise.

Let $y[n]$ denote the output of a discrete linear time-invariant system

$$y[n] = x[n] \star h[n], \tag{4.40}$$

in which $h[n]$ is a fixed impulse response sequence and input $x[n]$ is a zero-mean wide-sense stationary discrete random process. The output process $y[n]$ will also be WSS and can be shown to have the following relationships among the autocorrelations of the input and output process and the cross correlation between the input and output

$$r_{yx}[m] = r_{xx}[m] \star h[m] = \sum_{k=-\infty}^{\infty} r_{xx}[k-m]h[k]$$

$$r_{xy}[m] = r_{xx}[m] \star h^*[-m]$$

$$r_{yy}[m] = r_{xy}[m] \star h[m] = r_{xx}[m] \star (h^*[-m] \star h[m])$$

$$= r_{xx}[m] \star \left(\sum_{k=-\infty}^{\infty} h[k+m]h^*[k] \right). \tag{4.41}$$

Denoting the z-transforms of the various correlations as $P_{xx}(z) = \mathcal{Z}\{r_{xx}[m]\}$, $P_{xy}(z) = \mathcal{Z}\{r_{xy}[m]\}$, and $P_{yy}(z) = \mathcal{Z}\{r_{yy}[m]\}$ and the system function as $H(z) = \mathcal{Z}\{h[m]\}$, then the convolution theorem applied to relationships (4.41) will yield

$$P_{xy}(z) = P_{xx}(z)H^*(1/z^*)$$

$$P_{yy}(z) = P_{xy}(z)H(z)$$

$$P_{yy}(z) = P_{xx}(z)H(z)H^*(1/z^*), \tag{4.42}$$

where the property $\mathcal{Z}\{h^*[-m]\} = H^*(1/z^*)$ has been used. If $h[m]$ is real, then $H^*(1/z^*) = H(1/z)$. Based on relationship (2.52), definition (4.33), and Eq. (4.42), the power spectral density $P_{yy}(f)$ of the output process is related to the power spectral density $P_{xx}(f)$ of the input process as follows

$$P_{yy}(f) = T P_{yy}(z)\big|_{z=\exp(j2\pi fT)} = P_{xx}(f) |H(\exp[j2\pi fT])|^2, \tag{4.43}$$

where the time domain is discrete and the frequency domain is continuous.

One special case will serve as the basis for the spectral estimation techniques of Chaps. 6–10. If the input process is white noise as defined in (4.38), so that $P_{xx}(z) = \mathcal{Z}\{\rho_w \delta[m]\} = \rho_w$, and the rational system function (2.17) is selected and adjusted for unity gain (set $b[0] = 1$), then the z and DTFT transform relationships between input and output have the respective forms, based on (4.42) and (4.43),

$$P_{yy}(z) = \rho_w H(z)H^*(1/z^*), \quad P_{yy}(f) = T\rho_w |H(f)|^2 \tag{4.44}$$

where

$$H(z) = \frac{1 + \displaystyle\sum_{k=1}^{q} b[k]z^{-k}}{1 + \displaystyle\sum_{k=1}^{p} a[k]z^{-k}}, \quad H(f) = \frac{1 + \displaystyle\sum_{k=1}^{q} b[k]\exp(-j2\pi f kT)}{1 + \displaystyle\sum_{k=1}^{p} a[k]\exp(-j2\pi f kT)} \tag{4.45}$$

are, respectively, the discrete system function and the frequency response function.

Two important random processes used in this text are the Gaussian random process and the sinusoidal random process. If a random process is stationary and Gaussian, then the PDF of $x[n]$ for any time index n is the Gaussian PDF of either Eq. (4.10) or Eq. (4.15), depending on whether $x[n]$ is real or complex, respectively. The sequence samples $x[n], x[n+1], \ldots, x[n+M]$ will have a joint PDF given by either Eq. (4.11) or Eq. (4.16), depending on whether $x[n]$ is real or complex. If complex, for example, assuming a zero-mean process, accounting for the successive times that the samples are taken, and defining $\mathbf{x} = (x[n]\, x[n+1]\ldots x[n+M])^T$, then the joint Gaussian PDF is

$$p(\mathbf{x}) = [\pi^{M+1} \det \mathbf{R}_M]^{-1} \exp[-\mathbf{x}^H \mathbf{R}_M^{-1} \mathbf{x}], \tag{4.46}$$

in which the covariance matrix

$$\mathbf{R}_M = \mathbf{C}_M = \begin{pmatrix} r_{xx}[0] & r_{xx}^*[1] & \cdots & r_{xx}^*[M] \\ r_{xx}[1] & r_{xx}[0] & \cdots & r_{xx}^*[M-1] \\ \vdots & \vdots & \ddots & \vdots \\ r_{xx}[M] & r_{xx}[M-1] & \cdots & r_{xx}[0] \end{pmatrix} \tag{4.47}$$

is a Hermitian Toeplitz autocorrelation matrix of order M and dimension $(M+1) \times (M+1)$. This matrix will be encountered frequently in the chapters ahead.

Consider a deterministic real sinusoid process

$$x[n] = A \sin(2\pi f nT + \theta),$$

in which T is the sample interval and the amplitude A, frequency f, and phase θ are all fixed values. The mean and autocorrelation sequences for this case are

$$\bar{x}[n] = A \sin(2\pi f n T + \theta)$$

$$r_{xx}[n+m, n] = A \sin(2\pi[n+m]T + \theta)A \sin(2\pi f n T + \theta)$$

$$= \frac{A^2}{2}[\cos(2\pi f m T) - \cos(2\pi f[2n+m]T + 2\theta)]. \qquad (4.48)$$

The mean is not constant and the ACS is not strictly a function of the time difference m. A completely deterministic sinusoid cannot be modeled, therefore, as a wide-sense stationary random process. If the phase is now assumed to be a random variable that is uniformly distributed within the interval 0 to 2π, then the sinusoidal process will be stationary because the mean

$$\bar{x} = \int_0^{2\pi} A \sin(2\pi f n T + \theta)\frac{1}{2\pi} d\theta = 0$$

and autocorrelation sequence

$$r_{xx}[m] = \int_0^{2\pi} A \sin(2\pi f[n+m]T + \theta)A \sin(2\pi f m T + \theta)\frac{1}{2\pi} d\theta$$

$$= \frac{A^2}{2} \cos(2\pi f m T) \qquad (4.49)$$

are both independent of the time origin at index n. If there are L real sinusoids

$$x[n] = \sum_{l=1}^{L} A_l \sin(2\pi f_l n T + \theta_l),$$

each of which has a phase that is uniformly distributed on the interval 0 to 2π and independent of the other phases, then the mean of the L sinusoids is zero and the ACS is

$$r_{xx}[m] = \sum_{l=1}^{L} \frac{A_l^2}{2} \cos(2\pi f_l m T) \qquad (4.50)$$

and the companion PSD is

$$P_{xx}(f) = \sum_{l=1}^{L} \frac{A^2}{4} \{\delta(f - f_l) + \delta(f + f_l)\}$$

based on the transform properties found in Table 2.2. If the process consists of L complex sinusoids

$$x[n] = \sum_{l=1}^{L} A_l \exp[j(2\pi f_l n T + \theta_l)],$$

then the ACS will have the form

$$r_{xx}[m] = \sum_{l=1}^{L} A_l^2 \exp(j2\pi f_l m T) \tag{4.51}$$

in which

$$P_{xx}(f) = \sum_{l=1}^{L} A^2 \delta(f - f_l)$$

is the corresponding PSD.

If an independent white noise process $w[n]$, with variance ρ_w, is added to the complex sinusoids with random phases, then the combined process $y[n] = x[n] + w[n]$ will have the combined ACS components

$$r_{yy}[m] = r_{xx}[m] + r_{ww}[m]$$

$$= \sum_{l=1}^{L} A_l^2 \exp(j2\pi f_l m T) + \rho_w \delta[m]. \tag{4.52}$$

The autocorrelation matrix (4.47) for the random process of complex sinusoids with additive white noise can be succinctly expressed, based on Eq. (4.53), as

$$\mathbf{R}_{yy} = \sum_{l=1}^{L} A_l^2 \mathbf{e}_M(f_l)\mathbf{e}_M^H(f_l) + \rho_w \mathbf{I}_{M+1}, \tag{4.53}$$

where $\mathbf{I}_{M+1}$ is an $(M+1) \times (M+1)$ identity matrix and

$$\mathbf{e}_M(f_l) = \begin{pmatrix} 1 \\ \exp(j2\pi f_l T) \\ \vdots \\ \exp(j2\pi f_l M T) \end{pmatrix}$$

is a complex sinusoid vector of $M+1$ elements composed of complex sinusoidal terms that are functions of frequency f_l.

4.4 SUBSTITUTING TIME AVERAGES FOR ENSEMBLE AVERAGES

Up to this point in this chapter, only statistical ensemble averaging has been used to define the mean, correlation, covariance, and power spectral density descriptors of random processes. However, in practice one does not typically have an ensemble of waveforms from which to evaluate these statistical descriptors. It is desirable to *estimate* all these statistical properties from a *single* sample waveform $x(t)$ by substituting *time* averages for *ensemble* averages. The property required to accomplish this is *ergodicity*. A random process is said to be ergodic if, with probability 1, all its statistics can be predicted from a single waveform of the process ensemble via time averaging; that is, the time average statistics of almost all possible single waveforms are equal to the same statistics computed via ensemble averaging. The mathematical analysis of random processes is greatly simplified with the property of ergodicity.

The concept of ergodicity for first and second moment statistics (i.e., the mean and variance) requires the assumption that the data be stationary up to fourth moments. For a process to be stationary, the statistics must be independent of the time origin selected. This seems intuitively appropriate since a time-averaged quantity is independent of time and therefore cannot approximate a statistically nonstationary parameter that depends on time. Thus, if one were to look at a single waveform $x[n]$ at time points n_1, n_2, and so forth, then the value seen on average should be $\bar{x}$. In the limit as one looks at all time points, one would expect the average to be

$$\lim_{M\to\infty} \frac{1}{2M+1} \sum_{n=-M}^{M} x[n] = \mathcal{E}\{x[n]\} = \bar{x}. \qquad (4.54)$$

This limit, if it exists, can be shown to be true if and only if the variance of time average (4.55)

$$\lim_{M\to\infty} \frac{1}{2M+1} \sum_{m=-2M}^{2M} \left(1 - \frac{|m|}{2M+1}\right) c_{xx}[m] = 0 \qquad (4.55)$$

approaches zero, where $c_{xx}[m]$ is the true ensemble covariance of the random process $x[n]$. In this case, $x[n]$ is said to be *ergodic in the mean*.

In a similar manner, if one were to look in several places at the product of the sample values at two instants of time that are separated by lag m, say $x[n_1 + m]x[n_1]$, $x[n_2 + m]x[n_2]$, ..., then the product should yield on average for a stationary process the value of the ACS term $r_{xx}[m]$ (assuming the process is zero mean). In the limit as one averages over all lag products, one would expect the average to be

$$\lim_{M\to\infty} \frac{1}{2M+1} \sum_{n=-M}^{M} x[n+m]x^*[n] = \mathcal{E}\{x[n+m]x^*[n]\} = r_{xx}[m]. \qquad (4.56)$$

This can be shown to be true if and only if the variance of the above time average approaches zero in the limit, that is,

$$\lim_{M\to\infty} \frac{1}{2M+1} \sum_{m=-2M}^{2M} \left(1 - \frac{|m|}{2M+1}\right) c_{zz}[m] = 0 \qquad (4.57)$$

where $c_{zz}[m]$ is the true ensemble autocovariance (lag product z may not have zero mean although process x itself is zero mean) of the quadratic lag product random process $z_m[n] = x[n+m]x^*[n]$. Note that computation of $c_{zz}[m]$ involves the fourth-order statistical moment of the process $x[n]$. If the above is true, then $x[n]$ is said to be *autocorrelation ergodic*. If the stationary process $x[n]$ is zero-mean Gaussian, then it is straightforward to show that $x[n]$ is both mean ergodic and autocorrelation ergodic [Papoulis, 1977]. Although it is more difficult to prove ergodicity for non–Gaussian processes, suffice it to say that almost all practical data processes that are stationary are also ergodic. *Hereafter, it shall be assumed that sampled data sequences are stationary so that time average equivalent definitions of the mean and autocorrelation can replace the ensemble average definitions.*

The assumption of ergodicity not only permits substitution of time average definitions for the mean (4.55) and autocorrelation (4.56), but also permits substitution of the following equivalent time average definition for the power spectral density

$$P_{xx}(f) = \lim_{M \to \infty} \mathcal{E} \left\{ \frac{1}{(2M+1)T} \left| T \sum_{n=-M}^{M} x[n] \exp(-j2\pi f nT) \right|^2 \right\}. \quad (4.58)$$

This equivalent PSD, called the *limit spectrum*, is obtained by an ensemble average of the squared magnitude of the DTFT of a windowed data set divided by its data record length, in which the number of data points grows to infinity. The ensemble average is required because the DTFT is a random variable itself, varying for each realization of $x[n]$ used. In order to prove that Eq. (4.58) is equivalent to the Wiener-Khintchine theorem Eq. (4.33), expand the squared magnitude DTFT and exchange the order of the summation and the expectation operations,

$$P_{xx}(f) = \lim_{M \to \infty} \mathcal{E} \left\{ \frac{T}{2M+1} \sum_{m=-M}^{M} \sum_{n=-M}^{M} x[m]x^*[n] \exp(-j2\pi f[m-n]T) \right\}$$

$$= \lim_{M \to \infty} \frac{T}{2M+1} \sum_{m=-M}^{M} \sum_{n=-M}^{M} r_{xx}[m-n] \exp(-j2\pi f[m-n]T).$$

$$(4.59)$$

Using the simply proven fact

$$\sum_{m=-M}^{M} \sum_{n=-M}^{M} r_{xx}[m-n] = \sum_{m=-2M}^{2M} (2M+1-|m|)r_{xx}[m], \quad (4.60)$$

then Eq. (4.59) can be simplified to the form

$$P_{xx}(f) = \lim_{M \to \infty} \frac{T}{2M+1} \sum_{m=-2M}^{2M} (2M+1-|m|)r_{xx}[m] \exp(-j2\pi f mT)$$

$$= \lim_{M \to \infty} T \sum_{m=-2M}^{2M} \left(1 - \frac{|m|}{2M+1}\right) r_{xx}[m] \exp(-j2\pi f mT)$$

$$= T \sum_{m=-\infty}^{\infty} r_{xx}[m] \exp(-j2\pi f mT). \quad (4.61)$$

An assumption has been made in the last step of Eq. (4.61) that the ACS decays such that

$$\sum_{m=-\infty}^{\infty} |m|\, r_{xx}[m] < \infty \qquad (4.62)$$

remains finite, otherwise it would not be true that the discrete-time Fourier transform of the temporal ACS will tend to the true ensemble PSD as the data record length increases, even if the ACS is itself ergodic. For example, the condition is violated for random processes with nonzero means or processes with sinusoidal components, although the use of impulse functions can handle even this case. Thus, the two expressions Eqs. (4.33) and (4.58) for the PSD are equivalent under condition (4.62).

The two equivalent definitions of the PSD are summarized by the triangle diagram shown in Fig. 4.1. Wiener-Khintchine PSD definition (4.33) that uses the autocorrelation approach is termed an *indirect* method because the random process $x[n]$ is not used directly to estimate the PSD. Limit spectrum PSD definition (4.58) is termed the *direct* method because the PSD is defined directly in terms of the time sequence $x[n]$.

Disregarding the limit and ensemble averaging operations of Eq. (4.58) yields the PSD estimator for finite number of samples

$$\tilde{P}_{xx}(f) = \frac{1}{(2M+1)T} \left| T \sum_{n=-M}^{M} x[n] \exp(-j2\pi f n T) \right|^2, \qquad (4.63)$$

termed the *sample spectrum*. This is also the original Schuster *periodogram*, as described in Chap. 1. The sample spectrum is shown in Appendix 4.A to be a statistically inconsistent estimator of the true PSD as M grows in the limit to infinity. Although the mean of the sample spectrum will tend to the true PSD value in the limit, the variance will not tend to zero in the limit of infinite data, but it will in fact be comparable in magnitude to the mean value of the sample spectrum. Early users of the periodogram failed to regard the expectation operation and obtained, as a

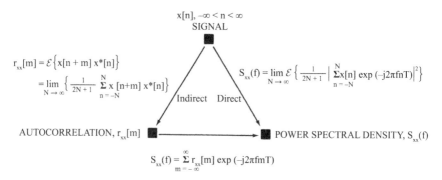

Figure 4.1. Equivalent definitions of the power spectral density under the assumption of ergodicity.

result, unreliable spectral estimates, even though Schuster himself (see Chap. 1) had warned of the need to perform some sort of averaging or smoothing operation when using the periodogram. As a result, the periodogram method fell into disfavor by data processing experimentalists until its revival in the early 1950s, when smoothing techniques in a well-understood statistical framework were introduced.

4.5 ENTROPY CONCEPTS

Claude Shannon introduced the concept of a statistical information measure. For example, an event that has M possibilities x_i, with $p[x_i]$ representing the probability of occurrence of the ith possibility, has a *self-information* of

$$I[x_i] = \ln(1/p[x_i]) = -\ln p[x_i]. \qquad (4.64)$$

The expected value, or mean, of the information is the *entropy*

$$H[x] = \sum_{i=1}^{M} p[x_i] I[x_i] = -\sum_{i=1}^{M} p[x_i] \log p[x_i]. \qquad (4.65)$$

Entropy is a measure of "uncertainty" associated with the outcome of an event. The higher the entropy, the higher the uncertainty as to which possibility will occur. The discrete event entropy will be maximized when the distribution is uniform, that is, all outcomes are equally likely.

In an analogous way, a continuous multivariate random variable $\mathbf{x}$ with a continuous probability density function $p(\mathbf{x})$ has an entropy

$$H(\mathbf{x}) = -\int_{-\infty}^{\infty} p(\mathbf{x}) \ln p(\mathbf{x}) \, d\mathbf{x}. \qquad (4.66)$$

In the discrete case, $H[x]$ was strictly nonnegative. In the continuous case, however, entropy may be negative. Substituting the zero-mean multivariate Gaussian PDF Eq. (4.46) into Eq. (4.66) yields the entropy for a Gaussian process [Smylie et al., 1973]

$$H = \frac{1}{2} \ln\left[\det \mathbf{R}_{xx}\right]. \qquad (4.67)$$

As the dimension M of $\mathbf{R}_{xx}$ grows, the Gaussian entropy may diverge. In such a case, the entropy rate

$$h = \lim_{M \to \infty} \frac{1}{M+1} H \qquad (4.68)$$

may be used. It can be shown [Smylie et al., 1973] that the entropy rate for a zero-mean Gaussian process of bandwidth $B = 1/2T$ can be expressed as

$$h = \frac{1}{2} \ln(2B) + \frac{1}{4B} \int_{-B}^{B} \ln[P_{xx}(f)] \, df. \qquad (4.69)$$

This will be used in Chap. 7 in the discussion of the maximum entropy method.

4.6 LIMIT SPECTRA OF TEST DATA

The MATLAB demonstration script `plot_psd.m` at the website cited in the preface will generate the theoretical limit spectrum (∞ number of data samples) related to the synthesized data sets `doppler_radar.dat` and `test1987.dat`, as shown in Figs. 4.2 and 4.3. These two data sets represent mixed spectral types in which the filtered noise component has a continuous PSD structure and the complex sinusoidal components each have an impulse function structure. The procedure selected for this text for such mixed spectra types is to place a vertical line at the frequency location of each delta function with a height proportional to the sinusoidal power associated with that delta function. For the mixed spectra of the two synthetic data cases, the height in dB above or below the noise power spectral density (PSD) curve is determined by the total power (i.e., the integrate the PSD across all frequencies) in the noise process relative to the sinusoidal power.

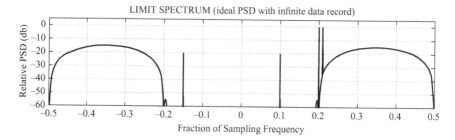

Figure 4.2. Limit spectrum of the 1987 test data.

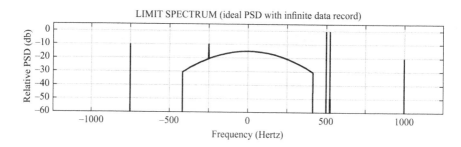

Figure 4.3. Limit spectrum of the Doppler radar data.

4.7 EXTRA: BIAS AND VARIANCE OF THE SAMPLE SPECTRUM

Statistical analysis of the periodogram for the general case is mathematically complicated. The analysis that follows is for the special case of a zero-mean white Gaussian process; the results may then be used as guidelines to the behavior of the periodogram spectral estimator in more general situations [Jenkins and Watts, 1968].

The sample spectrum of the finite-length sequence $x[n]$ for $n = 0$ to $n = N - 1$ is

$$\tilde{P}_{xx}(f) = \frac{1}{NT}|X(f)|^2,$$

where $X(f)$ is the discrete-time Fourier transform

$$X(f) = T \sum_{n=0}^{N-1} x[n] \exp(-j2\pi f nT).$$

This is identical to the discrete-time Fourier transform of the estimated autocorrelation sequence

$$\tilde{P}_{xx}(f) = T \sum_{m=-(N-1)}^{N-1} \hat{r}_{xx}[m] \exp(-j2\pi f mT),$$

where the biased autocorrelation estimate

$$\hat{r}_{xx}[m] = \frac{1}{N} \sum_{n=0}^{N-m-1} x[n+m]x^*[n]$$

has been used for the maximum possible number of lag estimates $[m = 0, \ldots, N - 1]$.

The mean of the sample spectrum is simply

$$\mathcal{E}\{\tilde{P}_{xx}(f)\} = T \sum_{m=-(N-1)}^{N-1} \mathcal{E}\{\hat{r}_{xx}[m]\} \exp(-j2\pi f mT)$$

$$= T \sum_{m=-(N-1)}^{N-1} \left[\frac{N - |m|}{N}\right] r_{xx}[m] \exp(-j2\pi f mT)$$

$$= W(f) \star S_{xx}(f),$$

where $W(f)$ is the DTFT of the Bartlett window,

$$w[m] = \begin{cases} 1 - |m|/N & \text{for } |m| \leq N - 1 \\ 0 & \text{for } |m| > N - 1 \end{cases}.$$

Thus, the mean of the sample spectrum is the convolution of the true PSD with the Fourier transform of the Bartlett window. The sample spectrum is, therefore, *biased* for finite N, but *unbiased* as $N \to \infty$ because

$$\lim_{N \to \infty} \mathcal{E}\{\tilde{P}_{xx}(f)\} = P_{xx}(f),$$

thus yielding the true spectrum.

The autocorrelation of the periodogram is given by

$$\mathcal{E}\{\tilde{P}_{xx}(f_1)\tilde{P}_{xx}(f_2)\} = \left(\frac{T^2}{N^2}\right) \sum_{k=0}^{N-1} \sum_{l=0}^{N-1} \sum_{m=0}^{N-1} \sum_{n=0}^{N-1} \mathcal{E}\{x[k]x[l]x[m]x[n]\}$$
$$\times \exp\{j2\pi T[f_1(k-l)+f_2(m-n)]\}.$$

The variance of the sample spectrum involves *fourth-order* moments, which are difficult to evaluate in the general case. However, because $x[k]$ is white and Gaussian, the fourth-order statistical moment reduces to a sum of second-order moments:

$$\mathcal{E}\{x[k]x[l]x[m]x[n]\} = \begin{cases} \rho_w^2 & \text{if } k=l \text{ and } m=n, \text{ or } k=m \text{ and } l=n, \\ & \text{or } k=n \text{ and } l=m \\ 0 & \text{otherwise} \end{cases},$$

where ρ is the variance of the Gaussian process. This means that

$$\mathcal{E}\{\tilde{P}_{xx}(f_1)\tilde{P}_{xx}(f_2)\}$$
$$= (T\rho_w)^2 \left[1 + \left(\frac{\sin \pi NT(f_1+f_2)}{N \sin \pi T(f_1+f_2)}\right)^2 + \left(\frac{\sin \pi NT(f_1-f_2)}{N \sin \pi T(f_1-f_2)}\right)^2 \right].$$

The covariance is then

$$\text{cov}\{\tilde{P}_{xx}(f_1)\tilde{P}_{xx}(f_2)\} = \mathcal{E}\{\tilde{P}_{xx}(f_1)\tilde{P}_{xx}(f_2)\} - \mathcal{E}\{\tilde{P}_{xx}(f_1)\}\mathcal{E}\{\tilde{P}_{xx}(f_2)\}.$$

Noting that $\mathcal{E}\{\tilde{P}_{xx}(f)\} = T\rho_w$ for a white Gaussian process,

$$\text{cov}\{\tilde{P}_{xx}(f_1)\tilde{P}_{xx}(f_2)\}$$
$$= (T\rho_w)^2 \left[\left(\frac{\sin \pi T(f_1+f_2)N}{N \sin \pi T(f_1+f_2)}\right)^2 + \left(\frac{\sin \pi T(f_1-f_2)N}{N \sin \pi T(f_1-f_2)}\right)^2 \right].$$

Looking at the variance at a particular frequency

$$\text{var}\left\{\tilde{P}_{xx}(f)\right\} = \text{cov}\{\tilde{P}_{xx}(f)\tilde{P}_{xx}(f)\} = (T\rho_w)^2 \left[1 + \left(\frac{\sin 2\pi f TN}{N \sin 2\pi fT}\right)^2 \right]$$
$$= P_{xx}^2(f) \left[1 + \left(\frac{\sin 2\pi f TN}{N \sin 2\pi fT}\right)^2 \right].$$

Note that $\text{var}\{\tilde{P}_{xx}(f)\}$ does not tend to zero no matter how large N may get. Thus, the sample spectrum is *not* a *consistent* estimator of the PSD. The magnitude of the variance is proportional to $P_{xx}^2(f)$ for any N, i.e., the standard deviation is as large as the mean of the PSD to be estimated and, therefore, the sample spectrum is not a reliable PSD estimator because its variance does not decrease as the data record increases!

The selection of harmonic frequencies $f_1 = m/NT$ and $f_2 = n/NT$ (frequencies which are integer multiples of $1/NT$), such that $m \neq n$, results in a zero covariance

$$\text{cov}\{\tilde{P}_{xx}(f_1)\tilde{P}_{xx}(f_2)\} = 0,$$

that is, the values of the periodogram separated in frequency by integer multiples of $1/(NT)$ Hz are uncorrelated. As the number of data samples N increases, these uncorrelated periodogram samples come closer together in frequency. This feature, coupled with the nondecreasing variance for increasing N, suggests that the periodogram (sample spectrum) will appear to have increasingly rapid fluctuations as the record length increases (see Fig. 1.3 for an example).

In general, any nonwhite process $x[n]$ can be generated by passing white noise of variance ρ_w through a linear filter of magnitude-squared response $|H(f)|^2$. Because the true PSD $P_{xx}(f)$ is related by [see Eq. (4.44)]

$$P_{xx}(f) = T\rho_w |H(f)|^2,$$

then it could be argued that the following is approximately true

$$\text{var}\left\{\tilde{P}_{xx}(f)\right\} \approx |H(f)|^4 \rho_w^2 \left[1 + \left(\frac{\sin 2\pi f T N}{N \sin 2\pi f T}\right)^2\right] \geq \mathcal{E}\{\tilde{P}_{xx}(f)\}^2.$$

Thus, as N increases, the sample spectrum variance does not approach zero, but becomes proportional to the square of the expected value of the PSD (i.e., the square of the true PSD), making the sample spectrum an *inconsistent* estimator of the PSD. That is, $\tilde{P}_{xx}(f)$ does not converge in the mean to $P_{xx}(f)$ at any value f where $P_{xx}(f) > 0$.

References

Brillinger, D. R., Fourier Analysis of Stationary Processes, *Proc. IEEE*, vol. 62, pp. 1628–1643, December 1974.

Gardner, W. A., *Introduction to Random Processes*, 2nd ed., McGraw Hill Publishing Company, New York, 1990.

Jenkins, G. M., and D. G. Watts, *Spectral Analysis and Its Applications*, Holden-Day, Inc., San Francisco, 1968.

Kay, S. M., *Modern Spectral Estimation*, Prentice-Hall, Inc., Englewood Cliffs, N. J., 1987.

Monzingo, R. A., and T. W. Miller, *Introduction to Adaptive Arrays*, John Wiley and Sons, Inc., New York, 1980.

Papoulis, Athanasios, *Signal Analysis*, McGraw-Hill Book Company, New York, 1977.

Papoulis, Athanasios, *Probability and Statistics*, Prentice-Hall, Inc., Englewood Cliffs, N. J., 1990.

Papoulis, Athanasios, *Probability, Random Variables, and Stochastic Processes*, 3rd ed., McGraw-Hill Book Company, New York, 1991.

Smylie, D. E., G. K. C. Clarke, and T. J. Ulrych, Analysis of Irregularities in the Earth's Rotation, in *Methods in Computational Physics*, R. Alder et al., eds., vol. 13, pp. 391–430, Academic Press,Inc., New York, 1973.

Therrien, C., *Discrete Random Signals & Statistical Signal Processing*, Prentice-Hall, Englewood Cliffs, N. J., 1992.

5

CLASSICAL SPECTRAL
ESTIMATION

5.1 INTRODUCTION

Two asymptotically equivalent definitions of power spectral density (PSD) were introduced in Chap. 4. The *indirect* approach, based on the Wiener-Khintchine theorem, used an infinite sequence of data values to compute the autocorrelation sequence, which in turn was Fourier transformed to yield the PSD. The *direct* approach generated the PSD by taking the squared magnitude of the Fourier transform of an infinite data sequence with appropriate statistical averaging. If this averaging was disregarded, then the resultant function, termed a *sample spectrum*, was shown to be unsatisfactory due to the statistically erratic spectral variations that it produces, that is, the standard deviation of the PSD estimate was as large as the mean of the PSD estimate. In this chapter, averaging techniques for both indirect and direct approaches that provide smooth and statistically stable spectral estimates from a finite number of data samples are summarized. PSD estimators that are based on direct transformation of the data followed by averaging are collectively termed *periodograms*. PSD estimators that transform finite correlation sequences estimated from the data are collectively termed *correlogram* methods of spectral estimation. Use of the terminology "periodogram" and "correlogram" evoke the historic roots for these methods cited in Chap. 1.

Unlike the continuous-time developments of spectral estimation often found in other texts, all classical methods covered in this chapter are developed in terms of discrete-time sample sequences and correlation sequences. For each PSD estimation method, the user is faced with many trade-offs in an effort to produce statistically reliable spectral estimates of highest possible resolution achievable with a finite data sample record. Trade-offs among data windowing, lag windowing, time-domain averaging, and frequency-domain averaging are required to balance the need to reduce sidelobes, to perform effective ensemble averaging, and to ensure adequate spectral resolution. Stable results (small spectral fluctuations) with good fidelity (small bias from the true spectral value at all frequencies) are achievable only if the product $T_e B_e$, where T_e is the total data record interval and B_e is the effective frequency resolution, is much larger than unity. These trade-offs can be quantified analytically for Gaussian processes and used as rough guidelines for non–Gaussian processes where analytic treatments are not normally possible. Despite detailed ana-lytical studies of the statistical behavior of classical spectral estimators for Gaussian processes, the ultimate selection of a particular method to use with non–Gaussian processes often relies only on experimental evidence. The selection of a sidelobe-suppressing window is also very frequently based more on experimental observation then on analytical studies.

To adequately describe the statistical performance of each method presented would require several chapters, so only highlights are provided here. References that provide more detailed development of the classical spectral estimation tech-niques include Bendat and Piersol [1986], Blackman and Tukey [1958], Brillinger [1981], Gardner [1988], Jenkins and Watts [1968], Koopmans [1974], Priestley [1981], and Yuen and Fraser [1979]. The three sections following this chapter sum-mary introduce some basic topics needed for a proper discussion of the classical spectral estimators. Windowing, an important component common to all classi-cal spectral estimators, is presented in Sec. 5.3. The time-bandwidth product of Chap. 2 must be augmented with a third term in order to account for and to quan-tify the statistical stability of classical spectral estimators. Section 5.4 discusses the augmented stability-time-bandwidth product. Estimation of the correlation func-tion, required for the correlogram method of spectral estimation, is presented in Sec. 5.5. Following these preliminary sections, Sections 5.6–5.8 summarize the correlogram method of spectral estimation (often called the Blackman-Tukey spec-tral estimator), the periodogram method of spectral estimation, and the combined correlogram-periodogram method of PSD estimation. The chapter concludes in Sec. 5.9 with an application of the classical spectral estimators to sunspot data.

All of the classical methods presented in this chapter have similar statistical behavior and similar gross appearance, although some visual differences in the fine detail of the spectral shape will be apparent. Because all classical spectral estima-tors yield similar statistical behavior, the specific method selected is often the one that may be computed most efficiently. Before the extensive availability of com-puters, the correlogram method as outlined by Blackman and Tukey [1958] was predominant. The advent of the FFT algorithm and the introduction of signal pro-cessing integrated circuits have promoted a shift to methods based on periodogram techniques [Jones, 1965].

5.2 SUMMARY

There are two basic classical PSD estimation approaches for finite data records, cor-responding to the two asymptotically equivalent definitions of the PSD presented in Chapter 4. The *direct method*, or periodogram, operates directly on the data set to yield a PSD estimate. The *indirect method* must first estimate a finite record autocor-relation sequence (ACS) [correlogram when plotted], and then Fourier transforms the estimated ACS to obtain a PSD estimate.

The periodogram spectral estimator is described in Sec. 5.7. A demonstra-tion MATLAB script `spectrum_demo_per.m` based on the Welch periodogram

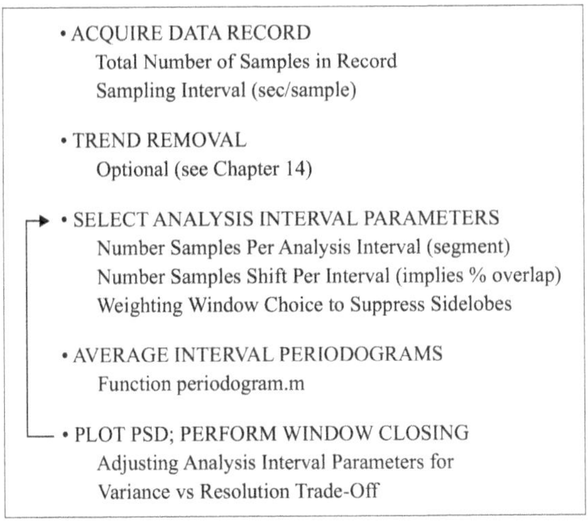

```
 • ACQUIRE DATA RECORD
    Total Number of Samples in Record
    Sampling Interval (sec/sample)

 • TREND REMOVAL
    Optional (see Chapter 14)

→ • SELECT ANALYSIS INTERVAL PARAMETERS
    Number Samples Per Analysis Interval (segment)
    Number Samples Shift Per Interval (implies % overlap)
    Weighting Window Choice to Suppress Sidelobes

 • AVERAGE INTERVAL PERIODOGRAMS
    Function periodogram.m

└── • PLOT PSD; PERFORM WINDOW CLOSING
    Adjusting Analysis Interval Parameters for
    Variance vs Resolution Trade-Off
```

Figure 5.1. Periodogram spectral estimator flowchart that will produce the test 1987 spectrum estimate shown in Figure 5.2.

estimator (see Sec. 5.7.3) is available at a Dover Publications web page. By inputting a response to the script questions, a periodogram spectral estimate for one of the three data sets (test1987, doppler_radar, sunspot_numbers) will be calculated, and a plot such as shown in Fig. 5.2 will be generated.

The Welch method of periodogram spectral estimation involves three parameter choices. An analysis interval (number of segment samples) is divided into possibly overlapping segments. Each segment is windowed to reduce the bias effect of sidelobes, transformed with an FFT, and the squared magnitude of the transform values taken to produce a single segment periodogram. The segment periodograms are then averaged to produce the final periodogram estimate. The "tuning knobs" for periodogram spectral estimators that control the resolution and statistical stability are the segmentation and overlap parameter selections. The selected segment sample size involves a trade-off between the degree of smoothness required in the spectral estimate versus the desired frequency resolution. Setting the segment size to smaller values will permit more segments to be averaged (which decreases spectral variance), but the smaller segment size also means decreased spectral resolution. Increasing the segment size will generate fewer segments for averaging (which increases the spectral variance), but the spectral resolution is improved due to a larger segment aperture.

The determination of the proper parameter selections to tune the periodogram estimator to the spectrum of the signal under analysis is not often apparent. John Tukey suggested an investigative approach called *window closing* to discover appropriate trade-offs among the parameter settings. The essence of the window closing concept is the trial generation of a sequence of periodogram spectral estimates at different frequency resolution/stability settings to learn more about the unknown

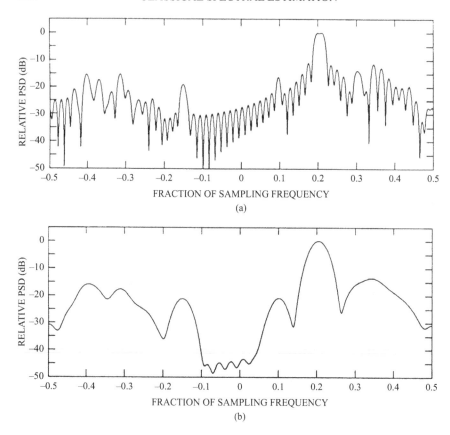

Figure 5.2. Periodogram spectral estimate of 64-sample test1987 data. (a) One segment of 64 samples, sampling interval T = 1 sec, no windowing, 0% overlap (0 samples). (b) Three 32-sample segments, Hamming window, 50% overlap (16 samples).

process being analyzed. The procedure starts with parameter settings that yield low resolution and high stability and works progressively to periodogram estimates of higher resolution and lower stability (less smoothing). If there is much detail in the spectrum, it may be necessary to give up stability (fewer segments averaged) for improved resolution (longer segment sizes). Although there are no exact rules for when to stop the window closing procedure, generally it should be continued until the resolution captures the narrowest *important* detail in the spectrum. Some prior knowledge of anticipated signals that may be present in the spectrum might be required in order to determine what is important and what is not of importance. Lacking that, one should then stop when there are no major changes in the key spectral estimate features as the data segment length increases.

The correlogram method of spectral estimation is described in Sec. 5.6. A demonstration script spectrum_demo_bt based on the Blackman-Tukey technique described in Sec. 5.6 is available at Dover Publications website. By inputting responses to the dialogue of this script, a Blackman-Tukey correlogram-based estimate of one of the three data sets will be generated. Script spectrum_demo_bt will graph the negative sidelobes that are often artifacts of the Blackman-Tukey estimation procedure because it disregards the sign of the PSD estimate values. Shading has been added to the spectral estimate of Fig. 5.4 to highlight sidelobes that were actually negative in value. A valid PSD was shown in Sec. 4.3 to be real and positive, yet the Blackman-Tukey spectral estimator can produce estimates with negative PSD artifact values in some regions of the spectrum.

Figure 5.3 depicts the three basic steps of the Blackman-Tukey method of correlogram spectral estimation. An ACS estimate is computed from the analysis interval data record, windowing is applied to the ACS estimates (rather than the data) to suppress the sidelobe effects, and an FFT of the windowed ACS sequence is computed to yield the PSD estimate.

The "tuning knob" for the Blackman-Tukey PSD estimator that controls the resolution and statistical stability trade-off of this estimator is the maximum lag index, which sets the range of the ACS estimates. The selection of the maximum lag determines the amount of smoothness versus the degree of resolution in the spectral estimate. A maximum lag of 16 was selected for the script dialogue of Fig. 5.3, which means the total time aperture of the ACS estimate is ± 16 samples in durations, or $\pm 16T$ (T is the sample interval). A window closing technique similar to that described for the periodogram can be used to perform a stability versus resolution trade-off investigation. As the number of estimated ACS terms (determined by the maximum lag parameter that was selected) progressively increases, the resolution increases but the stability decreases. The resolution/smoothness trade-off is apparent when comparing the spectral estimate of Fig. 5.4(b), which uses a smaller number of estimated ACS terms to produce a smoother but lower resolution spectral estimate, with the spectral estimate of Fig. 5.4(a).

Figure 5.3. Correlogram-based spectral estimator flowchart to produce spectral estimate shown in Fig. 5.4.

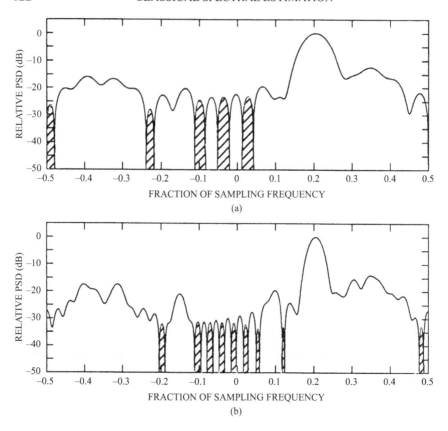

Figure 5.4. Blackman-Tukey method PSD estimate of `test1987` data. Shading indicates regions where the PSD estimate was negative. (a) Spectrum estimate based maximum lag 16 with Hamming window. (b) Spectrum estimate based on maximum lag 32 with Hamming window.

5.3 WINDOWS

An important component of all classical spectral estimation methods is windowing. Windowing is used to control the effect of sidelobes in classical spectral estimators. These effects are discussed in this section. The term *weighting* is used synonymously for windowing in this text, although some would reserve the term window only for the transform of a time-domain weighting function. The function $W(f)$ designates a *frequency window* if it is the discrete-time Fourier transform of a *data window* $w[n]$ applied to a discrete-time signal sequence. Data windows have also been called *tapering functions*. The function $\Omega(f)$ designates a *spectral window* if it is the discrete-time Fourier transform of a *lag window* $\omega[m]$ applied to a discrete-time correlation sequence. The primary effect of a data window is to reduce the bias in periodogram spectral estimates. The primary effect of a lag window is to reduce the PSD estimator variance of the correlogram method of PSD estimation.

It is convenient to consider a finite data record or finite correlation sequence (known or estimated) as a portion of an infinite-duration sequence seen through a window. For example, the observed data sequence $x_0[n]$ of N points may be mathematically expressed as the product with a unit amplitude rectangle function

$$\text{rect}[n] = \begin{cases} 1 & 0 \leq n < N \\ 0 & \text{otherwise} \end{cases} \tag{5.1}$$

and the infinite sequence $x[n]$,

$$x_0[n] = x[n] \cdot \text{rect}[n]. \tag{5.2}$$

An implicit assumption has been made that all unobserved samples are effectively zero, whether or not this is true. The discrete-time Fourier transform of the windowed sequence, expressed in terms of the transforms of $x[n]$ and $\text{rect}[n]$, is the convolution of transforms

$$X_0(f) = X(f) \star D_N(f), \tag{5.3}$$

in which

$$D_N(f) = \text{DTFT}\{\text{rect}[n]\} = T \exp(-j\pi fT[N-1]) \frac{\sin(\pi fTN)}{\sin(\pi fT)}. \tag{5.4}$$

The function $D_N(f)$, known as a *digital sinc* or the *Dirichlet kernel*, is the discrete-time Fourier transform of the rectangle function. The transform of the observed finite sequence is a distorted version of the infinite sequence transform. Figure 5.5 illustrates the effect of rectangular windowing on a discrete-time sinusoid of frequency f_0. The sharp impulsive discrete-time Fourier transform of the infinite sinusoidal sequence has been broadened by the replication of the shape of the window transform. Thus, the narrowest spectral response of a windowed sequence is limited to that set by the mainlobe of the window transform, independent of the data. Sidelobes of the window transform, called *leakage*, will bias the amplitudes of adjacent frequency responses. Because the discrete-time Fourier transform has a periodic replication, sidelobe aliasing from adjacent spectral replications can also contribute additional bias. Increasing the sample rate can mitigate the effect of aliased sidelobes by separating the periodic spectral response. Similar distortion will also occur for nonsinusoidal signals.

Leakage not only produces an amplitude bias in a discrete signal spectral response, but it may also mask the presence of weak signals and prevent their detection by classical spectral methods. Consider the discrete-time Fourier transform of

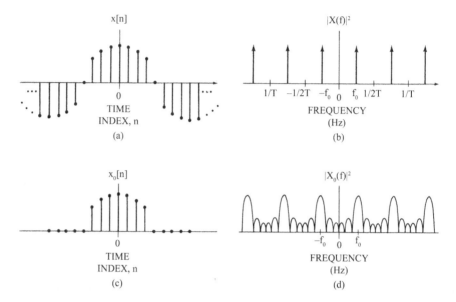

Figure 5.5. Illustration of discrete-time Fourier transform (DTFT) bias due to leakage of windowed data. (a) Original discrete-time sinusoidal sequence. (b) Magnitude of periodic DTFT of sinusoidal sequence. (c) Windowed sinusoidal sequence. (d) Magnitude of DTFT of windowed sequence.

16 samples of the two-sinusoid signal shown in Fig. 5.6. The weaker sinusoid has approximately 17 dB less power than the stronger sinusoid. In this extreme case, the spectral sidelobes of the stronger sinusoid have almost canceled the mainlobe response of the weaker sinusoid.

Other window functions have been devised to achieve better control over the sidelobe levels than obtainable with the rectangular window. Decreasing the sidelobe levels will reduce the bias. However, this can be achieved only by broadening the window's mainlobe frequency response, resulting in a reduction of spectral resolution. A trade-off must, therefore, be made between mainlobe width and level of sidelobe suppression.

Several figures of merit have been devised to categorize window functions. The mainlobe frequency bandwidth is indicative of the frequency resolution. Two measures are used to quantify the mainlobe bandwidth. The traditional measure is the *half-power bandwidth*, which is 3 dB down from the peak response. An alternative measure is the *equivalent bandwidth*, defined in Chap. 2. Two measures are used to quantify the sidelobe behavior. The *peak sidelobe level* is one indicator of how well a window suppresses leakage. The *sidelobe decay rate* indicates the rate at which the sidelobes closest to the mainlobe fall off. The sidelobe decay rate is actually a function of the number of sample points N, but it tends to approach an asymptotic value as N is increased. The asymptotic value is normally expressed in dB per octave of bandwidth.

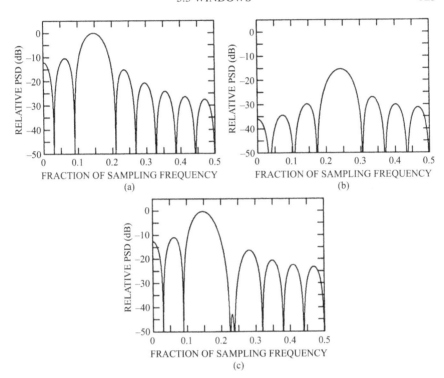

Figure 5.6. Illustration of windowing effect of weak signal masking by an adjacent strong signal. (a) Log of DTFT magnitude of strong sinusoid (16 samples, amplitude 0.1, fractional sample frequency 0.15, initial phase 45°). (b) Log of DTFT magnitude of weak sinusoid (16 samples, amplitude 0.19, fractional sample frequency 0.24, initial phase 162°). (c) Log of DTFT magnitude of combined signals. Weak signal response is nearly obliterated.

Table 5.1 defines some common discrete-time windows among many that have been proposed for use in spectral estimation (e.g., see Harris [1978], Nuttall [1981a], Geçkinli and Yavuz [1983]). Figure 5.7 illustrates some of the more common windows (51-point windows are illustrated) and their frequency responses, obtained by taking the magnitude of the discrete-time Fourier transform of each window. Table 5.1 provides mathematical definitions of N-point symmetric windows with two different origin points. For data windows, the origin point is $n = 0$. For lag windows, the origin point is $-(N-1)/2$. Odd-length windows are always used for lag windows (so that the center element will always be located at time index $n = 0$). An odd-length window has a point of symmetry located at the middle element of the window. This middle element will always have a value of one if it is a lag window. An even-length window will have a point of symmetry between the two window elements at the window center. Other subtle variations on window definitions are

Table 5.1. Definitions of typical N-point discrete-time windows

Window Name	Discrete-Time Function $w[n]$ or $w[m]$	Frequency Response $W(f)$ or $\Omega(f)$
Rectangle (Uniform)	1	$D_n(f)$
Triangle (Bartlett)	$1 - 2|t[n]|$	$\frac{2}{N} D_N^2(f/2)$
Squared Cosine (Hann)	$\cos^2(\pi t[n]) = .50 + .50\cos(2\pi t[n])$	$.50 D_N(f) + .25\left[D_N\left(f - \frac{1}{NT}\right) + D_N\left(f + \frac{1}{NT}\right)\right]$
Raised Cosine (Hamming)	$.54 + .46\cos(2\pi t[n])$	$.54 D_N(f) + .23\left[D_N\left(f - \frac{1}{NT}\right) + D_N\left(f + \frac{1}{NT}\right)\right]$
Weighted Cosine (Nuttall, R = 3)	$\sum_{r=0}^{R} a_r \cos(2\pi rt[n])$	$\sum_{r=0}^{R} .5 a_r \left[D_N\left(f - \frac{r}{NT}\right) + D_N\left(f + \frac{r}{NT}\right)\right]$
Truncated Gaussian ($\alpha = 2.5$)	$\exp\left[-\frac{1}{2}(2\alpha t[n])^2\right]$	$(\sqrt{2\pi}/2\alpha) \exp\left[-\frac{1}{2}(2\alpha fT/\alpha)^2\right] \star D_N(f)$
Equiripple (Chebyshev) ($\beta = 50$ dB)	Inverse DTFT of frequency response	$\cos[(N - 1)\cos^{-1}\{\alpha \cos(\pi fT)\}]/\cosh[(N - 1)\cosh^{-1}(\alpha)]$ $\alpha = \cosh[\cosh^{-1}(10^{(\beta/20)})/(N - 1)]$ β = ratio of mainlobe to sidelobe level (dB)

T is the sample interval.

Data window $w[n]$ is defined over the index range $0 \le n \le N - 1$ and N may be even or odd.

Lag window $\omega[m]$ is defined over the index range $-(N - 1)/2 \le m \le (N - 1)/2$ and N is always odd.

Function $t[n] = (n - [N - 1]/2)/[N - 1]$ for $w[n]$ and $t[m] = m/N$ for $\omega[m]$.

Function $D_N(f) = T\text{sinc}(fNT)$ for $w[n]$ and $D_N(f) = T \exp(j2\pi fT)\text{sinc}(fNT)$ for $\omega[m]$.

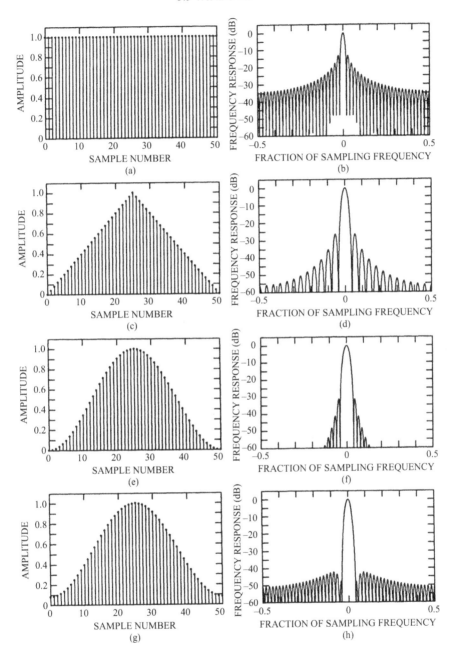

Figure 5.7. Typical discrete-time windows (left side) and the log of their DTFT magnitude (right side). (a–b) Rectangular. (c–d) Triangular. (e–f) Hann. (g–h) Hamming.

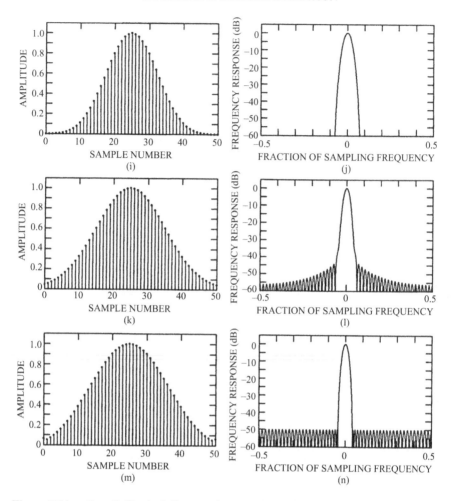

Figure 5.7 (continued). Typical discrete-time windows (left side) and the log of their DTFT magnitude (right side). (i–j) Nuttall. (k–l) Gaussian. (m–n) Chebyshev.

often found in the literature. For example, some authors prefer to omit the zero-value endpoints on windows. For operations with the DTFS involving no zero padding, the highest time index point is usually dropped so that a periodic replication of the window is correctly generated [Harris, 1978]. Table 5.2 lists the characteristics of the windows described in Table 5.1. The bandwidth entries are normalized with respect to a discrete-time Fourier series frequency bin of $1/NT$ Hz.

The frequency response of the rectangular window has the narrowest mainlobe among the windows illustrated, but it has the highest sidelobe level of any window. The triangular window was first described by Bartlett in connection with his analysis of the periodogram spectral estimate and, therefore, often carries his name.

Table 5.2. Characteristics of Table 5.1 Windows

Window Name	Highest Sidelobe Level	Asymptotic Sidelobe Decay Rate	Equiv. BW (DTFS bins)	1/2-Power BW (DTFS bins)
Rectangle	−13.3 dB	−6 dB/octave	1.00	0.89
Triangle	−26.5 dB	−12 dB/octave	1.33	1.28
Hann	−31.5 dB	−18 dB/octave	1.50	1.44
Hamming	−43 dB	−6 dB/octave	1.36	1.30
Nuttall ($r = 3$)	−98 dB	−6 dB/octave	1.80	1.70
Gaussian	−42 dB	−6 dB/octave	1.39	1.33
Equiripple	−50 dB	0 dB/octave	1.39	1.33

The autocorrelation of a rectangular window is a triangular window of twice the width of the rectangular window. The squared cosine window is named after the Austrian meteorologist Julius Van Hann. It is often mistakenly termed a *hanning* window. The historic appeal of the Hann window is that it may be easily implemented in the frequency domain with only three adds and two shift operations that effectively multiply by factors of $1/2$ and $1/4$ at each frequency point. The raised cosine window was introduced by R.W. Hamming, so it often bears his name. The factors 0.53836 and 0.46164 of the Hamming window were selected to essentially eliminate the highest sidelobe of the $D_N(f)$ function. This approach can be generalized to a whole family of weighted cosine windows [Nuttall, 1981a; Harris, 1978; Blackman and Tukey, 1958]. The weights of the cosines may be selected to be optimal with respect to a number of conditions, including minimum sidelobes, maximum sidelobe decay rate, or maximum smoothness (highest number of derivatives without a discontinuity). For example, the window of Fig. 5.7(i) is a third-order ($r = 3$) cosine window with minimum relative sidelobe levels of −98 dB, obtained by selecting $a_0 = 0.3635819, a_1 = 0.4891775, a_2 = 0.1365995$, and $a_3 = 0.0106411$ [Nuttall, 1981a]. The Gaussian window has the smallest time-bandwidth product of all continuous-time functions, although this is only approximately true when truncated to a finite window. The Gaussian window of Fig. 5.7(k) used a parameter setting of $\alpha = 2.5$. The Chebyshev (equiripple) window has all of its sidelobes at a constant level [Dolph 1947, Stegan 1953, Nuttall 1974]. It was first used in antenna array theory. It has the property of having, among all N-point discrete windows with sidelobes at or below a given peak sidelobe level, the narrowest mainlobe. A parameter setting of $\beta = 50$ dB was used to define the Chebyshev window of Fig. 5.7(m).

The strategy for window selection is dictated by a trade-off between bias due to interferers in nearby sidelobes versus bias due to interferers in distant sidelobes. For example, if there are moderate signal components both close and distant in frequency to a weaker signal component, then a window should be selected with equiripple sidelobes around the mainlobe in order to keep the bias small, as illustrated in

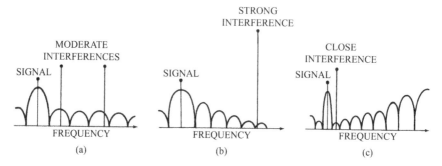

Figure 5.8. Window selection strategy. (a) Equiripple sidelobes for situations with both close and distant interferences of comparable level. (b) Rapidly decaying sidelobes for distant strong interference. (c) Nontraditional special window with low adjacent sidelobes for very close interference and increasing sidelobes.

Fig. 5.8(a). The decay rate is not important in this situation. If there is a distant strong component, as illustrated in Fig. 5.8(b), then a window with rapidly decaying sidelobes should be used. The sidelobe level in the immediate vicinity of the mainlobe is not important in this situation. In situations where high resolution between very close signal components is desired, and there are no distant signal components, then it is even conceivable that a window with *increasing* sidelobes, but with very narrow mainlobe, could be designed, as illustrated by Fig. 5.8(c). If the range of signal levels is limited so that sidelobe characteristics are not a significant consideration, then a window may be selected for its ease of computation. If the spectrum is relatively flat, then no windowing at all may be needed. Windows that can adjust, or "adapt," themselves in some sense to the data conditions, such as that of Fig. 5.8(c), are an important facet of minimum variance spectral estimation discussed in Chap. 13.

5.4 RESOLUTION AND THE STABILITY-TIME-BANDWIDTH PRODUCT

In Sec. 2.11, the time-bandwidth product (TBP) relationship for deterministic signals was introduced. It was shown that, for a temporal (spatial) signal of duration (aperture) $T_e = NT$ seconds (meters), the time-bandwidth product satisfies the equality

$$T_e B_e = 1,$$

where B_e is the equivalent bandwidth in Hz. The resolution of the *energy* spectral density made from a finite sequence of samples of a *deterministic* signal was said to have a resolution of B_e Hz, given as the reciprocal of the data record interval. In order to determine the resolution of the *power* spectral density made from a finite sequence of samples of a *random* signal, the time-bandwidth product must first be modified to account for the random nature of the signal that affects the statistical "quality" of the spectral estimate.

An appropriate modification is the *stability-time-bandwidth product* $QT_e B_s$, where Q is the statistical *quality ratio*, defined as the ratio of the variance of the PSD estimate to the squared mean of the PSD estimate

$$Q = \frac{\text{var}\{\hat{P}(f)\}}{[\mathcal{E}\{\hat{P}(f)\}]^2}. \tag{5.5}$$

The quality ratio Q is effectively an inverted signal-to-noise ratio (SNR) that relates directly to the statistical stability of the spectral estimate. Values of Q much less than unity would indicate smooth spectral estimates with small fluctuations (low variance). Values of Q much greater than unity would indicate very noisy spectral estimates with large fluctuations (large variance). The equivalent bandwidth B_e in the TBP has been replaced with B_s, the *effective statistical bandwidth* [Blackman and Tukey, 1958], defined for the spectral window $\Omega(f)$ as

$$B_s = \frac{\left[\displaystyle\int_{-1/2T}^{1/2T} \Omega(f)\,df\right]^2}{\displaystyle\int_{-1/2T}^{1/2T} \Omega^2(f)\,df}. \tag{5.6}$$

B_s represents the bandwidth of an equivalent rectangular window with an identical ratio of output variance to output squared mean as that of window $\Omega(f)$ when a white noise process is the input (see Sec. B.8 of Blackman and Tukey [1958] for a discussion). The effective statistical bandwidth B_s, rather than the equivalent deterministic bandwidth B_e, must be used due to the random nature of the signal. However, the two bandwidths may be related as $B_s = \alpha B_e$, where α is approximately 0.8 for a rectangular window, 1.3 for a Hann window, and 1.4 for a Hamming window. The bandwidth B_s may be used to roughly indicate the resolution of the spectral estimate, although the bandwidth B_e is closer to the actual resolution in Hz.

It is shown in Sec. 5.8 that, for a Gaussian noise process, the inequality

$$QT_e B_s \geq 1 \tag{5.7}$$

is applicable to all the classical spectral estimation procedures presented in this chapter. If $T_e B_s \approx 1$, then generally $QT_e B_s > 1$. On the other hand, if the time-bandwidth product $T_e B_s$ is selected to be much greater than unity, then the approximation

$$QT_e B_s \approx 1 \tag{5.8}$$

will hold for all classical spectral estimators. Even though all the classical spectral estimation procedures have similar statistical stability (small fluctuations), each estimator will manifest some slight visual differences on an estimate by estimate basis when processing a given data record.

The stability-time-bandwith product epitomizes the interrelationship of the three fundamental parameters that affect the performance of spectral estimators. For a given record length T_e, it is not possible simultaneously to have fine resolution (small values of B_s) and highly stable spectral estimates (small Q). For example, if the desired variance of the spectral estimate is selected to be one-tenth of its mean, then $Q = 0.1$ and the achievable resolution will be $B_s \approx 10/T_e$ Hz. This resolution bandwidth is ten times larger than the rule-of-thumb $1/T_e$ for deterministic signal resolution given in Chap. 2. For this value of Q, only variations in the PSD estimate greater than $\sqrt{Q} \times$ mean PSD may be considered confidently as spectral features rather than noise variations. This example emphasizes the compromise required to strike a balance between statistical stability and resolution that applies to all spectral estimation methods. Note that increasing the number of samples by increasing the sample rate, but holding the interval T_e constant, will have no effect on the ultimate resolution; only the total time aperture of the data record has an effect, not the number of samples. Much of the literature on classical spectral estimation is concerned with procedures that attain the ultimate stability (smallest Q) for a given record interval and specified resolution (e.g., Jenkins and Watts [1968] and Nuttall [1981b]).

Analytic determination of the stability factor Q for arbitrary PSD estimators operating on arbitrary random processes is generally mathematically intractable. Analytic investigations, such as the one leading to Eq. (5.7), are usually restricted for this reason to stationary Gaussian processes under the assumptions of a large $T_e B_s$ product and a constant PSD level within a resolution bin width. Because fourth-order statistical moments of Gaussian processes can be expressed as sums of products of second-order variances, the study of the variance of PSD estimators is greatly simplified. Although the inequality of Eq. (5.7) depends on the Gaussian assumption, it still serves as a reasonable approximation in practice for non–Gaussian processes. Thus, it may be used as a guideline to structure spectral estimation procedures that attain good statistical stability. Some trial settings of B_s and Q may still be necessary to "tune" a spectral estimator to yield the best spectral estimation performance in any given data situation. As Blackman and Tukey [1958, p. 21] state, "We must rely on observed changes from trial to trial as basically the safest measure of the lack of stability of our spectral density estimates."

5.5 AUTOCORRELATION AND CROSS CORRELATION ESTIMATION

The autocorrelation sequence $r_{xx}[m]$ of an ergodic process was defined previously by Eq. (4.56) as the limit of a time average. In practice, the autocorrelation sequence is rarely known, so it must be estimated with the available finite data record. Assuming N data samples $x[n]$ indexed from $n = 0$ to $n = N - 1$, an obvious discrete-time autocorrelation estimate is

$$\hat{r}_{xx}[m] = \frac{1}{(N-m)T} \sum_{n=0}^{N-m-1} x[n+m]x^*[n] \, T$$

$$= \frac{1}{N-m} \sum_{n=0}^{N-m-1} x[n+m]x^*[n]. \tag{5.9}$$

This estimate follows from Eq. (4.56) if $(N-m)$ is substituted for $(2M+1)$ and the summation index range is adjusted for the available data range. Equation (5.9) is applicable only for the positive lag indices $0 \le m \le N-1$ (the lags constitute a time interval of mT seconds). Autocorrelation estimates for lags greater than $N-1$ are not possible due to the finite data available. For negative lag indices $-(N-1) \le m < 0$, the estimate must use the modified form

$$\hat{r}_{xx}[m] = \frac{1}{(N+m)} \sum_{n=-m}^{N-1} x[n+m]x^*[n]$$

$$= \frac{1}{N-|m|} \sum_{n=0}^{N-|m|-1} x^*[n+|m|]x[n]. \tag{5.10}$$

It may be readily shown that this autocorrelation estimate is conjugate symmetric because it satisfies $\hat{r}_{xx}[m] = \hat{r}_{xx}^*[-m]$, a property held in common with the known autocorrelation sequence. When computing the estimate $\hat{r}_{xx}[m]$, only the ACS terms with positive lag indices need be computed, as the ACS terms with negative lag indices are obtained by simple conjugate symmetry. A total of $2N-1$ lags of the autocorrelation sequence may be estimated from N data samples. A plot of the $\hat{r}_{xx}[m]$ estimates versus m is historically called a "correlogram," a term in use since the early 1900s.

The discrete sequence $\hat{r}_{xx}[m]$ forms *unbiased* estimates of the true autocorrelation sequence because

$$\mathcal{E}\{\hat{r}_{xx}[m]\} = \frac{1}{N-m} \sum_{n=0}^{N-m-1} \mathcal{E}\{x[n+m]x^*[n]\} = r_{xx}[m]. \tag{5.11}$$

The variance of this autocorrelation estimate for a real zero-mean complex Gaussian random process as a function of the number of data samples has been shown by Therrien [1992] to be approximately

$$\text{var}\{\hat{r}_{xx}[m]\} \approx \frac{1}{N-m} \sum_{k=-(N-|m|-1)}^{N-|m|-1} \left(1 - \frac{|k|}{(N-m)}\right) |r_{xx}^2[k]|^2$$

and for a real Gaussian process

$$\text{var}\{\hat{r}_{xx}[m]\} \approx \frac{1}{N-m} \sum_{k=-(N-|m|-1)}^{N-|m|-1} \left(1 - \frac{|k|}{(N-m)}\right)$$
$$\cdot \left(r_{xx}^2[k] + r_{xx}[k-|m|]r_{xx}[k+|m|]\right) \qquad (5.12)$$

for $N \gg m$, that is, the lag index is much smaller than the number of data samples. Observe that the variance increases as the lag index increases. Because there are fewer data values to average at the larger lags, the statistical uncertainty of the autocorrelation estimate is increased. The variance tends to zero as N increases, so that $\hat{r}_{xx}[m]$ is a statistically consistent estimate of the discrete-time autocorrelation sequence. For non–Gaussian random processes, Eq. (5.12) is a good approximation to the variance for $N \gg |m|$ [Jenkins and Watts, 1968].

An alternative autocorrelation estimate is

$$\check{r}_{xx}[m] = \begin{cases} \dfrac{1}{N} \displaystyle\sum_{n=0}^{N-m-1} x[n+m]x^*[n] & \text{for } 0 \le m \le N-1 \\[4mm] \dfrac{1}{N} \displaystyle\sum_{n=0}^{N-|m|-1} x^*[n+|m|]x[n] & \text{for } -(N-1) \le m < 0 \end{cases} \qquad (5.13)$$

This estimate differs from $\hat{r}_{xx}[m]$ only in the normalizing factor, yielding the following relationship

$$\check{r}_{xx}[m] = \frac{N-|m|}{N}\hat{r}_{xx}[m]. \qquad (5.14)$$

This is a *biased* estimate for finite N because

$$\mathcal{E}\{\check{r}_{xx}[m]\} = \left(1 - \frac{|m|}{N}\right)r_{xx}[m], \qquad (5.15)$$

but it is asymptotically unbiased as $N \to \infty$. The bias is centered about the zero lag and forms a triangular window on the true autocorrelation. The variance of this estimate is approximately

$$\text{var}\{\check{r}_{xx}[m]\} = \frac{N-|m|}{N}\text{var}\{\hat{r}_{xx}[m]\}$$
$$\approx \frac{1}{N}\sum_{k=-\infty}^{\infty}\left(r_{xx}^2[k] + r_{xx}[k+m]r_{xx}[k-m]\right). \qquad (5.16)$$

The variance asymptotically approaches zero as the number of samples increases, for fixed lag index m. For typical applications, the mean square error (sum of variance and squared bias) tends to be higher for estimator $\hat{r}_{xx}[m]$ than for estimator $\check{r}_{xx}[m]$, especially as the highest lag index approaches the number of data samples

($m \approx N$). The unbiased autocorrelation estimate can also yield autocorrelation estimates that are not valid autocorrelation sequences. Consider, for example, the data sequence $x[1] = 1$, $x[2] = 1.1$, and $x[3] = 1$. The three lags of the unbiased autocorrelation estimate yield $\hat{r}_{xx}[0] = 1.07$, $\hat{r}_{xx}[1] = 1.1$, and $\hat{r}_{xx}[2] = 1$. Observe that $\hat{r}_{xx}[0] < \hat{r}_{xx}[1]$, which violates property Eq. (4.30). The unbiased autocorrelation estimates can thus lead to autocorrelation matrices (to be studied in Chap. 6) that are not positive semidefinite, which means certain matrix equations have no solution, whereas the autocorrelation matrices formed from the biased autocorrelation estimate will always be positive semidefinite. For these reasons, the biased autocorrelation estimator is often the preferred estimator.

Both autocorrelation estimates yield identical values at zero lag

$$\hat{r}_{xx}[0] = \check{r}_{xx}[0] = \frac{1}{N} \sum_{n=0}^{N-1} |x[n]|^2 = \frac{1}{NT} \left(\sum_{n=0}^{N-1} |x[n]|^2 T \right). \tag{5.17}$$

This represents the total estimated power of the measured signal (integrated energy divided by the total time interval). Thus, both ACS estimates preserve the power in the measured signal.

The summation in Eqs. (5.9) and (5.13) may be expressed as a linear convolution

$$\sum_{n=0}^{N-m-1} x[n+m]x^*[n] = x[n] \star x^*[-n]. \tag{5.18}$$

Computationally efficient procedures that perform fast convolutions with FFT algorithms may therefore be used to efficiently compute the discrete autocorrelation estimates. Consult references such as Otnes and Enochson [1978], Rader [1970], Chap. 6 in Rabiner and Gold [1975], or Oppenheim and Schafer [1989] for details. MATLAB function `correlation_sequence.m` uses a FFT-based fast computational technique to generate either the unbiased or the biased ACS estimate from a data record:

```
function r=correlation_sequence(max_lag,bias,x,y)
% Computes the unbiased/biased auto correlation sequence (ACS) or cross
% correlation sequence (CCS) estimate.  Most computationally efficient if
% data vector length is a power of 2.  Returns length 2*max_lag + 1
% correlation sequence in a column vector.  Zero lag of correlation falls
% in middle of the sequence at element (row) max_lag + 1.  'biased'
% estimates scaled by 1/length(x); 'unbiased' estimates scaled by
% 1/[length(x) - lag index].
%
%      ACS:   r = correlation_sequence(max_lag,bias,x)
%      CCS:   r = correlation_sequence(max_lag,bias,x,y)
%
% max_lag -- maximum time index for estimated ACS or CCS
% bias     -- select: 'unbiased' or 'biased' estimates
% x        -- vector of data samples from channel x
% y        -- vector of data samples from channel y (CCS only)
```

```
% r        -- ACS or CCS vector ordered from -max_lag to +max_lag;
%             zero lag is element max_lag + 1
```

Note that a MATLAB demonstration script `plot_correlation.m` has been provided to display the estimated biased or unbiased autocorrelation estimates of the `test1987` and `doppler_radar` test data sets.

The cross correlation estimate may similarly be defined. For example, assuming the two sequences $x[n]$ and $y[n]$ have been measured for $n = 0$ to $n = N - 1$, then the biased cross correlation estimate is

$$\check{r}_{xy}[m] = \begin{cases} \dfrac{1}{N} \displaystyle\sum_{n=0}^{N-m-1} x[n+m]y^*[n] & \text{for } 0 \le m \le N - 1 \\ \dfrac{1}{N} \displaystyle\sum_{n=0}^{N-|m|-1} x[n]y^*[n+|m|] & \text{for } -(N-1) \le m < 0 \end{cases} . \quad (5.19)$$

The unbiased cross correlation estimate uses $1/(N - |m|)$ rather than $1/N$. Unlike the autocorrelation estimate, the cross correlation estimate for negative lag indices must be computed separately because $\check{r}_{xy}[m] \ne \check{r}_{xy}^*[-m]$. The biased cross correlation estimate may be shown to have essentially identical bias and variance behavior as that of the biased autocorrelation. Care must be exercised when using the biased CCS estimator because the bias may obscure the true maximum CCS lag term since the zero lag is not necessarily the largest CCS term, as was the case for the ACS. MATLAB function `correlation_sequence` will also generate unbiased and biased CCS estimates if two data records are indicated as input parameters.

5.6 CORRELOGRAM METHOD PSD ESTIMATORS

The power spectral density was shown, based on the Wiener-Khintchine theorem of Chap. 4, to be the discrete-time Fourier transform of the autocorrelation sequence

$$P_{xx}(f) = T \sum_{m=-\infty}^{\infty} r_{xx}[m]\exp(-j2\pi fmT). \quad (5.20)$$

The *correlogram method* of PSD estimation simply substitutes a finite sequence of unbiased or biased autocorrelation estimates (the *correlogram*) for the infinite sequence of unknown true autocorrelation values. For example, substitution of the unbiased autocorrelation estimates $\hat{r}_{xx}[m]$ that have been computed out to maximum lag indices $\pm L$ yields one possible PSD estimator

$$\hat{P}_{xx}(f) = T \sum_{m=-L}^{L} \hat{r}_{xx}[m]\exp(-j2\pi fmT), \quad (5.21)$$

defined for $-1/2T \leq f \leq 1/2T$. The maximum lag index L is typically much less than the number of data samples N. A maximum of $L \approx N/10$ was suggested by Blackman and Tukey [1958], who were the first to extensively study and popularize the discrete-time correlogram method of PSD estimation. The motivation for this maximum is to avoid the greater statistical variance associated with the estimation of higher lag terms of the autocorrelation sequence. Selection of L closer to N would use ACS terms with increased variance, which would yield a more unstable PSD estimate.

The expected (mean) value of estimator $\hat{P}_{xx}(f)$ is

$$\mathcal{E}\{\hat{P}_{xx}(f)\} = T \sum_{m=-L}^{L} \mathcal{E}\{\hat{r}_{xx}[m]\} \exp(-j2\pi fm)$$

$$= T \sum_{m=-L}^{L} r_{xx}[m] \exp(-j2\pi fmT)$$

$$= P_{xx}(f) \star D_L(f), \tag{5.22}$$

where $D_L(f)$ is the Dirichlet kernel of Eq. (5.4). Even though the correlogram-based PSD estimator is computed with unbiased autocorrelation estimates, it will still be a biased, or smeared, estimate of the true PSD. The windowing effect of a finite autocorrelation sequence will produce an estimate that is effectively the convolution of the true PSD with the transform of the discrete-time rectangular window.

Another correlogram-based PSD estimator may also be formed by using the biased autocorrelation estimate $\check{r}_{xx}[m]$, yielding

$$\check{P}_{xx}(f) = T \sum_{m=-L}^{L} \check{r}_{xx}[m] \exp(-j2\pi fmT), \tag{5.23}$$

defined for $-1/2T \leq f \leq 1/2T$. An alternative matrix formulation of this PSD estimator is

$$\check{P}_{xx}(f) = \frac{T}{N} \mathbf{e}_L^H(f) \tilde{\mathbf{R}}_L \mathbf{e}_L(f), \tag{5.24}$$

in which $\mathbf{e}_L(f)$ is the complex sinusoid vector

$$\mathbf{e}_L(f) = \begin{pmatrix} 1 \\ \exp(j2\pi fT) \\ \vdots \\ \exp(j2\pi fLT) \end{pmatrix}$$

and $\tilde{\mathbf{R}}_L$ is a Hermitian Toeplitz matrix of unweighted correlation estimates products

$$\tilde{\mathbf{R}}_L = \begin{pmatrix} \tilde{r}_{xx}[0] & \tilde{r}_{xx}^*[1] & \cdots & \tilde{r}_{xx}^*[L] \\ \tilde{r}_{xx}[1] & \tilde{r}_{xx}[0] & \cdots & \tilde{r}_{xx}^*[L-1] \\ \vdots & \vdots & \ddots & \vdots \\ \tilde{r}_{xx}[L] & \tilde{r}_{xx}[L-1] & \cdots & \tilde{r}_{xx}[0] \end{pmatrix} \qquad (5.25)$$

in which $\tilde{r}_{xx}[m] = \sum x[n+m]x^*[n]$. The expected (mean) value of estimator $\check{P}_{xx}(f)$ is

$$\mathcal{E}\left\{\check{P}_{xx}(f)\right\} = T \sum_{m=-L}^{L} \left(1 - \frac{|m|}{N}\right) r_{xx}[m] \exp(-j2\pi f mT)$$

$$= P_{xx}(f) \star \frac{2}{L} D_L^2(f/2). \qquad (5.26)$$

The PSD estimate in this case will have a bias as a result of the convolution of the true PSD with the transform of the discrete-time triangular window. If the special case $L = N - 1$ is selected, i.e., the maximum lag is set equal to the number of available data samples less one, then the estimator $\check{P}_{xx}(f)$ is identical to the sample spectrum computed from the N data samples. Because the sample spectrum is known to be statistically unstable, this is another indication that one must choose $L \ll N$ in order to produce a statistically reliable spectral estimator.

To reduce the effect of leakage of the implicit rectangular or triangular window, which therefore reduces the bias in the estimate, an odd-length $(2L + 1)$-point lag window $\omega[m]$ over the interval $-L \leq m \leq L$ and symmetric about the origin is normally used. The most general form of the estimator for the correlogram method of PSD estimation has the form

$$\hat{P}_{BT}(f) = T \sum_{m=-L}^{L} \omega[m]\hat{r}_{xx}[m] \exp(-j2\pi f mT), \qquad (5.27)$$

where the unbiased autocorrelation estimate $\hat{r}_{xx}[m]$ is used. A subscript BT has been added to designate this PSD estimate as the one promoted by Blackman and Tukey. The window is normalized so that $\omega[0] = 1$. Therefore, $\hat{r}_{xx}[0]$ will be unbiased, the estimated signal power (Eq. 5.17) is preserved, and estimator $\hat{P}_{BT}(f)$ will be scaled correctly as a PSD. If one desires the *peaks* in the BT PSD estimator to be proportional to power (due perhaps to impulses in the spectrum generated by sinusoidal signals), rather than the integrated *area* under the PSD to be proportional to power (the proper PSD interpretation), then Eq. (5.27) should be scaled by $F = 1/NT$.

The expected (mean) value of the BT spectral estimate is

$$\mathcal{E}\left\{\hat{P}_{BT}(f)\right\} = T \sum_{m=-L}^{L} \omega[m]r_{xx}[m] \exp(-j2\pi f mT)$$

$$= P_{xx}(f) \star \Omega(f), \qquad (5.28)$$

where $\Omega(f)$ is the discrete-time Fourier transform of the lag window. Lag windows with regions in which $\Omega(f) < 0$ should be avoided, as this generates negative PSD estimates, contrary to the property that a PSD should be positive real. For instance, a rectangular window should be avoided as it will generate regions of negative PSD values as illustrated already in the summary to this chapter. The correlogram method of PSD estimation is asymptotically unbiased as the number of lags increases. Blackman and Tukey have suggested that the maximum lag of the ACS estimate should result in a standard deviation of the PSD that is no more than one-third of the PSD mean value, i.e., $Q = (1/3)^2$. Based on the stability-time-bandwidth product, $1/B_s \approx T_e Q = T_e/9$. Assuming that the lag interval of $T_L = (2L+1)T$ seconds is approximately the reciprocal of the B_s bandwidth, then $T_L \approx T_e/9$, that is, the number of lags of the discrete-time autocorrelation to be estimated should be about 10% of the available data samples. This is the basis of the earlier statement that the maximum lag L should be limited to one-tenth of the number of samples in the data record.

The FFT algorithm may be used to evaluate the PSD estimate of Eq. (5.27) on a grid of $K+1$ frequency points $f_i = i/KT$ for $0 \le i \le K$. The value of K is arbitrary, but generally $K \gg L$, so that sharp details in the PSD estimate will not be missed. Zero padding of data samples will be required for time indices from $L+1$ to K. MATLAB function correlogram_psd.m will compute the Blackman-Tukey method of correlation-based PSD estimation, Eq. (5.27):

```
function psd=correlogram_psd(num_psd,window,max_lag,T,x,y)
% Classical Blackman-Tukey auto/cross spectral estimate based on the Fourier
% transform of the auto/cross correlogram ( ACS  or  CCS  estimate).
%
%       Auto:   psd = correlogram_psd(num_psd,window,max_lag,T,x)
%       Cross:  psd = correlogram_psd(num_psd,window,max_lag,t,x,y)
%
% num_psd -- number of psd vector elements (must be power of 2); frequency
%            spacing between elements is F = 1/(num_psd*T) Hz, with psd(1)
%            corresponding to frequency  -1/(2*T) Hz, psd(num_psd/2 + 1)
%            corresponding to  0  Hz, and psd(num_psd) corresponding to
%            1/(2*T) - F  Hz.
% window   -- window selection:  0 -- none, 1 -- Hamming, 2 -- Nuttall
% max_lag -- maximum time lag (in samples) of ACS or CCS estimate
% T         -- sample interval in seconds
% x         -- vector of data samples
% y         -- vector of data samples (cross PSD only); must have length(y)
%            equal length(x)
% psd       -- vector of num_psd auto/cross PSD values
```

The Blackman-Tukey PSD estimator for the cross power spectral density has the form

$$\hat{P}_{xy}(f) = T \sum_{m=-L}^{L} \omega[m]\hat{r}_{xy}[m]\exp(-j2\pi f mT). \qquad (5.29)$$

This PSD estimate can have a large bias if the cross correlation sequence is not maximum at or near zero lag, where the lag window is centered. This bias can be minimized by prealigning the two discrete-time sequences so that the cross correlation peaks approximately at zero lag. If this is done, then PSD estimator (5.27) will have similar statistical properties as the autospectral estimator (5.25). MATLAB function `correlogram_psd` will also compute the Blackman-Tukey method of cross power spectral density estimation.

5.7 PERIODOGRAM PSD ESTIMATORS

The alternative formal definition of the PSD, based on ergodicity, was discussed in Chap. 4 and had the discrete-time form

$$P_{xx}(f) = \lim_{N \to \infty} \mathcal{E} \left\{ \frac{1}{(2N+1)T} \left| T \sum_{n=-N}^{N} x[n] \exp(-j2\pi f n T) \right|^2 \right\}. \quad (5.30)$$

Ignoring the expectation operation and assuming a finite data set of N samples, $x[0], \ldots, x[N-1]$, the sample spectrum

$$\tilde{P}_{xx}(f) = \frac{1}{NT} \left| T \sum_{n=0}^{N-1} x[n] \exp(-j2\pi f n T) \right|^2 = \frac{T}{N} \left| \sum_{n=0}^{N-1} x[n] \exp(-j2\pi f n T) \right|^2$$

$$(5.31)$$

may be computed from the finite data sequence. This was the original unmodified periodogram PSD estimator. It was shown in Chap. 4 that the sample spectrum will yield statistically inconsistent (unstable) PSD estimates because the expectation operation of Eq. (5.30) has been ignored. One must resort to pseudo ensemble averaging in order to smooth the periodogram PSD estimate. Three types of basic averaging schemes are cited here. The method of Daniell [1946] smooths the periodogram by averaging over adjacent frequency bins. The method of Bartlett [1948] averages multiple periodograms produced from segments of the original data sequence. The method of Welch [1967] extends the Bartlett approach by overlapping the segments and introducing data windows to improve the bias caused by leakage. Other optimal methods of smoothing the periodogram have appeared in the literature, but the methods are usually optimal only with respect to some restrictive class of signals. The three methods presented in this section may not always be optimum, but they have been proven by usage to be robust for many classes of signals.

5.7.1 Daniell Periodogram

Daniell [1946] suggested that a method to smooth the rapid fluctuations of the sample spectrum was to average over adjacent spectral frequencies. If an FFT is used to evaluate the sample spectrum $\tilde{P}_{xx}(f)$ over a uniform grid of frequencies $f_k = k/KT$ for $0 \leq k \leq K - 1$, then a modified periodogram estimate at frequency f_i could be obtained by averaging the P points on either side of this frequency

$$\hat{P}_{\mathrm{D}}[f_i] = \frac{1}{2P+1} \sum_{n=i-P}^{i+P} \tilde{P}_{xx}[f_n].$$
(5.32)

The generalization of this concept is to pass the sample spectrum through a low-pass filter with frequency response $H(f)$. The Daniell periodogram may therefore be expressed more generally as the convolution of the sample spectrum with a low-pass filter

$$\hat{P}_{\mathrm{D}}(f) = \tilde{P}_{xx}(f) \star H(f),$$
(5.33)

where subscript D will distinguish this periodogram estimator from other PSD estimators.

5.7.2 Bartlett Periodogram

Bartlett's scheme [1948] to smooth the sample spectrum was to create a pseudo ensemble by dividing the data sequence of N samples into P nonoverlapping segments of D samples each, so that $DP \leq N$. The pth segment will then consist of the samples

$$x^{(p)}[n] = x[pD + n]$$
(5.34)

for $0 \leq n \leq D - 1$, where the superscript $^{(p)}$ denotes the segment number. An independent sample spectrum

$$\tilde{P}_{xx}^{(p)}(f) = \frac{1}{DT} \left| T \sum_{m=0}^{D-1} x^{(p)}[m] \exp(-j2\pi f m T) \right|^2$$
(5.35)

for each segment $0 \leq p \leq P - 1$ is computed over the range $-1/2T \leq f \leq 1/2T$, typically with an FFT over a uniform grid of frequency points. At each frequency of interest, the P individual unmodified periodograms are averaged to produce the Bartlett averaged periodogram

$$\hat{P}_{\mathrm{B}}(f) = \frac{1}{P} \sum_{p=0}^{P-1} \tilde{P}_{xx}^{(p)}(f).$$
(5.36)

The bias of the Bartlett average of modified periodograms is

$$\mathcal{E}\left\{ \hat{P}_{\mathrm{B}}(f) \right\} = \frac{1}{P} \sum_{p=0}^{P-1} \mathcal{E}\left\{ \tilde{P}_{xx}^{(p)}(f) \right\} = \mathcal{E}\left\{ \tilde{P}_{xx}(f) \right\}.$$
(5.37)

Thus, $\hat{P}_{\mathrm{B}}(f)$ suffers from the same window bias as that of a single-segment sample spectrum. If the P segment periodograms are statistically independent, which will be approximately true if autocorrelation $r_{xx}[m]$ is small for lags $m > D$, then $\hat{P}_B(f)$ can be interpreted as the sample mean of a set of P independent observations of $\tilde{P}_{xx}(f)$. The Bartlett averaged periodogram can be shown, using a generalized variance expression developed in Sec. 5.8, to have a variance inversely proportional to the number of segments

$$\text{var}\left\{\hat{P}_B(f)\right\} \propto \frac{P_{xx}^2(f)}{P}. \tag{5.38}$$

The Bartlett spectral estimator stability improves, therefore, roughly as the inverse of the number of segments P. The variance reduction with increasing P will be less if the segment periodograms are not statistically independent. The resolution has been degraded due to the segmentation into data lengths of D samples each, where $D < N$. The effective spectral window will then have a broader mainlobe bandwidth. For fixed $N = PD$, there is the usual trade-off between high spectral resolution (D is as large as possible) and minimizing estimation variance (P is as large as possible).

5.7.3 Welch Periodogram

Welch modified the basic Bartlett segment-and-average scheme by using data windows and by permitting the data segments to overlap. A data window is applied to the data in each segment prior to the computation of the segment periodogram. The purpose of the window is to reduce the effect of sidelobes and to decrease the estimation bias, at the price of a slight decrease in resolution. The purpose of overlapping segments is to increase the number of segments that are averaged for a given data record length and, therefore, to decrease the PSD estimate variance. Welch also provided a computationally efficient procedure using the FFT for his method of averaging periodograms of weighted and overlapped data segments. It is this latter feature that has made the Welch method the most frequently used periodogram estimation approach in use today.

If the complex data record $x[0], \ldots, x[N-1]$ of N samples is divided into P segments of D samples each, with a shift of S samples between adjacent segments ($S \leq D$), then the maximum number of segments P is given by the integer part of $(N - D)/S + 1$. The pth segment will consist of the windowed samples $x[pS], \ldots, x[pS + D - 1]$ which have been weighted to produce the segment sequence

$$x^{(p)}[n] = w[n]x[n + pS] \tag{5.39}$$

indexed over the range $0 \leq n \leq D - 1$. The segments are indexed over the range $0 \leq p \leq P - 1$. The sample spectrum (raw periodogram) of the weighted pth segment is given by

$$\tilde{P}_{xx}^{(p)}(f) = \frac{1}{E_w}\frac{1}{DT}X^{(p)}(f)\left[X^{(p)}(f)\right]^* = \frac{1}{E_w}\frac{1}{DT}\left|X^{(p)}(f)\right|^2 \tag{5.40}$$

over the frequency range $-1/2T \le f \le 1/2T$, where $X^{(p)}(f)$ is the discrete-time Fourier transform of the pth segment

$$X^{(p)}(f) = T \sum_{n=0}^{D-1} x^{(p)}[n] \exp(-j2\pi f nT) \tag{5.41}$$

and E_w is the discrete-time window energy

$$E_w = T \sum_{n=0}^{D-1} |w[n]|^2. \tag{5.42}$$

The factor E_w is required to prevent the window from biasing the sample spectrum estimate and to preserve the estimated signal power in each segment

$$\int_{-1/2T}^{1/2T} \tilde{P}_{xx}^{(p)}(f)\,df = \frac{1}{DT}\left(\sum_{n=0}^{D-1} |x[n+pS]|^2 T\right)$$

regardless of the shape of the selected window. The average of the windowed segment periodograms yields the Welch periodogram estimator,

$$\hat{P}_W(f) = \frac{1}{P}\sum_{p=0}^{P-1} \tilde{P}_{xx}^{(p)}(f) = \frac{1}{PE_w}\frac{1}{DT}\sum_{p=0}^{P-1}\left|X^{(p)}(f)\right|^2. \tag{5.43}$$

The average value of the Welch periodogram may be shown (Nuttall [1981]) to be

$$\mathcal{E}\left\{\hat{P}_W(f)\right\} = \frac{1}{P}\sum_{p=0}^{P-1}\mathcal{E}\left\{\tilde{P}_{xx}^{(p)}(f)\right\} = \frac{1}{E_w}(P_{xx}(f) \star |W(f)|^2), \tag{5.44}$$

where

$$W(f) = T\sum_{n=0}^{D-1} w[n]\exp(-j2\pi f nT) \tag{5.45}$$

is the discrete-time Fourier transform of the data window and $P_{xx}(f)$ is the true PSD. The DTFT property $|X(f) \star W(f)|^2 = |X(f)|^2 \star |W(f)|^2$ has been used to establish (5.44). The inverse discrete-time Fourier transform of Eq. (5.44) yields the expected value of the effective autocorrelation sequence of the Welch periodogram

$$\mathcal{E}\left\{\hat{r}_{xx}^W[m]\right\} = \frac{1}{E_w}(r_{xx}[m]\cdot\phi[m]), \tag{5.46}$$

in which $r_{xx}[m] = \text{DTFT}^{-1}\{P_{xx}(f)\}$ and

$$\phi[m] = \text{DTFT}^{-1}\left\{\left|W(f)\right|^2\right\} = w[m] \star w^*[-m] = T \sum_{n=0}^{D-m-1} w[n+m]w^*[n]$$

(5.47)

is the autocorrelation of the data window. Note that $E_w = \phi[0]$ and therefore $\phi[m]/\phi[0]$ is effectively an ACS with unit value at lag 0. The total estimated power in the original sample data sequence is given by the zero lag term of the ACS estimate; because

$$\mathcal{E}\left\{\hat{r}_{xx}^W[0]\right\} = r_{xx}[0]\frac{\phi[0]}{E_w} = r_{xx}[0],$$

(5.48)

this implies $\hat{r}_{xx}^W[0]$ is an unbiased estimate of the power.

Welch specifically proposed the use of the Hann window and 50% overlap of the segments. This particular amount of overlap leads to a very efficient implementation with the FFT algorithm, which evaluates Eq. (5.43) over a uniform grid of frequency points. A 50% overlap also uses all data samples twice, except for the $D/2$ samples on each end of the original N-point data sequence. This tends to equalize the treatment of most of the data samples, giving a sample which received a small weight in one segment a larger weight in the next segment. A study of the variance of the Welch periodogram for Gaussian processes indicates, based on expressions given in the next section, that the PSD estimator variance for a Hann window is minimized when using an overlap of 65%; the PSD estimator variance increases approximately 8% if a 50% overlap is used. Like the Bartlett periodogram, the variance of the Welch periodogram is approximately proportional to the inverse of the number of segments

$$\text{var}\left\{\hat{P}_W(f)\right\} \propto \frac{P_{xx}^2(f)}{P},$$

(5.49)

assuming segment independence (overlapping, though, will tend to introduce some segment dependence). Due to overlapping, more segments are constructed from a fixed data record for use by the Welch method than for the Bartlett method. As a result, the Welch periodogram estimator will tend to have slightly less variance than the Bartlett periodogram estimator.

Unlike the lag window of the correlogram-based PSD estimators, the merits of the use of data windows in periodogram spectral estimators have occasionally been the subject of debate in the literature. It has been shown that the primary use of data windows is to reduce the bias produced by sidelobes from large amplitude signals in one part of the spectrum that affect weaker signal spectral responses in another part of the spectrum. Examples of such spectra that provide compelling motivation for data windowing have been provided by Thompson [1977] and Van Schooneveld and Frijling [1981]. The importance of data windowing for sidelobe control is diminished as the spectrum becomes flatter (has decreased signal amplitude range). There have been published studies that claim data windows give some data samples more importance than others and only serve to decrease resolution (broadened mainlobe of frequency window) without a compensating decrease in variance. These claims are certainly true for a sample spectrum produced by using all the data samples of a signal with a relatively flat spectral response. However, the claims are not true for the Welch periodogram, because overlapping segments equalize data treatment and increase the number of segments, which decreases PSD estimator variance. The claims are also not true for spectra involving large amplitude differences among signals where sidelobe effects need to be controlled.

The Welch periodogram estimator `periodogram_psd.m` has been implemented in a MATLAB function available through the text website:

```
function psd=periodogram_psd(num_psd,window,overlap,seg_size,T,x,y)
% Classical periodogram auto/cross power spectral density estimate by Welch
% procedure.
%
%    Auto:    psd = periodogram_psd(num_psd,window,overlap,seg_size,T,x)
%    Cross:   psd = periodogram_psd(num_psd,window,overlap,seg_size,T,x,y)
%
% num_psd    -- number of psd vector elements (must be power of 2);
%               frequency spacing between elements is F = 1/(num_psd*T)
%               Hz, with psd(1) corresponding to frequency  -1/(2*T)
%               Hz, psd(num_psd/2 + 1) corresponding to  0  Hz, and
%               psd(num_psd) corresponding to 1/(2*T) - F  Hz.
% window     -- window selection: 0 -- none, 1 -- Hamming, 2 -- Nuttall
% overlap    -- number of overlap samples between segments
% seg_size   -- number of samples per segment (must be even)
% T          -- sample interval in seconds
% x          -- vector of data samples
% y          -- vector of data samples (cross PSD only);
%               note that length(y) must equal length(x)
% psd        -- vector of num_psd auto/cross PSD values
```

The Welch procedure for estimation of the cross power spectral density is very similar to the auto power spectral density estimator. The N-sample sequences $x[n]$ and $y[n]$ are first segmented and weighted

$$x^{(p)}[n] = w[n]x[n + pS]$$
$$y^{(p)}[n] = w[n]y[n + pS] \tag{5.50}$$

for $0 \le n \le D - 1$ and $0 \le p \le P - 1$. Next, the sample cross spectrum is computed

$$\tilde{P}_{xy}^{(p)}(f) = \frac{1}{E_w}\frac{1}{DT}X^{(p)}(f)\left[Y^{(p)}(f)\right]^*, \tag{5.51}$$

where $X^{(p)}(f)$ and $Y^{(p)}(f)$ are the discrete-time Fourier transforms

$$X^{(p)}(f) = T\sum_{n=0}^{D-1} x^{(p)}[n]\exp(-j2\pi f nT)$$

$$Y^{(p)}(f) = T\sum_{n=0}^{D-1} y^{(p)}[n]\exp(-j2\pi f nT) \tag{5.52}$$

and E_w is identical to Eq. (5.42). Averaging over the segment periodograms yields the final cross periodogram estimate

$$\hat{P}_{xy}(f) = \frac{1}{P} \sum_{p=0}^{P-1} \tilde{P}_{xy}^{(p)}(f).$$

(5.53)

The mean and variance of the cross periodogram spectral estimator is very similar to that of the auto periodogram spectral estimator when the two processes are Gaussian. MATLAB function `periodogram_psd` will also generate the cross spectral density estimate for complex or real data [see also Nuttall, 1975].

5.8 COMBINED PERIODOGRAM/CORRELOGRAM ESTIMATORS

Schemes that combine aspects of both the periodogram and correlogram-based PSD estimators have frequently been promoted in the literature in an attempt to unify the classical PSD estimation methods (e.g., see Otnes and Enochson [1978] or Van Schoeneveld and Frijling [1981]). One of the more definitive generalizations has been presented by Nuttall and Carter [1980, 1982, 1983]. It includes all the methods in Sections 5.6 and 5.7 as special cases. The Nuttall-Carter generalization led to a discovery of an alternative to the popular Welch periodogram method with comparable statistical performance but half the computational burden.

The combined periodogram/correlogram method requires four steps. First, the usual Welch periodogram $\hat{P}_W(f)$ is computed according to Eqs. (5.39), (5.40), and (5.43). Arbitrary overlap S and data window $w[n]$ are permitted. Second, an inverse discrete-time Fourier transform of $\hat{P}_W(f)$ is computed to produce a symmetric autocorrelation estimate $\hat{r}_{xx}^W[m]$ of $2D + 1$ lags. The autocorrelation $\hat{r}_{xx}^W[m]$ spans the lag interval $|m| \leq D$ due to the broadening effect of the window autocorrelation $\phi[m] = w[n] \star w[n]$, as illustrated by Fig. 5.9(a) and (b) for the rectangular data window. Third, an odd-length real symmetric lag window $\omega[m]$ of $2L + 1$ lags is applied to produce the windowed autocorrelation $\hat{r}_{xx}^C[m] = \omega[m]\hat{r}_{xx}^W[m]$ for the combined method. The lag window is selected to be smaller than the width of the autocorrelation $\phi[m]$ of the data window ($L < D$). Fourth, a DTFT of $\hat{r}_{xx}^C[m]$ will yield the final PSD estimate. The final combined PSD estimator is given by

$$\hat{P}_C(f) = \hat{P}_W(f) \star \Omega(f),$$

(5.54)

where $\Omega(f)$ is the transform of the lag window $\omega[m]$.

The combined method achieves sidelobe control (leakage suppression) by a mixture of both temporal weighting and lag weighting. Stability is achieved by a combination of segment averaging and frequency smoothing (lag weighting). By restricting windows and overlaps appropriately, either the Welch periodogram PSD estimator or the Blackman-Tukey PSD estimator are obtained as special cases.

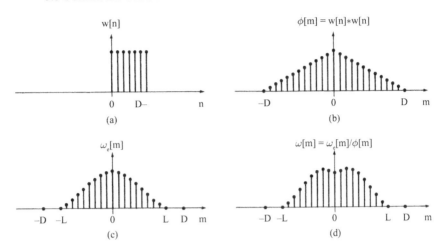

Figure 5.9. Illustration of lag reshaping. (a) Original discrete-time data window of length D. (b) Autocorrelation of discrete-time data window of length $2D + 1$. (c) Desired discrete-time effective lag window of length $2L + 1$, $L < D$. (d) Actual discrete-time lag window applied which incorporates the effect of $\phi[m]$.

The mean of the combined method PSD estimate and the mean of the combined method autocorrelation are easily shown to satisfy

$$\mathcal{E}\left\{\hat{P}_C(f)\right\} = P_{xx}(f) \star \Omega_e(f) \tag{5.55}$$

$$\mathcal{E}\left\{\hat{r}_{xx}^C[m]\right\} = r_{xx}[m]\omega_e[m], \tag{5.56}$$

where the *effective ACS lag window* $\omega_e[m]$ and its transform $\Omega_e(f)$ are given by

$$\Omega_e(f) = \Omega(f) \star |W(f)|^2 \tag{5.57}$$

$$\omega_e[m] = \omega[m]\frac{\phi[m]}{E_w} = \omega[m]\frac{\phi[m]}{\phi[0]}. \tag{5.58}$$

The only assumption required to obtain these is that $x[n]$ be stationary. In order that $\hat{r}_{xx}^C[0]$ be an unbiased estimate of the total power, one must select $\omega_e[0] = 1$. Selecting a lag window $\omega[m]$ such that $\omega[0] = 1$ will satisfy the requirement. An interesting observation can be made at this point. Suppose an ACS lag window is selected in advance, such as the one illustrated in Fig. 5.9(c). If $D > L$ and an arbitrary data window $w[n]$ are selected, then its effect may be incorporated into the lag window $\omega[m]$ as

$$\omega[m] = \frac{\omega_e[m]}{\phi[m]/E_w} \tag{5.59}$$

for $|m| \leq L$. This effect, called *lag window reshaping*, is illustrated in Fig. 5.9(d) for the case of a rectangular data window $w[n]$. Other data windows could also have been used.

The variance of the combined approach has been shown by Nuttall [1971, 1981b] to be approximately

$$\operatorname{var}\left\{\hat{P}_C(f)\right\} \approx P^2(f)\frac{1}{P} \sum_{p=-(P-1)}^{P-1} \left(1 - \frac{|p|}{P}\right) T \sum \omega^2[m]\psi[m, pS], \quad (5.60)$$

where the third-order correlation sequence $\psi[m, n]$ of the data window for even $m + n$ is

$$\psi[m, n] = T \sum_l w\left[l + \frac{n - m}{2}\right] w\left[l + \frac{-n - m}{2}\right]$$
$$\times w\left[l + \frac{n + m}{2}\right] w\left[l + \frac{-n + m}{2}\right]. \quad (5.61)$$

Note that this correlation sequence is symmetric, $\psi[m, n] = \psi[-m, -n]$. The only assumptions required to make this approximation are that $x[n]$ be a complex Gaussian process and that the true PSD $P_{xx}(f)$ be relatively constant within the frequency band $f \pm B_s/2$, representing the bandwidth of the effective lag window centered at f. This latter assumption yields the following approximation for the mean of the PSD

$$\mathcal{E}\left\{\hat{P}_C(f)\right\} \approx P_{xx}(f) \int_{-1/2T}^{1/2T} \Omega_e(f)\,df = P_{xx}(f)w_e[0] = P_{xx}(f). \quad (5.62)$$

The equivalent statistical bandwidth B_s may be explicitly expressed as a function of the data and lag window functions

$$B_s = \frac{\left[\int \Omega_e(f)\,df\right]^2}{\int \Omega_e^2(f)\,df} = \frac{\omega_e^2[0]}{T \sum\limits_{m=-L}^{L} \omega_e^2[m]} = \frac{1}{T \sum_{m=-L}^{L} \phi^2[m]\omega^2[m]}. \quad (5.63)$$

The stability-time-bandwidth product for the combined approach is obtained by substituting the variance equation (5.60), squared mean equation (5.62), and equivalent statistical bandwidth equation (5.63) to yield

$$QT_e B_s \approx \frac{\frac{1}{P}\sum_{p=-(P-1)}^{P-1}\left(1 - \frac{|p|}{P}\right) T \sum_{m=-L}^{L} \omega^2[m]\psi[m, pS]}{\sum_{m=-L}^{L} \omega^2[m]\phi^2[m]}. \quad (5.64)$$

This triple product illustrates that the PSD estimator performance is a function of the data record length T_e, the number of segments P, the amount of overlap S, the data window $w[n]$, and the lag window $\omega[m]$.

The Blackman-Tukey PSD estimator is obtained as a special case by the selection of one segment ($P = 1$), a lag window much less than the number of data points

- ACQUIRE DATA RECORD
  Total Number of Samples in Record
  Sampling Interval (sec/sample)

- COMPUTE BARTLETT PERIODOGR AM
  No segment overlap
  Rectangular data window

- INVERSE FFT OF PERIODOGRAM
  Yields correlation estimate

- WINDOW CORRELATION ESTIMATE
  Lag window reshaping per Section 5.8

- FFT OF WINDOWED CORRELATION
  Final PSD estimate

Figure 5.10. Flowchart of the Nuttall-Carter combined PSD estimator.

$(L \ll D = N)$, and a rectangular data window ($w[n] = 1$ for $n = 0$ to $n = N - 1$). Nuttall [1981b] has shown that $QT_eB_s \approx 1$ in this case. The Welch periodogram PSD estimator is obtained as a special case by the selection of $P = 2N/D - 1$ segments, no lag weighting ($\omega[m] = 1$ for $|m| \leq L = D$), 62% overlap ($S = 0.38D$), and Hann weighting for the data window. It may similarly be shown in this case that $QT_eB_s \approx 1$. If 50% overlap is selected, then $QT_eB_s \approx 1.08$. A special combined PSD estimator is obtained by the selection of a rectangular data window ($w[n] = 1$), no overlap ($S = D$) (thus, only half the number of segments as the Welch periodogram are generated), and lag window reshaping to obtain an effective lag window $\omega_e[m]$ which is a Hann weighting, as illustrated by Fig. 5.9. A flowchart of the four steps of the Nuttall-Carter combined PSD estimator for these special conditions is provided in Fig. 5.10. Nuttall and Carter [1982] have shown, using Eq. (5.64), that these conditions will yield $QT_eB_s \approx 1$. This special case of the combined approach will therefore have the same statistical behavior for a given time-bandwidth product T_eB_s as the popular Welch periodogram. However, it requires about half the computational effort of the Welch method because it uses the FFT for half as many segments as the Welch method and it involves no multiplications for data windowing since this operation is not required.

5.9   APPLICATION TO SUNSPOT NUMBERS

An interesting application of classical spectral estimation that has a long historical basis has been the search for periodicities in sunspot numbers, described in Sec. (1.2). A preliminary periodogram analysis is shown in Fig. 5.11. The 1728 monthly sunspot numbers from `sunspot_numbers.dat` for the years 1841 to 1984 were processed using 24-year segments (288 samples per segment), 12 year overlap

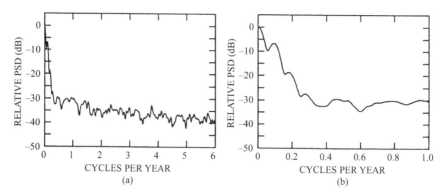

Figure 5.11. (a) Periodogram of monthly sunspot numbers for years 1841–1984. (b) Expansion of the frequency interval within 0 to 1 cycle/year.

(144 samples), and a Hamming data window for each segment. It is apparent from Fig. 5.11 that there are no significant features at frequencies higher than 1 sunspot cycle per year. Although not shown, this lack of features is also borne out by the Blackman-Tukey spectral estimator and by the autoregressive spectral estimator (this method is presented in Chaps. 7 and 8). A large 0-Hz component, due to the positive numerical property of the sunspot numbers, is apparent in Fig. 5.11. The classical 11-year sunspot period (frequency of 0.09 cycles per year) is also apparent in this figure.

In order to study the details in the spectrum between 0 and 1 cycle per year, the monthly sunspot numbers were filtered by a low-pass filter, decimated by a factor of 6 to reduce the sample rate to twice yearly, and finally filtered by a high-pass filter. Linear-phase filter designs using the MATLAB built-in filter functions were used to create the low- and high-pass filters. The low-pass filter had 49 coefficients and a transition zone from 0.5 to 1.0 cycle per year, as shown in Fig. 5.12(a). This eliminated the high-frequency noise in the monthly sunspot numbers. The output of the low-pass filter was decimated by 6 to yield a twice-yearly sample rate. The decimated samples were high-pass filtered to remove the biasing effect of the large DC component. The high-pass filter had 51 coefficients and a transition zone from 0 to 0.0455 cycles per year, as shown in Fig. 5.12(b). This transition zone was selected to permit the detection of a possible 22-year harmonic period in the sunspot numbers, due to the sunspot magnetic polarity, which is known to switch roughly every 22 years. The resultant bandpass-filtered data record consisted of 230 samples of biannual data running from July 1855 to January 1970 (the decimated data samples fell on January and July of each year).

Spectral estimates made from the filtered sunspot data by the classical methods are shown in Fig. 5.13. The periodogram used four segments of 46-year intervals (92 samples per segment) and 23-year overlap (46 samples) between segments. A Hamming window was applied to each segment. The Blackman-Tukey PSD estimate was based on a sequence of unbiased autocorrelation estimates from lag

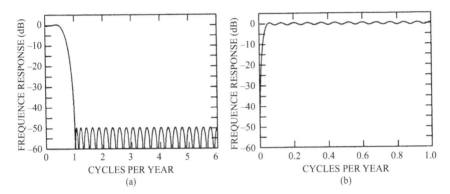

Figure 5.12. Log of frequency response magnitude. (a) 49-term linear-phase low-pass filter. (b) 51-term linear-phase high-pass filter.

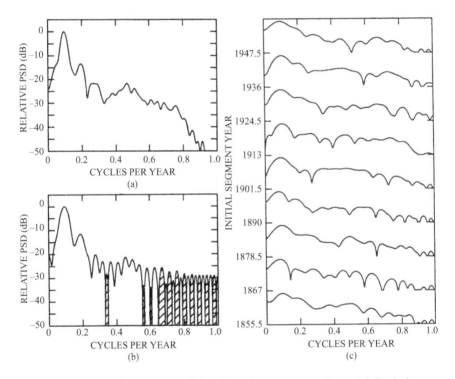

Figure 5.13. Spectral estimates of the filtered sunspot numbers. (a) Periodogram PSD estimate using 92 samples per segment, 5 segments, and the Hamming window. (b) Blackman-Tukey PSD estimate (negative PSD values shown shaded) using ACS estimates to lag 45 and a Hamming window. (c) Stack of single-segment periodograms; only the top 50 dB of each plot is shown.

0 to lag 45 that had been weighted with a Hamming window. The periodogram indicates two strong peaks at 0.095 cycles per year (10.5 years per cycle) and 0.194 cycles per year (5.16 years per cycle). The Blackman-Tukey PSD estimate positions these peaks at 0.095 cycles per year and 0.188 cycles per year (5.32 years per cycle). Thus, the second peak is probably a harmonic of the 10.5-year primary period of the sunspot activity. Other common features between the spectra of Fig. 5.13 are also observed around 0.25 and 0.45 cycle per year.

A strong indicator of the nonstationary characteristics of sunspot numbers is seen in Fig. 5.13(c). Single-segment periodograms of 23-year intervals each are stacked one above the other. A 50-dB plotting range is depicted for each periodogram. The starting year for each segment is indicated along the abscissa. Note that an overlap of 11.5 years between segments has been used. The two strongest spectral peaks do not stay at a fixed frequency, but instead tend to wander about the mean peak locations near 0.1 and 0.2 cycles per year seen in Fig. 5.13(a) and (b). The use of spectral estimation methods, which are based on the assumption that the data is stationary, must therefore be used cautiously when applied to the sunspot number sequence.

5.10 CONCLUSION

Classical spectral estimation techniques that have been outlined in this chapter are among the most robust of the spectral estimation methods presented in this text. They may be employed for just about any signal and noise classes that have stationary properties, whereas alternative high-resolution methods, such as those to be presented in Chaps. 6–14, are robust only for restricted classes of stationary signals. The classical methods typically employ the FFT algorithm and are, therefore, the most computationally efficient spectral estimation methods available. The classical PSD estimators are linearly proportional to the power of sinusoids present in the data; the methods of Chaps. 6–14 generally do not have a linear relationship between spectral peaks and sinusoidal power.

The key disadvantage of the classical spectral estimation techniques is the distorting impact of sidelobe leakage due to the inherent windowing caused by finite data intervals. Window weighting was shown to mitigate the effect of sidelobes at the expense of decreasing the spectral resolution. The resolution of the classical methods can be no better than the reciprocal of the data record duration, independent of the characteristics of the data. In order to improve the statistical stability of the classical spectral estimators, pseudo ensemble averaging over data segments was introduced at a price of a further decrease in the resolution. For a finite data record, trade-offs among resolution, stability (minimizing the estimate variance), and leakage suppression are necessary. Very short data records are often forced upon us and a practical need for the highest resolution may require that the entire data record be used in a single-segment analysis which, of course, sacrifices statistical stability. Alternative spectral analysis methods that may improve or maintain high resolution without sacrificing as much stability are the subject of the next chapter.

References

Bartlett, M. S., Smoothing Periodograms from Time Series with Continuous Spectra, *Nature*, London, vol. 161, pp. 686–687, May 1948.

Bendat, Julius S., and Allan G. Piersol, *Random Data: Analysis and Measurement Procedures (Revised and Expanded)*, 2nd ed., Wiley Interscience, 1986.

Blackman, R. B., and J. W. Tukey, *The Measurement of Power Spectra from the Point of View of Communication Engineering*, Dover Publications, Inc., New York, 1958; also appeared in the January 1958 and March 1958 *Bell Syst. Tech. J.*

Brillinger, D. R., *Time Series Data Analysis and Theory*, Expanded Edition, Holden Day, 1981.

Carter, G. C., and A. H. Nuttall, A Brief Summary of a Generalized Framework for Power Spectral Estimation (with reply by C. K. Yuen), *Signal Processing*, vol. 12, pp. 387–390, October 1980.

Carter, G. C., and A. H. Nuttall, Analysis of a Generalized Framework for Spectral Estimation. Part 1: The Technique and Its Mean Value, pp. 239–241; Part 2: Reshaping and Variance Results, pp. 242–245, *IEE Proc.*, vol. 130, pt. F, April 1983.

Daniell, P. J., Discussion of "On the Theoretical Specification and Sampling Properties of Autocorrelated Time-Series," *J. R. Stat. Soc.*, ser. B, vol. 8, pp. 88–90, 1946.

Dolph, C.L., A Current Distribution for Broadside Arrays Which Optimizes the Relationship Between Beam Width and Side-Lobe Level, *Proc. IRE*, vol. 34, pp. 335–348, June 1946.

Gardner, William A., *Statistical Spectral Analysis: A Non-Probabilistic Theory*, Prentice Hall, Englewood Cliffs, NJ, 1988.

Geçkinli, N. C., and D. Yavuz, *Discrete Fourier Transformation and Its Applications to Power Spectra Estimation*, Elsevier Scientific Publishing Company, Amsterdam, 1983.

Harris, F. J., On the Use of Windows for Harmonic Analysis with the Discrete Fourier Transform, *Proc. IEEE*, vol. 66, pp. 51–83, January 1978.

Jenkins, G. M., and D. G. Watts, *Spectral Analysis and Its Applications*, Holden-Day, Inc., San Francisco, 1968.

Jones, R. H., A Reappraisal of the Periodogram in Spectral Analysis, *Technometrics*, vol. 7, pp. 531–542, November 1965.

Koopmans, L. H., *The Spectral Analysis of Time Series*, Academic Press, Inc., New York, 1974.

Nuttall, Albert H., Spectral Estimation by Means of Overlapped Fast Fourier Transform Processing of Windowed Data, *Naval Underwater Systems Center Technical Report 4169*, New London, Conn., 13 October 1971.

Nuttall, Albert H., Generation of Dolph-Chebyshev Weights Via a Fast Fourier Transform, *Proc. IEEE*, vol. 62, pp. 1396, October 1974.

Nuttall, Albert H., Estimation of Cross Spectra Via Overlapped Fast Fourier Transform Processing, *Naval Underwater Systems Center Technical Report 4169-S*, New London, CT, 11 July 1975.

Nuttall, Albert H., Some Windows with Very Good Sidelobe Behavior, *IEEE Trans. Acoust. Speech Signal Process.*, vol. ASSP-29, pp. 84–91, February 1981a.

Nuttall, Albert H., Spectral Analysis via Quadratic Frequency-Smoothing of Fourier-Transformed, Overlapped, Weighted Data Segments, *Naval Underwater Systems Center Technical Report 6459*, New London, CT, 1 June 1981b.

Nuttall, A. H., and G. C. Carter, Spectral Estimation Using Combined Time and Lag Weighting, *Proc. IEEE*, vol. 70, pp. 1115–1125, September 1982.

Oppenheim, A. V., and R. W. Schafer, *Discrete-Time Signal Processing*, Prentice-Hall, Inc., Englewood Cliffs, N. J., 1989.

Otnes, R. K., and L. Enochson, *Applied Time Series Analysis*, John Wiley & Sons, Inc., New York, 1978.

Priestley, M. B., *Spectral Analysis and Time Series, Vol. I–Univariate Series*, Academic Press, London, 1981.

Rabiner, L. R., and Ben Gold, *Theory and Application of Digital Signal Processing*, Chapter 6, Prentice-Hall, Inc., Englewood Cliffs, N. J., 1975.

Rader, C. M., An Improved Algorithm for High Speed Autocorrelation with Applications to Spectral Estimation, *IEEE Trans. Audio Electroacoust.*, vol. AU-18, pp. 439–442, December 1970.

Stegen, R. J., Excitation Coefficients and Beamwidths of Tschebyscheff Arrays, *Proc. IRE*, vol. 41, pp. 1671–1674, November 1953.

Therrien, C., *Discrete Random Signals and Statistical Signal Processing*, Prentice-Hall, Englewood Cliffs, N. J., 1992.

Thompson, D. J., Spectrum Estimation Techniques for Characterization and Development of WT4 Waveguide-II, *Bell Syst. Tech. J.*, vol. 56, pp. 1983–2005, December 1977; Part I, vol. 50, pp. 1769–1815, November 1977.

Van Schooneveld, C., and D. J. Frijling, Spectral Analysis: On the Usefulness of Linear Tapering for Leakage Suppression, *IEEE Trans. Acoust. Speech Signal Process.*, vol. ASSP-29, pp. 323-329, April 1981.

Welch, P. D., The Use of Fast Fourier Transform for the Estimation of Power Spectra: A Method Based on Time Averaging over Short Modified Periodograms, *IEEE Trans. Audio Electroacoust.*, vol. AU-15, pp. 70–73, June 1967.

Yuen, C. K., and D. Fraser, *Digital Spectral Analysis*, Pitman Publishing Ltd., London, 1979.

6

PARAMETRIC MODELS OF
RANDOM PROCESSES

6.1 INTRODUCTION

The power spectral density (PSD) was defined in Chap. 4 as the discrete-time Fourier
transform of an autocorrelation sequence (ACS) defined over an infinite time index.
This transform relationship between the PSD and ACS is considered to be a *nonparametric* spectral description of the second-order statistics of a random process
because it is a function of only the ACS. Other nonparametric spectral estimators
that are functions of the ACS are considered in Chaps. 12 and 13. A *parametric* description of the second-order statistics may also be devised by assuming a
time-series model for the random process. The PSD of the time-series model can
therefore be formulated as a function dependent on the model parameters rather than
dependent on the ACS. A special class of linear filter models, driven by white noise
processes and possessing rational system functions, is described in this chapter. This
class includes the autoregressive (AR) process model, the moving average (MA)
process model, and the autoregressive–moving average (ARMA) process model.
The output processes of this class of models have power spectral densities that are
totally described in terms of the model parameters and the variance of the driving white noise process. The parameters and the white noise variance are obtained
from the autocorrelation sequence through relationships described in this chapter.
Chapters 8–10 describe methods to directly estimate the model parameters and
the white noise variance from a finite data record without first having to estimate
the ACS.

The motivation for parametric models of random processes is the possibility
to achieve better PSD estimators based upon the model than achievable by classical
spectral estimators. Better spectral resolution is one key improvement area. Both
the periodogram and correlogram methods of Chap. 5 yield PSD estimates from a
windowed finite record of data or ACS estimates. The unavailable data or ACS values outside the window are implicitly zero, normally an unrealistic assumption that
leads to distortions in the spectral estimate. Some knowledge about the process from
which the data samples are taken is often available. This information may be used to
select a time series model that approximates the process that generated the observed
data record. Such models can be used to make more realistic assumptions about the
unmeasured data outside the window other than the assumption that the unmeasured

data values are zero. Thus, the need for window functions to mitigate sidelobe effects is eliminated, along with their distorting impact on the PSD estimate. The degree of improvement in resolution and spectral fidelity, if any, is determined by the appropriateness of the selected model and the ability to fit the measured data or the ACS (known or estimated from the data) with a reasonable number of model parameters.

6.2 SUMMARY

The parametric approach to spectral estimation involves three steps. In step one, an appropriate parametric time-series model is selected to represent the measured data record. This chapter is concerned with three types of parametric time-series models: autoregressive, moving average, and combined autoregressive–moving average. In step two, an estimate of the parameters of the model is made. In step three, the estimated parameters are inserted into the theoretical power spectral density expression appropriate for that model, yielding a PSD estimate.

A summary of the key equation relationships for AR, MA, and ARMA time-series models that are developed in this chapter are provided in Table 6.1. The selection of one of the three models from Table 6.1 requires some knowledge concerning the likely spectral shape. If spectra with sharp peaks, but no deep nulls, are expected, then the AR model is appropriate. If spectra with deep nulls, but no sharp peaks, are expected, then the MA model is appropriate. The ARMA model can theoretically represent both kinds of extremes. If the spectral shape is reasonably modeled by any two or more of the three models, then the model that results in the fewest number of parameters or the least computations to estimate should be selected. This is the principle of parsimony promoted by Box and Jenkins [1970]. The reasoning for this principle is that better PSD estimator statistics are likely if the number of parameters to be estimated is minimized. However, the computational burden to estimate the parameter of an AR process is often significantly less than that required to estimate the parameters of an MA or ARMA process. Thus, an AR process may still be the model of choice, even though it may not be the model of least number of parameters. The question of how to select the number of parameters, once a model is selected, is left to Chaps. 8 and 10.

Table 6.1. Summary of Key Equation Relationships
for Parametric Models

Key Relationships	ARMA	AR	MA
Time Series Definition	6.1	6.13	6.11
Autocorrelation Sequence	6.29	6.31	6.35
PSD (based on parameters)	6.8	6.14	6.12
PSD (based on ACS)	—	6.33	6.36
AR (∞) Equivalence	6.17	—	6.17
MA (∞) Equivalence	6.24	6.24	—

6.3 AR, MA, AND ARMA RANDOM PROCESS MODELS

A time-series model that approximates many discrete-time deterministic and stochastic processes encountered in practice is represented by the linear filter difference
equation of complex coefficients

$$x[n] = -\sum_{k=1}^{p} a[k]x[n-k] + \sum_{k=0}^{q} b[k]u[n-k] \tag{6.1}$$

$$= \sum_{k=0}^{\infty} h[k]u[n-k], \tag{6.2}$$

in which $x[n]$ is the output sequence of a causal filter ($h[k] = 0$ for $k < 0$), which is
assumed to model the observed data, and $u[n]$ is an input sequence. The assumption
$b[0] = 1$ can be made without loss of generality because the input $u[n]$ can always
be scaled to account for any filter gain. The system function $H(z)$ between input
and output was shown [see Eq. (2.17)] to have the rational form

$$H(z) = \frac{B(z)}{A(z)}, \tag{6.3}$$

for which the polynomials are

$$A(z) = 1 + \sum_{k=1}^{p} a[k]z^{-k} \tag{6.4}$$

$$B(z) = 1 + \sum_{k=1}^{q} b[k]z^{-k} \tag{6.5}$$

$$H(z) = 1 + \sum_{k=1}^{\infty} h[k]z^{-k}. \tag{6.6}$$

If polynomials $A(z)$ and $B(z)$ have all of their zeros within the unit circle in the
z-plane, this will guarantee that $H(z)$ is a stable minimum-phase causal filter. If
only $A(z)$ has its zeros within the unit circle, then $H(z)$ is stable but not necessarily
minimum phase.

The z-transform of the autocorrelation of an output sequence $x[n]$ is related to
the z-transform of the autocorrelation of the input random process $u[n]$ by

$$P_{xx}(z) = P_{uu}(z)H(z)H^*(1/z^*) = P_{uu}(z)\frac{B(z)B^*(1/z^*)}{A(z)A^*(1/z^*)}, \tag{6.7}$$

as shown by Eq. (4.42). The input process $u[n]$ that drives the linear filter is not
usually available for purposes of spectral analysis. Many things could be assumed
for the driving process. It could be, for example, a unit impulse, an impulse train,
or white noise. If an impulse is assumed, then the methods of Chap. 11 could be

used to determine the model parameters by observing only the filter output. Here it will be assumed that the driving sequence is a white noise process of zero mean and variance ρ_w, so that $P_{uu}(z) = \rho_w$. Equation (6.1) represents an *autoregressive–moving average* (ARMA) model for the time series $x[n]$ when $u[n]$ is selected to be a white noise sequence. An ARMA process time-series model is illustrated in Fig. 6.1(a). The $a[k]$ parameters form the autoregressive portion of the ARMA model. The $b[k]$ parameters form the moving average portion of the ARMA model. Both $a[0]$, $b[0]$ are assumed to be unity so that any filter gain will be absorbed into ρ_w. The ARMA power spectral density is obtained by substituting $z = \exp(j2\pi fT)$ into Eq. (6.7) (see Eq. 4.43), yielding

$$P_{\text{ARMA}}(f) = T\rho_w \left| \frac{B(f)}{A(f)} \right|^2 = T\rho_w \frac{\mathbf{e}_q^H(f)\mathbf{b}\mathbf{b}^H\mathbf{e}_q(f)}{\mathbf{e}_p^H(f)\mathbf{a}\mathbf{a}^H\mathbf{e}_p(f)}, \qquad (6.8)$$

in which the polynomials $A(f)$ and $B(f)$ are defined as

$$A(f) = 1 + \sum_{k=1}^{p} a[k]\exp(-j2\pi fkT) \qquad (6.9)$$

$$B(f) = 1 + \sum_{k=1}^{q} b[k]\exp(-j2\pi fkT),$$

and the complex sinusoid vectors $\mathbf{e}_q(f)$ and $\mathbf{e}_p(f)$ and parameter vectors $\mathbf{a}$ and $\mathbf{b}$ are defined as

$$\mathbf{e}_p(f) = \begin{pmatrix} 1 \\ \exp(j2\pi fT) \\ \vdots \\ \exp(j2\pi fpT) \end{pmatrix}, \quad \mathbf{a} = \begin{pmatrix} 1 \\ a[1] \\ \vdots \\ a[p] \end{pmatrix},$$

$$\mathbf{e}_q(f) = \begin{pmatrix} 1 \\ \exp(j2\pi fT) \\ \vdots \\ \exp(j2\pi fqT) \end{pmatrix}, \quad \mathbf{b} = \begin{pmatrix} 1 \\ b[1] \\ \vdots \\ b[q] \end{pmatrix}. \qquad (6.10)$$

The ARMA power spectral density is evaluated over the range $-1/2T \le f \le 1/2T$.

Function `arma_psd.m` computes a set of power spectral density values for either an ARMA, an AR, or an MA process, given the arrays of process parameters (*a* or *b* or both) and the white noise variance (*RHO*). An FFT is used to efficiently evaluate the polynomials $A(f)$ and $B(f)$ of the process system function over a uniform frequency grid:

```
function psd=arma_psd(num_psd,T,rho,a,b)
% Generates the AR, MA, or ARMA power spectral density values.
% ARMA psd: eq (6.8); MA psd: eq (6.12); AR psd: eq (6.14)
%
%       psd = arma_psd(num_psd,T,rho,a,b)
```

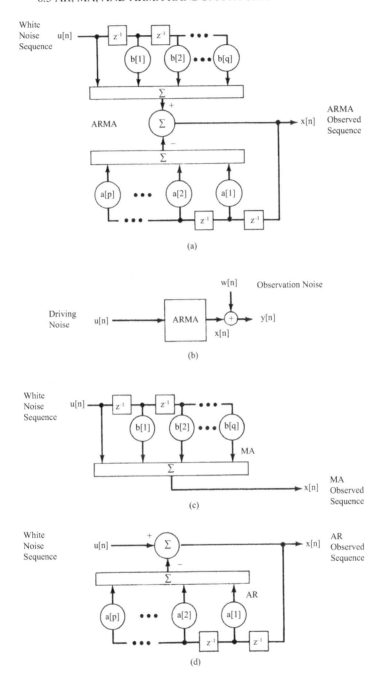

Figure 6.1. Time-series models that have rational system functions. (a) Autoregressive-moving average filter of order (p, q). (b) ARMA with observation noise. (c) Moving average filter of order q. (d) Autoregressive filter of order p.

```
%
% num_psd  -- size of psd vector (must be power of two)
% T        -- sample interval in seconds
% rho      -- variance of driving white noise process
% a        -- vector of autoregressive parameters
% b        -- vector of moving average parameters
% psd      -- vector of power spectral density values
```

The notation ARMA(p, q) is often used for convenience to indicate an ARMA process of *order p* autoregressive parameters and *order q* moving average parameters. Specification of the AR parameters, the MA parameters, and the white noise variance ρ_w entirely characterizes the ARMA PSD of process $x[n]$. Any additive *observation* noise present in the measured sequence must be modeled *separately* and should be *distinguishable* from the *driving* noise source of the ARMA process, as illustrated in Fig. 6.1(b). Effects of observation noise are considered in Chap. 8.

If all the autoregressive parameters are zero, except $a[0] = 1$, then

$$x[n] = \sum_{k=1}^{q} b[k]u[n-k] + u[n] \tag{6.11}$$

is strictly a moving average (MA) process of order q, or simply an MA(q). The MA power spectral density, obtained by setting $p = 0$ in Eq. (6.8), is

$$P_{\text{MA}}(f) = T\rho_w |B(f)|^2 = T\rho_w e_q^H(f)\mathbf{bb}^H e_q(f). \tag{6.12}$$

The MA process time-series model is illustrated in Fig. 6.1(c).

If all the moving average parameters are zero, except $b[0] = 1$, then

$$x[n] = -\sum_{k=1}^{p} a[k]x[n-k] + u[n] \tag{6.13}$$

is strictly an autoregressive (AR) process of order p, or simply an AR(p). The AR power spectral density, obtained by setting $q = 0$ in Eq. (6.8), is

$$P_{\text{AR}}(f) = \frac{T\rho_w}{|A(f)|^2} = \frac{T\rho_w}{e_p^H(f)\mathbf{aa}^H e_p(f)}. \tag{6.14}$$

The AR process time-series model is illustrated in Fig. 6.1(d). The ARMA, MA, and AR power spectral densities may be evaluated, given the process parameters and driving noise variance ρ_w, with MATLAB function arma_psd.

Figure 6.2 illustrates typical ARMA, AR, and MA spectra. The sharp peaks that characterize autoregressive spectra and the deep nulls that characterize moving average spectra are apparent in this figure. The ARMA spectrum shown in Fig. 6.2(c) is the result of combining the AR and MA of Fig. 6.2(a) and (b). The ARMA

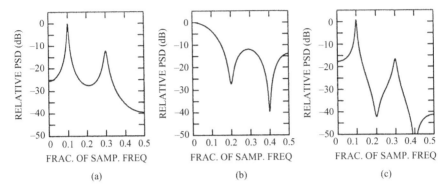

Figure 6.2. Typical parametric process spectra. (a) AR(4) spectrum. (b) MA(4) spectrum. (c) ARMA(4,4) spectrum.

spectrum is able to model both sharp peaks and deep nulls. Other insight into the spectral shape characteristics of these parametric models can be found in Gutowski et al. [1978] and Kay [1987].

6.4 RELATIONSHIPS AMONG AR, MA, AND ARMA PROCESS PARAMETERS

Given an AR, MA, or ARMA process of a finite number of parameters, it is possible to express the parameters of one process in terms of the parameters of the other two processes. An ARMA or MA process can be represented by a unique AR process of generally infinite order. An ARMA or AR process can be represented by a unique MA process of generally infinite order. This concept is important because any one of the three processes may be selected and its parameters used to obtain the parameters of the other two processes. There are often computational benefits to estimating the parameters of one process from the data and using these parameters to estimate the parameters of another process. The AR process, in particular, has many computationally efficient estimation algorithms. As will be shown in Chap. 10, a large order AR estimate is often the first step in an MA or an ARMA estimation algorithm.

Let

$$C(z) = 1 + \sum_{k=1}^{\infty} c[k]z^{-k} \qquad (6.15)$$

represent the denominator polynomial of an AR(∞) process. The $c[k]$ parameters of the AR(∞) process that are equivalent to a finite ARMA(p, q) process are obtained by equating

$$\frac{B(z)}{A(z)} = \frac{1}{C(z)}, \qquad (6.16)$$

or by forming the inverse z-transform of $C(z)B(z) = A(z)$. This will yield

$$c[n] = \begin{cases} 1 & \text{for } n = 0 \\ -\displaystyle\sum_{k=1}^{q} b[k]c[n-k] + a[n] & \text{for } 1 \le n \le p \\ -\displaystyle\sum_{k=1}^{q} b[k]c[n-k] & \text{for } 1 \le n \le p \end{cases} \tag{6.17}$$

with initial conditions $c[-1] = \cdots = c[-q] = 0$. Conversely, if given the parameters of an AR(∞) process that is known to be equivalent to an ARMA(p, q) process, then the MA parameters may be recovered by solving

$$\begin{pmatrix} c[p] & c[p-1] & \cdots & c[p-q+1] \\ c[p+1] & c[p] & \cdots & c[p-q+2] \\ \vdots & \vdots & \ddots & \vdots \\ c[p+q-1] & c[p+q-2] & \cdots & c[p] \end{pmatrix} \begin{pmatrix} b[1] \\ b[2] \\ \vdots \\ b[q] \end{pmatrix} = - \begin{pmatrix} c[p+1] \\ c[p+2] \\ \vdots \\ c[p+q] \end{pmatrix} \tag{6.18}$$

for the $b[k]$ terms using relationship (6.17) over the range $p+1 \le n \le p+q$. The matrix of $c[n]$ parameters in Eq. (6.18) is Toeplitz. Because Eq. (6.18) has a Toeplitz matrix, it may be solved using techniques of Sec. 3.8.3 in a number of operations proportional to p^2. Once the MA parameters have been found, then the AR parameters of the ARMA process may be recovered by convolution

$$a[n] = c[n] + \sum_{k=1}^{q} b[k]c[n-k], \tag{6.19}$$

for $1 \le n \le p$. Fast convolution algorithms based on the FFT are readily available. Equation (6.19) is derivable from Eq. (6.17).

    Note that a finite range of autoregressive parameters $c[1], \ldots, c[p+q]$ from the AR(∞) process are used by Eqs. (6.18) and (6.19). If the parameters $c[k]$ for $k > p+q$ are truncated to zero, then the resulting truncated AR($p+q$) process can only *approximate* the ARMA(p, q) process from which it came. It approximates the ARMA in the sense that the inverse ARMA polynomial

$$G(z) = \frac{1}{H(z)} = \frac{A(z)}{B(z)} = 1 + \sum_{k=1}^{\infty} g[k]z^{-k} \tag{6.20}$$

and the truncated AR polynomial

$$C(z) = 1 + \sum_{k=1}^{p+q} c[k]z^{-k} \tag{6.21}$$

match, i.e., $g[k] = c[k]$, only over the finite range $1 \le k \le p+q$. The higher the orders p and q are selected, the more terms that are matched and the better will

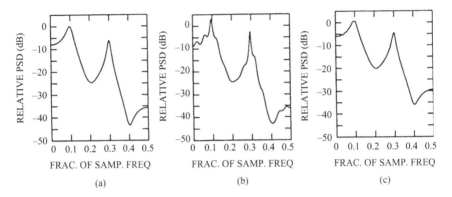

Figure 6.3. High-order AR approximation to a low-order ARMA process. (a) ARMA(4,4) spectrum. (b) Spectrum of AR(25) approximation. (c) Spectrum of AR(50) approximation.

be the approximation. This approximation of a rational function of finite-order polynomials with a polynomial of higher order is the *Padé approximation* problem [Padé, 1892; Weiss and McDonough, 1963]. This procedure will be useful in Chap. 10 for deriving approximations of ARMA process from high-order AR process. Figure 6.3 illustrates typical high-order AR approximations to a low-order ARMA process.

In a similar manner, let

$$D(z) = 1 + \sum_{k=1}^{\infty} d[k]z^{-k} \qquad (6.22)$$

represent the numerator polynomial of an MA(∞) process. The $b[k]$ parameters of the MA(∞) process that are equivalent to an ARMA(p, q) process are obtained by equating

$$\frac{B(z)}{A(z)} = D(z), \qquad (6.23)$$

or by forming the inverse z-transform of $D(z)A(z) = B(z)$. This will yield

$$d[n] = \begin{cases} 1 & \text{for } n = 0 \\ -\sum_{k=1}^{p} a[k]d[n-k] + b[n] & \text{for } 1 \le n \le q \\ -\sum_{k=1}^{p} a[k]d[n-k] & \text{for } n > q \end{cases} . \qquad (6.24)$$

Figure 6.4 illustrates high-order MA approximations to a low-order ARMA process.

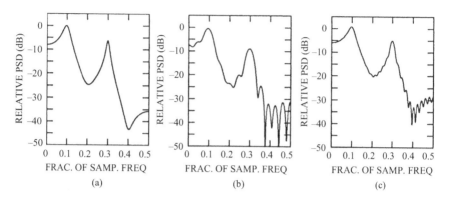

Figure 6.4. High-order MA approximation to a low-order ARMA process. (a) ARMA(4,4) spectrum. (b) Spectrum of MA(25) approximation. (c) Spectrum of MA(50) approximation.

6.5 RELATIONSHIP OF AR, MA, AND ARMA PARAMETERS TO THE AUTOCORRELATION SEQUENCE

This section is concerned with the determination of the AR, MA, and ARMA process parameters when the autocorrelation sequence is known. If Eq. (6.1) is multiplied by $x^*[n-m]$ and the expectation taken, then the result is

$$
\mathcal{E}\{x[n]x^*[n-m]\} = -\sum_{k=1}^{p} a[k]\mathcal{E}\{x[n-k]x^*[n-m]\}
$$
$$
+ \sum_{k=0}^{q} b[k]\mathcal{E}\{u[n-k]x^*[n-m]\} \qquad (6.25)
$$

or

$$
r_{xx}[m] = -\sum_{k=1}^{p} a[k]r_{xx}[m-k] + \sum_{k=0}^{q} b[k]r_{ux}[m-k]. \qquad (6.26)
$$

The cross correlation $r_{ux}[i]$ between the input and the output can be expressed in terms of the $h[k]$ parameters of Eq. (6.6) by using Eq. (6.2), yielding

$$
r_{ux}[i] = \mathcal{E}\{u[n+i]x^*[n]\} = \mathcal{E}\left\{ u[n+i]\left[u^*[n] + \sum_{k=1}^{\infty} h^*[k]u^*[n-k] \right] \right\}
$$
$$
= r_{uu}[i] + \sum_{k=1}^{\infty} h^*[k]r_{uu}[i+k]. \qquad (6.27)
$$

As $u[k]$ was assumed to be a white noise process, then

$$r_{ux}[i] = \begin{cases} 0 & \text{for } i > 0 \\ \rho_w & \text{for } i = 0 \\ \rho_w h^*[-i] & \text{for } i < 0 \end{cases} \quad (6.28)$$

The final relationship between the ARMA parameters and the autocorrelation of process $x[k]$ is therefore

$$r_{xx}[m] = \begin{cases} r_{xx}^*[-m] & \text{for } m < 0 \\ -\sum_{k=1}^{p} a[k] r_{xx}[m-k] + \rho_w \sum_{k=m}^{q} b[k] h^*[k-m] & \text{for } 0 \le m \le q \\ -\sum_{k=1}^{p} a[k] r_{xx}[m-k] & \text{for } m > q \end{cases} \quad (6.29)$$

where the reader is reminded that $h[0] = 1$ by definition [see Eq. (6.2)]. Note that the relationship is highly nonlinear for lag indices $0 \le m \le q$.

The autoregressive parameters of an ARMA process are related by a set of linear equations to the autocorrelation sequence. The linear equations of (6.29) for the p lag indices $q+1 \le m \le q+p$ can be grouped into the matrix expression

$$\begin{pmatrix} r_{xx}[q] & r_{xx}[q-1] & \cdots & r_{xx}[q-p+1] \\ r_{xx}[q+1] & r_{xx}[q] & \cdots & r_{xx}[q-p+2] \\ \vdots & \vdots & \ddots & \vdots \\ r_{xx}[q+p-1] & r_{xx}[q+p-2] & \cdots & r_{xx}[q] \end{pmatrix} \begin{pmatrix} a[1] \\ a[2] \\ \vdots \\ a[p] \end{pmatrix} \quad (6.30)$$
$$= -\begin{pmatrix} r_{xx}[q+1] \\ r_{xx}[q+2] \\ \vdots \\ r_{xx}[q+p] \end{pmatrix}.$$

If given, therefore, the autocorrelation sequence for lags $q - p + 1$ to $q + p$, then the autoregressive parameters may be found separately from the moving average parameters as the solution to the simultaneous equations (6.30). This relationship is known as the *ARMA Yule-Walker normal equations*, sometimes referenced as the *modified Yule-Walker equations* or the *high-order Yule-Walker equations*. The autocorrelation matrix of Eq. (6.30) is Toeplitz and may be solved in a number of computations proportional to p^2 using the techniques of Sec. 3.8.3. Unfortunately, the moving average parameters of an ARMA process cannot be expressed as the solution to a set of linear equations. The MA parameters are convolved with the impulse response coefficients $h[k]$, as indicated by Eq. (6.29), which means the relationship with the ACS is nonlinear.

The relationship between the autocorrelation sequence and a pure autoregressive process may be found by setting $q = 0$ in Eq. (6.29), yielding

$$r_{xx}[m] = \begin{cases} -\sum_{k=1}^{p} a[k]r_{xx}[m-k] & \text{for } m > 0 \\ -\sum_{k=1}^{p} a[k]r_{xx}[-k] + \rho_p & \text{for } m = 0 \\ r_{xx}^{*}[-m] & \text{for } m < 0 \end{cases} \quad (6.31)$$

This relationship may be evaluated for the $p + 1$ lag indices $0 \le m \le p$ and formed into the matrix expression

$$\begin{pmatrix} r_{xx}[0] & r_{xx}[-1] & \cdots & r_{xx}[-p] \\ r_{xx}[1] & r_{xx}[0] & \cdots & r_{xx}[-p+1] \\ \vdots & \vdots & \ddots & \vdots \\ r_{xx}[p] & r_{xx}[p-1] & \cdots & r_{xx}[0] \end{pmatrix} \begin{pmatrix} 1 \\ a[1] \\ \vdots \\ a[p] \end{pmatrix} = \begin{pmatrix} \rho_w \\ 0 \\ \vdots \\ 0 \end{pmatrix} \quad (6.32)$$

or

$$\begin{pmatrix} 1 & a[1] & \cdots & a[p] \end{pmatrix} \begin{pmatrix} r_{xx}[0] & r_{xx}[1] & \cdots & r_{xx}[p] \\ r_{xx}[-1] & r_{xx}[0] & \cdots & r_{xx}[p-1] \\ \vdots & \vdots & \ddots & \vdots \\ r_{xx}[-p] & r_{xx}[-p+1] & \cdots & r_{xx}[0] \end{pmatrix} = \begin{pmatrix} \rho_w & 0 & \cdots & 0 \end{pmatrix}.$$

If given the autocorrelation sequence for lags 0 to p, then the AR parameters may be found as the solution to Eq. (6.32). This relationship forms the *AR Yule-Walker normal equations*. The autocorrelation matrix of Eq. (6.32) is both Toeplitz and Hermitian because $r_{xx}[-k] = r_{xx}^{*}[k]$. MATLAB function levinson_recursion.m may be used to obtain the solution for ρ_w, $a[1]$, ..., $a[p]$, given the ACS over the lag range $0 \le m \le p$. The number of computations is proportional to p^2.

Using the autocorrelation sequence given by Eq. (6.31), an alternative expression for the autoregressive PSD is

$$P_{AR}(f) = \frac{T\rho_w}{|A(f)|^2} = T \sum_{k=-\infty}^{\infty} r_{xx}[k] \exp(-j2\pi fkT). \quad (6.33)$$

Note that autocorrelation lags 0 to p uniquely describe the autoregressive process of order p, as the lags for $|k| > p$ are obtained recursively from

$$r_{xx}[m] = -\sum_{k=1}^{p} a[k]r_{xx}[m-k], \quad (6.34)$$

as indicated by Eq. (6.29) with $q = 0$.

The relationship between the ACS and a moving average process may be found by specializing Eq. (6.29) for $p = 0$ and noting that $h[k] = b[k]$ for $1 \leq k \leq q$, yielding

$$
r_{xx}[m] = \begin{cases}
0 & \text{for } m > q \\[2mm]
\rho_w \displaystyle\sum_{k=m}^{q} b[k]b^*[k-m] & \text{for } 0 \leq m \leq q \ . \\[2mm]
r_{xx}^*[-m] & \text{for } m < 0
\end{cases}
\tag{6.35}
$$

This is a convolutional relationship between the ACS and the MA parameters. Using the autocorrelation sequence given by Eq. (6.35), an alternative expression for the moving average PSD is

$$
P_{\text{MA}}(f) = T\rho_w |B(f)|^2 = T \sum_{k=-q}^{q} r_{xx}[k] \exp(-j2\pi f k T).
\tag{6.36}
$$

Note that Eq. (6.36) has a finite summation range. This simply reflects the fact that a moving average process of order q is uncorrelated for lags $|k| > q$. Equation (6.36) is identical in form to the classical Blackman-Tukey method of PSD estimation

$$
\hat{P}_{\text{BT}}(f) = T \sum_{k=-q}^{q} \hat{r}_{xx}[k] \exp(-j2\pi f k T),
\tag{6.37}
$$

if autocorrelation estimates $\hat{r}_{xx}[k]$ up to lag q are used. The distinction between the two spectral methods is the manner in which the data is used. The correlogram method uses the data to estimate the autocorrelation sequence directly. The moving average method uses the data to estimate the MA parameters (see Chap. 10 for MA estimation procedures), then uses Eq. (6.12) to compute the PSD.

6.6   SPECTRAL FACTORIZATION

Consider the z-transform for an AR(p) process, defined by Eq. (6.7) as

$$
P_{\text{AR}}(z) = \frac{\rho_w}{A(z)A^*(1/z^*)},
\tag{6.38}
$$

in which $A(z)A^*(1/z^*)$ is a polynomial in z of order $2p$. The poles of Eq. (6.38), which are simply the zeros of $A(z)A^*(1/z^*)$, will occur in conjugate reciprocal pairs. For example, if z_k is a root of $A(z)$, then $(1/z_k)^* = 1/z_k^*$ will be a root of $A^*(1/z^*)$. If z_k is inside the unit circle, then $1/z_k^*$ will be outside the unit circle. If z_k is on the unit circle, then $1/z_k^*$ will also be on the unit circle. Furthermore, the roots of $A(z)$ will also come in complex conjugate pairs if the polynomial coefficients are all real.

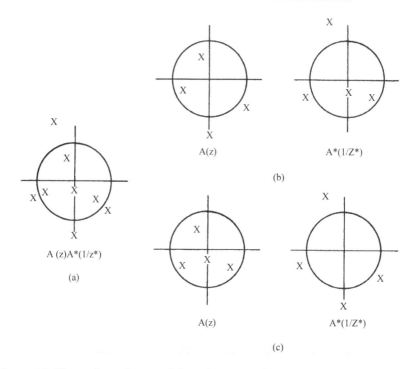

Figure 6.5. Illustration of spectral factorization and the stability issue. (a) Combined zeros of AR system function product. (b) Nonminimum phase factorization between $A(z)$ and $A^*(1/z^*)$. (c) Minimum phase factorization between $A(z)$ and $A^*(1/z^*)$.

Figure 6.5(a) illustrates a typical plot of the zero locations of the product polynomial $A(z)A^*(1/z^*)$. There are 2^P possible combinations of p zeros for $A(z)$ that would yield the identical polynomial $A(z)A^*(1/z^*)$. Figures 6.5(b) and (c) show two possible *spectral factorizations* of the zeros indicated in Fig. 6.5(a) for $A(z)A^*(1/z^*)$. A unique factorization can be imposed by requiring that the time-series model be *stable, causal,* and *minimum phase*, which was the assumption in Eq. (6.1). Such requirements (see Chap. 2) force $A(z)$ to have all of its roots *inside* the unit z-plane circle, as indicated in Fig. 6.5(c). The polynomial $A^*(1/z^*)$ will then have all of its roots *outside* the unit circle. The polynomial $A^*(1/z^*)$ is associated with the *stable anticausal* autoregressive process

$$x[n] = -\sum_{k=1}^{p} a^*[k]x[n+k] + e[n] \qquad (6.39)$$

defined for $n \leq 0$, rather than the causal AR process index range of $n \geq 0$.

The reader should not confuse the filter stability issue of spectral factorization with the statistical stability issue of PSD estimation. Filter stability is concerned with the need to choose AR or MA parameters that form valid autocorrelation sequences, as given by Eq. (6.29) or (6.31). Statistical stability is concerned with methods that reduce the spectral estimator variance for a given finite data record. Chapter 8 will present linear prediction estimation techniques that have good statistical stability, but which do not necessarily guarantee a minimum-phase estimate of $A(z)$. From the standpoint of a PSD estimate, it does not matter whether or not $A(z)$ is minimum phase. The PSD can be created from any $A(z)$ with arbitrary zero locations because the product $A(z)A^*(1/z^*)$ used in the PSD computation does not distinguish minimum phase $A(z)$ from nonminimum phase, as illustrated in Fig. 6.5. If a filter implementation using the estimated coefficients of $A(z)$ is required, then stability becomes an issue. In this case, a minimum-phase filter can be created by reflecting to the inside any zeros of $A(z)$ that are outside the unit circle. That is, one can simply form $A(z)A^*(1/z^*)$ and then decompose as a minimum-phase factorization.

References

Box, G. E. P., and G. M. Jenkins, *Time Series Analysis Forecasting and Control*, Holden-Day, Inc., San Francisco, 1970.

Friedlander, B., Lattice Algorithms for Factoring the Spectrum of a Moving Average Process, *IEEE Trans. Autom. Contr.*, vol. 18, pp. 1051–1055, November 1983.

Gutowski, P. R., E. A. Robinson, and S. Treitel, Spectral Estimation: Fact or Fiction?, *IEEE Trans. Geosci. Electron.*, vol. GE-16, pp. 80–84, April 1978.

Kay, S. M., *Modern Spectral Estimation*, Prentice-Hall, Inc., Englewood Cliffs, N. J., 1987.

Padé, H. E., Sur la representation approchée d'une function par des fractions rationnelles, *Ann. Sci. Ec. Norm. Sup.* (Paris), no. 3, vol. 9, pp. 1–98 (suppl.), 1892.

Weiss, L., and R. N. McDonough, Prony's Method, z-Transforms, and Padé Approximation, *SIAM Rev.*, vol. 5, pp. 145–149, April 1963.

7

AUTOREGRESSIVE PROCESS AND SPECTRUM PROPERTIES

7.1 INTRODUCTION

The autoregressive (AR) spectral estimator has received the most attention in the technical literature of the three linear-filter-based time-series models introduced in Chap. 6. The interest stems from two reasons. First, autoregressive spectra tend to have sharp peaks, a feature associated with high-resolution spectral estimators. Second, estimates of the AR parameters can be obtained as solutions to linear equations as was shown by Eq. (6.32), for example, that related the AR parameters and the autocorrelation sequence. Estimates of MA and ARMA parameters, however, require the solution of nonlinear equations, such as those of Eqs. (6.29) and (6.35).

The large body of material on autoregressive spectral estimation has necessitated a division of the presentation of the AR method into three chapters. Properties and features of the AR process and the AR spectrum are covered in this chapter under the assumption that the ACS is known, which is not generally the case in practice. These properties have led to alternative names for AR spectral analysis, such as the maximum entropy method (MEM) or linear prediction (LP) spectral estimation. Some of these properties have been utilized in the development of algorithms to estimate the autoregressive process parameters from data samples. These algorithms are described in the two sequel chapters. Chapter 8 describes AR parameter estimation algorithms and programs based on processing the data as a block. Chapter 9 describes AR parameter estimation algorithms based on processing the data sequentially one sample at a time.

7.2 SUMMARY

This chapter has been divided into two main sections: properties of autoregressive processes and properties of autoregressive spectra. These properties are developed under the assumption that the autocorrelation sequence is known. Table 7.1 summarizes the key AR process properties and the section where each property is discussed. A major result of these properties is the ability to represent the AR process as any one of three unique sequences, as summarized in Sec. 7.3.4.

Table 7.2 summarizes the key AR spectrum properties and the section where each property is discussed. The maximum entropy interpretation is appropriate

171

Table 7.1. Summary of Autoregressive Process Properties

Topic	Property	Section
Linear Prediction Filter	Whitens AR Process	3.7.1
Levinson Algorithm	Fast Computational Algorithm	3.7.2
Reflection Coefficients	AR Process as a Lattice Filter	3.7.3

Table 7.2. Summary of Autoregressive Power
Spectral Density Properties

Property	Section
Maximum Entropy Interpretation	7.4.1
Basis for High Resolution	7.4.2
Sinusoidal Power Estimation	7.4.3

when the process is Gaussian. If the process consists of sinusoids in noise, then the content of Sec. 7.4.3 will be of interest to the reader.

7.3 AUTOREGRESSIVE PROCESS PROPERTIES

7.3.1 Relationship to Linear Prediction Analysis

The equations of linear prediction analysis will be shown in this section to be structurally identical to the Yule-Walker equations of an autoregressive process. A close relationship, therefore, exists between a linear prediction filter and an AR process. This relationship is exploited by several of the algorithms presented in Chaps. 8 and 9.

Consider the *forward* linear prediction estimate

$$\hat{x}^f[n] = -\sum_{k=1}^{m} a^f[k]x[n-k] = -\mathbf{x}^T[n-1]\mathbf{a}^f \tag{7.1}$$

of the data sequence at sample $x[n]$, in which the vectors are

$$\mathbf{x}[n-1] = \begin{pmatrix} x[n-1] \\ \vdots \\ x[n-p] \end{pmatrix} \qquad \mathbf{a}^f = \begin{pmatrix} a^f[1] \\ \vdots \\ a^f[m] \end{pmatrix}$$

and $a^f[k]$ designates the forward linear prediction coefficient at time delay index k. The hat ˆ is used to denote an estimate and the superscript f is used to denote that this is a forward estimate. The prediction is forward in the sense that the estimate at time index n is based on m samples indexed earlier in time. The complex forward linear prediction error

$$e^f[n] = x[n] - \hat{x}^f[n] \tag{7.2}$$

has a real-valued, positive variance

$$\rho^f = \mathcal{E}\{|e^f[n]|^2\}. \tag{7.3}$$

Substituting Eqs. (7.1) and (7.2) into the above expression will yield

$$\rho^f = r_{xx}[0] + \sum_{k=1}^{m} a^f[k]r_{xx}[-k] + \sum_{j=1}^{m}(a^f[j])^*r_{xx}[j] \tag{7.4}$$

$$+ \sum_{k=1}^{m}\sum_{j=1}^{m}a^f[k](a^f[j])^*r_{xx}[j-k]$$

$$= r_{xx}[0] + \mathbf{r}_m^H\mathbf{a}^f + (\mathbf{a}^f)^H\mathbf{r}_m + (\mathbf{a}^f)^H\mathbf{R}_{m-1}\mathbf{a}^f$$

in which the vector $\mathbf{r}_m$ and matrix $\mathbf{R}_{m-1}$ are

$$\mathbf{r}_m = \begin{pmatrix} r_{xx}[1] \\ \vdots \\ r_{xx}[m] \end{pmatrix} \qquad \mathbf{R}_{m-1} = \begin{pmatrix} r_{xx}[0] & \cdots & r_{xx}^*[m-1] \\ \vdots & & \vdots \\ r_{xx}[m-1] & \cdots & r_{xx}[0] \end{pmatrix}.$$

The assumption that $x[n]$ is a wide-sense stationary process has been made, which means $r_{xx}[-k] = r_{xx}^*[k]$. This expression is identical in form to the quadratic equation (3.114). Therefore, the forward linear prediction coefficient vector $\mathbf{a}^f$ that minimizes the variance ρ^f is found as the solution to the normal equations given by

$$\begin{pmatrix} r_{xx}[0] & \mathbf{r}_m^H \\ \mathbf{r}_m & \mathbf{R}_{m-1} \end{pmatrix}\begin{pmatrix} 1 \\ \mathbf{a}^f \end{pmatrix} = \begin{pmatrix} \rho^f \\ \mathbf{0}_m \end{pmatrix}, \tag{7.5}$$

which follows directly from Eq. (3.115). It also follows from (7.5) that

$$\rho^f = \begin{pmatrix} 1 & (\mathbf{a}^f)^H \end{pmatrix}\mathbf{R}_m\begin{pmatrix} 1 \\ \mathbf{a}^f \end{pmatrix}.$$

The Hermitian Toeplitz structure of normal equations (7.5) becomes apparent if rewritten as

$$\begin{pmatrix} r_{xx}[0] & r_{xx}^*[1] & \cdots & r_{xx}^*[m] \\ r_{xx}[1] & r_{xx}[0] & \cdots & r_{xx}^*[m-1] \\ \vdots & \vdots & \ddots & \vdots \\ r_{xx}[m] & r_{xx}[m-1] & \cdots & r_{xx}[0] \end{pmatrix}\begin{pmatrix} 1 \\ a^f[1] \\ \vdots \\ a^f[m] \end{pmatrix} = \begin{pmatrix} \rho^f \\ 0 \\ \vdots \\ 0 \end{pmatrix} \tag{7.6}$$

or

$$\left(1 \; a^f[1] \; \cdots \; a^f[m]\right) \begin{pmatrix} r_{xx}[0] & r_{xx}[1] & \cdots & r_{xx}[m] \\ r_{xx}^*[1] & r_{xx}[0] & \cdots & r_{xx}[m-1] \\ \vdots & \vdots & \ddots & \vdots \\ r_{xx}^*[m] & r_{xx}^*[m-1] & \cdots & r_{xx}[0] \end{pmatrix} = \left(\rho^f \; 0 \; \cdots \; 0\right).$$

This matrix expression is structurally identical to the Yule-Walker equations (6.32) of an autoregressive process. If one rearranges Eq. (7.2) as

$$x[n] = -\sum_{k=1}^{m} a^f[k]x[n-k] + e^f[n], \tag{7.7}$$

then the similarity with Eq. (6.13), which describes an autoregressive process, may be seen. There are, however, two distinctions between a forward linear prediction process of order m and an AR(m) process. The sequence $u[n]$ of Eq. (6.13) is a *white noise* process used as the *input* to an autoregressive filter. The sequence $x[n]$ is the *output* of the autoregressive filter. The error sequence $e^f[n]$ of Eq. (7.7) is the *output* of a forward linear prediction error filter, as depicted in Fig. 7.1. The sequence $x[n]$ is the *input* to the prediction error filter. The linear prediction error sequence will be uncorrelated with the linear prediction estimate $\hat{x}^f[n]$, but it is not, in general, a white process *unless $x[n]$ was generated as an AR(p) process and $m = p$*. The error sequence in this case will be a white process, the forward linear prediction coefficients will be identical to the AR parameters ($a^f[k] = a[k]$) and, therefore, the linear prediction error filter may be regarded as a whitening filter for an AR process [Papoulis, 1981].

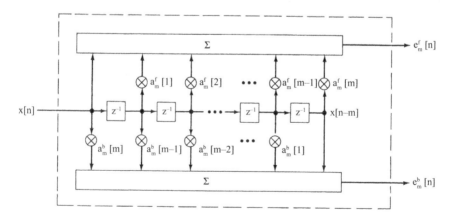

Figure 7.1. Linear prediction error filter of order m implemented using a transversal filter structure. If $x[n]$ is an AR(m) process, then $e^f[n]$ and $e^b[n]$ are white processes of variance $\rho_w = \rho^f = \rho^b$. The boxes with z^{-1} denote a delay of one sample.

A *backward* linear prediction error estimate

$$\hat{x}^b[n] = -\sum_{k=1}^{m} a^b[k]x[n+k] = -\mathbf{x}^T[n-1]\mathbf{Ja}^b \qquad (7.8)$$

may also be formed, in which $\mathbf{a}^{bT} = (a^b[1]\cdots a^b[m])$ and $a^b[k]$ designates the backward linear prediction coefficient at time delay index k. A superscript b is used to tag elements associated with the backward linear prediction estimate. The prediction is backward in the sense that the estimate at time index n is based on m samples indexed future in time. The backward linear prediction error is

$$e^b[n] = x[n-m] - \hat{x}^b[n-m]. \qquad (7.9)$$

The backward error has been intentionally indexed for notational convenience with respect to n, rather than $n - m$, so that $e^f[n]$ and $e^b[n]$ will be functions of the same set of data samples (namely, $x[n], x[n-1], \ldots, x[n-m]$) that are present in the linear prediction filter, as may be seen by examining Fig. 7.1. The complex backward linear prediction error has a real-valued, positive variance

$$\rho^b = \mathcal{E}\{|e^b[n]|^2\}. \qquad (7.10)$$

Substituting Eqs. (7.8) and (7.9) into (7.10) will yield

$$\rho^b = r_{xx}[0] + \sum_{k=1}^{m} a^b[k]r_{xx}[k] + \sum_{j=1}^{m}(a^b[j])^* r_{xx}[-j] \qquad (7.11)$$

$$+ \sum_{k=1}^{m}\sum_{j=1}^{m} a^b[k](a^b[j])^* r_{xx}[k-j]$$

$$= r_{xx}[0] + \mathbf{r}_m^T \mathbf{a}^b + (\mathbf{a}^b)^H \mathbf{r}_m^* + (\mathbf{a}^b)^H \mathbf{R}_{m-1}^* \mathbf{a}^b$$

where $\mathbf{r}_m$ and $\mathbf{R}_{m-1}$ have been previously defined. Based on Eq. (3.115), the backward linear prediction coefficient vector $\mathbf{a}^b$ that minimizes the variance ρ^b is found as the solution of the normal equations given by

$$\begin{pmatrix} r_{xx}[0] & \mathbf{r}_m^T \\ \mathbf{r}_m^* & \mathbf{R}_{m-1}^* \end{pmatrix} \begin{pmatrix} 1 \\ \mathbf{a}^b \end{pmatrix} = \begin{pmatrix} \rho^b \\ \mathbf{0}_m \end{pmatrix}.$$

Rewriting the normal equations to explicitly show the structure produces

$$\begin{pmatrix} r_{xx}[0] & r_{xx}[1] & \cdots & r_{xx}[m] \\ r_{xx}^*[1] & r_{xx}[0] & \cdots & r_{xx}[m-1] \\ \vdots & \vdots & \ddots & \vdots \\ r_{xx}^*[m] & r_{xx}^*[m-1] & \cdots & r_{xx}[0] \end{pmatrix} \begin{pmatrix} 1 \\ a^b[1] \\ \vdots \\ a^b[m] \end{pmatrix} = \begin{pmatrix} \rho^b \\ 0 \\ \vdots \\ 0 \end{pmatrix}. \qquad (7.12)$$

By conjugating both sides of Eq. (7.12), the following relationship will hold

$$\begin{pmatrix} r_{xx}[0] & r_{xx}^*[1] & \cdots & r_{xx}^*[m] \\ r_{xx}[1] & r_{xx}[0] & \cdots & r_{xx}^*[m-1] \\ \vdots & \vdots & \ddots & \vdots \\ r_{xx}[m] & r_{xx}[m-1] & \cdots & r_{xx}[0] \end{pmatrix} \begin{pmatrix} 1 \\ a^{b*}[1] \\ \vdots \\ a^{b*}[m] \end{pmatrix} = \begin{pmatrix} \rho^b \\ 0 \\ \vdots \\ 0 \end{pmatrix} \qquad (7.13)$$

in which the property $\rho^b = \rho^{b*}$ has been used (variance is real-valued). It can now be seen that Eqs. (7.6) and (7.13) both involve the identical Hermitian Toeplitz autocorrelation matrix and it may therefore be concluded that

$$\rho^b = \rho^f \qquad (7.14)$$

and

$$a^b[k] = a^{f*}[k] \qquad (7.15)$$

for $1 \le k \le m$. Thus, the backward and forward linear prediction error variances are identical. The backward linear prediction coefficients are simply the complex conjugates of the forward linear prediction coefficients when both prediction filters have the same length m. If the forward linear prediction error filter is a whitening filter of an AR process, then the backward linear prediction error filter is a whitening filter for the anticausal realization of the AR process.

7.3.2 Levinson Algorithm

The solution of the Hermitian Toeplitz equations (7.6) and (7.13) can be computed with Levinson's algorithm, as described in Sec. 3.8.1. If the associations $t[k] = r_{xx}[k]$, $a_m[k] = a_m^f[k] = a_m^{b*}[k]$, and $\rho_m = \rho_m^f = \rho_m^b$ are made with the parameters of Sec. 3.8.1, then the solution for a linear prediction filter of M coefficients is given by the recursion

$$\Delta_m = \begin{cases} r_{xx}[1] & \text{if } m = 1 \\ \displaystyle\sum_{k=1}^{m-1} a_{m-1}^f[k]r_{xx}[m-k] + r_{xx}[m] & \text{if } m > 1 \end{cases} \qquad (7.16)$$

$$a_m^f[k] = \begin{cases} -\Delta_m/\rho_{m-1}^f & \text{for } k = m \\ a_{m-1}^f[k] + a_m^f[m](a_{m-1}^f[m-k])^* & \text{for } 1 \le k \le m-1 \\ & \text{and } m > 1 \end{cases} \qquad (7.17)$$

$$\rho_m^f = \rho_{m-1}^f\left(1 - |a_m^f[m]|^2\right) \qquad (7.18)$$

for $m = 1$ to $m = M$. The single initial condition required is

$$\rho_0^f = r_{xx}[0]. \qquad (7.19)$$

The backward linear prediction coefficients are obtained simply by conjugating the forward linear prediction coefficients. A subscript m or $m - 1$ has been added to the elements to distinguish the order of the linear prediction filter. Observe that all linear prediction coefficients from order 1 to order M that minimize the prediction error variance are obtained automatically with the Levinson algorithm. It is possible, then, to produce all lower-order linear prediction filters with no additional computational effort. A MATLAB function `levinson_recursion.m` was previously discussed in Sec. (3.8.3).

If the Hermitian Toeplitz matrix of Eq. (7.6) is nonsingular and positive definite, then the polynomial

$$A_m^f(z) = 1 + \sum_{k=1}^{m} a_m^f[k]z^{-k} \qquad (7.20)$$

formed from the linear prediction coefficients of the Levinson algorithm solution will have its roots inside the unit circle [Markel and Gray, 1976]. This will yield a *minimum-phase* linear prediction error filter. A variety of proofs have been developed to prove this property [Kay, 1987; Pakula and Kay, 1983; Lang and McClellan, 1979; Burg, 1975]. The typical proof demonstrates that the variance ρ^f may be further reduced by reflecting any root outside the unit circle to the inside. The minimum-phase property ensures that the solution of the linear prediction equations (7.12) yields a stable linear prediction error filter $A(z)$, and eventually a stable autoregression filter $1/A(z)$. A stable filter means the ACS recursion of Eq. (6.31) will generate autocorrelation sequence values that remain finite and bounded in magnitude by the value of $r_{xx}[0]$, as required for a stationary process.

If the $x[n]$ process is actually an AR(p) process, then the linear prediction error filter parameters given as the solution of (7.6) for known ACS by the Levinson algorithm at order $m = p$ yields

$$a_p[k] = a_p^f[k] = a_p^{b*}[k] \qquad \text{for } 1 \leq k \leq p$$

$$\rho_p = \rho_p^f = \rho_p^b, \qquad (7.21)$$

where $a_p[k]$ are the AR parameters and ρ_p is the driving white noise variance. Because the AR process ACS must satisfy Eq. (6.35), any attempt to fit a linear prediction error filter to match the ACS for an order $k > p$ will find that $a_p^f[k] = 0$ for $k > m$ and the variance remains unchanged at a value of ρ_p^f. This may be used as a basis for determination of the order of an unknown AR process for which only the ACS is known. The Levinson algorithm is used to generate successively higher-order linear prediction error filters until the variance reaches a point at which it does not change, which is an indicator of the correct autoregressive model order.

It will be shown in the next section that $|a_p[p]| \leq 1$. The prediction error variance ρ_p^f, as given by Eq. (7.18), must therefore be a monotonically decreasing function, reaching a minimum at $k < p$ if the input process to the linear prediction error filter is an AR(p) process. This will also be a useful property for model order selection, as discussed in Chap. 8. In practice where the ACS must be estimated,

the order selected for an unknown AR process is the one in which the variance
reaches a point where it changes very little from order to order.

7.3.3 Reflection Coefficients

The linear prediction coefficients $a_1^f[1], \ldots, a_p^f[p]$ are often termed the *reflection coefficients* due to their interpretation as physical parameters of an acoustical
tube model of speech [Makhoul, 1975, 1981] or in seismic layered earth models
[Robinson and Treitel, 1980]. A special symbol $k_m = a_m^f[m]$ for $1 \leq m \leq p$ is typically applied to distinguish these particular linear prediction coefficients from the
remaining coefficients. The negated reflection coefficient $-k_m$ has been called a
partial correlation (PARCOR) in the statistical literature [Box and Jenkins, 1970]
because it represents the normalized correlation between $x[n]$ and $x[n-m]$ with
the correlative components of $x[n-1], \ldots, x[n-m+1]$ removed. The reflection
coefficients also have an interpretation as the negative normalized correlation coefficient between the forward and backward linear prediction errors with one unit of
delay

$$k_m = \frac{-\mathcal{E}\{e_{m-1}^f[n]e_{m-1}^{b*}[n-1]\}}{\sqrt{\mathcal{E}\{|e_{m-1}^f[n]|^2\}}\sqrt{\mathcal{E}\{|e_{m-1}^b[n-1]|^2\}}} . \tag{7.22}$$

This may be derived by using definition equations (7.2), (7.4), and (7.9), yielding

$$\mathcal{E}\{e_{m-1}^f[n]e_{m-1}^{b*}[n-1]\}$$

$$= \mathcal{E}\left\{\left(x[n]+\sum_{k=1}^{m-1}a_{m-1}^f[k]x[n-k]\right)\left(x^*[n-m]+\sum_{k=1}^{m-1}a_{m-1}^{b*}[k]x^*[n-m+k]\right)\right\}$$

$$= r_{xx}[m]+\sum_{k=1}^{m-1}a_{m-1}^f[k]r_{xx}[m-k] = \Delta_m, \tag{7.23}$$

and noting that

$$\rho_{m-1}^f = \rho_{m-1}^b = \mathcal{E}\{|e_{m-1}^f[n]|^2\} = \mathcal{E}\{|e_{m-1}^b[n]|^2\} = \mathcal{E}\{|e_{m-1}^b[n-1]|^2\}, \tag{7.24}$$

which is a consequence of the stationary property and the Hermitian Toeplitz property of the autocorrelation matrix. Substituting Eqs. (7.23) and (7.24) into the
definition

$$k_m = a_m[m] = -\frac{\Delta_m}{\rho_{m-1}^f} = -\frac{\Delta_m}{\sqrt{\rho_{m-1}^f}\sqrt{\rho_{m-1}^b}} \tag{7.25}$$

will yield Eq. (7.22). It is straightforward to show, using Eq. (7.22), that $|k_m| \leq 1$
for $1 \leq m \leq p$. Having k_m bounded by unit magnitude is a necessary and sufficient
condition for the autocorrelation matrix to be positive semidefinite.

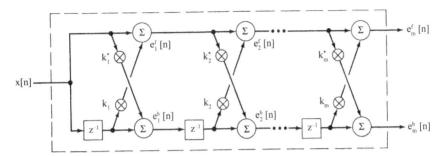

Figure 7.2. Linear prediction error filter implemented using a lattice filter structure. The boxes with z^{-1} denote a delay of one sample.

If the recursion of (7.17) is substituted for $a_m^f[k] = a_m^{b*}[k]$ in definitions (7.2) and (7.9) for the forward and backward linear prediction errors, then it is simple to show that

$$e_m^f[n] = e_{m-1}^f[n] + \mathsf{k}_m e_{m-1}^b[n-1] \qquad (7.26)$$

$$e_m^b[n] = e_{m-1}^b[n-1] + \mathsf{k}_m^* e_{m-1}^f[n] \qquad (7.27)$$

with initial conditions $x[n] = e_0^f[n] = e_0^b[n]$. Relationships (7.26) and (7.27) yield an alternative interpretation of the prediction error filter as a lattice structure, depicted in Fig. 7.2. The parameters at each stage of the lattice filter are the reflection coefficients. Note that both the forward and backward linear prediction errors are propagated simultaneously in the lattice filter. The lattice filter has been found to offer implementation advantages over tapped delay line filters such as Fig. 7.1 because they suffer from less round-off noise and less sensitivity to coefficient value perturbations [Friedlander, 1982]. The lattice filter is also an orthogonalizing filter because the backward prediction errors (but not the forward prediction errors) at each stage are mutually orthogonal [Makhoul, 1978].

7.3.4 Summary of Equivalent Representations for Autoregressive Processes

An AR(p) process has three equivalent representations: as an infinite-duration autocorrelation sequence, as a finite sequence of autoregressive parameters, or as a finite sequence of reflection coefficients. These equivalences are depicted by the triangle diagram of Fig. 7.3. Recursive mappings exist to transform from one representation into either of the other representations. Although the ACS for an AR(p) process is infinite in duration, the entire ACS is determined uniquely from the finite autocorrelation sequence $r_{xx}[0], \ldots, r_{xx}[p]$. The remainder of the ACS can be obtained from recursion Eq. (6.34). The Levinson algorithm of Sec. 7.3.2 provides both AR parameters and reflection coefficients, given the finite ACS from lag 0 to lag p. The use of Eq. (7.17) and the initial condition $a_1[1] = \mathsf{k}_1$ forms a simple recursion that

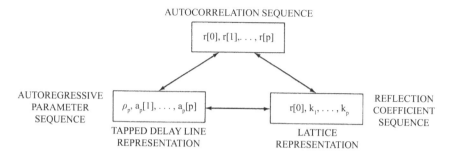

Figure 7.3. The three equivalent representations of an autoregressive process in terms of the ACS sequence, the AR parameter sequence, or the reflection coefficient sequence.

yields the AR parameters of all orders from $m = 1$ to $m = p$, based on knowledge of just the reflection coefficient sequence k_1 to k_p. This forms a mapping from the reflection coefficients to the autoregressive parameters. The recursions of the Levinson algorithm may also be run in reverse to provide descending-order recursions from order m to order $m - 1$, using a procedure called a *step-down Levinson recursion*. This enables one to compute the autocorrelation sequence recursively from lag p down to lag 0 using either the AR parameter set $\{\rho_p, a_p[1], \ldots, a_p[p]\}$ or the reflection coefficient set $\{\rho_p, k_1, \ldots, k_p\}$. A mapping from the AR parameters to the reflection coefficients is also possible. The following three equivalent statements represent the required conditions for the mappings depicted in Fig. 7.3 to be feasible:

- $r_{xx}[0], \ldots, r_{xx}[p]$ is a positive-definite sequence.
- $A(z) = 1 + \sum_{m=1}^{p} a_p[m] z^{-m}$ is minimum phase and $\rho_p > 0$.
- $|k_m| < 1$ for $1 \leq m \leq p$ and $\rho_p > 0$.

7.4 AUTOREGRESSIVE POWER SPECTRAL DENSITY PROPERTIES

The AR power spectral density function Eq. (6.14) was shown in Chap. 6 to have the form

$$P_{\mathrm{AR}}(f) = \frac{T\rho_w}{|A(f)|^2}, \quad A(f) = 1 + \sum_{m=1}^{p} a_p[m] \exp(-j2\pi f m T)$$

when expressed in terms of the AR parameter sequence used in the tapped delay line filter of Fig. 7.1. It can be shown that an equivalent representation of the AR

PSD, in terms of the reflection coefficient sequence of the lattice filter of Fig. 7.2, is

$$P_{AR}(f) = \frac{T\rho_w}{|E^f(f)|^2} = \frac{T\rho_w}{|E^b(f)|^2}$$

in which the forward prediction error system function $E_p^f(f)$ and backward prediction error system function $E_p^b(f)$ for an AR(p) process are determined through the expression

$$\begin{pmatrix} E_p^f(f) \\ E_p^b(f) \end{pmatrix} = \prod_{m=1}^{p} \begin{pmatrix} 1 & k_m \exp(-j2\pi f m T) \\ k_m^* & \exp(-j2\pi f m T) \end{pmatrix} \begin{pmatrix} 1 \\ 1 \end{pmatrix}. \tag{7.28}$$

Two interesting relationships exist between the AR parameters and the reflection coefficients of real-valued processes

$$1 + \sum_{m=1}^{p} a_p[m] = \prod_{m=1}^{p} \left(1 + k_m\right) > 0$$

$$1 + \sum_{m=1}^{p} (-1)^m a_p[m] = \prod_{m=1}^{p} \left(1 + (-1)^m k_m\right) > 0$$

which can be derived from Eq. (7.28) [Chan, 1990].

7.4.1 Maximum Entropy Interpretation

The estimate of the autoregressive power spectral density has frequently been termed the *maximum entropy method* (MEM) of spectral analysis in the literature. This interpretation was introduced by Burg [1975]. If only the finite autocorrelation sequence $r_{xx}[0], \ldots, r_{xx}[p]$ is assumed known, then a question may be posed as to how the remaining unknown lags $r_{xx}[p+1], r_{xx}[p+2], \ldots$ should be specified in order to guarantee that the entire autocorrelation sequence is positive semidefinite. There are an infinite number of possible extrapolations that will yield valid autocorrelation sequences. Burg argued that the extrapolation should be made in such a way as to maximize the entropy of the time series characterized by the extrapolated ACS. This time series would then be the most random, in an entropy sense, of all the series that match the known ACS for lags 0 to p. A spectral estimate produced from this extrapolated ACS would correspond to the process that had maximum entropy. In particular, the entropy rate for a Gaussian random process was shown in Sec. 4.5 to be proportional to

$$\int_{-1/2T}^{1/2T} \ln P_{MEM}(f)\, df,$$

where $P_{MEM}(f)$ is the PSD of the process whose entropy rate is to be maximized. $P_{MEM}(f)$ is found by maximizing the above expression, subject to the constraints that it satisfy the inverse Wiener-Khintchine relationship (4.34) for the $p + 1$ known autocorrelation sequence values

$$\int_{-1/2T}^{1/2T} P_{\text{MEM}}(f) \exp(j2\pi f m T) \, df = r_{xx}[m] \tag{7.29}$$

over $0 \leq m \leq p$. The solution, found by the Lagrange multiplier technique (see, for example, Haykin and Kesler [1983]), is

$$P_{\text{MEM}}(f) = \frac{T \rho_w}{\left| 1 + \sum_{m=1}^{p} a_p[m] \exp(-j2\pi f m T) \right|^2}, \tag{7.30}$$

for which $\rho_w, a_p[1], \ldots, a_p[p]$ are found via the solution of the Yule-Walker equations. Thus, the MEM and AR spectral estimators are identical for *Gaussian random processes* of *known* autocorrelation sequence of uniform spacing from lag 0 to lag p. It is incorrect to say that PSD estimators of the form of Eq. (7.30) are maximum entropy spectral estimators (1) if the process is not Gaussian or (2) if the ACS terms are not known but must be estimated from the data. An MEM spectral analysis can also be performed over a nonuniform spacing of the ACS. In this case, the AR and MEM power spectral densities will not be identical. Other maximum entropy interpretations may be found in Jaynes [1982], Van Den Bos [1971], and Nikias and Raghuveer [1985].

7.4.2   Basis for High Resolution

The autoregressive PSD estimator, as expressed by Eqs. (6.14) and (6.33), was shown to have the two equivalent representations [DuBroff, 1975]

$$P_{\text{AR}}(f) = \frac{T \rho_w}{\left| 1 + \sum_{k=1}^{p} a[p] \exp(-j2\pi f k T) \right|^2} \tag{7.31}$$

$$= T \sum_{k=-\infty}^{\infty} r_{xx}[k] \exp(-j2\pi f k T). \tag{7.32}$$

Given the autocorrelation sequence for lags 0 to p, one may solve the Yule-Walker equations for the white noise variance ρ_w and the autoregressive parameters $a[1], \ldots, a[p]$ and then use Eq. (7.33) to evaluate the AR PSD. Alternatively, the extrapolation

$$r_{xx}[n] = -\sum_{k=1}^{p} a[k] r_{xx}[n-k] \tag{7.33}$$

for $n > p$, which follows from Eq. (6.34), may be used to extend the autocorrelation sequence so that Eq. (7.34) may be used to compute the AR PSD (although this would not be very practical). As a contrast, consider the classical Blackman-Tukey correlogram-based method of power spectral density estimation presented in

Chap. 5. This method would use the given $p + 1$ autocorrelation sequence values to form the spectral estimator

$$P_{BT}(f) = T \sum_{k=-p}^{p} r_{xx}[k] \exp(-j2\pi f kT). \qquad (7.34)$$

The unavailable ACS terms are tacitly assumed to be zero. Comparing Eqs. (7.34) and (7.36), it may be seen that the AR PSD estimator matches the autocorrelation sequence over the summation range of the Blackman-Tukey PSD estimator up to lag p, but then uses a nonzero ACS extrapolation given by Eq. (7.35), rather than the zero extension assumed by the Blackman-Tukey method, to generate the unavailable lags. This is depicted in Fig. 7.4. Even though Eq. (7.33) is used to compute the AR PSD, this is equivalent to computing (7.34) with the implied ACS extrapolation of (7.35). This implied extrapolation of the autocorrelation sequence when computing the AR PSD is the basis for the high-resolution property attributed to the AR PSD estimator. There is no windowing of the autocorrelation sequence as there is with

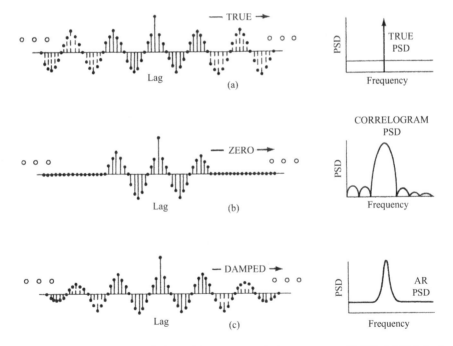

Figure 7.4. Implied extrapolation of the autocorrelation sequence (ACS). (a) Original infinite ACS and true spectrum of a single real sinusoid in white noise. (b) Zero ACS extrapolation implied by the Blackman-Tukey PSD estimator and associated spectral estimate. (c) Autoregressive ACS extrapolation and associated AR spectral estimate.

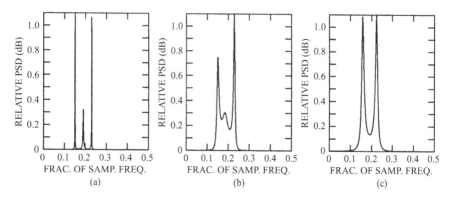

Figure 7.5. Autoregressive PSD estimates based on nine ACS terms for three equal-power sinusoids in white noise. The sinusoids have fractional sampling frequencies of 0.15, 0.19, and 0.23. (a) SNR of 30 dB. (b) SNR of 20 dB. (c) SNR of 10 dB.

the classic spectral estimators. Autoregressive power spectral density estimators, for this reason, do not possess the sidelobe artifacts of classic spectral estimators.

The degree of improved resolution of the AR PSD estimator for processes consisting of sinusoids in white noise is, however, a function of signal-to-noise ratio (SNR). Consider the AR(8) PSD estimates for three equal-power sinusoids in white noise shown in Fig. 7.5. The estimates were produced using lags 0 through 8 of the exact autocorrelation sequence for sinusoids in noise [Eq. (4.53)]. As the SNR decreases from 30 dB to 10 dB, the number of resolved sinusoids decreases from three to two; the frequency estimates given by the peak locations are also biased when the SNR is 10 dB. A measure of the sinusoidal resolution of equal-power sinusoids has been analytically determined in the case of known ACS [Marple, 1982] to be roughly

$$F = \frac{1.03}{Tp[\text{SNR}(p+1)]^{0.31}},\qquad(7.35)$$

where F is the resolution in Hz, T is the sample interval in sec, p is the highest lag of the autocorrelation sequence, and SNR is the signal-to-noise ratio of a single sinusoid expressed in linear (rather than dB) units. Note also the variance estimates of Baggeroer [1976].

7.4.3 Sinusoidal Power Estimation

Autoregressive PSD estimation is often used to detect the presence of sinusoidal components in the data. The *power* associated with components in the any PSD estimate (including the AR PSD) is obtainable only by *integrating* the area under the *power spectral density* (power spectral density × frequency = power). This may be computationally expensive. As an alternative, it is tempting to use the height of each spectral peak in an AR PSD estimate as an indicator of relative sinusoidal power levels. The PSD height was found to be a reasonable indicator of relative power when classical spectral estimators are used because their peaks are linearly related to the sinusoidal power when the input consists of sinusoids in white noise. This unfortunately will *not* work with AR spectral estimators. Consider the autocorrelation sequence for a single complex sinusoid in white noise

$$r_{xx}[k] = P \exp(j2\pi f_0 kT) + \rho_w \delta[k], \qquad (7.36)$$

as given by Eq. (4.53). Lacoss [1971] has used Eq. (7.38) to prove analytically that the mth-order autoregressive PSD estimator for a single sinusoid in white noise process can be expressed as

$$P_{AR}(f) = \frac{T\rho_w[1 + p/(mP + \rho_w)]}{\left|1 - \frac{P}{mP+\rho_w} \sum_{k=1}^{m} \exp(-j2\pi k[f - f_0]T)\right|^2}. \qquad (7.37)$$

The function $P_{AR}(f)$ reaches its maximum at $f = f_0$,

$$P_{AR}(f_0) = \rho_w T[1 + m\,\text{SNR}][1 + (m+1)\,\text{SNR}] \approx T(m\,\text{SNR})^2, \qquad (7.38)$$

where $\text{SNR} = P/\rho_w$ and the assumption $m \cdot \text{SNR} >> 1$ (high SNR) has been made. Thus, the peak is proportional to the *square* of the power; the *area* under the peak is, however, *linearly* proportional to the power.

One method to estimate the power of multiple real sinusoids from the peaks in an AR spectrum was suggested by Johnsen and Andersen [1978]. The method generally works only for well-separated peaks of high-SNR sinusoidal components in the data. If the AR PSD is expressed in z-transform notation

$$\mathsf{P}_{AR}(z) = \frac{T\rho_w}{\mathsf{A}(z)\mathsf{A}^*(1/z^*)}, \qquad (7.39)$$

then a peak observed at f_k in the AR spectrum will have an estimated power of

$$\text{Power}(f_k) = 2\,\text{Re}\left\{\text{Residue of } \frac{\mathsf{P}_{AR}(z)}{z} \text{ evaluated at } z_k = \exp(j2\pi f_k T)\right\}. \qquad (7.40)$$

The residue is defined by Eq. (2.20). Negative power estimates can occur with this technique if peaks are very close together. The power estimation technique of the Prony method (Chap. 11) is probably a more reliable estimate if the data is known to consist solely of sinusoids in noise.

References

Baggeroer, A. B., Confidence Intervals for Regression (MEM) Spectral Estimates, *IEEE Trans. Inf. Thy.*, vol. IT-22, pp. 534–545, September 1976.

Box, G. E. P., and G. M. Jenkins, *Time Series Analysis Forecasting and Control*, Holden-Day, Inc., San Francisco, 1970.

Burg, J. P., Maximum Entropy Spectral Analysis, Ph.D. dissertation, Department of Geophysics, Stanford University, Stanford, Calif., 1975.

Chan, J.-K., Two Autoregressive Identities, *Signal Processing*, vol. 19, pp. 337–342, April 1990.

DuBroff, R. E., The Effective Autocorrelation Function of Maximum Entropy Spectra, *Proc. IEEE*, vol. 63, pp. 1622–1623, November 1975.

Friedlander, B., Lattice Filters for Adaptive Processing, *Proc. IEEE*, vol. 70, pp. 829–867, August 1982.

Haykin, S. S., and S. Kesler, Prediction-Error Filtering and Maximum-Entropy Spectral Estimation, Chapter 2 in *Nonlinear Methods of Spectral Analysis*, 2nd ed., S. Haykin, ed., Springer-Verlag, New York, 1983.

Jaynes, E. T., On the Rationale of Maximum-Entropy Method, *Proc. IEEE*, vol. 70, pp. 939–952, September 1982.

Johnsen, S. J., and N. Andersen, On Power Estimation in Maximum Entropy Spectral Analysis, *Geophysics*, vol. 43, pp. 681–690, June 1978.

Kay, S. M., *Modern Spectral Estimation*, Prentice-Hall, Inc., Englewood Cliffs, N. J., 1987.

Lacoss, R. T., Data Adaptive Spectral Analysis Methods, *Geophysics*, vol. 36, pp. 661–675, August 1971.

Lang, S. W., and J. H. McClellan, A Simple Proof of Stability for All-Pole Linear Prediction Models, *Proc. IEEE*, vol. 67, pp. 860–861, May 1979.

Makhoul, J., Linear Prediction: A Tutorial Review, *Proc. IEEE*, vol. 63, pp. 561–580, April 1975; corrections in vol. 64, p. 285, February 1976.

Makhoul, J., A Class of All-Zero Lattice Digital Filters: Properties and Applications, *IEEE Trans. Acoust. Speech Signal Process.*, vol. ASSP-26, pp. 304–314, August 1978.

Makhoul, J., Lattice Methods in Spectral Estimation, in *Applied Time Series Analysis II*, D. F. Findley, ed., Academic Press, Inc., New York, pp. 301–324, 1981.

Markel, J. D., and A. H. Gray, *Linear Prediction of Speech*, Springer-Verlag, New York, 1976.

Marple, S. L., Jr., Frequency Resolution of Fourier and Maximum Entropy Spectral Estimates, *Geophysics*, vol. 47, pp. 1303–1307, September 1982.

Nikias, C. L., and M. Raghuveer, Discussion on Higher Order Autospectra by MEM, *Geophysics*, vol. 50, pp. 165–166, January 1985.

Pakula, L., and S. M. Kay, Simple Proofs of the Minimum Phase Property of the Prediction Error Filter, *IEEE Trans. Acoust. Speech Signal Process.*, vol. ASSP-31, p. 501, April 1983.

Papoulis, A., Maximum Entropy and Spectral Estimation: A Review, *IEEE Trans. Acoust. Speech Signal Process.*, vol. ASSP-29, pp. 1176–1186, December 1981.

Robinson, E. A., and S. Treitel, *Geophysical Signal Analysis*, Prentice-Hall, Inc., Englewood Cliffs, N. J., 1980.

Satorius, E. H., and J. Zeidler, Maximum Entropy Spectral Analysis of Multiple Sinusoids in Noise, *Geophysics*, vol. 43, pp. 1111–1118, October 1978.

Van Den Bos, A., Alternative Interpretation of Maximum Entropy Spectral Analysis, *IEEE Trans. Inf. Theory*, vol. IT–17, pp. 493–494, July 1971.

8

AUTOREGRESSIVE SPECTRAL ESTIMATION: BLOCK DATA ALGORITHMS

8.1 INTRODUCTION

Basic properties and relationships of the autoregressive (AR) model and its associated PSD function have been presented in Chaps. 6 and 7. The underlying assumption in these chapters was the availability of the exact autocorrelation function of the random process. In practice, though, the autocorrelation is usually not available, so one must make an AR spectral estimate based on the available data. Both this chapter and the next chapter are concerned with algorithmic techniques for producing AR spectral estimates from data samples. These techniques actually make estimates of the AR model parameters and, from these, the AR PSD function may be evaluated.

The techniques fall into two categories: algorithms for *block* data and algorithms for *sequential* data. This chapter presents block data methods, that is, algorithms that process an entire fixed-size block of accumulated data samples. Chapter 9 presents sequential data methods, that is, algorithms that process the data sequentially as each sample becomes available. The block techniques in this chapter may be succinctly described as fixed-time, recursive-in-order algorithms in the sense that they operate on a fixed block of time samples and recursively yield higher-order AR model parameters estimates based on lower-order AR parameter estimates. This is an advantage in situations where the appropriate AR model order is not known and many different orders must be tried and compared in order to select a suitable order. The sequential techniques of Chap. 9, on the other hand, may be described as fixed-order, recursive-in-time algorithms in the sense that they process the data sequentially to update the estimates of a fixed-order AR model. This is an advantage in situations in which tracking, or adapting to, a slowly time varying spectrum is required.

Conceptually, the simplest procedure to obtain an AR spectral estimate from data samples would be to produce estimates of the autocorrelation sequence from the data using MATLAB function `correlation_sequence` in Chap. 5. These autocorrelation estimates would then be used, in lieu of the unavailable true autocorrelation sequence, in the Yule-Walker equations of Chap. 6 to yield the

AR coefficients, and from these the AR PSD function. However, better results are obtained, particularly for short data segments, by algorithms that yield the AR model parameters directly from the data, without the need for autocorrelation estimates, as examples in this chapter will illustrate.

The block data techniques considered in this chapter fall into three categories, with a section devoted to each. In Chap. 7, it was shown that the AR parameters could be obtained through the equivalent representations of either the autocorrelation sequence or the reflection coefficient sequence. The so-called *Yule-Walker method* of AR parameter estimation from autocorrelation sequence estimates is covered in Sec. 8.3. Two methods of AR parameter estimation from reflection coefficient sequence estimates are provided in Sec. 8.4, including the popular *Burg algorithm*. An important category of AR parameter estimation methods is based on a least squares linear prediction approach. Techniques in this category, which are discussed in Sec. 8.5, are further distinguished by the type of linear prediction estimate used. One class of techniques performs *separate* minimization of the forward and backward linear prediction squared errors. Included in this class are the *autocorrelation* and *covariance* methods. Another class of techniques performs combined minimization of the forward and backward linear prediction squared errors. Included in this class is the *modified covariance* method. Other sections in this chapter present methods for AR model order selection and cover modifications required to handle observation noise in data measurements.

8.2 SUMMARY

In order to produce an autoregressive PSD estimate directly from a block of data samples, MATLAB functions discussed in this chapter may be found at the text website to provide the reader with a choice of four possible algorithms. Figure 8.1 is a flowchart of the steps required to estimate the AR power spectral density. One of four subroutines for estimating the AR parameters must be selected. The Yule-Walker method (Sec. 8.3) produces AR spectra for short data records with the least resolution among the four methods. The Burg (Sec. 8.4.2) and covariance (Sec. 8.5.1) methods produce comparable AR spectral estimates. The modified covariance method (Sec. 8.5.2) is best for sinusoidal components in the data. The Yule-Walker and Burg methods make direct AR parameter estimates $\hat{a}[n]$ from data. The covariance and modified covariance methods actually make estimates of the linear prediction coefficients, which are then used as the AR parameter estimates by equating the AR parameters to the forward linear prediction coefficients, $\hat{a}[n] = a^f[n]$, or to the complex conjugated backward linear prediction coefficients, $\hat{a}[n] = (a^b[n])^*$. Once the AR parameters have been estimated, then the AR spectral estimate is computed as

$$\hat{P}_{AR}(f) = \frac{T\hat{\rho}_w}{\left|1 + \sum_{n=1}^{p} \hat{a}[n]\exp(-j2\pi f nT)\right|^2}, \qquad (8.1)$$

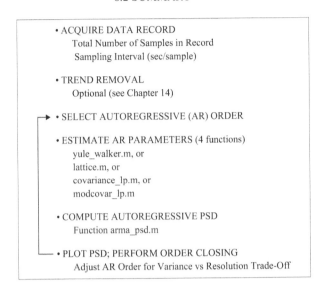

Figure 8.1. Flowchart for four AR PSD block estimation algorithms implemented in this chapter.

in which $\hat{\rho}_w$ is an estimate of the driving noise variance, which is also provided by all four methods of Fig. 8.1. Note that the AR model order must also be provided as an input parameter, in addition to the input data. The order determines the trade-off between resolution and estimate variance in AR spectra, analogous to the effect of windows in classical spectral estimates. Many different orders should be tried in a fashion similar to the window-closing approach of classical spectral estimation in order to determine the most suitable order to use. This will be called *order closing*, in analogy with the window closing scheme of classical spectral estimation. Section 8.7 presents methods for suggested order that could be incorporated into the program code of any algorithm, thereby relieving the need to select an order parameter.

MATLAB demonstration script `spectrum_demo_ar.m` available through the text website provides spectral estimation demonstration plots for all four AR estimation techniques presented in this chapter. By inputting a response to the script questions, an AR spectral estimate for a selected one of the three data sets `test1987`, `doppler_radar`, or `sunspot_numbers` will be calculated and a plot made. Examples of the four AR PSD estimates made from the 64-point `test1987` data sequence are shown in Fig. 8.2. A 15th-order AR process was selected for all four methods. An inset on the covariance method also shows an 8th-order spectral estimate to illustrate the resolution/smoothing trade-off that order selection permits. These estimates should be compared with the true limit spectrum depicted in Fig. 8.1. The spectra of Fig. 8.2 exhibit very sharp spectral lines. To ensure that the PSD function has been adequately sampled in frequency, so that no sharp peaks are missed, one need only sum the PSD values over all frequencies. This sum should be equal to the total

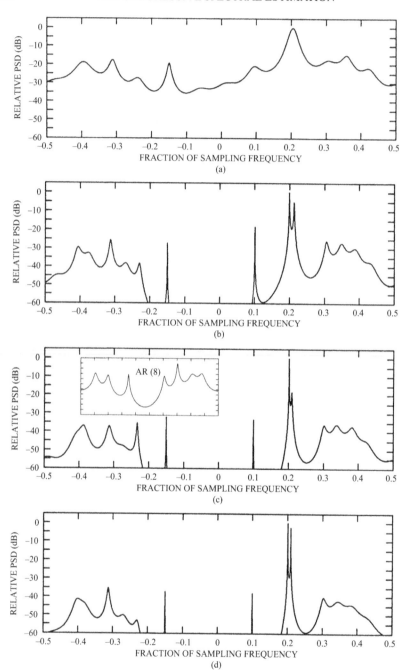

Figure 8.2. Examples of four AR(15) spectral estimates made from the 64-point test data sequence. (a) Yule-Walker method. (b) Burg method. (c) Covariance method. (d) Modified covariance method.

power in the process and, therefore, should agree closely in value with the zero-lag autocorrelation term, which also represents the total power in the process. If the values do not agree, then a denser frequency grid (requires the input parameter num_psd) is necessary in order to find one or more missed sharp peaks.

8.3 CORRELATION FUNCTION ESTIMATION METHOD

The most obvious approach to AR PSD estimation from data samples is to use autocorrelation function estimates in lieu of the unknown autocorrelation function for the solution of the Yule-Walker equations. This will be termed the *Yule-Walker algorithm*. If the unbiased autocorrelation estimate is selected, it may result in a nonpositive-definite autocorrelation matrix, so that a stable AR filter will not be assured. The biased correlation estimates will always yield positive-semidefinite autocorrelation matrices, so that a stable AR filter will be assured. For long data records, this AR parameter estimator can produce reasonable spectral estimates. For short data records, the use of the Yule-Walker approach produces poor-resolution spectral estimates relative to other AR methods; other techniques can produce better spectral estimates, as illustrated in Fig. 8.2. Available through the text website is the function yule_walker.m to compute AR estimates from data:

```
function [rho,a]=yule_walker(p,x)
% Yule-Walker autoregressive parameter estimation algorithm.
%
%      [rho,a] = yule_walker(p,x)
%
% p   -- order of autoregressive (AR) process
% x   -- vector of data samples
% rho -- noise variance estimate
% a   -- vector of autoregressive parameters
```

8.4 REFLECTION COEFFICIENT ESTIMATION METHODS

The Levinson recursive solution to the Yule-Walker equations relates the order p AR parameters to the order $p - 1$ parameters as

$$a_p[n] = a_{p-1}[n] + k_p a_{p-1}^*[p-n] \tag{8.2}$$

for $n = 1$ to $n = p - 1$, where the reflection coefficient k_p is found by using the known autocorrelation function for lags 0 to $p - 1$,

$$k_p = a_p[p] = \frac{-\sum_{n=0}^{p-1} a_{p-1}[n] r_{xx}[p-n]}{\rho_{p-1}}. \tag{8.3}$$

The recursion for the driving white noise variance was given by

$$\rho_p = \rho_{p-1}(1 - |\mathsf{k}_p|^2), \tag{8.4}$$

where $\rho_0 = r_{xx}[0]$. Among the quantities in Eqs. (8.2)–(8.4), only the reflection coefficient depends directly on the autocorrelation function. This suggests that one procedure for yielding an AR PSD estimate, when only a block of data samples is available, is to use the data samples to estimate the relection coefficient at each stage of the Levinson recursion. Three methods have evolved from this concept, one based on the geometric mean of the forward and backward linear prediction squared errors, one based on the arithmetic mean of these squared errors, and one based on an approximate maximum likelihood approach.

8.4.1 Geometric Algorithm

The forward linear prediction error is given by

$$e_p^f[n] = x[n] + \sum_{m=1}^{p} a_p^f[m]x[n-m] \tag{8.5}$$

and the backward linear prediction error is given by

$$e_p^b[n] = x[n-p] + \sum_{m=1}^{p} a_p^{f\,*}[m]x[n+m-p]. \tag{8.6}$$

Substitution of Eq. (8.2) into these definitions yields the recursive relationships

$$e_p^f[n] = e_{p-1}^f[n] + \mathsf{k}_p e_{p-1}^b[n-1]$$
$$e_p^b[n] = e_{p-1}^b[n-1] + \mathsf{k}_p^* e_{p-1}^f[n] \tag{8.7}$$

that relate the order $p-1$ prediction errors to the order p prediction errors. In Chap. 7, it was shown that k_p has an interpretation as a partial correlation coefficient between the forward and backward linear prediction errors, with the property

$$\mathsf{k}_p = -\frac{\mathcal{E}\{e_{p-1}^f[n]e_{p-1}^{b\,*}[n-1]\}}{\mathcal{E}\{|e_{p-1}^f[n]|^2\}^{1/2}\mathcal{E}\{|e_{p-1}^b[n-1]|^2\}^{1/2}}. \tag{8.8}$$

This represents the "geometric" mean [the geometric mean between two numbers a and b is $\sqrt{ab}$, the arithmetic mean is $(a+b)/2$, and the harmonic mean is $2ab/(a+b)$] between $\mathcal{E}\{e_{p-1}^f[n]e_{p-1}^{b\,*}[n-1]\}/\mathcal{E}\{|e_{p-1}^f[n]|^2\}$, the forward prediction partial correlation, and $\mathcal{E}\{e_{p-1}^f[n]e_{p-1}^{b\,*}[n-1]\}/\mathcal{E}\{|e_{p-1}^b[n-1]|^2\}$, the backward prediction partial correlation. By substituting the prediction error autocorrelation and cross correlation estimates

$$\frac{1}{N} \sum_{n=p+1}^{N} e_{p-1}^{f}[n] e_{p-1}^{b*}[n-1]$$

$$\frac{1}{N} \sum_{n=p+1}^{N} |e_{p-1}^{f}[n]|^2 \tag{8.9}$$

$$\frac{1}{N} \sum_{n=p+1}^{N} |e_{p-1}^{b}[n-1]|^2,$$

the reflection coefficient may be approximated as

$$\hat{k}_p = \frac{-\sum_{n=p+1}^{N} e_{p-1}^{f}[n] e_{p-1}^{b*}[n-1]}{\left(\sum_{n=p+1}^{N} |e_{p-1}^{f}[n]|^2\right)^{1/2} \left(\sum_{n=p+1}^{N} |e_{p-1}^{b}[n-1]|^2\right)^{1/2}}. \tag{8.10}$$

Note that it did not matter whether the biased or unbiased correlation estimator was used, as the scale factor divides out. It has been assumed in Eq. (8.10) that there are N samples $x[1], \ldots, x[N]$, and prediction errors $e_p^f[n]$ and $e_p^b[n]$ only in the index range from $n = p+1$ to $n = N$ are formed, because this uses only the available data samples. It is easy to show that $\hat{k}_p$ has magnitude less than unity, so that a stable AR filter (poles within the unit circle) will be ensured.

The geometric algorithm, therefore, will use the Levinson algorithm, substituting estimate $\hat{k}_p$ for the usual reflection coefficient computation based on known correlation values. The basic Levinson algorithm is supplemented with the prediction error recursions given by Eq. (8.7), which are initialized at order zero by

$$e_0^f[n] = e_0^b[n] = x[n] \tag{8.11}$$

for $1 \leq n \leq N$. The linear prediction error variance is initialized by the estimate

$$\rho_0 = \frac{1}{N} \sum_{n=1}^{N} |x[n]|^2, \tag{8.12}$$

which represents the sample power in the N data samples. A multichannel version of the geometric algorithm is developed in Chap. 15.

8.4.2 Harmonic (Burg) Algorithm

One of the earliest and best known algorithms for AR spectral estimation was introduced by John Burg [1967], and later was described in detail by Andersen [1974],

Datta [1977], Haykin and Kesler [1976], and Ulrych et al. [1973]. This algorithm has been called a maximum entropy algorithm, but it should be considered as a separate and distinct contribution from Burg's maximum entropy concept [1975].

The Burg algorithm is identical to the geometric algorithm, except that an alternative estimate for the reflection coefficient is determined, based on a least squares criteria. At each order p, the arithmetic mean of the forward and backward linear prediction error power (sample prediction error variance)

$$\rho_p^{fb} = \frac{1}{2} \left[\frac{1}{N} \sum_{n=p+1}^{N} |e_p^f[n]|^2 + \frac{1}{N} \sum_{n=p+1}^{N} |e_p^b[n]|^2 \right] \qquad (8.13)$$

is minimized, subject to the recursion given by Eq. (8.7). Note again that the summation ranges only over the available data. Thus, ρ_p^{fb} is a function of a single parameter, namely the complex-valued reflection coefficient k_p, because the prediction errors from order $p-1$ will be known. Setting the complex derivative of Eq. (8.13) to zero

$$\frac{\partial \rho_p^{fb}}{\partial \operatorname{Re}\{k_p\}} + j \frac{\partial \rho_p^{fb}}{\partial \operatorname{Im}\{k_p\}} = 0$$

and solving for k_p yields the least squares estimate

$$\hat{k}_p = \frac{-2\sum_{n=p+1}^{N} e_{p-1}^f[n] e_{p-1}^{b*}[n-1]}{\sum_{n=p+1}^{N} |e_{p-1}^f[n]|^2 + \sum_{n=p+1}^{N} |e_{p-1}^b[n-1]|^2}. \qquad (8.14)$$

This estimate of the reflection coefficient represents the "harmonic" mean between the forward and backward partial correlation coefficients. It is easy to prove that this estimate of the reflection coefficient has magnitude less than or equal to unity, thereby ensuring a stable all-pole filter. A further recursion that simplifies the evaluation of the denominator term in Eq. (8.14) was discovered by Andersen [1978]. Define

$$\text{DEN}_p = \sum_{n=p+1}^{N} \left(|e_{p-1}^f[n]|^2 + |e_{p-1}^b[n-1]|^2 \right); \qquad (8.15)$$

then

$$\text{DEN}_p = \left(1 - |\hat{k}_{p-1}|^2 \right) \text{DEN}_{p-1} - |e_{p-1}^f[p]|^2 - |e_{p-1}^b[N]|^2. \qquad (8.16)$$

A MATLAB function lattice.m of the geometric and Burg algorithms is available through the text website:

```
function [rho,a]=lattice(method,p,x)
% Lattice autoregressive parameter estimation algorithms based on the geometric
% or Burg algorithms that estimate the reflection coefficients, then apply
% the Levinson algorithm to get the AR parameters from the reflec. coeffs.
```

```
%
%      [rho,a] = lattice(method,p,x)
%
% method -- selection choices: 'geometric', 'burg'
% p        -- order of autoregressive (AR) process
% x        -- vector of data samples
% rho      -- noise variance estimate
% a        -- vector of autoregressive parameters
```

It is annotated with equation references to the text. A computational complexity analysis (counting only second-order terms) will show that $3Np - p^2$ complex adds and multiplies, as well as p real divisions, are required. Storage of $3N + p$ complex values is also required.

The harmonic method exhibits some bias in the estimation of sinusoidal frequencies, as will be illustrated in Sec. 8.6. This has led some researchers to propose variations of the harmonic method in order to mitigate these biases. The basic variation is a weighting of the squared prediction errors,

$$
\rho_p^{fb} = \frac{1}{2}\left[\frac{1}{N} \sum_{n=p+1}^{N} w_p[n]\left\{ |e_p^f[n]|^2 + |e_p^b[n]|^2 \right\} \right],
$$

leading to the reflection coefficient estimate

$$
\hat{k}_p = \frac{-2\sum_{n=p+1}^{N} w_{p-1}[n] e_{p-1}^f[n] e_{p-1}^{b*}[n-1]}{\sum_{n=p+1}^{N} w_{p-1}[n]\left(|e_{p-1}^f[n]|^2 + |e_{p-1}^b[n-1]|^2 \right)}, \qquad (8.17)
$$

where $w_{p-1}[n]$ represents the weighting function. As long as $w[n] \geq 0$, then $|\hat{k}_p| \leq 1$. The Burg algorithm selects the uniform weighting $w_{p-1}[n] = 1/N$. Swingler [1979b, 1980] provided simulations that indicated a Hamming taper function reduced the frequency bias. Kaveh and Lippert [1983] found that a special quadratic window could also achieve a reduction in the sinusoidal frequency bias over most of the frequency range. Helme and Nikias [1984] propose the data-adaptive weighting

$$
w_{p-1}[n] = \sum_{k=n-p+1}^{n-1} |x[k]|^2 \quad \text{for} \quad p \geq 2,
$$

which represents the common data energy in the forward and backward linear prediction errors $e_{p-1}^f[n]$ and $e_{p-1}^b[n-1]$ at time index n.

8.4.3 Recursive MLE Method

The recursive MLE [Kay, 1983] derives its name based upon the MLE principles that are used to develop an estimate of the reflection coefficients and the order update of the autoregressive coefficients provided by the Levinson recursion. The derivation details are quite extensive; only a summary of the algorithm is given here. The reader may consult Kay [1983] for a derivation assuming real data. The extension to the complex data case is not yet available.

If the probability density function of a zero-mean Gaussian autoregressive process at order p is minimized with respect to the reflection coefficient k_p and the driving noise variance ρ_p, assuming the order $p-1$ autoregressive coefficients have previously been computed, then the k_p and ρ_p estimates that minimize the PDF for a given data sequence $x[0], x[1], \ldots, x[N-1]$ are given as the solution of the cubic equation

$$\hat{k}_p^3 + \left(\frac{N-2p}{N-p}\right)\left(\frac{\alpha_p}{\beta_p}\right)\hat{k}_p^2 - \left(\frac{p\delta_{p-1}+N\alpha_p}{[N-p]\beta_p}\right)\hat{k}_p - \left(\frac{N}{N-p}\right)\left(\frac{\alpha_p}{\beta_p}\right) = 0,$$

for which the real root within the interval $[-1, 1]$ is selected, and the linear equations

$$\delta_p = \delta_{p-1} + 2\alpha_p\hat{k}_p + \beta_p\hat{k}_p^2$$
$$\hat{\rho}_p = \delta_p/N$$

are used. The real coefficients α_p and β_p are determined from the quadratic forms

$$\alpha_p = \begin{pmatrix} \mathbf{a}_{p-1}^T & 0 \end{pmatrix} \mathbf{S}_p \begin{pmatrix} \mathbf{Ja}_{p-1} \\ 0 \end{pmatrix}$$

$$\beta_p = \begin{pmatrix} 0 & \mathbf{a}_{p-1}^T\mathbf{J} \end{pmatrix} \mathbf{S}_p \begin{pmatrix} \mathbf{Ja}_{p-1} \\ 0 \end{pmatrix},$$

for which $\mathbf{a} = (1 \quad a_{p-1}[1] \quad \ldots \quad a_{p-1}[p-1])^T$ is the autoregressive coefficient vector, $\mathbf{J}$ is a $p \times p$ reflection matrix, and $\mathbf{S}_p$ is a $(p+1) \times (p+1)$ matrix with elements

$$s_p[i, j] = \sum_{n=p}^{N-1+i+j-p} x[n-i]x[n-j]$$

defined for $0 \le i, j \le p$. Once $\hat{k}_p$ has been estimated, then the usual Levinson recursion [Eq. (8.2)] is used to update the order $p-1$ autoregressive coefficients to the order p coefficients. The initial condition

$$\delta_0 = \sum_{n=0}^{N-1} x^2[n]$$

is required to initialize the recursive MLE algorithm. A computer program that implements this algorithm may be found in Kay [1987].

8.5 LEAST SQUARES LINEAR PREDICTION ESTIMATION METHODS

By constraining the AR coefficients to satisfy the Levinson recursive relationship, Eq. (8.2), Burg was able to perform a least squares optimization with respect to a single parameter, namely, the reflection coefficient. An alternative approach would be to perform a least squares minimization with respect to *all* of the linear prediction coefficients, entirely removing the Levinson recursion constraint. Such an approach will yield some improvement to the spectral estimate, as Sec. 8.6 will demonstrate. Two types of least squares estimates will be considered. The first type utilizes separate forward or backward linear prediction for the estimate. The second type employs a combination of forward and backward linear prediction.

8.5.1 Separate Forward and Backward Linear Prediction Algorithms

Assume the N-point data sequence $x[1], \ldots, x[N]$ is to be used to estimate the pth-order AR parameters. The forward linear predictor $\hat{x}[n]$ for sample $x[n]$ will have the usual form

$$\hat{x}^f[n] = -\sum_{k=1}^{p} a_p^f[k]x[n-k], \qquad (8.18)$$

where the $a_p^f[k]$ are the forward linear prediction coefficients for order p. The prediction is "forward" in the sense that the prediction for the current data sample is a weighted sum of p prior samples. The forward linear prediction error is

$$e_p^f[n] = x[n] - \hat{x}^f[n] = x[n] + \sum_{k=1}^{p} a_p^f[k]x[n-k]. \qquad (8.19)$$

This expression is identical in form to the recursion that describes a pth-order autoregressive process, except $e_p^f[n]$ is not a driving white noise process. The linear prediction error derived from a finite data sequence may, or may not, be a white process. For purposes of fitting an autoregressive model, it is *assumed* that the prediction error is a whitened process, permitting one to equate the autoregressive parameters to the linear prediction coefficients.

The forward linear prediction error may be defined over the range from $n = 1$ to $n = N + p$, if one assumes that the data prior to the first sample and after the last sample are zero ($x[n] = 0$ for $n < 1$ and $n > N$). The $N + p$ forward linear prediction error terms represented by Eq. (8.19) may be summarized, using matrix-vector notation, as

$$
\begin{pmatrix}
e_p^f[1] \\
\vdots \\
e_p^f[p+1] \\
\vdots \\
e_p^f[N-p] \\
\vdots \\
e_p^f[N] \\
\vdots \\
e_p^f[N+p]
\end{pmatrix}
=
\underbrace{
\begin{pmatrix}
x[1] & \cdots & 0 \\
\vdots & \ddots & \vdots \\
x[p+1] & & x[1] \\
\vdots & \ddots & \vdots \\
x[N-p] & & x[p+1] \\
\vdots & \ddots & \vdots \\
x[N] & & x[N-p] \\
\vdots & \ddots & \vdots \\
0 & \cdots & x[N]
\end{pmatrix}
}_{\mathbf{X}_p}
\begin{pmatrix}
1 \\
a_p^f[1] \\
\vdots \\
a_p^f[p]
\end{pmatrix},
\tag{8.20}
$$

in which $\mathbf{X}_p$ is an $(N+p) \times (p+1)$ rectangular Toeplitz data matrix. The upper right and lower left corners of the data matrix are zeros, clearly emphasizing the implied windowing of the data sequence. The data matrix may be partitioned into three components

$$
\mathbf{X}_p = \begin{pmatrix} \mathbf{L}_p \\ \mathbf{T}_p \\ \mathbf{U}_p \end{pmatrix},
\tag{8.21}
$$

where the lower triangular $p \times (p+1)$ matrix $\mathbf{L}_p$, rectangular $(N-p) \times (p+1)$ matrix $\mathbf{T}_p$, and upper triangular $p \times (p+1)$ matrix $\mathbf{U}_p$ are defined as

$$
\mathbf{L}_p = \begin{pmatrix}
x[1] & \cdots & 0 & 0 \\
\vdots & \ddots & \vdots & \vdots \\
x[p] & \cdots & x[1] & 0
\end{pmatrix}
$$

$$
\mathbf{T}_p = \begin{pmatrix}
x[p+1] & \cdots & x[1] \\
\vdots & \ddots & \vdots \\
x[N-p] & & x[p+1] \\
\vdots & \ddots & \vdots \\
x[N] & \cdots & x[N-p]
\end{pmatrix}
\tag{8.22}
$$

$$
\mathbf{U}_p = \begin{pmatrix}
0 & x[N] & \cdots & x[N-p+1] \\
\vdots & \vdots & \ddots & \vdots \\
0 & 0 & \cdots & x[N]
\end{pmatrix}.
$$

The forward linear prediction squared error magnitude to be minimized is simply

$$
\rho_p^f = \sum_n |e_p^f[n]|^2.
\tag{8.23}
$$

Dividing ρ_p^f by N will yield the sample variance. The summation range for ρ_p^f has purposely not been specified, because there are many possible range cases that one may select. Three cases have been used for spectral estimation. Selecting the full range from $e_p^f[1]$ to $e_p^f[N+p]$ is the *windowed* case, because it involves pre- and postwindowing of the data (i.e., setting unavailable data values to zero). Selecting the range $e_p^f[1]$ to $e_p^f[N]$ is the *prewindowed* case because it requires data prior to $x[1]$ to be set to zero. Selecting the range $e_p^f[p+1]$ to $e_p^f[N]$ is the *nonwindowed* case because only available data samples are used. The windowed case has been termed the *autocorrelation* method of linear prediction. The nonwindowed case has been termed the *covariance* method of linear prediction. This terminology is based on historical usage in speech processing [Makhoul, 1975] and is not justified by the standard statistical definitions for these terms (i.e., the covariance is the correlation with the mean removed).

Consider first the least squares linear prediction estimate based on minimization of ρ_p^f for the *autocorrelation* (windowed) method. The relationship between the forward linear prediction errors and the linear prediction coefficients may be succinctly expressed as

$$\mathbf{e}_p^f = \mathbf{X}_p \begin{pmatrix} 1 \\ \mathbf{a}_p^f \end{pmatrix}, \tag{8.24}$$

in which the $(N+p)$-element error vector $\mathbf{e}_p^f$ and p-element linear prediction coefficient vector $\mathbf{a}_p^f$ are defined as

$$\mathbf{e}_p^f = \begin{pmatrix} e_p^f[1] \\ \vdots \\ e_p^f[N+p] \end{pmatrix}, \quad \mathbf{a}_p^f = \begin{pmatrix} a_p^f[1] \\ \vdots \\ a_p^f[p] \end{pmatrix} \tag{8.25}$$

and $\mathbf{X}_p$ is the $(N+p) \times (p+1)$ data matrix of Eq. (8.20). Equation (8.24) has the same form as that of Eq. (3.62), so that the normal equations that minimize the squared error

$$\rho_p^f = \sum_{n=1}^{N+p} |e_p^f[n]|^2 = (\mathbf{e}_p^f)^H \mathbf{e}_p^f \tag{8.26}$$

at order p are determined simply by inspection of Eq. (3.67) to be

$$\mathbf{X}_p^H \mathbf{X}_p \begin{pmatrix} 1 \\ \mathbf{a}_p^f \end{pmatrix} = \begin{pmatrix} \rho_p^f \\ \mathbf{0}_p \end{pmatrix}. \tag{8.27}$$

The product $\mathbf{X}_p^H \mathbf{X}_p$ forms a square $(p+1) \times (p+1)$ Hermitian matrix $\mathbf{R}_p$ of the form

$$\mathbf{R}_p = \mathbf{X}_p^H \mathbf{X}_p = \begin{pmatrix} r_p[0,0] & \cdots & r_p[0,p] \\ \vdots & & \vdots \\ r_p[p,0] & \cdots & r_p[p,p] \end{pmatrix}, \tag{8.28}$$

in which $r_p[i, j] = r_p^*[j, i]$. Elements of $\mathbf{R}_p$ have the lagged-product form

$$r_p[i, j] = \sum_{n=1}^{N+p} x^*[n-i]x[n-j] = \sum_{k=1}^{N-(i-j)} x[k+(i-j)]x^*[k] \qquad (8.29)$$

for $0 \leq i - j \leq p$. Because $r_p[i, j]$ depends only on the difference $(i - j)$, $\mathbf{X}_p^H\mathbf{X}_p$ forms a Toeplitz matrix [see Eq. (3.35)]. If $r_p[i, j]$ is divided by N, these terms will then be identical to the biased autocorrelation sequence estimate $\check{r}_{xx}[i - j] = r_p[i, j]/N$; in this case, normal equations (8.27) will be identical to the Yule-Walker equations (6.32). Thus, the autocorrelation method of least squares linear prediction analysis is equivalent to the Yule-Walker method of autoregressive parameter estimation in which biased autocorrelation estimates are used. The windowing of the autocorrelation method is responsible for the apparent reduction in resolution relative to the other linear-prediction-based estimation methods to be discussed (see Fig. 8.2 for an example). Thus, this approach is rarely used in practice for short data segments because better performance is obtained by the other least squares methods.

The relationship between the forward linear prediction errors and the linear prediction coefficients for the *covariance* (nonwindowed) method may be succinctly expressed as

$$\mathbf{e}_p^f = \mathbf{T}_p \begin{pmatrix} 1 \\ \mathbf{a}_p^f \end{pmatrix}, \qquad (8.30)$$

in which the $N \times (p + 1)$ Toeplitz data matrix $\mathbf{T}_p$ was defined in Eq. (8.22), the p-element linear prediction coefficient vector $\mathbf{a}_p^f$ was defined in Eq. (8.25), and the $(N - p)$-element error vector $\mathbf{e}_p^f$ now has the structure

$$\mathbf{e}_p^f = \begin{pmatrix} e_p^f[p+1] \\ \vdots \\ e_p^f[N] \end{pmatrix}. \qquad (8.31)$$

The normal equations that minimize the squared error

$$\rho_p^f = \sum_{n=p+1}^{N} |e_p^f[n]|^2 = (\mathbf{e}_p^f)^H\mathbf{e}_p^f \qquad (8.32)$$

at order p are determined, again, simply by inspection of Eq. (3.67) to be

$$\mathbf{T}_p^H\mathbf{T}_p \begin{pmatrix} 1 \\ \mathbf{a}_p^f \end{pmatrix} = \begin{pmatrix} \rho_p^f \\ \mathbf{0}_p \end{pmatrix}. \qquad (8.33)$$

The $(p + 1) \times (p + 1)$ Hermitian matrix $\mathbf{R}_p = \mathbf{T}_p^H \mathbf{T}_p$ has lagged-product elements of the form

$$r_p[i, j] = \sum_{n=p+1}^{N} x^*[n - i]x[n - j] \tag{8.34}$$

for $0 \leq i, j \leq p$. The elements of $\mathbf{R}_p$ in the covariance method cannot be expressed as functions of the difference $(i - j)$, which means the product $\mathbf{T}_p^H \mathbf{T}_p$ is not Toeplitz. However, the fact that $\mathbf{R}_p$ is the product of two Toeplitz matrices still makes it possible to develop a fast algorithm similar to that of the Levinson algorithm. Thus, the solution of Eq. (8.33) in the covariance case may be obtained with an algorithm of computational complexity proportional to p^2 operations, as compared to p^3 operations that would be required if the Cholesky algorithm had been used. Available through the text website is a MATLAB function `covariance_lp.m` that implements the fast algorithm solution of Eq. (8.30). Details of the fast algorithm may be found in Sec. (8.10):

```
function [rho_f,a_f,rho_b,a_b]=covariance_lp(p,x)
% Covariance least squares autoregressive parameter estimation algorithm using
% a fast QR-decomposition type of linear prediction computational solution.
%
%      [rho_f,a_f,rho_b,a_b] = covariance_lp(p,x)
%
% p      -- order of linear prediction/autoregressive filter
% x      -- vector of data samples
% rho_f  -- least squares estimate of forward linear prediction variance
% a_f    -- vector of forward linear prediction/autoregressive parameters
% rho_b  -- least squares estimate of backward linear prediction variance
% a_b    -- vector of backward linear prediction/autoregressive parameters
```

A necessary, but not sufficient, condition for $\mathbf{R}_p$ to be a nonsingular matrix is that $N - p \geq p$, or $p \leq N/2$. The selected order must, therefore, be no greater than half the data length. The covariance method will be encountered again in Chap. 11 as part of the Prony method. It can be shown [Kay, 1987] that the covariance method is an approximate maximum likelihood estimate for the AR parameters when the process is Gaussian.

Based on a development similar to that used for the windowed and non-windowed cases, it is simple to show that the normal equation matrix $\mathbf{R}_p$ in the *prewindowed* case may be expressed as the data matrix product

$$\mathbf{R}_p = \begin{pmatrix} \mathbf{L}_p \\ \mathbf{T}_p \end{pmatrix}^H \begin{pmatrix} \mathbf{L}_p \\ \mathbf{T}_p \end{pmatrix} = \mathbf{L}_p^H \mathbf{L}_p + \mathbf{T}_p^H \mathbf{T}_p = \mathbf{X}_p^H \mathbf{X}_p - \mathbf{U}_p^H \mathbf{U}_p \tag{8.35}$$

with elements of the form

$$r_p[i, j] = \sum_{n=1}^{N} x^*[n - i]x[n - j] \tag{8.36}$$

for $0 \leq i, j \leq p$. The prewindowed method is of most interest for the sequential least squares linear prediction methods. This method will be treated in more detail in Chap. 9.

The windowed (autocorrelation), nonwindowed (covariance), and prewindowed cases may also be developed for the case of the *backward* linear predictor

$$\hat{x}^b[k] = -\sum_{k=1}^{p} a_p^b[k]x[k+p], \tag{8.37}$$

where the $a_p^b[k]$ are the backward linear prediction coefficients for order p. The prediction is "backward" in the sense that the prediction for the current data sample is a weighted sum of p future samples. For a finite data set, the backward linear prediction parameters determined by least squares analysis are not, in general, identical to the forward prediction parameters. The backward linear prediction error is

$$e_p^b[n] = x[n-p] - \hat{x}^b[n-p] = x[n-p] + \sum_{k=1}^{p} a_p^b[k]x[n-p+k]. \tag{8.38}$$

With the assumption that $x[n] = 0$ for $n < 0$ and $n > N$, the backward linear prediction error terms may be summarized by the matrix-vector product

$$\mathbf{e}_p^b = \mathbf{X}_p \begin{pmatrix} \mathbf{a}_p^b \\ 1 \end{pmatrix}, \tag{8.39}$$

where the backward linear prediction error vector $\mathbf{e}_p^b$ and the backward linear prediction coefficient vector $\mathbf{a}_p^b$ are defined as

$$\mathbf{e}_p^b = \begin{pmatrix} e_p^b[N+p] \\ \vdots \\ e_p^b[1] \end{pmatrix} \qquad \mathbf{a}_p^b = \begin{pmatrix} a_p^b[p] \\ \vdots \\ a_p^b[1] \end{pmatrix} \tag{8.40}$$

and $\mathbf{X}_p$ is the same Toeplitz data matrix defined by Eq. (8.21).

The backward linear prediction squared error is given by

$$\rho_p^b = \sum_{n} |e_p^b[n]|^2. \tag{8.41}$$

The summation range is 1 to $N + p$ for the windowed (autocorrelation) case, $p + 1$ to N for the nonwindowed (covariance) case, and 1 to N for the prewindowed case. Using the same arguments as used for the forward linear prediction cases, the normal equations for the backward linear prediction cases have the form

$$\mathbf{R}_p \begin{pmatrix} \mathbf{a}_p^b \\ 1 \end{pmatrix} = \begin{pmatrix} \mathbf{0}_p \\ \rho_p^b \end{pmatrix},$$ (8.42)

where $\mathbf{R}_p$ equals $\mathbf{X}_p^H \mathbf{X}_p$ for the windowed case, $\mathbf{T}_p^H \mathbf{T}_p$ for the nonwindowed case, and Eq. (8.35) in the prewindowed case. The solutions for the forward and backward normal equations are coupled because both have the same $\mathbf{R}_p$ matrix. The fast algorithms for both the covariance and prewindowed methods will simultaneously solve the normal equations for both the forward and backward linear prediction coefficients at all intermediate orders, because the solutions for each are intertwined. Thus, both sets of coefficients are available for analysis at no additional computational cost.

The forward and backward linear prediction coefficients generated by the covariance method do not guarantee a stable filter. This is not generally a problem if the use of the coefficients is only for spectral estimation purposes. In fact, spectral estimates produced with the autoregressive coefficients estimated by the covariance method usually have less distortion [Nuttall, 1976; Lang and McClellan, 1980] than spectral estimates produced by methods that ensure the filter stability, such as the autocorrelation method. If an actual linear prediction filter is synthesized for nonspectral estimation purposes, then the filter stability issue is of concern.

One variation to the covariance method was proposed by Nikias and Scott [1983]. Their method selects the forward and backward linear prediction coefficients based on the weighted squared errors

$$\rho_p^f = \sum_{n=p+1}^N w_p[n]|e_p^f[n]|^2 \qquad \rho_p^b = \sum_{n=p+1}^N w_p[n]|e_p^b[n]|^2,$$ (8.43)

in which the weights represent the energy of the data used to form the prediction errors within the window,

$$w_p[n] = \sum_{k=n-p+1}^n |x[k]|^2.$$

No fast algorithm is known for the general weighted covariance method, so that general matrix solution routines must be used, requiring a number of computations proportional to p^3 in order to solve the associated normal equations.

8.5.2 Combined Forward and Backward Linear Prediction Algorithms

It was shown in Chap. 7 that the forward and backward autoregressive coefficients for a stationary random process are simply complex conjugates of one another, so that the backward linear prediction error may then by expressed as

$$e_p^b[n] = x[n-p] + \sum_{k=1}^{p} a_p^{f*}[k]x[n-p+k].$$ (8.44)

Because both directions have similar statistical information, it seems reasonable to combine the linear prediction error statistics of both forward and backward errors in order to generate more error points. The net result should be an improved estimate of the autoregressive parameters. The $(N-p)$ forward and the $(N-p)$ backward linear prediction errors of the nonwindowed case may be summarized concisely by the matrix-vector product

$$\mathbf{e}_p = \begin{pmatrix} \mathbf{e}_p^f \\ \mathbf{e}_p^{b*} \end{pmatrix} = \begin{pmatrix} \mathbf{T}_p \\ \mathbf{T}_p^*\mathbf{J} \end{pmatrix} \begin{pmatrix} 1 \\ \mathbf{a}_p^{fb} \end{pmatrix}.$$ (8.45)

The error vector $\mathbf{e}_p$ has $2(N-p)$ elements, consisting of $\mathbf{e}_p^f$, an $(N-p)$-element forward linear prediction error vector, and $\mathbf{e}_p^b$, an $(N-p)$-element backward linear prediction error vector, as follows:

$$\mathbf{e}_p^f = \begin{pmatrix} e_p^f[p+1] \\ \vdots \\ e_p^f[N] \end{pmatrix}, \quad \mathbf{e}_p^b = \begin{pmatrix} e_p^b[p+1] \\ \vdots \\ e_p^b[N] \end{pmatrix}.$$

The linear prediction coefficient vector $\mathbf{a}_p^{fb}$ in Eq. (8.45) is given by

$$\mathbf{a}_p^{fb} = \begin{pmatrix} a_p[1] \\ \vdots \\ a_p[p] \end{pmatrix}.$$

The $(N-p) \times (p+1)$ data matrix $\mathbf{T}_p$ in Eq. (8.45) was defined by Eq. (8.22). Noting that $\mathbf{J}$ is a $(p+1) \times (p+1)$ reflection matrix, then

$$\mathbf{T}_p^*\mathbf{J} = \begin{pmatrix} x^*[1] & \cdots & x^*[p+1] \\ \vdots & \ddots & \vdots \\ x^*[p+1] & \cdots & x^*[N-p] \\ \vdots & \ddots & \vdots \\ x^*[N-p] & \cdots & x^*[N] \end{pmatrix}$$

forms a Hankel matrix of conjugated data elements. Observe that the superscript f has been omitted from the linear prediction coefficients $a_p[k]$ because these apply now to both forward *and* backward prediction errors.

Minimizing the *average* of the forward *and* backward linear prediction squared errors over the available data

$$\rho_p^{fb} = \frac{1}{2}\left[\sum_{n=p+1}^{N}|e_p^f[n]|^2 + \sum_{n=p+1}^{N}|e_p^b[n]|^2\right]$$

$$= \frac{1}{2}\mathbf{e}_p^H\mathbf{e}_p = \frac{1}{2}\left[(\mathbf{e}_p^f)^H\mathbf{e}_p^f + (\mathbf{e}_p^b)^H\mathbf{e}_p^b\right] \qquad (8.46)$$

leads to the set of normal equations

$$\mathbf{R}_p\mathbf{a}_p^{fb} = \begin{pmatrix} 2\rho_p^{fb} \\ \mathbf{0}_p \end{pmatrix}, \qquad (8.47)$$

in which

$$\mathbf{R}_p = \begin{pmatrix} \mathbf{T}_p \\ \mathbf{T}_p^*\mathbf{J} \end{pmatrix}^H \begin{pmatrix} \mathbf{T}_p \\ \mathbf{T}_p^*\mathbf{J} \end{pmatrix} = \mathbf{T}_p^H\mathbf{T}_p + \mathbf{J}\mathbf{T}_p^T\mathbf{T}_p^*\mathbf{J} \qquad (8.48)$$

and $\mathbf{0}_p$ is an p-element all-zeros vector. Equation (8.47) follows by inspection from Eq. (3.67) because the error equation (8.45) is of the same form as Eq. (3.62). Individual elements of $\mathbf{R}_p$ have the form

$$r_p[i, j] = \sum_{n=p+1}^{N}\left(x^*[n-i]x[n-j] + x[n-p+i]x^*[n-p+j]\right) \qquad (8.49)$$

for $0 \le j, k \le p$. Because the summation range of squared errors in Eq. (8.46) is identical to that of the covariance method, this forward and backward least squares error approach has been called the *modified covariance method*. Burg [1967], Ulrych and Clayton [1976], and Nuttall [1976] have independently proposed this method. Ulrych and Clayton gave it the name *least squares* (LS) method, even though this is not the only least squares method of linear prediction analysis used for spectral estimation. Other summation ranges for Eq. (8.46) are possible, but are not considered here because they generally do not yield better performance than the modified covariance method.

The modified covariance algorithm and Burg harmonic algorithm are both based on the minimization of the forward and backward squared prediction errors. The modified covariance method is based on minimization with respect to all the prediction coefficients, whereas the Burg method performs a constrained least squares minimization with respect to only a single prediction coefficient $a_p[p]$ (i.e., the reflection coefficient k_p). Several problems with the Burg method, including spectral line splitting and bias of the frequency estimate, appear to be eliminated when the modified covariance method is used. This suggests that the constrained least squares approach creates problems.

Direct solution of Eq. (8.47) by the Cholesky algorithm would require a number of computations proportional to p^3 and element storage proportional to p^2.

However, matrix $\mathbf{R}_p$ of Eq. (8.48) has much structure, including centrosymmetry, that can be exploited to develop a fast computational solution of (8.47). Available through the text website is a MATLAB function modcovar_lp.m that implements the fast algorithm solution of Eq. (8.51). Details of the fast algorithm may be found in Sec. (8.11):

```
function [rho,a]=modcovar_lp(p,x)
% Modified covariance least squares autoregressive parameter estimation
% algorithm using fast QR-decomposition linear prediction solution.
%
%     [rho,a] = modcovar_lp(p,x)
%
% p   -- order of linear prediction/autoregressive filter
% x   -- vector of data samples
% rho -- least squares estimate of linear prediction variance
% a   -- vector of linear prediction/autoregressive parameters
```

For a pth-order model and N data samples, the algorithm requires $Np + 6p^2$ computational operations (adds/multiplies) and element storage of $N + 4p$. This is computationally comparable with the Burg algorithm; for $p \ll N$, the modified covariance algorithm is actually more efficient than the Burg algorithm when the same order is used for both algorithms. A necessary, but not sufficient, condition for $\mathbf{R}_p$ to be a nonsingular matrix is that $2(N - p) \geq p$, or $p \leq 2N/3$, so that the selected order must be no greater than two-thirds of the data length.

The modified covariance method, unlike the Burg algorithm, does not guarantee a stable linear prediction filter (poles inside the unit circle of the z-plane), although most of the time it will yield stable poles [Marple, 1980]. This is not a problem for purposes of spectral estimation, but it should be a consideration if the coefficients are to be used for filter synthesis. It is shown in Sec. (8.11) that the modified covariance method always yields reflection coefficients with magnitudes less than unity. Thus, stable lattice filters will result from use of the reflection coefficients computed from the modified covariance method.

8.6 ESTIMATOR CHARACTERISTICS

The behavior of spectra produced from many of the autoregressive parameter estimators presented in this chapter has been documented extensively in the literature. Because the Burg method is one of the earliest and most frequently used algorithms, it has also been the most extensively tested and compared. The introduction of new autoregressive estimation schemes has often been the result of a need to improve upon some anomalies in spectral estimates produced by the Burg harmonic algorithm. Anomalies to be discussed include spurious spectral peaks, frequency estimation biases, and spectral line splitting.

If a large model order is selected relative to the number of data samples, AR spectral estimates tend to exhibit spurious peaks. Due to estimation error, the normal equations of most AR methods will generally be full rank, even for high orders, so that solutions for the autoregressive parameters will be obtained even though the true model order may be much lower. The extra poles generated by the additional AR parameters give rise to the spurious peaks. To mitigate against spurious peaks, methods for determining model order should be used. Section 8.7 presents some order selection criteria. Lowering the selected order to prevent spurious peaks also reduces the resolution. Among the methods presented in this chapter, the Burg, covariance, and modified covariance methods tend to have better resolution for a given order than other AR algorithms, as Fig. 8.2 has illustrated. This is primarily due to windowing effects in these other algorithms. The autocorrelation method suffers the most degradation in resolution of all methods for this reason.

It has been observed in the geometric mean, harmonic mean (Burg), and autocorrelation methods that two or more closely spaced peaks in the spectral estimate can occur in certain situations, where only one peak should be present. This phenomenon, known as *spectral line splitting*, is illustrated in Fig. 8.3 with spectral estimates based on the Burg and modified covariance methods. Both spectra are order 25 estimates produced from 101 samples of a process consisting of a unit amplitude sinusoid, of $45°$ initial phase and frequency 7.25 Hz, with additive noise of variance 0.01 (SNR = 50 dB). The sample rate was 100 Hz. Figure 8.3(a) exhibits a Burg spectral estimate that has split into two peaks about 7.25 Hz, falsely indicating two sinusoidal frequencies. The modified covariance spectra, based on the same data sequence, yields a single peak at the correct frequency. Spectral line splitting was first documented by Fougere et al. [1976]. They observed that the Burg algorithm would most likely exhibit spectral line splitting when (1) the SNR is high, (2) the initial phase of sinusoidal components is some odd multiple of $45°$, (3) the time duration of the data sequence is such that sinusoidal components have an odd number of quarter cycles, and (4) the number of AR parameters estimated is a large percentage of the number of data samples used for the estimation. Both real and

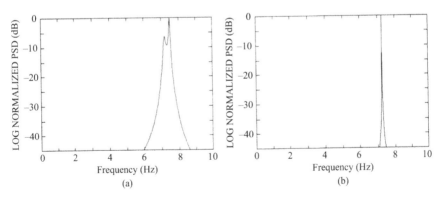

Figure 8.3. Response of two autoregressive spectral estimators to 101 samples of a 7.25-Hz sinusoid in white noise (SNR = 50). The sample rate was 100 Hz and the initial phase was $45°$. (a) Burg spectrum with line splitting. (b) Modified covariance spectrum of same data with no line splitting.

complex data sets have been found to produce line splitting. Many spurious spectral peaks often accompany spectra that exhibit line splitting. The phenomenon of line splitting tends to disappear as the data record increases. An analytical discussion of the reasons for line splitting in the Yule-Walker (autocorrelation) method has been presented by Kay and Marple [1979]. By using the unbiased autocorrelation estimate, line splitting in the Yule-Walker method can be reduced, if not eliminated. An analytical discussion of the reasons for line splitting in the Burg method has been presented by Herring [1980]. The use of weighted squared errors [Eq. (8.43)] seems to reduce the occurrence of line splitting in the Burg method, which appears to be caused by bias between positive and negative spectral components of real sinusoids. However, use of the modified covariance method seems to be the best cure because no evidence of line splitting has ever been observed, especially for those cases for which the Burg algorithm is known to have line splitting.

Chen and Stegen [1974], Ulrych and Clayton [1976], and Thorvaldsen [1981] observed, for a process consisting of one or two sinusoids in white noise, that the spectral peaks of the Burg AR spectral estimate shifted as a function of the initial phases of the sinusoids. In one of the Ulrych and Clayton experiments, Burg and modified covariance spectral estimates were produced from ensembles of 15 data samples of a real-valued unit amplitude 1 Hz sinusoid with 10% additive white noise (the sample interval was 0.05 sec). Figure 8.4 is a plot of the mean variation of the spectral estimate peak as a function of the initial phase of a sinusoid. Each plotted point represents the mean peak frequency over an ensemble of 50 independent data realizations. It is obvious from Fig. 8.4 that the modified covariance method in

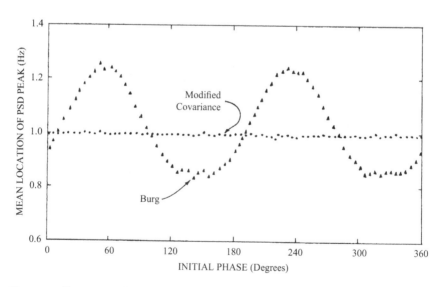

Figure 8.4. Frequency bias of two AR spectral methods. Plot represents the ensemble mean of spectral peak location as a function of initial phase for a cosine signal in white noise (SNR = 10) for Burg and modified covariance methods, using order 9.

the mean is fairly insensitive to the initial phase and is an accurate estimate of the sinusoid frequency. On the other hand, the Burg method has a rather severe sinusoidal-like bias in the frequency estimate as a function of the initial phase. Swingler [1979a, 1980] has demonstrated an analytical basis for the sinusoidal-like variation of the frequency bias; the bias can be as much as 16% of a resolution cell (a resolution cell is $1/NT$ Hz). The use of weighted squared errors, such as Eq. (8.17), has been found to reduce the phase dependence of the Burg method frequency estimate [Swingler, 1979b]. Use of the analytic (complex) signal also reduces the effect of the bias [Jackson et al., 1978].

Nuttall [1976] has made an extensive investigation of the power spectral density variance of many different AR methods, including those presented in this chapter, by use of ensemble averaging over a large number of data sets. He found, of all the methods, that the Burg and modified covariance methods tended to have the least PSD variance and the least frequency variance when tested with nonsinusoidal processes. Shon and Mehrotra [1984] also found this to be true. For example, Fig. 8.5 shows the overlapped spectral estimates for 50 data sequences of 40 samples each of the AR(4) process

$$x[k] = 2.7607x[k-1] - 3.8106x[k-2] + 2.6535x[k-3]$$
$$- 0.9238x[k-4] + w[k],$$

where $w[k]$ is a white Gaussian noise process. The model order for each PSD estimate was preselected to be four. One observation that can be made from Fig. 8.5 is that the spectral estimates of the linear-prediction-based methods tend to have slightly less variability in the skirts, but more spiky estimates near the peaks of the spectra, than that seen in the reflection coefficient estimation methods. That is, the linear prediction methods produce AR spectral estimates with slightly less frequency variance, but more PSD variance, than the reflection coefficient methods. The greater PSD variance may be attributed to the fact that the linear prediction methods do not restrict the poles from moving close to the unit circle, unlike the reflection coefficient methods, which are constrained to be stable (poles within the unit circle only). The Yule-Walker method does the poorest job of estimating the AR(4) spectrum.

Signals that contain large DC levels or a linear trend have been found to corrupt AR spectral estimates [Kane and Trivedi, 1979], particularly at the low-frequency end of the spectrum. These components should be estimated and removed prior to AR spectral estimation.

Although the emphasis in this section has been on short-time-sequence behavior of AR spectral estimates, additional references on the asymptotic statistical behavior of AR spectral estimation may also be found in the literature. Sakai [1979] has shown empirically that the frequency variance of an AR spectral estimate is inversely proportional to both the data length and the square of the SNR for sinusoids in noise. Keeler [1978] has provided empirical evidence that the variance is proportional to the inverse of both the duration and the SNR (rather than proportional to the square of the SNR) for nonsinusoidal processes. Akaike [1969],

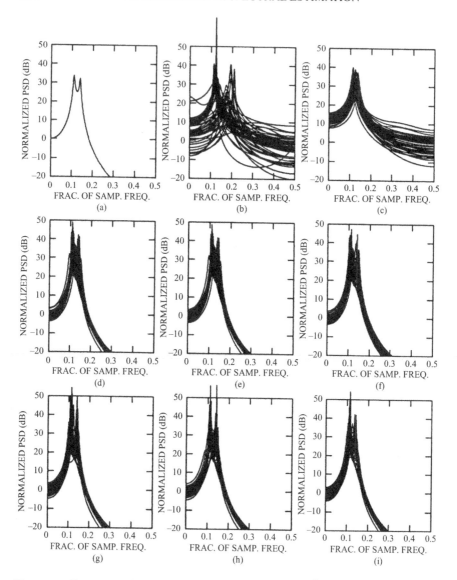

Figure 8.5. Response of five AR spectral estimators to a fourth-order AR process. Overlapped estimates of 50 realizations of 40-sample records are shown for each method. (a) True spectrum of AR(4) process. (b) Yule-Walker, unbiased autocorrelation estimates. (c) Yule-Walker, biased autocorrelation estimates. (d) Geometric method. (e) Burg method. (f) Kay method. (g) Covariance method, forward prediction coefficients. (h) Covariance method, backward prediction coefficients. (i) Modified covariance method.

Kromer [1970], and Berk [1974] have shown that asymptotic performance of the autoregressive PSD, based on the autocorrelation method, is distributed according to a Gaussian PDF as the number of data points N grows asymptotically; the mean of the AR PSD estimate is the true AR PSD and the variance is proportional to $(4p/N)P_{AR}^2(f)$. Confidence interval results for AR spectra have been given by Baggeroer [1976]. Other performance comparisons may be found in the paper by Kaveh and Cooper [1976].

8.7 MODEL ORDER SELECTION

Because the best choice of filter order p is not generally known a priori, it is usually necessary in practice to postulate several model orders. Based on these, one then computes some error criterion that indicates which model order to choose. Too low a guess for model order results in a highly smoothed spectral estimate. Too high an order increases the resolution and introduces spurious detail into the spectrum. Thus, model order selection represents for AR spectral estimation the classic trade-off of increased resolution and decreased variance. One intuitive approach would be to construct AR models of increasing order until the computed prediction error variance reaches a minimum. However, all the estimation procedures discussed in this chapter have prediction error (residual) variances that decrease monotonically with increasing order p. For example, the Burg, modified covariance, and Yule-Walker autocorrelation methods involve a relationship of the form

$$\rho_p = \rho_{p-1}(1 - |a_p[p]|^2) \tag{8.50}$$

for the mean square or least squared error. As long as $|a_p[p]|^2$ is nonzero (it must be ≤ 1), the prediction error variance decreases. Thus, the prediction error variance alone is usually not sufficient to indicate when to terminate the search, unless there is some definitive order after which the rate of change in error variance reduction suddenly decreases. Another simple check concerning proper order selection is to perform a periodogram analysis of the prediction error, or residual, sequence. The residuals should be approximately white, so that an approximately flat periodogram estimate should be obtained.

Many criteria have been proposed as allegedly objective functions for selection of the AR model order. Akaike [1969] has provided two criteria. His first criterion is the *final prediction error* (FPE). This criterion selects the order of the AR process so that the average error variance for a one-step prediction is minimized. He considers the error variance to be the sum of the power in the unpredictable part of the process and a quantity representing the inaccuracies in estimating the AR parameters. The FPE for an AR process is defined as

$$\text{FPE}[p] = \hat{\rho}_p \left(\frac{N + (p+1)}{N - (p+1)} \right), \tag{8.51}$$

where N is the number of data samples, p is the order, and $\hat{\rho}_p$ is the estimated white noise variance (the linear prediction error variance will be used for this estimate). Note that Eq. (8.51) assumes that the sample mean has been subtracted from the data. The term in parentheses increases as the order increases, reflecting the increase in the uncertainty of the estimate $\hat{\rho}_p$ of the prediction error variance. The order p selected is the one for which the FPE is minimum. The FPE has been studied for application by Landers and Lacoss [1977], Ulrych and Bishop [1975], and Nuttall [1976]. For a pure AR process, the FPE works fairly well in selecting the correct order. However, when applied against actual signals, the FPE seems to be conservative, often selecting too low an order. For example, in processing geophysical data, both Jones [1976] and Berryman [1978] found that the order selected tended to be too low.

Akaike [1974] also suggested another selection criterion using a maximum likelihood approach to derive a criterion termed the *Akaike information criterion* (AIC). The AIC determines the model order by minimizing an information theoretic function. Assuming the process has Gaussian statistics, the AIC for AR processes has the form

$$\text{AIC}[p] = N \ln(\hat{\rho}_p) + 2p. \tag{8.52}$$

For a development of this criterion, see Kay [1987]. The term $2p$ represents the penalty for the use of extra AR coefficients that do not result in a substantial reduction in the prediction error variance. Again, the order p selected is the one that minimizes the AIC. As $N \to \infty$, the AIC and FPE are asymptotically equivalent. The AIC has been studied for application to pure AR processes by Tong [1975, 1977] and Kaveh and Bruzzone [1979]. As with the FPE, many researchers have found that the order selected by AIC is often too low for practical non-AR data sets. Kaveh and Bruzzone suggest a variant of the AIC that accounts for the initial conditions in the maximum likelihood formulation of the estimator of the AR parameters. Note that Kashyap [1980] has found that the AIC is statistically inconsistent in that the probability of error in choosing the correct order does not tend to zero as $N \to \infty$; the result is a tendency to overestimate the order as the data record length increases. In response to this, Rissanen [1983] developed a variant information theoretic criterion to the AIC with the form

$$\text{MDL}[p] = N \ln(\hat{\rho}_p) + p \ln(N), \tag{8.53}$$

where MDL stands for *minimum description length*. It is said to be statistically consistent because $p \ln(N)$ increases with N faster than with p.

A third method was proposed by Parzen [1974] and is termed the *criterion autoregressive transfer* (CAT) function. The order p is selected to be the one in which the estimate of the difference between mean square errors of the true prediction error filter (which may be of infinite length) and the estimated filter is a minimum. Parzen showed that this difference can be calculated, without explicitly knowing the true prediction error filter, by

$$\text{CAT}[p] = \left(\frac{1}{N} \sum_{j=1}^{p} \bar{\rho}_j^{-1} \right) - \bar{\rho}_p^{-1} \qquad (8.54)$$

where $\bar{\rho}_j = [N/(N-j)]\hat{\rho}_j$. Again, p is chosen to minimize $\text{CAT}[p]$.

The results of spectra using the FPE, AIC, and CAT have been mixed, particularly when applied to actual data rather than simulated AR processes. Ulrych and Clayton [1976] have found that for short data segments, none of the criteria work well. For harmonic processes in noise, the FPE and AIC also tend to underestimate the order, especially if the SNR is high. Ulrych and Ooe [1983] experimentally determined, in the case of short data segments of harmonic processes, that an order selection between $N/3$ to $N/2$ often produced satisfactory results when using the covariance or modified covariance methods. Reinforcement for this experimental observation was given by Lang and McClellan [1980], who showed analytically for the modified covariance method that the spectral peak position variance for one sinusoid or multiple sinusoids well separated in frequency closely approach the Cramer-Rao lower bound when the model order is about $1/3$ of the data length. In the final analysis, subjective judgment, rather than science, is still required to select an order for data from actual signals of an unknown process. The criteria presented in this section may be used as guidelines for initial order selection, which are known to work well for computer-generated synthetic AR signals, but may or may not work well with actual data, depending on how well such data is modeled by an AR process.

8.8 AUTOREGRESSIVE PROCESSES WITH OBSERVATION NOISE

In many practical situations, observation noise corrupts the data samples, degrading the performance and resolution of AR spectral estimates [Kay, 1979]. The reason for this degradation is that the all-pole model assumed for the spectral analysis is no longer valid when observation noise is present. Assume $y[k]$ is the noise-corrupted AR process

$$y[k] = x[k] + n[k],$$

where $x[k]$ is the AR(p) process and $n[k]$ is the observation noise. If $n[k]$ is white noise of variance ρ_n and is uncorrelated with $x[k]$, then

$$\begin{aligned}
P_{yy}(z) &= \frac{\rho_w}{A(z)A^*(1/z^*)} + \rho_n = \frac{[\rho_w + \rho_n A(z)A^*(1/z^*)]}{A(z)A^*(1/z^*)} \\
&= \frac{\rho[B(z)B^*(1/z^*)]}{A(z)A^*(1/z^*)}.
\end{aligned}$$

Thus, the z-transform of the output process autocorrelation is characterized by poles and zeros, i.e., $y[k]$ is an ARMA(p, p) process. Processes of this type should be analyzed, therefore, by the ARMA methods of Chap. 10.

If the observation noise is white, then several concepts have been proposed to reduce the bias effect of this noise through noise power cancellation schemes (e.g., Marple [1978] or Kay [1980]). Noting that the observation noise power affects only the zero-lag autocorrelation term,

$$r_{yy}[k] = \begin{cases} r_{xx}[0] + \rho_n & \text{for } k = 0 \\ r_{xx}[k] & \text{for } k \neq 0 \end{cases},$$

then compensating the estimate of the $r_{yy}[0]$ autocorrelation term by removing an estimate of ρ_n serves to remove the effect of the observation noise. The Pisarenko harmonic decomposition method in Chap. 13 provides one such special scheme for estimating ρ_n for the case of sinusoids in white noise. Similar compensation schemes can be developed for the reflection coefficients. In general, the noise cancellation schemes can reduce the bias, but will tend to increase the variance of the spectral estimate. A serious deficiency is that one does not know how much noise power to remove; if the estimate of ρ_n is too large, the estimated AR spectrum will exhibit sharper peaks than the true spectrum.

8.9   APPLICATION TO SUNSPOT NUMBERS

The same filtered sequence of 230 samples of sunspot numbers used in Sec. 5.9 was used to produce autoregressive power spectral density estimates. The AIC expression, Eq. (8.5), was used to attempt an estimate of an appropriate AR model order. Figure 8.6 plots the AIC from order 0 to order 32. No obvious minimum is apparent. The AIC of the Burg method levels out approximately at order 12. Figure 8.7 has spectral plots of the AR PSD estimates at order 12 by the covariance and Burg methods. The spectra are essentially identical. The detail in the spectra, however, is disappointing. This illustrates the conservative nature of order selection

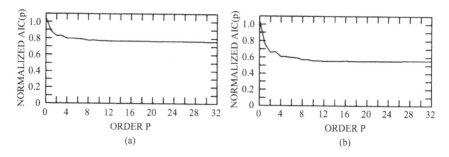

Figure 8.6. AIC function of filtered sunspot numbers, normalized to AIC[0]. (a) AIC of covariance method. (b) AIC of Burg method.

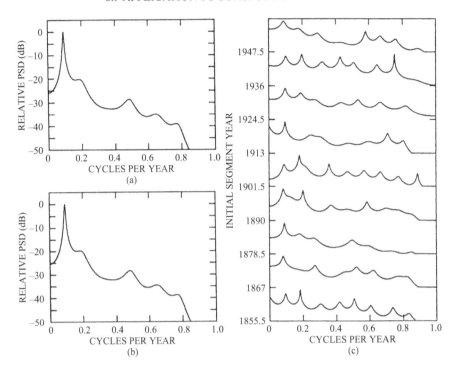

Figure 8.7. Autoregressive spectral estimates of filtered sunspot numbers. Order was $p = 12$. (a) Covariance method (forward coefficients). (b) Burg method. (c) Stack of AR(18) covariance method spectral estimates of sunspot number segments of 46 samples each.

rules like AIC. Figure 8.8(c) illustrates the short-term behavior of autoregressive spectral estimates using the covariance method for order 18. Each spectral estimate represents a segment of 46 filtered sunspot samples, beginning with the year shown on the absissca. A 50-dB range is plotted for each segment spectral estimate. A similar nonstationary behavior, as discovered in Fig. 5.12, is also obvious with the AR spectral estimates.

If the order is increased to 36, then the autoregressive spectral estimates of Fig. 8.8 begin to show more detail. Three spectral estimation methods were selected, including spectral estimates made from both the forward and the backward coefficients of the covariance linear prediction method. All four spectral estimates are very similar. A peak at approximately 22 years per cycle is present in all four spectra. The first three peaks of Fig. 8.8(a), converted to units of period, are located at 20.90, 10.72, and 5.20 years per cycle. The first three peaks of Fig. 8.8(b) are located at 19.88, 10.67, and 5.21 years per cycle. The first three peaks of Fig. 8.8(c) are located at 21.33, 10.56, and 5.25 years per cycle. The first three peaks of Fig. 8.8(d) are

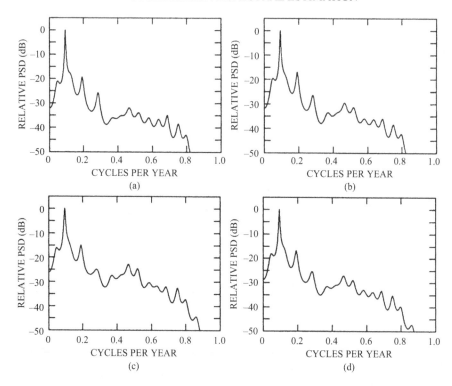

Figure 8.8. Autoregressive spectral estimates of filtered sunspot numbers. Order was $p = 36$. (a) Covariance method (forward coefficients). (b) Burg method. (c) Covariance method (backward coefficients). (d) Modified covariance method.

located at 20.90, 10.67, and 5.21 years per cycle. These sunspot cycle components are very similar to those estimated by Currie [1973].

8.10 EXTRA: COVARIANCE LINEAR PREDICTION FAST ALGORITHM

8.10.1 Introduction

This section has further computational reductions for the published version of this fast algorithm [Marple, 1981]. The algorithm computes all lower-order least squares solutions for free as part of the recursive structure of the algorithm which is useful if searching for a correct order solution. The implemented MATLAB function does not save all of the intermediate lower-order solutions, although the program could be easily modified to do so.

The pth-order forward and backward linear prediction filter errors may be represented as the filter vector inner products

$$e_p^f[n] = \mathbf{x}_p^T[n]\mathbf{a}_p^f \tag{8.55}$$

$$e_p^b[n] = \mathbf{x}_p^T[n]\mathbf{J}\mathbf{a}_p^b \tag{8.56}$$

where the data vector $\mathbf{x}_p[n]$, forward linear prediction coefficient vector $\mathbf{a}_p^f$, and backward linear prediction coefficient vector $\mathbf{a}_p^b$ have the definitions

$$\mathbf{x}_p[n] = \begin{pmatrix} x[n] \\ x[n-1] \\ \vdots \\ x[n-p] \end{pmatrix}, \quad \mathbf{a}_p^f = \begin{pmatrix} 1 \\ a_p^f[1] \\ \vdots \\ a_p^f[p] \end{pmatrix}, \quad \mathbf{a}_p^b = \begin{pmatrix} 1 \\ a_p^b[1] \\ \vdots \\ a_p^b[p] \end{pmatrix}. \tag{8.57}$$

Observe that $\mathbf{a}_p^f$ and $\mathbf{a}_p^b$ as defined here include the unity element, in contrast to definitions (8.25) and (8.40). Based on measured complex data samples $x[1], \ldots, x[N]$, the covariance method of linear prediction *separately* minimizes the forward and backward linear prediction squared errors

$$\rho_p^f = \sum_{n=p+1}^{N} |e_p^f[n]|^2, \quad \rho_p^b = \sum_{n=p+1}^{N} |e_p^b[n]|^2, \tag{8.58}$$

resulting in the normal equations

$$\mathbf{R}_p\mathbf{a}_p^f = \begin{pmatrix} \rho_p^f \\ \mathbf{0}_p \end{pmatrix}, \quad \mathbf{R}_p\mathbf{a}_p^b = \begin{pmatrix} \mathbf{0}_p \\ \rho_p^b \end{pmatrix}, \tag{8.59}$$

where $\mathbf{0}_p$ is a $p \times 1$ all-zeros vector. Note that approximate variance estimates would be $\rho^f/(N-p)$ and $\rho^b/(N-p)$. Unlike the Levinson algorithm, $\mathbf{a}_p^f \neq \mathbf{a}_p^{b*}$ in general for the covariance algorithm. A vector outer product representation for $\mathbf{R}_p$, that is algebraically equivalent to that given by Eq. (8.33), is

$$\mathbf{R}_p = \sum_{n=p+1}^{N} \mathbf{x}_p^*[n]\mathbf{x}_p^T[n]. \tag{8.60}$$

If the error term $e[p+1]$ is not used, the normal equations that result from considering just the data sequence $x[2], \ldots, x[N]$ have the form

$$e_p^{f'}[n] = \mathbf{x}_p^T[n]\mathbf{a}_p^{f'}, \quad \rho_p^{f'} = \sum_{n=p+2}^{N} \left|e_p^{f'}[n]\right|^2,$$

$$\mathbf{R}_p'\mathbf{a}_p^{f'} = \begin{pmatrix} \rho_p^{f'} \\ \mathbf{0}_p \end{pmatrix}, \quad \mathbf{R}_p'\mathbf{a}_p^{b'} = \begin{pmatrix} \mathbf{0}_p \\ \rho_p^{b'} \end{pmatrix}, \tag{8.61}$$

where the prime denotes this as the solution for the reduced data sequence case. Similar to (8.60), it can be shown that

$$\mathbf{R}'_p = \sum_{n=p+2}^{N} \mathbf{x}^*_p[n]\mathbf{x}^T_p[n]. \tag{8.62}$$

Similarly, if the error term $e[N]$ is eliminated, the normal equations considering just the data sequence $x[1], \ldots, x[N-1]$ have the form

$$e^{b''}_p[n] = \mathbf{x}^T_p[n]\mathbf{J}\mathbf{a}^{b''}_p, \quad \rho^{b''}_p = \sum_{n=p+1}^{N-1} \left| e^{b''}_p[n] \right|^2,$$

$$\mathbf{R}''_p \mathbf{a}^{f''}_p = \begin{pmatrix} \rho^{f''}_p \\ \mathbf{0}_p \end{pmatrix}, \quad \mathbf{R}''_p \mathbf{a}^{b''}_p = \begin{pmatrix} \mathbf{0}_p \\ \rho^{b''}_p \end{pmatrix}, \tag{8.63}$$

where the double prime denotes these as the solutions for the reduced data sequence case. In this situation

$$\mathbf{R}''_p = \sum_{n=p+1}^{N-1} \mathbf{x}^*_p[n]\mathbf{x}^T_p[n]. \tag{8.64}$$

8.10.2 Special Partitions and Auxiliary Parameters

Crucial to the development of a fast algorithm are the following order-index partitions of the $(p+1) \times (p+1)$ matrix $\mathbf{R}_p$,

$$\mathbf{R}_p = \begin{pmatrix} r_p[0,0] & \mathbf{s}^H_p \\ \mathbf{s}_p & \mathbf{R}''_{p-1} \end{pmatrix}, \quad \mathbf{R}_p = \begin{pmatrix} \mathbf{R}'_{p-1} & \mathbf{r}_p \\ \mathbf{r}^H_p & r_p[p,p] \end{pmatrix}, \tag{8.65}$$

where the $p \times 1$ column vectors $\mathbf{r}_p$ and $\mathbf{s}_p$ are

$$\mathbf{s}_p = \begin{pmatrix} r_p[1,0] \\ \vdots \\ r_p[p,0] \end{pmatrix}, \quad \mathbf{r}_p = \begin{pmatrix} r_p[0,p] \\ \vdots \\ r_p[p-1,p] \end{pmatrix}. \tag{8.66}$$

Partitions (8.65) are very similar to partitions (3.124) of the Toeplitz matrix given in Chap. 3.

Also crucial are the following time-index partitions

$$\mathbf{R}'_p = \mathbf{R}_p - \mathbf{x}^*_p[p+1]\mathbf{x}^T_p[p+1] \tag{8.67}$$

$$\mathbf{R}''_p = \mathbf{R}_p - \mathbf{x}^*_p[N]\mathbf{x}^T_p[N]. \tag{8.68}$$

These may be considered to be the adjustments necessary to account for the very nearly Toeplitz structure of $\mathbf{R}_p$.

The pair of auxiliary $(p+1) \times 1$ vectors $\mathbf{c}_p$ and $\mathbf{d}_p$, which act as gain vectors, are also needed in the development of a fast algorithm. These are defined by

$$\mathbf{R}_p \mathbf{c}_p = \mathbf{x}_p^*[N] \qquad (8.69)$$
$$\mathbf{R}_p \mathbf{d}_p = \mathbf{x}_p^*[p+1] \qquad (8.70)$$

and have the structure

$$\mathbf{c}_p = \begin{pmatrix} c_p[0] \\ \vdots \\ c_p[p] \end{pmatrix}, \quad \mathbf{d}_p = \begin{pmatrix} d_p[0] \\ \vdots \\ d_p[p] \end{pmatrix}.$$

Considering vectors $\mathbf{c}_p$ and $\mathbf{d}_p$ as filter coefficients, then "gain error" terms similar to that of (8.55) and (8.56) can be formed

$$e_p^c[n] = \mathbf{x}_p^T \mathbf{c}_p, \quad e_p^d[n] = \mathbf{x}_p^T[n]\mathbf{d}_p. \qquad (8.71)$$

Similarly defined vectors $\mathbf{c}_p'$ and $\mathbf{d}_p'$ are associated with the matrix $\mathbf{R}_p'$, and $\mathbf{c}_p''$ and $\mathbf{d}_p''$ are associated with the matrix $\mathbf{R}_p''$. Similar gain error terms $e_p^{c'}[n]$ and $e_p^{d''}$ as (8.71) can be defined.

8.10.3 Order Update Recursions

The order update for the forward linear prediction vector is very similar in structure to the Levinson recursion. It is proposed, with hindsight, that the update has the structure

$$\mathbf{a}_p^f = \begin{pmatrix} \mathbf{a}_{p-1}^{f'} \\ 0 \end{pmatrix} + k_p^f \begin{pmatrix} 0 \\ \mathbf{Ja}_{p-1}^{b''} \end{pmatrix}, \qquad (8.72)$$

where $a_p^f[p]$ is given by

$$a_p^f[p] = k_p^f = -\frac{\Delta_p}{\rho_{p-1}^{b''}} \qquad (8.73)$$

in which

$$\Delta_p = \mathbf{r}_p^H \mathbf{a}_{p-1}^{f'}.$$

Parameter k_p^f can be interpreted as the forward linear predictor reflection coefficient. To verify (8.72), premultiply both sides by $\mathbf{R}_p$ and substitute the appropriate order partition (8.65). This will also yield the relationship for the forward prediction squared error

$$\rho_p^f = \rho_{p-1}^{f'} - \Delta_p \nabla_p / \rho_{p-1}^{b''} = \rho_{p-1}^{f'} \left(1 - k_p^f k_p^b \right), \qquad (8.74)$$

where
$$\nabla_p = \mathbf{s}_p^H \mathbf{a}_{p-1}^{b''}.$$

It may be shown that $\nabla_p = \Delta_p^*$, thus ensuring that (8.74) is real-valued. The order update for the backward linear prediction vector is

$$\mathbf{a}_p^b = \begin{pmatrix} \mathbf{a}_{p-1}^{b''} \\ 0 \end{pmatrix} + k_p^b \begin{pmatrix} 0 \\ \mathbf{J}\mathbf{a}_{p-1}^{f'} \end{pmatrix}, \tag{8.75}$$

where
$$a_p^b[p] = k_p^b = -\nabla_p / \rho_{p-1}^{f'} = -\Delta_p^* / \rho_{p-1}^{f'}. \tag{8.76}$$

Equation (8.75) may be verified by premultiplication of both sides by $\mathbf{R}_p$ and substitution of the appropriate order partitions (8.65). This will also yield the relationship for the backward prediction squared error

$$\rho_p^b = \rho_{p-1}^{b''} - |\Delta_p|^2 / \rho_{p-1}^{f'} = \rho_{p-1}^{b''} \left(1 - k_p^f k_p^b \right). \tag{8.77}$$

By multiplying Eqs. (8.72) and (8.75) on the left by $\mathbf{x}_p^T[n]$ will yield the following linear prediction error order-update recursions

$$e_p^f[n] = e_{p-1}^{f'}[n] + k_p^f e_{p-1}^{b''}[n-1] \tag{8.78}$$

$$e_p^b[n] = e_{p-1}^{b''}[n-1] + k_p^b e_{p-1}^{f'}[n] \tag{8.79}$$

for $p + 1 \leq n \leq N$.

The vectors $\mathbf{c}_p$ and $\mathbf{d}_p$ obey the following order update recursions

$$\mathbf{c}_p = \begin{pmatrix} \mathbf{c}_{p-1}' \\ 0 \end{pmatrix} + c_p[p] \mathbf{J}\mathbf{a}_p^b \tag{8.80}$$

$$\mathbf{d}_p = \begin{pmatrix} 0 \\ \mathbf{d}_{p-1}'' \end{pmatrix} + d_p[0] \mathbf{a}_p^f, \tag{8.81}$$

which, again, may be verified by premultiplying both sides by $\mathbf{R}_p$ and using either the partitions (8.65) and the expressions (8.59). The scalars $c_p[p]$ and $d_p[0]$ may be determined from the following scalar identities (which uses the fact that $\mathbf{R}_p$ is Hermitian)

$$\mathbf{d}_p^H \mathbf{R}_p \mathbf{a}_p^f = (\mathbf{a}_p^{f\,H} \mathbf{R}_p \mathbf{d}_p)^* \implies d_p[0] = (e_p^f[p+1])^* / \rho_p^f \tag{8.82}$$

$$\mathbf{c}_p^H \mathbf{R}_p \mathbf{a}_p^b = (\mathbf{a}_p^{b\,H} \mathbf{R}_p \mathbf{c}_p)^* \implies c_p[p] = (e_p^b[N])^* / \rho_p^b. \tag{8.83}$$

Define the gain scalars δ_p and γ_p to be

$$\delta_p = 1 - e_p^d[p+1] = 1 - \mathbf{x}_p^T[p+1]\mathbf{d}_p = 1 - \mathbf{x}_p^T[p+1]\mathbf{R}_p^{-1}\mathbf{x}_p^*[p+1] \tag{8.84}$$

$$\gamma_p = 1 - e_p^c[N] = 1 - \mathbf{x}_p^T[N]\mathbf{c}_p = 1 - \mathbf{x}_p^T[N]\mathbf{R}_p^{-1}\mathbf{x}_p^*[N]. \tag{8.85}$$

From inspection of the quadratic products (8.82) and (8.83), δ_p and γ_p may be seen to be real-valued. Similarly, the real-valued terms $\delta_p'' = 1 - e_p^{b''}[p+1]$ and

$\gamma'_p = 1 - e^{c'}_p[N]$ can be defined. If each equation in (8.79) is multiplied on the left by $\mathbf{x}^T_p[n]$, it will yield the order updates for the filtered gain errors

$$e^c_p[n] = e^{c'}_{p-1}[n] + c_p[p]e^b_b[n] \qquad (8.86)$$

$$e^d_p[n] = e^{d''}_{p-1}[n-1] + d_p[0]e^f_p[n] \qquad (8.87)$$

for $p + 1 \leq n \leq N$. An important relationship is

$$\Delta_p = \sum_{n=p+1}^{N} e^{f'}_{p-1}[n] \left(e^{b''}_{p-1}[n-1]\right)^* \qquad (8.88)$$

which shows that Δ_p is a cross correlation coefficient similar to Eq. (7.77).

8.10.4 Time Update Recursion

The time-index update of the forward linear prediction coefficient vector is given by

$$\mathbf{a}^{f'}_p = \mathbf{a}^f_p + \frac{e^f_p[p+1]}{\delta''_{p-1}} \begin{pmatrix} 0 \\ \mathbf{d}''_{p-1} \end{pmatrix}, \qquad (8.89)$$

which may be verified by premultiplying both sides by $\mathbf{R}'_p$ and substituting the time-index partition (8.67) and order-index partition (8.65) as appropriate. A by-product of this verification is the time-index update for the forward linear prediction squared error

$$\rho^{f'}_p = \rho^{f*}_p - e^f_p[p+1] \left(\frac{e^f_p[p+1]}{\delta''_{p-1}}\right). \qquad (8.90)$$

Similarly, the backward linear prediction coefficient vector is given by

$$\mathbf{a}^{b''}_p = \mathbf{a}^b_p + \frac{e^b_p[N]}{\gamma'_{p-1}} \begin{pmatrix} 0 \\ \mathbf{Jc}'_{p-1} \end{pmatrix}, \qquad (8.91)$$

which may be verified by multiplying both sides on the left by $\mathbf{R}''_p$ and substituting the time-index partition (8.68) and the order-index partition (8.65) as appropriate. A by-product of this verification is the update for the backward linear prediction squared error

$$\rho^{b''}_p = \rho^{b*}_p - e^b_p[N] \left(\frac{e^b_p[N]}{\gamma'_{p-1}}\right). \qquad (8.92)$$

If (8.89) and (8.91) are multiplied on the left by $\mathbf{x}_p^T[n]$, this will yield the following time-update recursions for the forward and backward linear prediction errors

$$e_p^{f'}[n] = e_p^f[n] + \left(\frac{e_p^f[p+1]}{\delta_{p-1}''}\right) e_{p-1}^{d''}[n-1] \tag{8.93}$$

$$e_p^{b''}[n] = e_p^b[n] + \left(\frac{e_p^b[N]}{\gamma_{p-1}'}\right) e_{p-1}^{c'}[n] \tag{8.94}$$

for $p+1 \leq n \leq N$. The auxiliary gain vectors $\mathbf{c}_p$ and $\mathbf{d}_p$ satisfy the following time-index update recursions

$$\mathbf{c}_p' = \mathbf{c}_p + \left(\frac{e_p^{d*}[N]}{1 - e_p^d[p+1]}\right) \mathbf{d}_p \tag{8.95}$$

$$\mathbf{d}_p'' = \mathbf{d}_p + \left(\frac{e_p^c[p+1]}{1 - e_p^c[N]}\right) \mathbf{c}_p. \tag{8.96}$$

The update for $\mathbf{c}_p$ may be verified by multiplying both sides of Eq. (8.92) on the left by $\mathbf{R}_p'$ and substituting partition (8.67); the update for $\mathbf{d}_p$ may be verified by multiplying both sides of Eq. (8.93) on the left by $\mathbf{R}_p'$ and substituting partition (8.67). The time-index recursions for the filtered gain errors $e_p^{c'}[n]$ and $e_p^{d''}$ may be found simply by forming $\mathbf{x}_p^T[N]\mathbf{c}_p'$ from Eq. (8.92) and $\mathbf{x}_p^T[p+1]\mathbf{d}_p''$ from Eq. (8.93), yielding

$$e_p^{c'}[n] = e_p^c[n] + \left(\frac{e_p^{d*}[N]}{1 - e_p^d[p+1]}\right) e_p^d[n] \tag{8.97}$$

$$e_p^{d''}[n] = e_p^d[n] + \left(\frac{e_p^c[p+1]}{1 - e_p^c[N]}\right) e_p^c[n] \tag{8.98}$$

for $p+1 \leq n \leq N$.

8.10.5   Initial Conditions

The initial conditions needed to commence the recursive solution starting at order zero are

$$\rho_0^f = \rho_0^b = \sum_{n=1}^{N} |x[n]|^2 \tag{8.99}$$

$$\rho_0^{f'} = \sum_{n=2}^{N} |x[n]|^2 = \rho_0^f - |x[1]|^2 \tag{8.100}$$

$$\rho_0^{b''} = \sum_{n=1}^{N-1} |x[n]|^2 = \rho_0^b - |x[N]|^2 \tag{8.101}$$

$$c_0[0] = x^*[N]/\rho_0^b \tag{8.102}$$

$$d_0[0] = x^*[1]/\rho_0^f \tag{8.103}$$

$$e_0^f[n] = e_0^{f'}[n] = e_0^b[n] = e_0^{b''}[n] = x[n], \quad 1 \le n \le N \tag{8.104}$$

$$e_0^c[n] = x[n]c_0[0] = x[n]x^*[N]/\rho_0^f, \quad 1 \le n \le N \tag{8.105}$$

$$e_0^d[n] = x[n]d_0[0] = x[n]x^*[1]/\rho_0^f, \quad 1 \le n \le N, \tag{8.106}$$

which follow from their definitions.

8.10.6 Simple Checks for Ill-Conditioning

Several simple checks are built into the algorithm for testing for numerical ill-conditioning or singularity of the normal equations. The squared prediction errors ρ^f, ρ^b, $\rho^{f'}$, and $\rho^{b''}$ are, by definition, positive real scalars. If these ever become negative, it is due to numerical ill-conditioning. If any become zero, it would indicate a singular normal equation matrix. The real scalars $e_p^c[N]$, $e_p^{c'}[N]$, $e_p^d[p+1]$, and $e_p^{d''}[p+1]$ must be bounded between zero and unity, as will be shown. If any of these scalars fall outside this range, it is indicative of numerical ill-conditioning.

To illustrate the bounds for the δ and γ scalars, consider the matrix inversion lemma, Eq. (3.53), applied to time-index partition equation (8.67),

$$\left(\mathbf{R}_p'\right)^{-1} = \mathbf{R}_p^{-1} - \left(1 - \mathbf{x}_p^T[p+1]\mathbf{R}_p^{-1}\mathbf{x}_p^*[p+1]\right)^{-1}$$
$$\times \left(\mathbf{R}_p^{-1}\mathbf{x}_p^*[p+1]\right)\left(\mathbf{x}_p^T[p+1]\mathbf{R}_p^{-1}\right)$$

or

$$\left(\mathbf{R}_p'\right)^{-1} = \mathbf{R}_p^{-1} - \mathbf{d}_p\mathbf{d}_p^H/\delta_p.$$

Form the quadratic product

$$\mathbf{x}_p^T[p+1]\left(\mathbf{R}_p'\right)^{-1}\mathbf{x}_p^*[p+1]$$
$$= \mathbf{x}_p^T[p+1]\mathbf{R}_p^{-1}\mathbf{x}_p^*[p+1] - \mathbf{x}_p^T[p+1]\mathbf{d}_p\mathbf{d}_p^H\mathbf{x}_p^*[p+1]/\delta_p$$
$$= (1 - \delta_p) - (1 - \delta_p)^2/\delta_p = 2\delta_p(1 - \delta_p).$$

If $\mathbf{R}_p'$ is an invertible positive-definite matrix, then this quadratic product must be positive,

$$2\delta_p(1 - \delta_p) > 0.$$

For this to be true, the real scalar δ_p must satisfy $0 < \delta_p < 1$.

8.10.7   Computation and Storage Requirements

In the main loop, there are $N + 12m$ computational operations (adds or multiplies) per pass that are functions of N or m. For $m = 1$ to $m = M$, this results in a total of $NM + 6M^2$ operations (counting only the quadratic terms). An examination of the vector dimensions in the program indicates memory requirements of $N + 5M$ are needed to store the contents of the vectors. Note that this storage is linear in order, rather than quadratic as would be the case had the entire $\mathbf{R}_p$ matrix been stored. All lower-order solutions are obtained as part of the recursive solution.

8.11   EXTRA: MODIFIED COVARIANCE LINEAR PREDICTION FAST ALGORITHM

8.11.1   Introduction

The original fast algorithm for the solution of the modified covariance normal equations was published by Marple [1980]. The presentation here contains further computational reductions over the original algorithm. The pth-order forward and backward linear prediction errors for the modified covariance method may be represented as the vector inner products

$$e_p^f[n] = \mathbf{x}_p^T[n]\mathbf{a}_p^{fb} \tag{8.107}$$

$$e_p^b[n] = \mathbf{x}_p^T[n]\mathbf{J}\mathbf{a}_p^{fb*}, \tag{8.108}$$

where the data vector $\mathbf{x}_p[n]$ and linear prediction coefficient vector $\mathbf{a}_p^{fb}$ are defined as

$$\mathbf{x}_p[n] = \begin{pmatrix} x[n] \\ x[n-1] \\ \vdots \\ x[n-p] \end{pmatrix}, \quad \mathbf{a}_p^{fb} = \begin{pmatrix} 1 \\ a_p[1] \\ \vdots \\ a_p[p] \end{pmatrix}, \tag{8.109}$$

and $\mathbf{J}$ is a $(p+1) \times (p+1)$ reflection matrix. Based on measured complex data samples $x[1], \ldots, x[N]$, the modified covariance method minimizes the average of the forward and backward linear prediction squared errors

$$\rho_p^{fb} = \sum_{n=p+1}^{N} \left(\left| e_p^f[n] \right|^2 + \left| e_p^b[n] \right|^2 \right), \tag{8.110}$$

resulting in the normal equations

$$\mathbf{R}_p \mathbf{a}_p^{fb} = \begin{pmatrix} \rho_p^{fb} \\ \mathbf{0}_p \end{pmatrix}, \tag{8.111}$$

where $\mathbf{0}_p$ is a $p \times 1$ all-zero vector. Note that an appropriate variance estimate is $\rho_p^{fb}/2(N-p)$ for spectral estimation. A dyadic vector definition for $\mathbf{R}_p$, equivalent to that given by Eq. (8.47), is

$$\mathbf{R}_p = \sum_{n=p+1}^{N} \left(\mathbf{x}_p^*[n]\mathbf{x}_p^T[n] + \mathbf{J}\mathbf{x}_p[n]\mathbf{x}_p^H[n]\mathbf{J} \right). \tag{8.112}$$

Note that $\mathbf{R}_p$ is persymmetric, because $\mathbf{J}\mathbf{R}_p^*\mathbf{J} = \mathbf{R}_p$, and also Hermitian, because $\mathbf{R}_p^H = \mathbf{R}_p$. These properties are the result of the assumption that $\mathbf{a}_p^{fb}$ is the forward linear prediction vector and the conjugated and inverted vector $\mathbf{J}\mathbf{a}_p^{fb*}$ is the backward linear prediction coefficient vector (these were different in the covariance method).

If the forward prediction error term $e_p^f[p+1]$ and the backward prediction error term $e_p^b[N]$ are both not used, the resulting squared error

$$\rho_p^{fb\prime} = \sum_{n=p+2}^{N} \left(|e_p^{f\prime}[n]|^2 + |e_p^{b\prime}[n-1]|^2 \right) \tag{8.113}$$

is minimized, where $e_p^{f\prime}[n] = \mathbf{x}_p^T[n]\mathbf{a}_p^\prime$ and $e_p^{b\prime}[n] = \mathbf{x}_p^T[n]\mathbf{J}\mathbf{a}_p^{\prime*}$. The linear prediction coefficients will then satisfy the normal equations

$$\mathbf{R}_p^\prime \mathbf{a}_p^{fb\prime} = \begin{pmatrix} \rho_p^{fb\prime} \\ \mathbf{0}_p \end{pmatrix}, \tag{8.114}$$

where the prime denotes the solution for this case of omitted error terms. The matrix $\mathbf{R}_p^\prime$ may be expressed as

$$\mathbf{R}_p^\prime = \sum_{n=p+2}^{N} \left(\mathbf{x}_p^*[n]\mathbf{x}_p^T[n] + \mathbf{J}\mathbf{x}_p[n-1]\mathbf{x}_p^H[n-1]\mathbf{J} \right). \tag{8.115}$$

8.11.2 Special Partitions and Auxiliary Parameters

Crucial to the development of a fast algorithm are the following order-index partitions of the $(p+1) \times (p+1)$ matrix $\mathbf{R}_p$,

$$\mathbf{R}_p = \begin{pmatrix} \mathbf{R}_{p-1}^\prime & \mathbf{r}_p \\ \mathbf{r}_p^H & r_p[p,p] \end{pmatrix} = \begin{pmatrix} r_p[0,0] & \mathbf{r}_p^T\mathbf{J} \\ \mathbf{J}\mathbf{r}_p^* & \mathbf{J}\mathbf{R}_{p-1}^{\prime*}\mathbf{J} \end{pmatrix}, \tag{8.116}$$

where the $p \times 1$ column vector $\mathbf{r}_p$ is

$$\mathbf{r}_p = \begin{pmatrix} r_p[0,p] \\ \vdots \\ r_p[p-1,p] \end{pmatrix} \tag{8.117}$$

and $\mathbf{J}$ is a $p \times p$ reflection matrix. This is very similar to the partition equation (3.124) of the Toeplitz matrix in Chap. 3. Also crucial is the following time-index partition

$$\mathbf{R}_p = \mathbf{R}'_p + \mathbf{x}_p^*[p+1]\mathbf{x}_p^T[p+1] + \mathbf{J}\mathbf{x}_p[N]\mathbf{x}_p^H[N]\mathbf{J} \qquad (8.118)$$

and its centrosymmetric variant

$$\mathbf{R}_p = \mathbf{J}\mathbf{R}_p'^*\mathbf{J} + \mathbf{J}\mathbf{x}_p[p+1]\mathbf{x}_p^H[p+1]\mathbf{J} + \mathbf{x}_p^*[N]\mathbf{x}_p^T[N]. \qquad (8.119)$$

The pair of $(p+1) \times 1$ auxiliary vectors $\mathbf{c}_p$ and $\mathbf{d}_p$ are also needed in the development of this fast algorithm, defined by the matrix equations

$$\mathbf{R}_p\mathbf{c}_p = \mathbf{J}\mathbf{x}_p[N] \qquad (8.120)$$

$$\mathbf{R}_p\mathbf{d}_p = \mathbf{x}_p^*[p+1], \qquad (8.121)$$

where

$$\mathbf{c}_p = \begin{pmatrix} c_p[0] \\ \vdots \\ c_p[p] \end{pmatrix}, \quad \mathbf{d}_p = \begin{pmatrix} d_p[0] \\ \vdots \\ d_p[p] \end{pmatrix}. \qquad (8.122)$$

Due to the Hermitian persymmetry of $\mathbf{R}_p$, it may easily be shown that

$$\mathbf{J}\mathbf{R}_p^*\mathbf{a}_p^{fb*} = \mathbf{R}_p\mathbf{J}\mathbf{a}_p^{fb*} = \begin{pmatrix} \mathbf{0}_p \\ \rho_p^{fb} \end{pmatrix} \qquad (8.123)$$

$$\mathbf{R}_p\mathbf{J}\mathbf{c}_p^* = \mathbf{x}_p^*[N] \qquad (8.124)$$

$$\mathbf{R}_p\mathbf{J}\mathbf{d}_p^* = \mathbf{J}\mathbf{x}_p[p+1]. \qquad (8.125)$$

Vectors $\mathbf{c}'_p$ and $\mathbf{d}'_p$ are defined for the matrix $\mathbf{R}'_p$ in a similar manner

$$\left(\mathbf{J}\mathbf{R}_p'^*\mathbf{J}\right)\mathbf{c}'_p = \mathbf{J}\mathbf{x}_p[N] \qquad (8.126)$$

$$\left(\mathbf{J}\mathbf{R}_p'^*\mathbf{J}\right)\mathbf{d}'_p = \mathbf{x}_p^*[p+1] \qquad (8.127)$$

and simply use inverted and conjugated $\mathbf{x}_p$ vectors. Considering $\mathbf{c}_p$ and $\mathbf{d}_p$ as filter coefficients, then the following eight forward and backward filtered terms

$$e_p^{cf}[n] = \mathbf{x}_p^T[n]\mathbf{c}_p \qquad e_p^{cf\prime}[n] = \mathbf{x}_p^T\mathbf{c}'_p \qquad (8.128)$$

$$e_p^{cb}[n] = \mathbf{x}_p^T[n]\mathbf{J}\mathbf{c}_p^* \qquad e_p^{cb\prime}[n] = \mathbf{x}_p^T\mathbf{J}\mathbf{c}_p'^* \qquad (8.129)$$

$$e_p^{df}[n] = \mathbf{x}_p^T[n]\mathbf{d}_p \qquad e_p^{df\prime}[n] = \mathbf{x}_p^T\mathbf{d}'_p \qquad (8.130)$$

$$e_p^{db}[n] = \mathbf{x}_p^T[n]\mathbf{J}\mathbf{d}_p^* \qquad e_p^{db\prime}[n] = \mathbf{x}_p^T\mathbf{J}\mathbf{d}_p'^* \qquad (8.131)$$

can be defined.

8.11.3 Order Update Recursions

The order update for the linear prediction coefficient vector is very similar to that found in the Levinson algorithm,

$$\mathbf{a}_p^{fb} = \begin{pmatrix} \mathbf{a}_{p-1}^{fb\prime} \\ 0 \end{pmatrix} + k_p \begin{pmatrix} 0 \\ \mathbf{Ja}_{p-1}^{fb\prime *} \end{pmatrix}, \tag{8.132}$$

where $a_p[p]$ is given by

$$a_p[p] = k_p = -\Delta_p / \rho_{p-1}^{fb\prime}, \tag{8.133}$$

k_p is the reflection coefficient, and

$$\Delta_p = \mathbf{r}_p^H \mathbf{a}_{p-1}^{fb\prime}. \tag{8.134}$$

To verify Eq. (8.132), premultiply both sides by $\mathbf{R}_p$ and substitute the appropriate order partition given in Eq. (8.116). Property (8.125) has also been used. The verification will also yield the order update relationship for the linear prediction squared error

$$\rho_p^{fb} = \rho_{p-1}^{fb\prime} - |\Delta_p|^2 / \rho_{p-1}^{fb\prime} = \rho_{p-1}^{fb\prime} \left(1 - |k_p|^2\right). \tag{8.135}$$

Observe that in order for ρ_p^{fb} to be positive and real, as it must be because it is a squared error, then it must be true that $|a_p[p]|^2 < 1$. Thus, the reflection coefficient of the modified covariance method has *magnitude less than unity*. By multiplying (8.132) by $\mathbf{x}_p^T[n]$ on the left or conjugating (8.132) and then multiplying, the following forward and backward linear prediction error order updates are obtained:

$$e_p^f[n] = e_{p-1}^{f\prime}[n] + k_p e_{p-1}^{b\prime}[n-1] \text{ for } p+2 \le n \le N \tag{8.136}$$

$$e_p^b[n] = e_{p-1}^{b\prime}[n-1] + k_p^* e_{p-1}^{f\prime}[n] \text{ for } p+1 \le n \le N-1. \tag{8.137}$$

The vectors $\mathbf{c}_p$ and $\mathbf{d}_p$ obey the following order update recursions

$$\mathbf{c}_p = \begin{pmatrix} 0 \\ \mathbf{c}_{p-1}' \end{pmatrix} + c_p[0]\mathbf{a}_p^{fb} \tag{8.138}$$

$$\mathbf{d}_p = \begin{pmatrix} 0 \\ \mathbf{d}_{p-1}' \end{pmatrix} + d_p[0]\mathbf{a}_p^{fb}, \tag{8.139}$$

which again may be verified by premultiplying both sides by $\mathbf{R}_p$ and substituting the appropriate order partitions from Eq. (8.116). The scalar terms $c_p[0]$ and $d_p[0]$ may be determined from the following scalar identities (which use the Hermitian property $\mathbf{R}_p^H = \mathbf{R}_p$):

$$\mathbf{c}_p^H \mathbf{R}_p \mathbf{a}_p^{fb} = \left(\mathbf{a}_p^{fbH} \mathbf{R}_p \mathbf{c}_p\right)^* \implies c_p[0] = e_p^b[N]/\rho_p^{fb} \tag{8.140}$$

$$\mathbf{d}_p^H \mathbf{R}_p \mathbf{a}_p^{fb} = \left(\mathbf{a}_p^{fbH} \mathbf{R}_p \mathbf{d}_p\right)^* \implies d_p[0] = e_p^{f*}[p+1]/\rho_p^{fb}. \tag{8.141}$$

Define the scalars δ_p and γ_p to be

$$\gamma_p = 1 - e_p^{cb}[N] = 1 - \mathbf{x}_p^H[N]\mathbf{Jc}_p = 1 - \mathbf{x}_p^H[N]\mathbf{JR}_p^{-1}\mathbf{Jx}_p[N] \tag{8.142}$$

$$\delta_p = 1 - e_p^{df} = 1 - \mathbf{x}_p^T[p+1]\mathbf{d}_p = 1 - \mathbf{x}_p^T[p+1]\mathbf{R}_p^{-1}\mathbf{x}_p^*[p+1]. \tag{8.143}$$

By inspection, these are real-valued due to the particular quadratic product definition assumed.

Multiply each side of the equations (8.138) and (8.139) by $\mathbf{x}_p^T[n]$ on the left. This yields the order update for the filter terms

$$e_p^{cf}[n] = e_{p-1}^{cf\prime}[n-1] + c_p[0]e_p^f[n] \tag{8.144}$$

$$e_p^{df}[n] = e_{p-1}^{df\prime}[n-1] + d_p[0]e_p^f[n]. \tag{8.145}$$

Order updates of the $\mathbf{c}$ and $\mathbf{d}$ filter terms defined in (8.120) and (8.121) are obtained as follows. Either multiply each side of the equations of (8.138) and (8.139) on the left by $\mathbf{x}_p^T[n]$ or conjugate the equations and multiply on the left by $\mathbf{x}_p^T[n]\mathbf{J}$ to yield

$$e_p^{cf}[n] = e_{p-1}^{cf\prime}[n-1] + c_p[0]e_p^f[n] \tag{8.146}$$

$$e_p^{cb}[n] = e_{p-1}^{cb\prime}[n] + c_p^*[0]e_p^b[n] \tag{8.147}$$

$$e_p^{df}[n] = e_{p-1}^{df\prime}[n-1] + d_p[0]e_p^f[n] \tag{8.148}$$

$$e_p^{db}[n] = e_{p-1}^{db\prime}[n] + d_p^*[0]e_p^b[n]. \tag{8.149}$$

Some identities are

$$\mathbf{d}_p^H \mathbf{R}_p \mathbf{c}_p = (\mathbf{c}_p^H \mathbf{R}_p \mathbf{d}_p)^* \implies e_p^{db}[N] = \left(e_p^{cf}[p+1]\right)^* \tag{8.150}$$

$$\mathbf{c}_p^H \mathbf{R}_p \mathbf{Ja}_p^* = \left(\mathbf{a}_p^T \mathbf{JR}_p \mathbf{c}_p\right)^* \implies c_p[p]\rho_p^{fb} = e_p^f[N] \tag{8.151}$$

$$\mathbf{d}_p^H \mathbf{R}_p \mathbf{Ja}_p^* = \left(\mathbf{a}_p^T \mathbf{JR}_p \mathbf{d}_p\right)^* \implies d_p[p]\rho_p^{fb} = e_p^b[p+1]. \tag{8.152}$$

One final order update is required for Δ_p,

$$\Delta_p = 2 \sum_{n=p+1}^{N} e_{p-1}^{f\prime}[n]e_{p-1}^{b\prime*}[n-1]. \tag{8.153}$$

8.11.4   Time Update Recursions

The time-index update of the linear prediction coefficient vector is given by

$$\mathbf{a}_p^{fb\prime} = \mathbf{a}_p^{fb} + \alpha_1 \begin{pmatrix} 0 \\ \mathbf{c}_{p-1}' \end{pmatrix} + \beta_1 \begin{pmatrix} 0 \\ \mathbf{d}_{p-1}' \end{pmatrix}, \tag{8.154}$$

where

$$\alpha_1 = \left(e_p^{b*}[N]\delta_{p-1}' + e_p^{f}[p+1]\lambda_{p-1}' \right)/\mathrm{DEN}_{p-1}' \tag{8.155}$$

$$\beta_1 = \left(e_p^{f}[p+1]\gamma_{p-1}' + e_p^{b*}[N]\lambda_{p-1}'^* \right)/\mathrm{DEN}_{p-1}' \tag{8.156}$$

$$\mathrm{DEN}_{p-1}' = \gamma_{p-1}'\delta_{p-1}' - |\lambda_{p-1}'|^2.$$

Equation (8.154) may be verified by premultiplying both sides by $\mathbf{R}_p'$ and substituting the time-index partition (8.118) and order-index partition (8.116), as appropriate. A by-product of this verification is the update for the linear prediction squared error

$$\rho_p^{fb\prime} = \rho_p^{fb} - \alpha_1 e_p^{b}[N] - \beta_1 \left(e_p^{f}[p+1] \right)^* \tag{8.157}$$

$$= \rho_p^{fb} - \frac{|e_p^{b}[N]|^2\delta_{p-1}' + |e_p^{f}[p+1]|^2\gamma_{p-1}' + 2\,\mathrm{Re}\left\{ e_p^{f}[p+1]e_p^{b}[N]\lambda_{p-1}' \right\}}{\mathrm{DEN}_{p-1}'}. \tag{8.158}$$

The time-index updates of the forward and backward linear prediction errors are obtained either by multiplying the equations of (8.154) on the left by $\mathbf{x}_{p}^{t}{}_{p}[n]$ or by multiplying the conjugated equations of (8.154) on the left by $\mathbf{x}_{p}^{t}{}_{p}[n]\mathbf{J}$ to yield

$$e_p^{f\prime}[n] = e_p^{f}[n] + \alpha_1 e_{p-1}^{cf\prime}[n-1] + \beta_1 e_{p-1}^{df\prime}[n-1] \tag{8.159}$$

$$e_p^{b\prime}[n] = e_p^{b}[n] + \alpha_1^* e_{p-1}^{cb\prime}[n] + \beta_1 e_{p-1}^{db\prime}[n]. \tag{8.160}$$

The auxiliary vectors $\mathbf{c}_p$ and $\mathbf{d}_p$ satisfy the following time-index recursions

$$\mathbf{c}_p' = \mathbf{c}_p + \alpha_2 \mathbf{J}\mathbf{c}_p^* + \beta_2 \mathbf{J}\mathbf{d}_p^* \tag{8.161}$$

$$\mathbf{d}_p' = \mathbf{d}_p + \alpha_3 \mathbf{J}\mathbf{c}_p^* + \beta_3 \mathbf{J}\mathbf{d}_p^*, \tag{8.162}$$

where

$$\alpha_2 = \left(e_p^{cb*}[p+1]e_p^{cf}[p+1] + e_p^{cf}[N]\delta_p \right)/\mathrm{DEN}_p \tag{8.163}$$

$$\beta_2 = \left(e_p^{cf}[N]e_p^{cf*}[p+1] + e_p^{cb*}[p+1]\gamma_p \right)/\mathrm{DEN}_p \tag{8.164}$$

$$\alpha_3 = \left(e_p^{db*}[p+1]e_p^{cf}[p+1] + e_p^{cb*}[p+1]\delta_p \right)/\mathrm{DEN}_p \tag{8.165}$$

$$\beta_3 = \left(e_p^{cb*}[p+1]e_p^{cf*}[p+1] + e_p^{db*}[p+1]\gamma_p \right)/\mathrm{DEN}_p \tag{8.166}$$

$$\mathrm{DEN}_p = \gamma_p\delta_p - |e_p^{cf}[p+1]|^2. \tag{8.167}$$

Equations (8.161) and (8.162) may be verified by premultiplying both sides of these equations by $\mathbf{JR}_p'^*\mathbf{J}$ and substituting time-index partition (8.118). The identity $\theta_p = \mathbf{x}_p^H[p+1]\mathbf{Jc}_p$ has been used to obtain recursions (8.161) and (8.162). This identity follows from the property $\mathbf{c}_p^H\mathbf{JR}_p\mathbf{d}_p^* = \left(\mathbf{d}_p^H\mathbf{R}_p\mathbf{Jc}_p^*\right)^*$.

The time-index recursions of the $\mathbf{c}$ and $\mathbf{d}$ filter terms may be obtained by forming $\mathbf{x}_p^T[n]\mathbf{c}_p'$ and $\mathbf{x}_p^H[n]\mathbf{Jc}_{p}'^*$ from (8.161) and $\mathbf{x}_p^T[n]\mathbf{d}_p'$ and $\mathbf{x}_p^H[n]\mathbf{Jd}_p'^*$ from (8.162), to yield

$$e_p^{cf'}[n] = e_p^{cf}[n] + \alpha_2 e_p^{cb}[n] + \beta_2 e_p^{db}[n] \tag{8.168}$$

$$e_p^{cb'}[n] = e_p^{cb}[n] + \alpha_2^* e_p^{cf}[n] + \beta_2^* e_p^{df}[n] \tag{8.169}$$

and

$$e_p^{df'}[n] = e_p^{df}[n] + \alpha_3 e_p^{cb}[n] + \beta_3 e_p^{db}[n] \tag{8.170}$$

$$e_p^{db'}[n] = e_p^{db}[n] + \alpha_3 e_p^{cf}[n] + \beta_3 e_p^{df}[n]. \tag{8.171}$$

8.11.5 Initial Conditions

The initial conditions needed to start this recursive algorithm at order zero are

$$\rho_0^{fb} = 2\sum_{n=0}^{N-1} |x[n]|^2 \tag{8.172}$$

$$\rho_0^{fb'} = \rho_0^{fb} - |x[0]|^2 - |x[N-1]|^2 \tag{8.173}$$

$$c_0[0] = x[N-1]/\rho_0^{fb} \tag{8.174}$$

$$d_0[0] = x^*[0]/\rho_0^{fb} \tag{8.175}$$

$$e_0^f[n] = e_0^b[n] = e_0^{f'}[n] = e_0^{b'}[n] = x[n] \tag{8.176}$$

$$e_0^{df}[n] = x[n]c_0[0] = x[n]x[N-1]/\rho_0^{fb} \tag{8.177}$$

$$e_0^{db}[n] = x[n]c_0^*[0] = x[n]x^*[N-1]/\rho_0^{fb} \tag{8.178}$$

$$e_0^{cf}[n] = x[n]d_0[0] = x[n]x^*[0]/\rho_0^{fb} \tag{8.179}$$

$$e_0^{cb}[n] = x[n]d_0^*[0] = x[n]x[0]/\rho_0^{fb} \tag{8.180}$$

which may be derived from the parameter definitions. Note that $e_0^{cf}[0] = |x[0]|^2/\rho_0^{fb} = 1 - \delta_0, e_0^{db}[N] = |x[N-1]|^2/\rho_0^{fb} = 1 - \gamma_0$ are real-valued and the complex-valued equality $e_0^{df}[0] = e_0^{cb}[N-1] = x[0]x[N-1]/\rho_0^{fb}$.

8.11.6 Simple Checks for Numerical Ill-Conditioning

As with the covariance algorithm it can be shown that $0 < \mathrm{DEN} \le 1$ and $0 < \mathrm{DEN}' \le 1$. Also, it can be shown that $0 < \gamma, \delta \le 1$. These are checked, as well as the positivity of the squared error terms, at each stage of the algorithm. Barrodale et al. [1983] have provided some improvements to the basic algorithm for both the high-SNR and the high-order cases.

8.11.7 Computation and Storage Count

In the main loop, there are $N + 12p$ computational operations (adds or multiplies) that are functions of N or p. For $p = 1$ to $p = M$, this yields a total of $13NM - 7M^2/2 + 5N$ operations (counting only the quadratic terms). An examination of the vector dimensions in the program listing reveals that memory requirements of $7N + 3M$ are needed.

References

Akaike, H., Power Spectrum Estimation through Autoregression Model Fitting, *Ann. Inst. Stat. Math.*, vol. 21, pp. 407–419, 1969.

Akaike, H., A New Look at the Statistical Model Identification, *IEEE Trans. Autom. Control*, vol. AC-19, pp. 716–723, December 1974.

Andersen, N. O., On the Calculation of Filter Coefficients for Maximum Entropy Spectral Analysis, *Geophysics*, vol. 39, pp. 69–72, February 1974.

Andersen, N. O., Comments on the Performance of Maximum Entropy Algorithms, *Proc. IEEE*, vol. 66, pp. 1581–1582, November 1978.

Baggeroer, A. B., Confidence Intervals for Regression (MEM) Spectral Estimates, *IEEE Trans. Inform. Theory*, vol. IT-22, pp. 534–545, September 1976.

Barrodale, I., L. M. Delves, R. E. Erickson, and C. A. Zala, Computational Experience with Marple's Algorithm for Autoregressive Spectrum Analysis, *Geophysics*, vol. 48, pp. 1274–1286, September 1983.

Berberidis, K., and S. Theodoridis, Efficient Symmetric Algorithms for the Modified Covariance Method of Autoregressive Spectral Analysis, *IEEE Trans. Signal Processing*, 1992.

Berk, K. N., Consistent Autoregressive Spectral Estimates, *Ann. Stat.*, vol. 2, pp. 489–502, 1974.

Berryman, J. G., Choice of Operator Length for Maximum Entropy Spectral Analysis, *Geophysics*, vol. 43, pp. 1384–1391, December 1978.

Burg, J. P., Maximum Entropy Spectral Analysis, *Proceedings of the 37th Meeting of the Society of Exploration Geophysicists*, 1967.

Burg, J. P., Maximum Entropy Spectral Analysis, Ph.D. dissertation, Department of Geophysics, Stanford University, Stanford, Calif., May 1975.

Cadzow, J. A., Spectral Analysis, Chapter 9 in *Handbook of Digital Signal Processing Engineering Applications*, D. F. Elliott, ed., Academic Press, 1987.

Chen, W. Y., and G. R. Stegen, Experiments with Maximum Entropy Power Spectra of Sinusoids, *J. Geophys. Res.*, vol. 79, pp. 3019–3022, 10 July 1974.

Currie, R., Fine Structure in the Sunspot Spectrum—2 to 70 Years, *Astrophys. Space Sci.*, vol. 20, pp. 509–518, 1973.

Datta, A. K., Comments on "The Complex Form of the Maximum Entropy Method for Spectral Estimation," *Proc. IEEE*, vol. 65, PP. 1219–1220, August 1977.

Demeure, C. J., and L. L. Scharf, Non Causal Linear Statistical Models and Related Algorithms, Proceedings of 22nd Asilomar Conference on Signals, Systems, and Computers, pp. 774–778, 1987.

Fougere, P. F., E. J. Zawalick, and H. R. Radoski, Spontaneous Line Splitting in Maximum Entropy Power Spectrum Analysis, *Phys. Earth Planet. Inter.*, vol. 12, pp. 201–207, August 1976.

Friedlander, B., and B. Porat, The Exact Cramer-Rao Bound for Gaussian Autoregressive Processes, *IEEE Trans. Aerospace and Electronic Systems*, vol. AES-25, pp. 3–7, January 1989.

Haykin, S., and S. Kesler, The Complex Form of the Maximum Entropy Method for Spectral Estimation, *Proc. IEEE*, vol. 64, pp. 822–823, May 1976.

Haykin, S., *Adaptive Filter Theory*, Prentice Hall, Englewood Cliffs, NJ, 1991.

Haykin, S., editor, *Advances in Spectrum Analysis and Array Processing*, Volume I, Prentice Hall, Englewood Cliffs, NJ, 1991.

Helme, B., and C. L. Nikias, Improved Spectrum Performance via a Data-Adaptive Weighted Burg Technique, *IEEE Trans. Acoust. Speech Signal Process.*, vol. ASSP-33, pp. 903–910, August 1985.

Herring, R. W., The Cause of Line Splitting in Burg Maximum-Entropy Spectral Analysis, *IEEE Trans. Acoust. Speech Signal Process.*, vol. ASSP-28, pp. 692–701, December 1980.

Jackson, L. B., D. W. Tufts, F. K. Soong, and R. M. Rao, Frequency Estimation by Linear Prediction, *Proceedings of the 1978 IEEE International Conference on Acoustics, Speech, and Signal Processing*, pp. 352–356, 1978.

Jones, R. H., Autoregression Order Selection, *Geophysics*, vol. 41, pp. 771–773, August 1976.

Kane, R. P., and N. B. Trivedi, Effects of Linear Trend and Mean Value on Maximum Entropy Spectral Analysis, *Institute de Pesquisas Espacials, Report INPE-1568-RPE/069*, San Jose dos Compos, Brasil, 1979 (available from NTIS, N79-33949).

Kashyap, R. L., Inconsistency of the AIC Rule for Estimating the Order of Autoregressive Models, *IEEE Trans. Autom. Control*, vol. AC-25, pp. 996–998, October 1980.

Kaveh, M., and S. P. Bruzzone, Order Determination for Autoregressive Spectral Estimation, *Record of the 1979 Rome Air Development Center Spectral Estimation Workshop*, pp. 139–145, 1979.

Kaveh, M., and G. R. Cooper, An Empirical Investigation of the Properties of the Autoregressive Spectral Estimator, *IEEE Trans. Inf. Theory*, vol. IT-22, pp. 313–323, May 1976.

Kaveh, M., and G. A. Lippert, An Optimum Tapered Burg Algorithm for Linear Prediction and Spectral Analysis, *IEEE Trans. Acoust. Speech Signal Process.*, vol. ASSP-31, pp. 438–444, April 1983.

Kay, S. M., The Effect of Noise on the Autoregressive Spectral Estimator, *IEEE Trans. Acoust. Speech Signal Process.*, vol. ASSP-27, pp. 478–485, October 1979.

Kay, S. M., Noise Compensation for Autoregressive Spectral Estimators, *IEEE Trans. Acoust. Speech Signal Process.*, vol. ASSP-28, pp. 292–303, June 1980.

Kay, S. M., Recursive Maximum Likelihood Estimation of Autoregressive Processes, *IEEE Trans. Acoust. Speech Signal Process.*, vol. ASSP-31, pp. 56–65, February 1983.

Kay, S. M., *Modern Spectral Estimation*, Prentice-Hall, Inc., Englewood Cliffs, N. J., 1987.

Kay, S. M., and S. L. Marple, Jr., Sources of and Remedies for Spectral Line Splitting in Autoregressive Spectrum Analysis, *Proceedings of the 1979 IEEE International Conference on Acoustics, Speech, and Signal Processing*, pp. 151–154, 1979.

Keeler, R. J., Uncertainties in Adaptive Maximum Entropy Frequency Estimators, *IEEE Trans. Acoust. Speech Signal Process.*, vol. ASSP-26, pp. 469–471, October 1978.

Kromer, R. E., Asymptotic Properties of the Autoregressive Spectral Estimator, Ph.D. dissertation, Department of Statistics, Stanford University, Stanford, Calif., 1969.

Landers, T. E., and R. T. Lacoss, Some Geophysical Applications of Autoregressive Spectral Estimates, *IEEE Trans. Geosci. Electron.*, vol. GE-15, pp. 26–32, January 1977.

Lang, S. W., and J. H. McClellan, Frequency Estimation with Maximum Entropy Spectral Estimators, *IEEE Trans. Acoust. Speech Signal Process.*, vol. ASSP-28, pp. 716–724, December 1980.

Makhoul, J., Linear Prediction: A Tutorial Review, *Proc. IEEE*, vol. 63, pp. 561–580, April 1975; correction in *Proc. IEEE*, vol. 64, p. 285, February 1976.

Algorithm, *IEEE Trans. Acoust. Speech Signal Process.*, vol. ASSP-28, pp. 441–454, August 1980.

Marple, S. L. Jr., High Resolution Autoregressive Spectrum Analysis Using Noise Power Cancellation, *Proceedings of 1978 IEEE International Conference on Acoustics, Speech, and Signal Processing*, Tulsa, Okla., pp. 345–348, April 1978.

Marple, S. L. Jr., Efficient Least Squares FIR System Identification, *IEEE Trans. Acoust. Speech Signal Process.*, vol. ASSP-29, pp. 62–73, February 1981.

Marple, S. L. Jr., A Fast Computational Algorithm for the Modified Covariance Method of Linear Prediction, *Digital Signal Processing: A Review Journal*, vol. 3, pp. 124–133, July 1991.

Morf, M., B. Dickinson, T. Kailath, and A. Vieira, Efficient Solution of Covariance Equations for Linear Prediction, *IEEE Trans. Acoust. Speech Signal Process.*, vol. ASSP-25, pp. 429–433, October 1977.

Nikias, C. L., and P. D. Scott, The Covariance Least-Squares Algorithm for Spectral Estimation of Processes of Short Data Length, *IEEE Trans. Geosci. Remote Sensing*, vol. GE-21, pp. 180–190, April 1983.

Nuttall, A. H., Spectral Analysis of a Univariate Process with Bad Data Points, via Maximum Entropy and Linear Predictive Techniques, *Naval Underwater Systems Center Technical Report TR-5303*, New London, Conn., March 1976.

Parzen, E., Some Recent Advances in Time Series Modeling, *IEEE Trans. Autom. Control*, vol. AC-19, pp. 723–730, December 1974.

Paulson, A. S., and G. R. Swope, Autoregressive Process Order Selection Via Model-Critical Methods, *IEEE J. Oceanic Engineering*, vol. OE-12, pp. 75–80, January 1987.

Rissanen, J., A Universal Prior for the Integers and Estimation by Minimum Description Length, *Ann. Stat.*, vol. 11, pp. 417–431, 1983.

Sakai, H., Statistical Properties of AR Spectral Analysis, *IEEE Trans. Acoust. Speech Signal Process.*, vol. ASSP-27, pp. 402–409, August 1979.

Shon, S., and K. Mehrotra, Performance Comparison of Autoregressive Estimation Methods, *Proceedings of the 1984 IEEE International Conference on Acoustics, Speech, and Signal Processing*, pp. 14.3.1–14.3.4, 1984.

Slock, D. T. M., and T. Kailath, Split-Levinson-Type Algorithms for LS Linear-Phase Filtering and FBLP Prediction, Proceedings 22nd Asilomar Conference on Signals, Systems, and Computers, pp. 323–327, 1987.

Strobach, P., *Linear Prediction Theory*, Springer-Verlag, 1990.

Swingler, D. N., A Comparison between Burg's Maximum Entropy Method and a Nonrecursive Technique for the Spectral Analysis of Deterministic Signals, *J. Geophys. Res.*, vol. 84, pp. 679–685, 10 February 1979a.

234 AUTOREGRESSIVE SPECTRAL ESTIMATION

Swingler, D. N., A Modified Burg Algorithm for Maximum Entropy Spectral Analysis, *Proc. IEEE*, vol. 67, pp. 1368–1369, September 1979b.

Swingler, D. N., Frequency Errors in MEM Processing, *IEEE Trans. Acoust. Speech Signal Process.*, vol. ASSP-28, pp. 257–259, April 1980.

Thorvaldsen, T., A Comparison of the Least Squares Method and the Burg Method for Autoregressive Spectral Analysis, *IEEE Trans. Antennas Propag.*, vol. AP-29, pp. 675–679, July 1981.

Tong, H., Autoregressive Model Fitting with Noisy Data by Akaike's Information Criterion, *IEEE Trans. Inf. Theory*, vol. IT-21, pp. 476–480, July 1975; see also the sequel in *IEEE Trans. Inf. Theory*, vol. IT-23, pp. 409–410, May 1977.

Ulrych, T. J., and T. N. Bishop, Maximum Entropy Spectral Analysis and Autoregressive Decomposition, *Rev. Geophys. Space Phys.*, vol. 13, pp. 183–200, February 1975.

Ulrych, T. J., and R. W. Clayton, Time Series Modeling and Maximum Entropy, *Phys. Earth Planet. Inter.*, vol. 12, pp. 188–200, August 1976.

Ulrych, T. J., and M. Ooe, Autoregressive and Mixed ARMA Models and Spectra, Chapter 3 in *Nonlinear Methods of Spectral Analysis*, 2nd ed., S. Haykin, ed., Springer-Verlag, New York, 1983.

Ulrych, T. J., D. E. Smylie, O. G. Jensen, and G. K. C. Clarke, Predictive Filtering and Smoothing of Short Records by Using Maximum Entropy, *J. Geophys. Res.*, vol. 78, pp. 4959–4964, August 1973.

Van den Bos, A., Alternative Interpretation of Maximum Entropy Spectral Analysis, *IEEE Trans. Inform. Theory*, vol. IT-17, pp. 493–494, July 1971.

Wax, M., and T. Kailath, Detection of Signals by Information Theoretic Criteria, *IEEE Trans. Acoust. Speech, Sig. Proc.*, vol. ASSP-33, pp. 387–394, April 1985.

Wu, C.-H., and A. E. Yagle, Numerical Performances of Autoregressive Spectrum Estimators Based on Three-Term Recurrences, *IEEE Trans. Signal Processing*, vol. 40, pp. 249–252, January 1992.

9

AUTOREGRESSIVE SPECTRAL ESTIMATION: SEQUENTIAL DATA ALGORITHMS

9.1 INTRODUCTION

Block estimation methods are appropriate for applications where data is very limited and the best possible performance with the available data is desired. For longer data sets, a variety of sequential estimation techniques are available for updating the autoregressive parameter estimates as each new data sample is received from a real-time data acquisition system. A new spectral density plot can then be made from these updated parameters as frequently as desired. These techniques are especially useful for tracking signals with slowly time varying characteristics. An example would be the doppler tone frequency of a sonar signal from a moving target. Time-sequential estimation methods have sometimes been called *adaptive* algorithms because they constantly adapt to the characteristics of the signal, even as it changes. This concept of adaptive should not be confused with the terminology *data adaptive* that has often appeared in the spectral estimation literature. Data adaptive is used to distinguish parametric methods of spectral estimation from nonparametric methods. Parametric methods are said to "adapt" their parameters to the nature of the data, whereas nonparametric methods have no parameters to adjust.

The sequential algorithms for adaptive AR parameter estimation fall into two categories. The simplest algorithms are based on a gradient approximation approach and include the well-known least-mean-square (LMS) algorithm. Recursive least squares (RLS) algorithms form a second category. The RLS algorithms provide better performance than the LMS algorithm, but at some additional computational expense. Both types of algorithms are presented in this chapter.

Algorithms for the sequential update of the parameters of a lattice filter are also available in the literature. Because the lattice filter parameters represent the reflection coefficients of an AR process, the Levinson algorithm must be used each time to convert the reflection coefficient sequence to an autoregressive parameter sequence for subsequent use in the AR PSD formula. The sequential algorithms presented in this chapter directly update the AR parameters, rather than the reflection coefficients, thereby avoiding the need to use the Levinson algorithm and saving on computations. For those readers interested in sequential lattice algorithms, consult the tutorial papers by Friedlander [1982a, 1982b].

9.2 SUMMARY

MATLAB demonstration script `spectrum_demo_rls.m` available at the text website provides spectral estimation demonstration plots for two possible time-sequential estimates (one gradient and one RLS) of the AR parameters for the `test1987` data record. Although the demonstration makes a plot after all 64 samples have been recursively processed, one could stop after any time update operation to compute an AR PSD estimate to that point in the recursion

$$\hat{P}_{\mathrm{AR}}(f) = \frac{T\hat{\rho}_w}{\left|1 + \sum_{n=1}^{p} \hat{a}[n]\exp(-j2\pi fnT)\right|^2}. \tag{9.1}$$

Figure 9.1 is a flowchart of the steps required to compute the AR PSD with the two sequential algorithms. Note that the model order must be provided as an input parameter. The adaptive time constant μ must also be provided if the LMS algorithm is selected, or decay factor ω if the RLS algorithm is selected. The model order must remain fixed once it is selected.

Figure 9.2 illustrates the relative performance of the two algorithms using the 64-point `test1987` test data set. A 15th-order AR model was selected. The fast RLS algorithm has produced the reasonable spectral estimate of Fig. 9.2(b) after

```
• ACQUIRE DATA RECORD
     Total Number of Samples in Record
     Sampling Interval (sec/sample)

• TREND REMOVAL
     Optional (see Chapter 14)

• SELECT AUTOREGRESSIVE (AR) ORDER

• INITIALIZE WITH 1ˢᵗ SAMPLE

• NEXT SAMPLE: UPDATE AR PARAMETERS
     lms.m, or
     fast_rls.m, or

• COMPUTE AUTOREGRESSIVE PSD
     Function arma_psd.m

• PLOT AR PSD. Continue?
```

Figure 9.1. Flowchart for two time-sequential algorithms for the estimation of the autoregressive PSD.

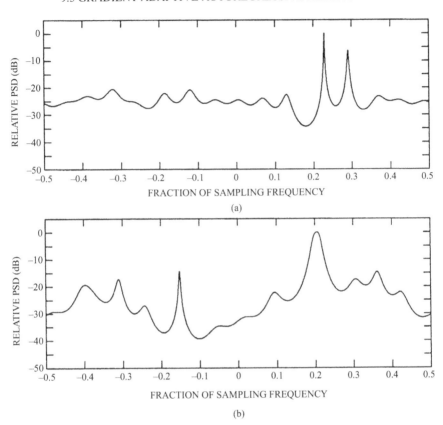

Figure 9.2. Examples of time-sequential AR spectral estimates made from the 64-point test data sequence. (a) LMS method. (b) Fast RLS method.

just 64 time updates using $\omega = 1$, which is to be expected because it is a fast-converging exact least squares method. Sections 9.4 and 9.5 discuss this method. The gradient LMS algorithm has done poorly, but this is to be expected because it is a slowly converging sequential algorithm and 64 time updates are not a sufficient record length for convergence. A time constant of $\mu = 0.02$ was selected for the LMS algorithm, based on a procedure given in Sec. 9.3.

9.3   GRADIENT ADAPTIVE AUTOREGRESSIVE METHODS

The gradient steepest descent adaptive procedure recursively estimates the p component AR parameter vector $\mathbf{a}_p^T = (a_p[1], \ldots, a_p[p])$ at time index $N + 1$ from the previous estimate at time index N

$$\hat{\mathbf{a}}_{p,N+1} = \hat{\mathbf{a}}_{p,N} - \mu \nabla \mathcal{E}\{|e^f_{p,N}[N+1]|^2\}, \tag{9.2}$$

where μ is a positive scalar adaptive step size that controls the adaptive time constant and other properties of the algorithm, ∇ denotes the gradient, and $e^f_{p,N}[N+1]$ is the forward linear prediction *filter residual*

$$e^f_{p,N}[N+1] = x[N+1] + \sum_{k=1}^{p} \hat{a}_{p,N}[k]x[N+1-k]$$

$$= x[N+1] + \mathbf{x}^T_{p-1}[N]\hat{\mathbf{a}}_{p,N}. \tag{9.3}$$

The data vector is defined as $\mathbf{x}_{p-1}[N] = (x[N], \dots, x[N-p+1])^T$. The filter residual $e^f_{p,N}[N+1]$ in Eq. (9.3) is not a prediction *error* because it is defined using the vector $\hat{\mathbf{a}}_{p,N}$ rather than the vector $\hat{\mathbf{a}}_{p,N+1}$. The mean value of $|e_{p,N}[N+1]|^2$ is given by

$$\mathcal{E}\{|e^f_{p,N}[N+1]|^2\} = r[0] + \hat{\mathbf{a}}^H_{p,N}\mathbf{R}_{p-1}\hat{\mathbf{a}}_{p,N} + 2\text{Real } \hat{\mathbf{a}}^H_{p,N}\mathbf{r}_p), \tag{9.4}$$

where

$$r[0] = \mathcal{E}\{x[N+1]x^*[N+1]\},$$

$$\mathbf{r}_p = \mathcal{E}\{x[N+1]\mathbf{x}^*_{p-1}[N]\},$$

and

$$\mathbf{R}_{p-1} = \mathcal{E}\{\mathbf{x}^*_{p-1}[N]\mathbf{x}^T_{p-1}[N]\}$$

are an autocorrelation element, vector, and matrix, respectively. The gradient is then

$$\nabla \mathcal{E}\{|e^f_{p,N}[N+1]|^2\} = 2\mathbf{r}_p + 2\mathbf{R}_{p-1}\hat{\mathbf{a}}_{p,N}. \tag{9.5}$$

When the gradient adaptive procedure converges, $\nabla \mathcal{E}\{|e^f_{p,N}[N+1]|^2\} = 0$ and, therefore, $\hat{\mathbf{a}}_{p,N} = \mathbf{a}_p$, yielding

$$\mathbf{R}_{p-1}\mathbf{a}_p = -\mathbf{r}_p, \tag{9.6}$$

which is identical to the Yule-Walker equation (6.32).

In practice, the autocorrelation sequence is not available, so the instantaneous estimates

$$\hat{\mathbf{r}}_p = x[N+1]\mathbf{x}^*_{p-1}[N], \quad \hat{\mathbf{R}}_{p-1} = \mathbf{x}^*_{p-1}\mathbf{x}^T_{p-1} \tag{9.7}$$

are substituted. This yields

$$\hat{\mathbf{a}}_{p,N+1} = \hat{\mathbf{a}}_{p,N} - e^f_{p,N}[N+1]2\mu\mathbf{x}^*_{p-1}[N]. \tag{9.8}$$

Expressions (9.3) and (9.8) together form the classic *least mean square* (LMS) adaptive algorithm [Widrow and Stearns, 1985]. Each time update then takes roughly $2p$ add or multiply computations.

Rules for selecting appropriate values of μ involve difficult analyses. Assuming the input process is a zero-mean Gaussian process, convergence of the mean of $\hat{\mathbf{a}}_p$ to $\mathbf{a}_p$ is guaranteed as long as μ is selected as a positive value between 0 and $1/\lambda_{max}$, where λ_{max} is the maximum eigenvalue of the matrix $\mathbf{R}_{p-1}$. The trace of a matrix was shown by Eq. (3.81) to be equal to the sum of the eigenvalues of the same matrix. Therefore, $\lambda_{max} \leq \operatorname{tr} \mathbf{R}_{p-1}$ and a more restrictive bound for the adaptive time constant is $0 < \mu < 1/\operatorname{tr} \mathbf{R}_{p-1}$. This bound is easier to apply because $\mathbf{R}_{p-1}$ is Toeplitz and, therefore, $\operatorname{tr} \mathbf{R}_{p-1} = p\, r[0]$. An estimate of $r[0]$, representing the sample signal power, is obtained simply from N samples of the data,

$$\hat{r}_{xx}[0] = \frac{1}{N} \sum_{k=1}^{N} |x[k]|^2 .$$

The mean convergence rate, or exponential convergence time constant τ_k along the kth eigenvector λ_k of $\mathbf{R}_p$, is roughly [Widrow and Stearns, 1985, p. 107]

$$\tau_k \approx \frac{1}{4\mu\lambda_k}. \tag{9.9}$$

The convergence rate is seen to be a function of both the adaptive factor μ and the spread in eigenvalues of the autocorrelation matrix. The convergence behavior thus depends on the combined effects of all the modes associated with the eigenvalues of $\mathbf{R}_p$. The convergence time is dominated by the smallest eigenvalue. For a large spread $\lambda_{max}/\lambda_{min} \gg 1$, such as that characteristic of a narrowband process in noise, the convergence time may be quite slow.

Because a gradient *estimate* is used, this has the effect of introducing self-noise into the adaptive process. This self-noise manifests itself as a steady-state variance or excess mean square error of the $\hat{\mathbf{a}}_p$ parameters about their means once convergence is achieved. This variance may be made arbitrarily small by choice of a sufficiently small μ, which will unfortunately cause a slow convergence. Selecting the value of μ therefore involves a trade-off between the rate of convergence and the accuracy of the AR parameter estimates (level of self-noise to be tolerated).

The LMS adaptive algorithm is able to track time variations of the input AR parameter signal statistics, provided they are sufficiently slowly varying with respect to the convergence rate of the adaptive algorithm. A time average of the error sequence generated by the LMS algorithm provides an estimate of the white noise variance needed by the AR PSD formula. Griffiths [1975] was an early proponent of this method of AR spectral estimation. The MATLAB functionn lms.m available through the text website performs the gradient algorithm:

```
function a=lms(mu,x,a)
% Least-mean-square (LMS) time-recursive adaptive AR algorithm.
%
%     a = lms(mu,x,a)
%
% mu -- adaptive time constant
```

```
% x  -- vector of data samples
% a  -- vector of linear prediction/autoregressive (AR) parameters
```

As illustrated by the example of Fig. 9.2, the gradient LMS adaptive algorithm has inferior performance and convergence rate on relatively short data records when compared to the RLS algorithms to be presented in the next section. However, the LMS is more robust in its behavior and less sensitive to numerical ill-conditioning and finite-word-length effects than the RLS algorithms. For applications in which data is plentiful and slower convergence is tolerable, the LMS method may be the more prudent choice of sequential algorithms.

9.4 RECURSIVE LEAST SQUARES (RLS) AUTOREGRESSIVE METHODS

The N-point data sequence $x[1], \ldots, x[N]$ can be used to estimate the pth-order AR parameters by forming the forward linear prediction error

$$e^f_{p,N}[n] = x[n] + \sum_{k=1}^{p} a_{p,N}[k]x[n-k] \qquad (9.10)$$

and solving for the set of $a_{p,N}[k]$ parameters that minimize the exponentially weighted squared error, given all available measured data up to time index N,

$$\rho^f_{p,N} = \sum_{n=1}^{N} \omega^{N-n}|e^f_{p,N}[n]|^2, \qquad (9.11)$$

where ω is a positive real scalar satisfying $0 < \omega \le 1$. This exponential window moves with the data, giving least reduction to current errors and most reduction to older error terms, thus permitting tracking of slowly varying signal parameters (nonstationary processes). The summation range in Eq. (9.11) for selection $\omega = 1$ was called the prewindowed case in Sec. 8.5, because it made the assumption $x[n] = 0$ for $n \le 0$. Following the same minimization procedure used in Sec. 8.5 and Sec. 8.10, it may be shown that the linear prediction/autoregressive coefficients that minimize ρ^f_p satisfy the $(p+1) \times (p+1)$ matrix normal equation

$$\mathbf{R}_{p,N}\bar{\mathbf{a}}_{p,N} = \begin{pmatrix} \rho^f_{p,N} \\ \mathbf{0}_p \end{pmatrix}, \qquad (9.12)$$

where

$$\mathbf{R}_{p,N} = \sum_{n=1}^{N} \omega^{N-n}\mathbf{x}^*_p[n]\mathbf{x}^T_p[n] \qquad (9.13)$$

and

$$
\mathbf{x}_p[n] = \begin{pmatrix} x[n] \\ \vdots \\ x[n-p] \end{pmatrix}, \quad \bar{\mathbf{a}}_{p,N} = \begin{pmatrix} 1 \\ a_{p,N}[1] \\ \vdots \\ a_{p,N}[p] \end{pmatrix} = \begin{pmatrix} 1 \\ \mathbf{a}_{p,N} \end{pmatrix} \tag{9.14}
$$

and $\mathbf{0}_p$ is an all-zero $p \times 1$ vector. Solution of Eq. (9.12) yields the linear predictor vector $\mathbf{a}_{p,N}$ and the least squared error $\rho_{p,N}^f$. The matrix $\mathbf{R}_{p,N}$ and parameter vector $\bar{\mathbf{a}}_{p,N}$ are indexed by N to indicate they are based on all data samples up to time index N. It can be shown that the elements of $\bar{\mathbf{a}}_{p,N}$ give rise to a stable prediction filter.

If a new sample $x[N+1]$ is acquired, then the least squares fit to the entire set of $N+1$ samples may be expressed as the matrix normal equation

$$
\mathbf{R}_{p,N+1}\bar{\mathbf{a}}_{p,N+1} = \begin{pmatrix} \rho_{p,N+1}^f \\ \mathbf{0}_p \end{pmatrix}, \tag{9.15}
$$

where

$$
\mathbf{R}_{p,N+1} = \sum_{n=1}^{N+1} \omega^{N+1-n}\mathbf{x}_p^*[n]\mathbf{x}_p^T[n]. \tag{9.16}
$$

An algorithm that, upon acquisition of a new sample $x[N+1]$, can update $\mathbf{a}_{p,N}$ to $\mathbf{a}_{p,N+1}$ without explicitly solving Eq. (9.15) is termed a *recursive least squares* procedure. Two RLS algorithms will be discussed. The traditional RLS algorithm requires a number of computations proportional to p^2 per time update. A new fast RLS algorithm exploits the structure in the normal equations to yield the identical solution with a procedure that only requires a number of computations proportional to p, a significant reduction in the computational load. The fast RLS algorithm is the sequential counterpart to the fast block-processing algorithms presented in Chap. 8.

9.4.1 Classical Recursive Least Squares Algorithm

The traditional RLS algorithm [Ljung and Söderström, 1983] provides an exact least squares recursive solution for the linear prediction coefficients as each data sample is made available. Noting that matrix $\mathbf{R}_{p,N}$ can be factored as

$$
\mathbf{R}_{p,N}\bar{\mathbf{a}}_{p,N} = \begin{pmatrix} r_{p,N}[0,0] & \mathbf{r}_{p,N}^H \\ \mathbf{r}_{p,N} & \mathbf{R}_{p-1,N-1} \end{pmatrix} \begin{pmatrix} 1 \\ \mathbf{a}_{p,N} \end{pmatrix} = \begin{pmatrix} \rho_{p,N}^f \\ \mathbf{0}_p \end{pmatrix}, \tag{9.17}
$$

where

$$\mathbf{R}_{p-1,N-1} = \sum_{n=1}^{N} \omega^{N-n} \mathbf{x}_{p-1}^{*}[n-1]\mathbf{x}_{p-1}^{T}[n-1]$$

$$= \sum_{n=1}^{N-1} \omega^{N-1-n} \mathbf{x}_{p-1}^{*}[n]\mathbf{x}_{p-1}^{T}[n]$$

$$\mathbf{r}_{p,N} = \sum_{n=1}^{N} \omega^{N-n} \mathbf{x}_{p-1}^{*}[n-1]x[n]$$

$$r_{p,N}[0,0] = \sum_{n=1}^{N} \omega^{N-n} |x[n]|^{2}, \qquad (9.18)$$

the equations for $\rho_{p,N}^{f}$ and $\mathbf{a}_{p,N}$ may then be written separately as

$$\mathbf{R}_{p-1,N-1}\mathbf{a}_{p,N} = -\mathbf{r}_{p,N} \qquad (9.19)$$

$$\rho_{p,N}^{f} = r_{p,N}[0,0] + \mathbf{r}_{p,N}^{H}\mathbf{a}_{p,N}. \qquad (9.20)$$

Using the definition

$$\mathbf{P}_{N-1} = \mathbf{R}_{p-1,N-1}^{-1}, \qquad (9.21)$$

the least squares solution at time index $N+1$ for $\mathbf{a}_{p,N+1}$ then has the form

$$\mathbf{a}_{p,N+1} = -\mathbf{P}_{N}\mathbf{r}_{p,N+1}. \qquad (9.22)$$

Noting the identity

$$\mathbf{r}_{p,N+1} = \omega\mathbf{r}_{p,N} + \mathbf{x}_{p-1}^{*}[N]x[N+1], \qquad (9.23)$$

which may be proved by inspection of definition (9.17), then

$$\mathbf{a}_{p,N+1} = -\mathbf{P}_{N}(\omega\mathbf{r}_{p,N} + \mathbf{x}_{p-1}^{*}[N]x[N+1])$$

$$= -\mathbf{P}_{N}(\omega\mathbf{P}_{N-1}^{-1}\mathbf{P}_{N-1}\mathbf{r}_{p,N} + \mathbf{x}_{p-1}^{*}[N]x[N+1])$$

$$= \mathbf{P}_{N}(\omega\mathbf{P}_{N-1}^{-1}\mathbf{a}_{p,N} - \mathbf{x}_{p-1}^{*}[N]x[N+1]). \qquad (9.24)$$

Using the identity

$$\omega\mathbf{P}_{N-1}^{-1} = \mathbf{P}_{N}^{-1} - \mathbf{x}_{p-1}^{*}\mathbf{x}_{p-1}^{T}[N],$$

which is easily proved using the definition of $\mathbf{R}_{p-1,N-1}$ in Eq. (9.17), then substituting into Eq. (9.23) yields

$$\mathbf{a}_{p,N+1} = \mathbf{a}_{p,N} - \mathbf{P}_{N}\mathbf{x}_{p-1}^{*}[N](\mathbf{x}_{p-1}^{T}[N]\mathbf{a}_{p,N} + x[N+1])$$

$$= \mathbf{a}_{p,N} - e_{p,N}^{f}[N+1]\mathbf{P}_{N}\mathbf{x}_{p-1}^{*}[N] \qquad (9.25)$$

$$= \mathbf{a}_{p,N} - e_{p,N}^{f}[N+1]\mathbf{c}_{p-1,N}, \qquad (9.26)$$

where the vector $\mathbf{c}_{p-1,N}$ has the definition

$$\mathbf{c}_{p-1,N} = \mathbf{P}_N \mathbf{x}_{p-1}^*[N]. \tag{9.27}$$

The scalar $e_{p,N}^f[N+1]$ is a *filtered residual*, rather than a *prediction error*, because the $\mathbf{a}_{p,N}$ parameter vector is used, rather than the $\mathbf{a}_{p,N+1}$ vector.

The distinction between the RLS and LMS adaptive algorithms may be seen by comparing the parameter update Eqs. (9.8) and (9.25). They are identical, except for the adaptive gains. In the LMS algorithm, the adaptive gain 2μ is a *constant scalar*. In the RLS algorithm, the adaptive gain $\mathbf{P}_N$ is a *time-varying matrix*. The time update for $\mathbf{P}_N$ may be determined by applying the matrix inversion lemma, Eq. (3.53), to expression (9.24), yielding

$$\mathbf{P}_N = (\omega \mathbf{P}_{N-1}^{-1} + \mathbf{x}_{p-1}^*[N] \mathbf{x}_{p-1}^T[N])^{-1}$$

$$= \omega^{-1} \left[\frac{\mathbf{P}_{N-1} - \mathbf{P}_{N-1} \mathbf{x}_{p-1}^*[N] \mathbf{x}_{p-1}^T[N] \mathbf{P}_{N-1}}{(\omega + \mathbf{x}_{p-1}^T[N] \mathbf{P}_{N-1} \mathbf{x}_{p-1}^*[N])} \right]. \tag{9.28}$$

It is simple to show from Eq. (9.28) that

$$\mathbf{c}_{p-1,N} = \mathbf{P}_{N-1} \mathbf{x}_{p-1}^*[N] / (\omega + \mathbf{x}_{p-1}^T[N] \mathbf{P}_{N-1} \mathbf{x}_{p-1}^*[N]), \tag{9.29}$$

so that Eq. (9.28) may be simplified to

$$\mathbf{P}_N = \omega^{-1} (\mathbf{I} - \mathbf{c}_{p-1,N} \mathbf{x}_{p-1}^T[N]) \mathbf{P}_{N-1}. \tag{9.30}$$

To complete the RLS algorithm, the update for $\rho_{p,N+1}^f$ is obtained by combining Eq. (9.19) with Eq. (9.22).

Equations (9.26), (9.29), and (9.30) form the basic RLS algorithm. They are similar in structure to the Kalman filter equations of random control theory. The vector $\mathbf{x}_{p-1}[N]$ and matrix $\mathbf{P}_{N-1}$, which are composed of data elements, are analogous to the correlation vector and correlation matrix of the Kalman formulation. The vector $\mathbf{c}_{p-1,N}$ is analogous to the *Kalman gain vector*. The RLS algorithm requires a number of computations per time update proportional to p^2 due to the matrix updates (9.29) and (9.30). This is in contrast to the LMS algorithm, which has computational complexity proportional to p.

Initial values must be assigned to AR parameter vector $\mathbf{a}_0$ and gain matrix $\mathbf{P}_0$ in order to start the recursion. The choice of $\mathbf{a}_0$ must be made carefully to avoid biases in $\mathbf{a}_N$, the AR parameter vector after N samples. Setting $\mathbf{R}_{p,0}$ to a diagonal matrix $\epsilon \mathbf{I}$ with small positive constant ϵ will ensure that it is invertible, so that $\mathbf{P}_0 = 1/\epsilon \mathbf{I}$. The effect of ϵ diminishes as N becomes large. The RLS algorithm is sensitive to numerical ill-conditioning caused by round-off noise that accumulates during the recursive computations. To partly remedy this instability, square root factorization algorithms based on triangular decomposition of the symmetric matrix $\mathbf{P}_N$ have been devised [Bierman, 1977].

9.4.2   Fast Recursive Least Squares Algorithm

The classical RLS algorithm was developed to handle general linear least squares minimization problems. Equation (9.16) has additional structure that may be exploited to yield a more efficient algorithm and still be able to yield the exact least squares solution. This structure is similar to that used to develop the fast block algorithms of Chap. 8. The fast block algorithms were recursive in order, but fixed in time (a block of N data samples). Using similar principles devised to create the fast block algorithms, a fast RLS algorithm can be developed. In contrast to the fast block algorithms, the fast RLS algorithm is recursive in time, but fixed in order. Section 9.7 provides development details of the fast algorithm. The key to a fast algorithm is an updating procedure for the gain vector $c_{p-1,N}$ in Eq. (9.26) that uses only vector operations rather than matrix operations such as Eqs. (9.29) and (9.30). The computational complexity of the entire RLS algorithm is reduced from p^2 to exactly $5p$ complex operations (multiplies/adds) per time update. This is two and one-half times the complexity of the LMS algorithm, which requires $2p$ operations per time update. Thus, the fast RLS algorithm is computationally competitive with the LMS algorithm. There has been a tendency to call the fast RLS algorithm a "fast Kalman" algorithm, but this seems inappropriate due to the nonrandom nature of the deterministic least squares solution.

The fast RLS algorithm has two additional virtues not obtained with the classical RLS algorithm. The backward linear prediction coefficients $a_{p,N}^b[k]$ that minimize the sum of squared backward linear prediction errors

$$\rho_{p,N}^b = \sum_{n=1}^{N} \omega^{N-n} |e_{p,N}^b[n]|^2,$$

where

$$e_{p,N}^b[n] = x[n-p] + \sum_{k=1}^{p} a_{p,N}^b[k]x[n-p+k],$$

are also computed and updated as part of the fast RLS algorithm, along with $\rho_{p,N}^b$. This may be useful for exponential determination methods such as those described in Sec. 11.9.

The matrix $\mathbf{R}_{p,N}$ requires at least $p+1$ data samples before it becomes non-singular (has rank $p+1$). The classical RLS algorithm handles this initialization difficulty by assigning arbitrary initial values to $\mathbf{P}_0$ and $\mathbf{a}_{p,0}$. This adds a bias to the exact least squares solution. The fast RLS algorithm in Appendix 9.C has been provided with a simultaneous order and time updating routine for initialization. This routine processes the first $p+1$ samples such that, at time update $N = p+1$, all parameters are set to their *exact* least squares solution values for those first $p+1$ data samples. Thus, the initialization bias of the classical RLS procedure is avoided with the fast RLS algorithm. MATLAB function `fast_rls.m` available through the text website implements the fast RLS algorithm:

```
function [init,rho_f,a_f,rho_b,a_b,gamma,c]=...
                 fast_rls(init,omega,x,rho_f,a_f,rho_b,a_b,gamma,c)
% Fast recursive-least-squares (RLS) sequential adaptive linear prediction/AR
% algorithm with two redundant variable computations to assure long-term
% algorithm numerical stability.
%
%    [init,rho_f,a_f,rho_b,a_b,gamma,c] =
%                 fast_rls(init,omega,x,rho_f,a_f,rho_b,a_b,gamma,c)
%
% init  -- set to 0 to initialize algorithm; function will
%                 auto increment this variable with each iteration
% omega -- exponential weighting factor (between 0 and 1)
% x     -- vector of most recent data samples in linear prediction error filter
% rho_f -- forward linear prediction squared error
% a_f   -- vector of forward linear prediction parameters
% rho_b -- backward linear prediction squared error
% a_b   -- vector of backward linear prediction parameters
% gamma -- gain factor
% c     -- vector of gain parameters
```

The fast RLS algorithm appears to have poor long-term numerical stability, but is quite accurate for short- to medium-length data records. The performance of RLS algorithms have been observed to diverge in long-term operation of the adaptive algorithm [Cioffi and Kailath, 1984]. This is due to error accumulation from finite-precision computations, especially when the $\mathbf{R}_{p,N}$ matrix is nearly singular, as it can be for narrowband inputs. One solution is to add low-level white noise to the input. Lin [1984] has introduced a rescue solution based upon reinitialization of the gain vector. Square-root-normalized versions of the fast RLS algorithm can increase dynamic range for limited-precision machines and improve performance [Fabre and Gueguen, 1986]. The normalized algorithms basically double the computational complexity in exchange for a halving of the required precision to store and compute the algorithm quantities. Setting $\omega \ll 1$ appears to introduce significant self-noise into the algorithm [Ling and Proakis, 1984a]. The RLS converges exponentially and uniformly, regardless of the eigenvalue spread of the signal autocorrelation matrix; convergence depends only on the weighting factor ω after a change in the input (the initial convergence does not depend on ω).

9.4.3 Comparison of RLS and LMS Performance

Typical convergence performance plots for RLS and LMS algorithms are depicted in Fig. 9.3. The RLS algorithm generally converges to steady state faster than the LMS algorithm, but it is never slower than LMS. In high noise, or where the eigenvalue disparity is low, the RLS and LMS have comparable convergence rates. The self-noise generated by the gradient estimate of the LMS algorithm manifests itself as a bias in the filter output variance and as misadjusted prediction filter parameters. The level of bias and misadjustment can be decreased by decreasing the value of the adaptive time constant μ. This, however, is at the expense of an increased

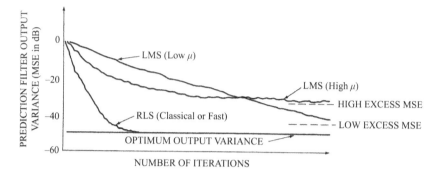

Figure 9.3. Typical convergence performance of gradient LMS algorithm and exact least squares RLS algorithm.

convergence time for the LMS algorithm, as Fig. 9.3 illustrates. Convergence rate and misadjustment noise are interdependent in the LMS algorithm, and one cannot be improved without adversely affecting the other.

If the LMS and RLS algorithms are implemented in hardware for real-time spectral estimation, the performance of the two adaptive algorithms will be affected by quantization error and round-off error due to small word lengths and fixed-point arithmetic [Lawrence and Tewksbury, 1983]. The round-off error may accumulate and increase with time, until it destroys the normal operation of the algorithm. Selecting $\omega < 1$ in the RLS algorithm will bound this error accumulation and keep the filter stable, although prediction parameter numerical accuracy will be affected. Ling and Proakis [1984b] found in one experiment that the fast RLS algorithm needed at least 10 bits to work properly, whereas the LMS algorithm could operate properly with as few as 7 bits. Both algorithms gave comparable steady-state behavior when at least 12 bits were used.

9.5 FAST LATTICE AUTOREGRESSIVE METHODS

An AR(p) process was shown in Chap. 7 to have an equivalent parameterization as either p prediction parameters or as p reflection coefficients. If parameterized by the prediction parameters, the autoregressive process is generated by passing white noise through a tapped delay line (transversal) filter with the prediction parameters forming the filter weights. If parameterized by the reflection coefficients, the equivalent autoregressive process is generated by passing white noise through a lattice filter with the reflection coefficients forming the lattice filter weights. The fast RLS algorithm may be reformulated into a mathematically equivalent fast lattice algorithm. This algorithm also yields the exact least squares solution as the transversal filter, but only in terms of the reflection coefficients rather than in terms of the linear prediction (AR) parameters. The fast lattice algorithm takes about twice the number of operations of the fast RLS algorithm. Interested readers may consult the papers of Friedlander [1982a, 1982b] or Satorius and Pack [1981] or the book by Honig and Messerschmitt [1984] for details of fast least squares lattice algorithms.

The prediction parameters from the fast RLS algorithm may be directly (instantaneously) used after any time update to evaluate the AR PSD function. However, the reflection coefficients from the fast least squares lattice algorithm must be converted by a Levinson-like recursion to the equivalent prediction parameters before the AR PSD function can be evaluated. This roughly requires an additional p^2 computational operations at any time step for which a spectral function evaluation is desired. This two-stage procedure and additional computational burden is a disadvantage of the lattice algorithm for AR spectral estimation, especially for frequent updating of the AR spectral estimate. However, the fast lattice algorithms have better numerical properties than the related fast RLS algorithms, which may be useful for hardware implementation. The performance of lattice algorithms is less sensitive to round-off noise and coefficient quantization than the fast RLS algorithm. The fast lattice algorithms also yield all lower-order least squares solutions, so that the filter order may be adjusted at any time step.

9.6 APPLICATION TO SUNSPOT NUMBERS

The 230 samples of the filtered sunspot data set are too few to achieve convergence with the LMS adaptive algorithm. Therefore, only the RLS algorithm is considered in this section. Figure 9.4 illustrates the tracking behavior of the fast RLS algorithm for various selections of the exponential weighting factor ω. A typical order 15 autoregressive spectral estimate is shown in Fig. 9.4(a). This spectrum is the result of sequentially processing the first 115 samples from year 1855.5 to year 1913.0 for

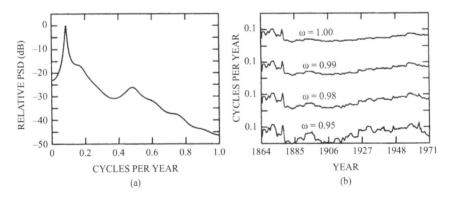

(a)   (b)

Figure 9.4. Demonstration of capability of the fast RLS algorithm to track the dominant sunspot cycle. (a) Typical AR(16) spectral estimate after 115 samples with $\omega = 0.98$. (b) Location of highest peak in spectrum as a function of year for four weighting factors. Each division along the vertical axis represents 0.02 cycle per year.

the selection $\omega = 0.98$. Figure 9.4(b) depicts the sequential frequency location of the dominant peak in the spectrum, obtained by searching the AR spectral response after each data point update. Tracking behavior for four choices of ω are shown. A rough estimate of the effective exponential window duration in samples is given by $1/(1 - \omega)$. The effective window durations used in Fig. 9.4(b) are therefore 20, 50, 100, and ∞ samples. The nonstationary behavior of the dominant sunspot cycle, previously detected in stacked periodogram plots, is clearly indicated by the variations in Fig. 9.4(b).

9.7 EXTRA: FAST RLS ALGORITHM FOR RECURSIVE LINEAR PREDICTION

9.7.1 Introduction

This algorithm is based on the fast recursive least squares adaptive filter developed by Ljung et al. [1978], Falconer and Ljung [1978], Carayannis et al. [1983], and Cioffi and Kailath [1984].

The pth-order forward and backward linear prediction errors based on all the data available up to time N may be represented as the vector inner products

$$e_{p,N}^f[n] = \mathbf{x}_p^T[n]\mathbf{a}_{p,N}^f, \tag{9.31}$$

$$e_{p,N}^b[n] = \mathbf{x}_p^T[n]\mathbf{a}_{p,N}^b, \tag{9.32}$$

where the data vector $\mathbf{x}_p[n]$, forward linear prediction coefficient vector $\mathbf{a}_{p,N}^f$, and backward linear prediction coefficient vector $\mathbf{a}_{p,N}^b$ have the definitions

$$\mathbf{x}_p[n] = \begin{pmatrix} x[n] \\ x[n-1] \\ \vdots \\ x[n-p] \end{pmatrix}, \quad \mathbf{a}_{p,N}^f = \begin{pmatrix} 1 \\ a_{p,N}^f[1] \\ \vdots \\ a_{p,N}^f[p] \end{pmatrix}, \quad \mathbf{a}_{p,N}^b = \begin{pmatrix} a_{p,N}^b[p] \\ \vdots \\ a_{p,N}^b[1] \\ 1 \end{pmatrix}. \tag{9.33}$$

Based on observed complex data samples $x[1], \ldots, x[N]$, the prewindowed method separately minimizes the forward and backward exponentially weighted linear prediction squared errors

$$\rho_{p,N}^f = \sum_{n=1}^{N} \omega^{N-n} |e_{p,N}^f[n]|^2, \quad \rho_{p,N}^b = \sum_{n=1}^{N} \omega^{N-n} |e_{p,N}^b[n]|^2, \tag{9.34}$$

resulting in the normal equations

$$\mathbf{R}_{p,N}\mathbf{a}_{p,N}^f = \begin{pmatrix} \rho_{p,N}^f \\ \mathbf{0}_p \end{pmatrix}, \quad \mathbf{R}_{p,N}\mathbf{a}_{p,N}^b = \begin{pmatrix} \mathbf{0}_p \\ \rho_{p,N}^b \end{pmatrix}, \tag{9.35}$$

where $\mathbf{0}_p$ is a $p \times 1$ all-zero vector. The effective window length is approximately $1/(1 - \omega)$. An equivalent dyadic vector definition for $\mathbf{R}_{p,N}$ is given by

$$\mathbf{R}_{p,N} = \sum_{n=1}^{N} \omega^{N-n} \mathbf{x}_p^*[n] \mathbf{x}_p^T[n] \tag{9.36}$$

and individual elements by

$$r_{p,N}[j,k] = \sum_{n=1}^{N} \omega^{N-n} x[n-k] x^*[n-j]$$

for $0 \le j, k \le p$.

9.7.2 Special Partitions and Auxiliary Parameters

Crucial to the development of a fast RLS algorithm are the following order-index partitions of the $(p+1) \times (p+1)$ matrix $\mathbf{R}_{p,N}$,

$$\mathbf{R}_{p,N} = \begin{pmatrix} \mathbf{R}_{p-1,N} & \mathbf{s}_{p,N} \\ \mathbf{s}_{p,N}^H & r_{p,N}[p,p] \end{pmatrix} = \begin{pmatrix} r_{p,N}[0,0] & \mathbf{r}_{p,N}^H \\ \mathbf{r}_{p,N} & \mathbf{R}_{p-1,N-1} \end{pmatrix}, \tag{9.37}$$

where the $p \times 1$ column vectors $\mathbf{r}_{p,N}$ and $\mathbf{s}_{p,N}$ are

$$\mathbf{r}_{p,N} = \begin{pmatrix} r_{p,N}[1,0] \\ \vdots \\ r_{p,N}[p,0] \end{pmatrix}, \quad \mathbf{s}_{p,N} = \begin{pmatrix} r_{p,N}[0,p] \\ \vdots \\ r_{p,N}[p-1,p] \end{pmatrix}. \tag{9.38}$$

Also crucial is the following time-index partition

$$\mathbf{R}_{p,N} = \omega \mathbf{R}_{p,N-1} + \mathbf{x}_p^*[N] \mathbf{x}_p^T[N], \tag{9.39}$$

obtained by inspection of Eq. (9.36).

The application of the matrix inversion lemma (3.53) to the two order partitions of Eq. (9.35) will yield, after some considerable matrix algebra,

$$\mathbf{R}_{p,N}^{-1} = \begin{pmatrix} \mathbf{R}_{p-1,N}^{-1} & \mathbf{0}_p \\ \mathbf{0}_p^H & 0 \end{pmatrix} + \mathbf{a}_{p,N}^b \mathbf{a}_{p,N}^{b \, H} / \rho_{p,N}^b \tag{9.40}$$

$$\mathbf{R}_{p,N}^{-1} = \begin{pmatrix} 0 & \mathbf{0}_p^H \\ \mathbf{0}_p & \mathbf{R}_{p-1,N-1}^{-1} \end{pmatrix} \mathbf{a}_{p,N}^f \mathbf{a}_{p,N}^{f \, H} / \rho_{p,N}^f. \tag{9.41}$$

A $(p+1) \times 1$ gain vector $\mathbf{c}_{p,N}$, defined as

$$\mathbf{R}_{p,N-1} \mathbf{c}_{p,N} = \omega^{-1} \mathbf{x}_p^*[N], \tag{9.42}$$

is also required by the fast algorithm. Note that $\mathbf{c}_{p,N}$ is defined in a different manner than the auxiliary vector required in the fast block algorithms of Chap. 8 because $\mathbf{R}_{p,N-1}$, rather than $\mathbf{R}_{p,N}$, is used. The column vector $\mathbf{c}_{p,N}$ has the structure

$$\mathbf{c}_{p,N} = \begin{pmatrix} c_{p,N}[0] \\ \vdots \\ c_{p,N}[p] \end{pmatrix}.$$

9.7.3 Time Update Recursions

The time-index update of the forward linear prediction coefficient vector is given by

$$\mathbf{a}^f_{p,N+1} = \mathbf{a}^f_{p,N} - e^f_{p,N+1}[N+1]\begin{pmatrix} 0 \\ \mathbf{c}_{p-1,N} \end{pmatrix}, \tag{9.43}$$

which may be verified by premultiplying both sides by $\mathbf{R}_{p,N+1}$ and substituting either the time-index partition (9.39) or order-index partitions (9.37) as appropriate. A by-product of this verification is the time update of the real forward linear prediction squared error

$$\begin{aligned} \rho^f_{p,N+1} &= \omega\rho^f_{p,N} + e^f_{p,N+1}[N+1](x^*[N+1] - \mathbf{r}^H_{p,N}\mathbf{c}_{p-1,N}) \\ &= \omega\rho^f_{p,N} + e^f_{p,N+1}[N+1]e^{f*}_{p,N}[N+1]. \end{aligned} \tag{9.44}$$

Multiplying both sides of Eq. (9.43) by $\mathbf{x}^T_p[N+1]$ to give the result

$$e^f_{p,N+1}[N+1] = e^f_{p,N}[N+1] - e^f_{p,N+1}[N+1](\mathbf{x}^T_{p-1}[N]\mathbf{c}_{p-1,N}), \tag{9.45}$$

or more concisely (relates filter residuals to prediction error)

$$e^f_{p,N+1}[N+1] = e^f_{p,N}[N+1]/\gamma_{p-1,N}, \tag{9.46}$$

where, by definition, $\gamma_{p-1,N}$ is a real, positive scalar

$$\gamma_{p-1,N} = 1 + \mathbf{x}^T_{p-1}[N]\mathbf{c}_{p-1,N} = 1 + \omega^{-1}\mathbf{x}^T_{p-1}[N]\mathbf{R}^{-1}_{p-1,N-1}\mathbf{x}^*_{p-1}[N]. \tag{9.47}$$

Substituting Eq. (9.43) into (9.44) yields

$$\rho^f_{p,N+1} = \omega\rho^f_{p,N} + |e^f_{p,N}[N+1]|^2/\gamma_{p-1,N}, \tag{9.48}$$

which upholds the real and positive nature of the squared error.

The time-index update of the backward linear prediction coefficient vector is given by

$$\mathbf{a}^b_{p,N+1} = \mathbf{a}^b_{p,N} - e^b_{p,N+1}[N+1]\begin{pmatrix} \mathbf{c}_{p-1,N+1} \\ 0 \end{pmatrix}, \tag{9.49}$$

which may be verified by premultiplying both sides by $\mathbf{R}_{p,N+1}$ and substituting either the time-index partition (9.39) or the appropriate order-index partition of

Eq. (9.37). A by-product of this verification is the time update of the real-valued backward linear prediction squared error

$$\rho_{p,N+1}^{b} = \omega\rho_{p,N}^{b} + e_{p,N+1}^{b}[N+1](x^{*}[N-p+1] - \mathbf{s}_{p,N}^{H}\mathbf{c}_{p-1,N+1})$$
$$= \omega\rho_{p,N}^{b} + e_{p,N+1}^{b}[N+1]e_{p,N}^{b*}[N+1]. \tag{9.50}$$

Multiply both sides of Eq. (9.49) by $\mathbf{x}_p^T[N+1]$ to give the result

$$e_{p,N+1}^{b}[N+1] = e_{p,N}^{b}[N+1] - e_{p,N+1}^{b}[N+1](\mathbf{x}_{p-1}^T[N+1]\mathbf{c}_{p-1,N+1}) \tag{9.51}$$

or, more concisely,

$$e_{p,N+1}^{b}[N+1] = e_{p,N}^{b}[N+1]/\gamma_{p-1,N+1}, \tag{9.52}$$

where, by definition, $\gamma_{p-1,N+1}$ is the real, positive scalar

$$\gamma_{p-1,N+1} = 1 + \mathbf{x}_{p-1}^T[N+1]\mathbf{c}_{p-1,N+1}$$
$$= 1 + \omega^{-1}\mathbf{x}_{p-1}^T[N+1]\mathbf{R}_{p-1,N}^{-1}\mathbf{x}_{p-1}^{*}[N+1]. \tag{9.53}$$

Substituting Eq. (9.52) into (9.50) yields

$$\rho_{p,N+1}^{b} = \rho_{p,N}^{b} + |e_{p,N}^{b}[N+1]|^2/\gamma_{p-1,N+1}, \tag{9.54}$$

which upholds the real and positive properties of the squared error.

A useful scalar relationship is obtained from the quadratic product

$$\mathbf{c}_{p,N+1}^{H}\mathbf{R}_{p,N}\mathbf{a}_{p,N}^{b} = (\mathbf{a}_{p,N}^{b}{}^{H}\mathbf{R}_{p,N}\mathbf{c}_{p,N+1})^{*}.$$

This reduces to

$$\mathbf{c}_{p,N+1}^{H}\begin{pmatrix} \mathbf{0}_p \\ \rho_{p,N}^{b} \end{pmatrix} = \mathbf{a}_{p,N}^{b}{}^{T}\omega^{-1}\mathbf{x}_p[N+1],$$

and this in turn reduces to

$$e_{p,N}^{b}[N+1] = \omega c_{p,N+1}^{*}[p]\rho_{p,N}^{b}. \tag{9.55}$$

9.7.4 Order Update Recursions

If the substitution (9.41) for $\mathbf{R}_{p,N}^{-1}$ in

$$\mathbf{c}_{p,N+1} = \omega^{-1}\mathbf{R}_{p,N}^{-1}\mathbf{x}_p^{*}[N+1] \tag{9.56}$$

is made, then the result is

$$\mathbf{c}_{p,N+1} = \omega^{-1}\begin{pmatrix} 0 & \mathbf{0}_p^H \\ \mathbf{0}_p & \mathbf{R}_{p-1,N-1}^{-1} \end{pmatrix}\mathbf{x}_p^{*}[N+1] + (\omega\rho_{p,N}^{f})^{-1}\mathbf{a}_{p,N}^{f}\mathbf{a}_{p,N}^{f}{}^{H}\mathbf{x}_p^{*}[N+1]$$

or

$$\mathbf{c}_{p,N+1} = \begin{pmatrix} 0 \\ \mathbf{c}_{p-1,N} \end{pmatrix} + (\omega \rho_{p,N}^f)^{-1} e_{p,N}^{f*}[N+1] \mathbf{a}_{p,N}^f. \qquad (9.57)$$

This forms a *step-up recursion* for $\mathbf{c}_{p,N+1}$ in the sense that the recursion is from order $p-1$ to order p. Premultiplying both sides of Eq. (9.57) by $\mathbf{x}_p^T[N+1]$, and then adding one to each side, yields the scalar order update

$$\gamma_{p,N+1} = \gamma_{p-1,N} + |e_{p,N}^f[N+1]|^2 / (\omega \rho_{p,N}^f). \qquad (9.58)$$

If the substitution (9.40) for $\mathbf{R}_{p,N}^{-1}$ in Eq. (9.56) is made, then

$$\mathbf{c}_{p,N+1} = \omega^{-1} \begin{pmatrix} \mathbf{R}_{p-1,N}^{-1} & \mathbf{0}_p \\ \mathbf{0}_p^H & 0 \end{pmatrix} \mathbf{x}_p^*[N+1] + (\omega \rho_{p,N}^b)^{-1} \mathbf{a}_{p,N}^b \mathbf{a}_{p,N}^{b\,H} \mathbf{x}_p^*[N+1]$$

or

$$\begin{pmatrix} \mathbf{c}_{p-1,N+1} \\ 0 \end{pmatrix} = \mathbf{c}_{p,N+1} - (\omega \rho_{p,N}^b)^{-1} e_{p,N}^{b*}[N+1] \mathbf{b}_{p,N}. \qquad (9.59)$$

This forms a *step-down recursion* for $\mathbf{c}_{p-1,N+1}$ in the sense that the recursion is from order p to order $p-1$. Premultiplying both sides of Eq. (9.59) by $\mathbf{x}_p^T[N+1]$ and then adding one to each side yields the scalar order update

$$\gamma_{p-1,N+1} = \gamma_{p,N+1} - |e_{p,N}^b[N+1]|^2 / (\omega \rho_{p,N}^b)$$
$$= \gamma_{p,N+1} - c_{p,N+1}[p] e_{p,N}^b[N+1]. \qquad (9.60)$$

9.7.5 Initialization

The steady-state fast RLS algorithm for a fixed-order p filter applies only when $N > p$; otherwise, the least squares normal equations are underdetermined. During the interval $1 \leq N \leq p$, *simultaneous order and time updating* can be developed to provide an $(N-1)$-order exact least squares solution to the available N data samples. With the arrival of time sample $x[p+1]$, a switch from the initialization *order and time* updating procedure to the steady-state *fixed-order time-update* RLS algorithm is made in order to continue the exact RLS solution for $N > p$.

Upon acquisition of the first time sample $x[1]$, the following parameters are initialized

$$\mathbf{a}_{0,1}^f = \mathbf{a}_{0,1}^b = 1$$
$$\gamma_{-1,1} = 1 \qquad (9.61)$$
$$\rho_{0,1}^f = |x[1]|^2.$$

For data samples in the range $2 \leq N \leq p+1$, the following order and time relationships hold:

$$e^f_{N-2,N-1}[N] = \mathbf{a}^f_{N-2,N-1}{}^T \mathbf{x}_{N-2}[N]$$

$$e^f_{N-2,N}[N] = e^f_{N-2,N-1}[N]/\gamma_{N-3,N-1}$$

$$\rho^f_{N-1,N} = \omega \rho^f_{N-2,N-1}$$

$$\gamma_{N-2,N} = \gamma_{N-3,N-1} + |e^f_{N-2,N-1}[N]|^2/\rho^f_{N-1,N} \qquad (9.62)$$

$$\mathbf{a}^f_{N-1,N} = \begin{pmatrix} \mathbf{a}^f_{N-2,N-1} \\ -e^f_{N-2,N-1}[N]/x[1] \end{pmatrix}$$

$$\mathbf{c}_{N-2,N} = \begin{pmatrix} 0 \\ \mathbf{c}_{N-3,N-1} \end{pmatrix} + \frac{e^{f*}_{N-2,N}[N]}{\rho^f_{N-1,N}} \mathbf{a}^f_{N-2,N-1}.$$

The values for $\mathbf{a}^b_{N-1,N}$ and $\rho^b_{N-1,N}$ need to be initialized *only* at time $N = p+1$,

$$\mathbf{a}^b_{p,p+1} = \begin{pmatrix} -x[1]\mathbf{c}_{p-1,p+1}/\gamma_{p-1,p+1} \\ 1 \end{pmatrix} \qquad (9.63)$$

$$\rho^b_{p,p+1} = |x[1]|^2/\gamma_{p+1,p}.$$

9.7.6 Checks for Ill-Conditioning

The invertibility of $\mathbf{R}_{p,N}$ and $\mathbf{R}_{p,N+1}$ is guaranteed as long as $\gamma_{p,N}$ and $\gamma_{p,N+1}$ are both positive and bounded between 0 and 1, and also that the squared errors $\rho^f_{p,N}$ and $\rho^b_{p,N}$ are positive. Violation of these positivity properties in the implementation given by subroutine `fast_rls.m` is related to the numerical deterioration of the algorithm due to round-off and finite-precision effects. The matrix $\mathbf{R}_{p,N}$ is positive definite, as is easily proved.

References

Bierman, G. J., *Factorization Methods for Discrete Sequential Estimation*, Academic Press, Inc., New York, 1977.

Carayannis, G., D. G. Manolakis, and N. Kalouptsidis, A Fast Sequential Algorithm for Least Squares Filtering and Prediction, *IEEE Trans. Acoust. Speech Signal Process.*, vol. ASSP-31, pp. 1394–1402, December 1983.

Cioffi, J. M., and T. Kailath, Fast, Recursive-Least-Squares, Transversal Filters for Adaptive Filtering, *IEEE Trans. Acoust. Speech Signal Process.*, vol. ASSP-32, pp. 304–337, April 1984.

Fabre, P., and C. Gueguen, Improvement of the Fast Recursive Least-Squares Algorithms via Normalization: A Comparative Study, *IEEE Trans. Acoust. Speech Signal Process.*, vol. ASSP-34, pp. 296–308, April 1986.

Falconer, D. D., and L. Ljung, Application of Fast-Kalman Estimation to Adaptive Equalization, *IEEE Trans. Commun.*, vol. COM-26, pp. 1439–1446, October 1978.

Friedlander, B., Lattice Filters for Adaptive Processing, *Proc. IEEE*, vol. 70, pp. 829–867, August 1982a.

Friedlander, B., Recursive Lattice Forms for Spectral Estimation, *IEEE Trans. Acoust. Speech Signal Process.*, vol. ASSP-30, pp. 920–930, December 1982b.

Griffiths, L. J., Rapid Measurements of Digital Instantaneous Frequency, *IEEE Trans. Acoust. Speech Signal Process.*, vol. ASSP-23, pp. 207–222, April 1975.

Honig, M., and D. Messerschmitt, *Adaptive Filters: Structures, Algorithms, and Applications*, Kluwer Academic Publishers, Hingham, Mass., 1984.

Lawrence, V. B., and S. K. Tewksbury, Microprocessor Implementation of Adaptive Digital Filters, *IEEE Trans. Commun.*, vol. COM-31, pp. 826–835, June 1983.

Lin, D. W., On Digital Implementation of the Fast Kalman Algorithm, *IEEE Trans. Acoust. Speech Signal Process.*, vol. ASSP-32, pp. 998–1006, October 1984.

Ling, F., and J. G. Proakis, Nonstationary Learning Characteristics of Least Squares Adaptive Estimation Algorithms, *Proceedings 1984 International Conference on Acoustics, Speech, and Signal Processing*, paper 3.7, 1984a.

Ling, F., and J. G. Proakis, Numerical Accuracy and Stability: Two Problems of Adaptive Estimation Algorithms Caused by Round-off Error, *Proceedings 1984 International Conference on Acoustics, Speech, and Signal Processing*, paper 30.3, 1984b.

Ljung, L., and T. Söderström, *Theory and Practice of Recursive Identification*, The MIT Press, Cambridge, Mass., 1983.

Ljung, L., M. Morf, and D. Falconer, Fast Calculation of Gain Matrices for Recursive Estimation Schemes, *Int. J. Control*, vol. 27, pp. 1–19, January 1978.

Satorius, E. H., and J. D. Pack, Application of Least Squares Lattice Algorithms to Adaptive Equalization, *IEEE Trans. Commun.*, vol. COM-29, pp. 136–142, February 1981.

Widrow, B., and S. D. Stearns, *Adaptive Signal Processing*, Prentice-Hall, Inc., Englewood Cliffs, N. J., 1985.

10

AUTOREGRESSIVE MOVING AVERAGE SPECTRAL ESTIMATION

10.1 INTRODUCTION

The autoregressive-moving average (ARMA) model has more degrees of freedom than the autoregressive model (AR). A greater latitude in its ability to generate diverse spectral shapes is therefore expected of ARMA PSD estimators. Unlike the extensive repertoire of linear algorithms available to produce AR PSD estimators, there have been few algorithms produced for ARMA PSD estimators. This is due primarily to the nonlinear nature required of algorithms that must *simultaneously* estimate the MA and AR parameters of the ARMA model. The nonlinear equation (6.29) demonstrates the difficulty of estimating the ARMA parameters, even when the autocorrelation sequence is exactly known. Iterative optimization techniques based on maximum likelihood estimation (MLE) and related concepts are often used to solve the nonlinear techniques [Kay, 1987]. These techniques involve significant computations and are not guaranteed to converge, or they may converge to the wrong solution, and are, therefore, not as practical for real-time processing as are AR algorithms. Suboptimum techniques have been developed to significantly reduce the computational complexity. These techniques are usually based on a least squares criterion and require solution of linear equations. These methods generally estimate the AR and MA parameters *separately*, rather than *jointly*, as required for optimal parameter estimation. The AR parameters are typically estimated first, independently of the MA parameters, by some variant of the modified Yule-Walker equation (6.30). The MA parameters are then estimated assuming the AR parameters are known or have been previously estimated.

Very little statistical characterization of ARMA spectral estimators for finite data records is available. Most characterizations have been made on empirical evidence of a few simulations. Some asymptotic results with MLE estimates of the ARMA parameters are available, but these do not lend much insight into the finite data record estimation behavior.

Section 10.3 is devoted to moving average parameter estimation. The methods of this section can then be used to generate an MA PSD estimate from the data or to develop the MA component of an ARMA model, as described in Sec. 10.4.

255

A suboptimum technique for separate estimation of the AR and MA parameters is developed in Sec. 10.4. Simultaneous estimation of the AR and MA parameters is discussed in Sec. 10.5. Sequential estimation methods are summarized in Sec. 10.6. A special ARMA model that fits a process consisting of sinusoids in white noise is described in Sec. 10.7.

10.2 SUMMARY

Figure 10.1 is a flowchart of an algorithm to estimate the MA parameters from a data record. A MATLAB demonstration script available through the text website, spectrum_demo_ma.m, provides a spectral estimation demonstration for one MA parameter estimator algorithm and a plot. Specifically, the MA parameter array and white noise variance estimate from this program are used to compute the MA spectral estimate

$$\hat{P}_{MA}(f) = T\hat{\rho}_w \left| 1 + \sum_{l=1}^{p} \hat{b}[l] \exp(-j2\pi f l T) \right|^2 . \qquad (10.1)$$

Figure 10.2 illustrates a typical MA(15) spectral plot made from the 64-point test1987 test sequence. As with the AR methods, the order selected controls the trade-off between resolution bias and PSD variance. The spectral estimate is similar to the correlogram method PSD of Fig. 5.4.

A flowchart of an algorithm to estimate the ARMA parameters from a sample sequence, based on a method that separately estimates the AR and the MA parameters, is illustrated in Fig. 10.3. MATLAB demonstration script

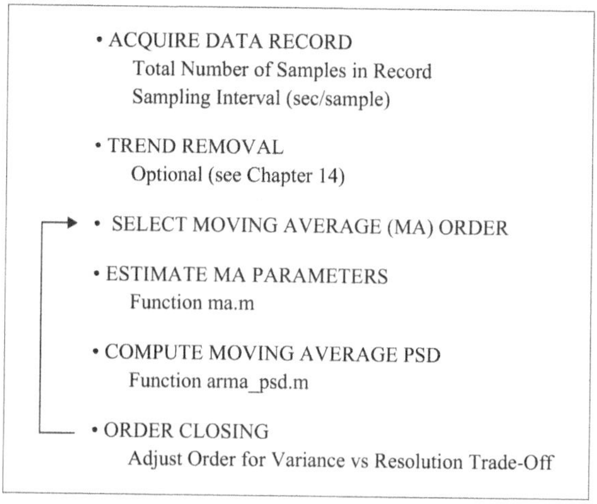

Figure 10.1. Flowchart of the MA PSD estimator.

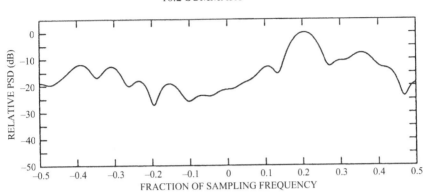

Figure 10.2. MA(15) spectral estimate of the 64-point test sequence using a long AR(30).

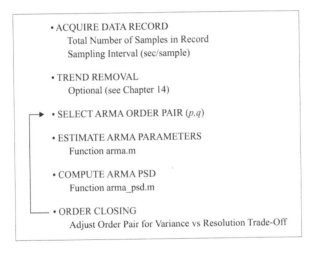

Figure 10.3. Flowchart of the ARMA PSD estimator.

`spectrum_demo_arma.m` available through the text website provides a spectral estimation demonstration for one ARMA parameter estimator algorithm and a plot. The AR parameter array, the MA parameter array, and the white noise variance estimate from this program are used to compute the ARMA spectral estimate

$$\hat{P}_{\text{ARMA}}(f) = T\hat{\rho}_w \left| \frac{1 + \sum_{k=1}^{q} \hat{b}[k]\exp(-j2\pi fkT)}{1 + \sum_{l=1}^{p} \hat{a}[l]\exp(-j2\pi flT)} \right|^2. \tag{10.2}$$

Figure 10.4 illustrates a typical ARMA(15,15) spectral estimate made from the 64-point `test1987` test sequence. Order selection will effect the resolution versus variance trade-off.

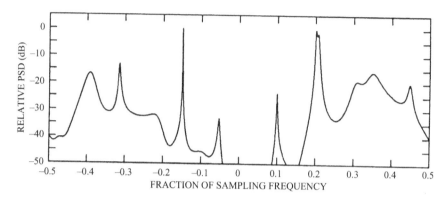

Figure 10.4. ARMA(15,15) spectral estimate of the 64-point test sequence using 30 autocorrelation lag estimates ($M = 30$).

Few guidelines exist for the selection of the ARMA model orders p and q, or for the MA model order q. Most often, it is assumed that $p = q$, although there is no reason for them to be the same. The AIC rule of AR processes has often been applied. Checking the rank of the modified Yule-Walker equations (6.30) can also give some indication of the AR model order p.

10.3 MOVING AVERAGE PARAMETER ESTIMATION

This section is generally applicable to all data sequences for which estimates of moving average parameters are desired, be it an original measured time sequence or the filtered residual sequence of the ARMA method described in Sec. 10.4. Recall that MA models are appropriate for processes that have broad peaks or sharp nulls in their spectra. The MA PSD does not model narrowband spectra well and is not, therefore, a "high-resolution" spectral estimator. This has tended to limit the research for means to estimate the MA parameters to be used in an MA power spectral density estimator.

The most obvious approach to estimate the MA parameters would be to solve the nonlinear equation (6.35) using autocorrelation estimates made from the data. Solutions of Eq. (6.35) often involve difficult spectral factorizations techniques [Box and Jenkins, 1970]. There is, however, an alternative approach based on a high-order AR approximation to the MA process that uses only linear operations. As the linear approach can yield satisfactory MA spectral estimates, it shall be the preferred algorithm for MA parameter estimation in this text.

Let

$$B(z) = 1 + \sum_{k=1}^{q} b[k]z^{-k} \tag{10.3}$$

denote the system function of an MA(q) process. Let $1/A_\infty(z)$, where

$$A_\infty(z) = 1 + \sum_{k=1}^{\infty} a[k] z^{-k}, \tag{10.4}$$

be the system function of an AR(∞) process that is equivalent to the MA(q) process, that is,

$$B(z) = \frac{1}{A_\infty(z)}, \tag{10.5}$$

or $B(z)A_\infty(z) = 1$. The inverse z-transform of the transform product $B(z)A_\infty(z)$ is the convolution of the MA parameters with the AR parameters, which may be deduced from Table 2.1. The inverse z-transform of the constant 1 is the sample function $\delta[m]$. The net result is

$$a[m] + \sum_{n=1}^{q} b[n] a[m-n] = \delta[m] = \begin{cases} 1 & \text{if } m = 0 \\ 0 & \text{if } m \neq 0 \end{cases}, \tag{10.6}$$

where, by definition, $a[0] = 1$ and $a[k] = 0$ for $k < 0$. Thus, the moving average parameters can be determined from the parameters of an equivalent infinite-order autoregressive model by solving any subset of q equations formed from Eq. (10.6). In practice, one can calculate a high-order AR(M) estimate, such that $M \gg q$. Based on the high-order autoregressive parameter estimates $1, \hat{a}_M[1], \ldots, \hat{a}_M[M]$, a set of equations of the form

$$e_{\text{MA}}[m] = \hat{a}_M[m] + \sum_{n=1}^{q} b[n] \hat{a}_M[m-n] \tag{10.7}$$

can be formed. Ideally, the error $e_{\text{MA}}[m]$ should be zero for all m except $m = 0$, as suggested by Eq. (10.6). Practically, the error will not be zero when dealing with finite data records, so estimates of the MA parameters are obtained by the minimization of the sum of squared errors

$$\hat{\rho}_q = \sum_m |e_{\text{MA}}[m]|^2, \tag{10.8}$$

where the index range will be defined. This estimation procedure can be shown [Kay, 1987] to be an approximate maximum likelihood estimate under the assumption that the process is Gaussian. Equation (10.7) is structurally identical to a forward linear prediction error, with the $\hat{a}[k]$ parameters taking the place of data samples. Two possible summation ranges for Eq. (10.8) are $0 \leq m \leq M + q$, corresponding to the *autocorrelation* method of linear prediction, and $q \leq m \leq M$, corresponding to the *covariance* method of linear prediction. Fast computational algorithms were presented in Chap. 8 for the solution of these two linear prediction methods.

MATLAB function ma.m implements the MA parameter estimator using this technique of fitting a high order AR into a low order MA:

```
function [rho,b]=ma(q,x,m)
% Moving average (MA) parameter estimation algorithm based on a high-order AR
% model estimated with the Yule-Walker algorithm.
%
%      [rho,b] = ma(q,x,m)
%
% q   -- order of moving average MA filter
% x   -- vector of data samples
% m   -- order of 'high-order' AR filter
% rho -- white noise variance estimate
% b   -- vector of moving average parameters
```

Note that an estimate of the driving noise variance of the MA process is simply the AR(M) least squares variance estimate.

Several methods may be used to determine an appropriate MA model order q. The AIC criterion for MA models has the same form as in the AR case

$$\text{AIC}[q] = N \ln(\hat{\rho}_q) + 2q$$

where $\hat{\rho}_q$ is an estimate of the white noise variance (use the least squares error variance for this). One could also filter the original data with the inverse MA filter and test for whiteness of the filtered residuals. [Chow, 1972a]. The residuals are white if the estimated autocorrelation sequence made from the residuals is approximately zero for all lags except lag zero.

10.4 SEPARATE AUTOREGRESSIVE AND MOVING AVERAGE PARAMETER ESTIMATION

Suboptimum procedures are most often used to produce ARMA parameters and PSD estimates. Most of these procedures make separate estimates of the AR parameters and the MA parameters. Typically, the AR parameters are estimated first, and then used to construct an inverse filter to apply to the original data. The residuals out of this filter should be representative of a moving average process, to which an MA parameter estimator can then be applied. This section describes one such ARMA algorithm involving three main steps.

It was shown in Chap. 6 [Eq. (6.29)] that the autocorrelation sequence for an ARMA(p, q) satisfies the relationship

$$r_{xx}[n] = -\sum_{k=1}^{p} a_p[k] r_{xx}[n-k] \qquad (10.9)$$

for $n > q$. If the autocorrelation sequence were known exactly, then the p relations for $q + 1 \le n \le q + p$ can be set up as a set of simultaneous equations to solve for the AR parameters. These were called the modified Yule-Walker equations, as formulated by Eq. (6.30). In a practical situation, only data samples are available, so that autocorrelation estimates must be substituted for the unknown autocorrelation sequence. Estimates of the AR parameters using the modified Yule-Walker equations are generally of poor quality (large variance) where the ARMA PSD normally has small values [Sakai and Tokumura, 1980] because the autocorrelation estimates are inaccurate, or the selected order is not correct, or both [Kay, 1987, Sec. 10.4].

An alternative approach examined by Mehra [1971], Cadzow [1981, 1982], and Porat and Friedlander [1985] is to use more than p equations for lags greater than q and minimize a squared error to fit p AR parameters. Such a procedure then becomes a *least squares modified Yule-Walker* technique. Assuming autocorrelation estimates $\hat{r}_{xx}[n]$ from lag 0 to lag M have been calculated, where M is the highest lag index that can be accurately estimated, the $M - q$ equations (such that $M - q > p$)

$$\hat{r}_{xx}[n] = -\sum_{k=1}^{p} a[k]\hat{r}_{xx}[n-k] + \epsilon[n] \qquad (10.10)$$

for $q + 1 \leq n \leq M$ are formed, where $\epsilon[n]$ is the estimation error. The *unbiased* autocorrelation estimates must be used to ensure that the bias of $\epsilon[n]$ will be zero. The sum of squared errors

$$\rho = \sum_{n=q+1}^{M} |\epsilon[n]|^2 \qquad (10.11)$$

is minimized with respect to the p autoregressive parameters $a[k]$. The normal equations that result are identical to the covariance method of linear prediction presented in Sec. 8.5.1, namely,

$$\left(\mathbf{T}_p^H \mathbf{T}_p\right) \begin{pmatrix} 1 \\ a[1] \\ \vdots \\ a[p] \end{pmatrix} = \begin{pmatrix} \rho \\ 0 \\ \vdots \\ 0 \end{pmatrix}, \qquad (10.12)$$

where $\mathbf{T}_p$ is a rectangular Toeplitz matrix of autocorrelation estimates

$$\mathbf{T}_p = \begin{pmatrix} \hat{r}_{xx}[q+1] & \cdots & \hat{r}_{xx}[q-p+1] \\ \vdots & \ddots & \vdots \\ \hat{r}_{xx}[M-p] & & \hat{r}_{xx}[q+1] \\ \vdots & \ddots & \vdots \\ \hat{r}_{xx}[M] & \cdots & \hat{r}_{xx}[M-p] \end{pmatrix}, \qquad (10.13)$$

rather than data elements as used in Chap. 8. Subroutine `covariance_lp.m` is a fast algorithm that may be used to solve Eq. (10.12), using $\hat{r}_{xx}[q-p+1], \ldots, \hat{r}_{xx}[M]$

as the input sequence, rather than the data values $x[1]$, ..., $x[N]$ as used in the covariance linear prediction method of Chap. 8.

A weighted squared error

$$\rho = \sum_{n=q+1}^{M} w[n]|\epsilon[n]|^2 \tag{10.14}$$

of decreasing weight sequence $w[n]$ with increasing n has been proposed by Fried-lander and Sharman [1985] to reduce the effect of the increased estimation variance associated with the autocorrelation estimates that are used to compute the higher-indexed error terms. As the lag index of the autocorrelation estimate increases, fewer data samples are used than at the lower lag indices, increasing the estimation variance. Cadzow [1982] has proposed a similar scheme, except errors are averaged in a neighborhood of error values.

The AR parameters estimated by the least squares modified Yule-Walker technique will not yield, in general, a minimum-phase autoregression. Spectral fac-torization may be used to create a minimum-phase autoregression by reflecting the poles outside the unit circle to the inside. No statistical results for finite data records are available for this technique. Kaveh and Bruzzone [1983] have made empirical observations that the performance is highly dependent upon the pole locations, giv-ing the best results when the AR poles are close to the unit circle and the worst results when the MA zeros are near the unit circle and the AR poles are farther from the unit circle. Both Izraelevitz and Lim [1985] and Friedlander and Sharman [1985] have provided some asymptotic analytical results of the least squares mod-ified Yule-Walker techniques as functions of the number of error terms used, the signal-to-noise ratio, and the spectral bandwidth of features in the ARMA spectra.

Selection of the order of the AR portion of the ARMA process can be based on similar order selection methods discused in Chap. 8 for pure AR processes. A high AR order will be required if there are sharp narrowband features in the ARMA spectra. Bruzzone and Kaveh [1980] have even shown overdetermination of the actual AR order in the ARMA model can improve the overall ARMA spectral estimate. One can also test the rank of $\mathbf{T}_p$ in Eq. (10.13) based on SVD analysis as an indicator of order [Cadzow, 1982].

Once the AR parameters have been determined, a residual time series that is approximately an MA process can be created by filtering the original data sequence with the filter that has the system function

$$\hat{A}(z) = 1 + \sum_{k=1}^{p} \hat{a}[k]z^{-k}, \tag{10.15}$$

in which $\hat{a}[k]$ are the estimated AR parameters determined by the least squares modified Yule-Walker technique. The system function of the ARMA process is

$B(z)/A(z)$, so that

$$\frac{B(z)}{A(z)}\hat{A}(z) \approx B(z). \tag{10.16}$$

Therefore, passing the measured data record through the filter with system function $\hat{A}(z)$ will yield an approximate moving average process at the filter output. The procedure of Sec. 10.3 may then be used to estimate the MA parameters. The filtered sequence of length $N - p$ is given by the convolution

$$z[n] = x[n] + \sum_{m=1}^{p} \hat{a}[m]x[n - m] \tag{10.17}$$

for $p + 1 \le n \le N$. Based on the above, MATLAB function arma.m estimates ARMA parameters based on the three step process: (1) estimate the AR parameters by the least squares modified Yule-Walker method, (2) filter the original data, and (3) estimate the MA parameters from the filter residuals:

```
function [rho,a,b]=arma(p,q,x,m)
% Estimates the AR and MA parameters of an ARMA process from sample data.
%
%    [rho,a,b] = arma(p,q,x,m)
%
% p   -- order of AR portion of ARMA model
% q   -- order of MA portion of ARMA model
% x   -- vector of data samples
% m   -- maximum ACS lag for modified Yule-Walker solution for AR (set m > p+q)
% rho -- estimated driving white noise variance
% a   -- vector of AR parameter estimates
% b   -- vector of MA parameter estimates
```

Although the MA parameters are needed for estimating the ARMA time-series model, the ARMA PSD estimator only needs an estimate of $T\rho_w|\hat{B}(f)|^2$. Explicit estimation of the MA parameters is not required [Kaveh, 1979]. The correlogram method PSD estimator could also be used to directly estimate the numerator function of the ARMA PSD estimator,

$$T\rho_w|\hat{B}(f)|^2 = T \sum_{m=-q}^{q} \hat{r}_{zz}[m]\exp(-j2\pi fmT), \tag{10.18}$$

in which $\hat{r}_{zz}[m]$ is the autocorrelation estimate produced from the filtered sequence $z[n]$. Spectral factorization can be used, if desired, to ensure a minimum-phase MA process.

Selecting the AR and MA orders of an ARMA time series model is not a simple procedure. There is not much in the literature on order selection except for simple cases. The AIC criterion for an ARMA has been one of the most used techniques.

It has the form

$$\text{AIC}[p, q] = N \ln(\hat{\rho}_{pq}) + 2(p + q)$$

in which $\hat{\rho}_{pq}$ is an estimate of the input white noise variance of the assumed ARMA(p, q) model. The actual value used is typically the squared error variance from the techniques described in this section or the next section. The minimum of the AIC$[p, q]$ over all orders p and q represents the appropriate order to use. One could also test the whiteness of the residual out of the inverse filter with systems function $\hat{B}(z)/\hat{A}(z)$ that uses the estimated AR and MA parameters. The estimated autocorrelation of the residual sequence should be approximately zero except at lag zero [Chow, 1972b].

10.5 SIMULTANEOUS AUTOREGRESSIVE AND MOVING AVERAGE PARAMETER ESTIMATION

A suboptimum method that makes a simultaneous estimate of both the autoregressive and the moving average parameters is motivated by a variant of the iterative prefiltering concept of Steiglitz and McBride [1965], one of a class of techniques that attempt to set up a linear problem by estimating the unknown white noise input to the ARMA filter. Other methods from this class of techniques have been described by Kay [1987]. The method to be described here is based on the improvement of a linear problems by a succession of these linear problems that iteratively refine the AR and MA parameter estimates.

Assuming for the moment that the input to the ARMA filter has been estimated as the sequence $\hat{u}[n]$, then this makes it possible to form the approximation error equations for an ARMA(p, q) process

$$e[n] = x[n] + \sum_{k=1}^{p} \hat{a}[k]x[n - k] + \sum_{k=1}^{q} \hat{b}[k]\hat{u}[n - k], \qquad (10.19)$$

in which the estimates of $\hat{u}[n]$ have been used in place of the unknown driving noise sequence values $u[n]$. The error becomes, therefore, linear in the unknown $a[k]$ and $b[k]$ parameters. The range for the time index is $1 \leq \alpha \leq n \leq \beta \leq N$, assuming data samples $x[n]$ from 1 to N are available. The actual values for α and β will depend on the specific algorithm used to estimate $u[n]$, to be discussed later. The set of $\beta - \alpha + 1$ error equations represented by (10.19) can be expressed in matrix form as

$$\mathbf{e} = \mathbf{X}\hat{\mathbf{a}} + \hat{\mathbf{U}}\hat{\mathbf{b}}, \qquad (10.20)$$

in which

$$\mathbf{e} = \begin{pmatrix} e[\alpha] \\ \vdots \\ e[\beta] \end{pmatrix}, \quad \mathbf{X} = \begin{pmatrix} x[\alpha] & \cdots & x[\alpha - p] \\ \vdots & & \vdots \\ x[\beta] & \cdots & x[\beta - p] \end{pmatrix}, \quad \hat{\mathbf{a}} = \begin{pmatrix} 1 \\ \hat{a}[1] \\ \vdots \\ \hat{a}[p] \end{pmatrix}$$

and

$$\hat{\mathbf{U}} = \begin{pmatrix} \hat{u}[\alpha - 1] & \cdots & \hat{u}[\alpha - q] \\ \vdots & & \vdots \\ \hat{u}[\beta - 1] & \cdots & \hat{u}[\beta - q] \end{pmatrix}, \hat{\mathbf{b}} = \begin{pmatrix} \hat{b}[1] \\ \vdots \\ \hat{b}[q] \end{pmatrix}.$$

Using the least squares solution described in Sec. 3.4 that minimizes

$$\rho_{ARMA} = \sum_{n=\alpha}^{\beta} |e[n]|^2,$$

the set of linear simultaneous equations, to be solved for the AR parameter estimates $\hat{\mathbf{a}}$ and the MA parameter estimates $\hat{\mathbf{b}}$, may easily be seen to take the form

$$\begin{pmatrix} \mathbf{X}^H \mathbf{X} & \mathbf{X}^H \hat{\mathbf{U}} \\ \hat{\mathbf{U}}^H \mathbf{X} & \hat{\mathbf{U}}^H \hat{\mathbf{U}} \end{pmatrix} \begin{pmatrix} \hat{\mathbf{a}} \\ \hat{\mathbf{b}} \end{pmatrix} = \begin{pmatrix} \rho_{ARMA} \\ \mathbf{0}_{p+q} \end{pmatrix}. \tag{10.21}$$

Once Eq. (10.21) has been solved, the new estimates of $\hat{\mathbf{a}}$ and $\hat{\mathbf{b}}$ may be used to filter the original time series $x[n]$ with the inverse ARMA filter $\hat{A}(z)/\hat{B}(z)$ to yield an updated estimate of the input sequence, $\hat{u}[n]$. The normal equations (10.21) are then solved iteratively in this fashion until convergence is achieved. The impulse response $h[k]$ of the filter $1/\hat{B}(z)$ will, in general, be infinite in duration and must be truncated to a finite interval in order to operate on the finite data sequence $x[n]$. This will be one source of error in this estimation procedure. The truncation interval selected and the accounting for the filter end effects will determine the index values α and β that can be used. The whole iterative process is started by making an estimate of $u[n]$ based on a long-order AR approximation using any AR algorithm from Chap. 8.

This iterative procedure is not guaranteed to converge and may, in fact, diverge. Conditions for convergence are not generally known. The inverse filter $\hat{A}(z)/\hat{B}(z)$ formed from the AR and MA parameter estimates is not guaranteed to be minimum phase, so the filter could be unstable. This usually requires some modification of the iterative algorithm to yield a minimum phase inverse filter, such as reflecting the poles/zeros outside of the unit z-plane circle to the inside. A one-shot version of this algorithm that terminates after the first iteration has been implemented as program MAYNE in the text by Kay [1987].

10.6 SEQUENTIAL APPROACH TO ARMA ESTIMATION

Time-sequential AR parameter estimation algorithms were introduced in Chap. 9. The ARMA process may be expressed in a form that permits similar time-sequential algorithms to be applied for the *simultaneous* estimation of the AR and MA parameters. The ARMA process may be set up by embedding it into a two-channel autoregressive process

$$e[n] = y[n] + \sum_{k=1}^{p} A[k]y[n-k], \qquad (10.22)$$

in which the two-element vectors $e[n]$ and $y[n]$ and 2×2 parameter matrix $A[k]$ are defined as

$$e[n] = \begin{pmatrix} e_x[n] \\ e_u[n] \end{pmatrix}, \quad y[n] = \begin{pmatrix} x[n] \\ u[n] \end{pmatrix}, \quad A[k] = \begin{pmatrix} a[k] & -b[k] \\ 0 & 0 \end{pmatrix}. \qquad (10.23)$$

The parameter matrix $A[k]$ is half zeros, reflecting the assumed white statistics of $u[n]$. Techniques of multichannel autoregressive analysis are presented in Chap. 15. A multichannel version of the single-channel fast RLS algorithm is available in the literature [Cioffi, 1984]. Assumptions have been made in Eq. (10.22) that the input driving noise sequence $u[n]$ is available, which is not the case, and that the MA order is identical to the AR order ($p = q$). Embedding methods can be modified to account for nonequal AR and MA orders [Lee et al., 1982; Friedlander, 1982]. The unknown process $u[n]$ is replaced in practice by the estimate

$$\hat{u}[n] = e_x[n] \qquad (10.24)$$

after each time update with a new data sample and treating the $\hat{u}[n]$ as "known" data for the next update, permitting the sequential algorithm to operate by "boot-strapping" to its own two-channel prediction error sequence. Some convergence conditions for this two-step method [Eqs. (10.22) and (10.24)] have been conjectured by Friedlander [1982], although detailed analysis still remains to be performed.

10.7 SPECIAL ARMA PROCESS FOR SINUSOIDS IN WHITE NOISE

The sinusoids in additive white noise process can be modeled as a special ARMA process. It is shown in Chap. 11 on the Prony method that a sample sequence consisting of M real sinusoids of arbitrary frequencies within the range $0 \le f \le 1/2T$ (T is the sample interval) may be generated by a symmetric $2M$-order linear difference equation of the form

$$\sum_{k=0}^{2M} a[k]x[n-k] = 0 \qquad (10.25)$$

such that $a[0] = a[2M] = 1$ and $a[2M-k] = a[k]$ for $0 \le k \le 2M$. The initial conditions $x[1], \ldots, x[2M]$ determine the amplitude and initial phase of the M

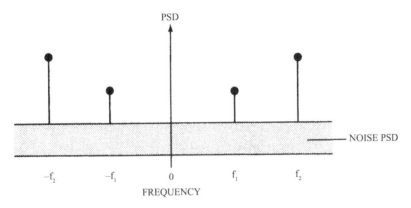

Figure 10.5. Spectrum of special ARMA PSD for sinusoids in white noise.

sinusoids. Adding white noise $w[n]$ produces the sinusoids in noise process

$$y[n] = x[n] + w[n]. \qquad (10.26)$$

Substituting $x[n - k] = y[n - k] - w[n - k]$ into Eq. (10.25) yields

$$y[n] = -\sum_{k=1}^{2M} a[k]y[n-k] + \sum_{k=0}^{2M} a[k]w[n-k]. \qquad (10.27)$$

This has the structure of an ARMA($2M, 2M$) process in which the AR and MA parameters are identical and the poles and zeros are on the unit circle [Ulrych and Clayton, 1976]. This forms an all-pass filter for which the system function $B(z)/A(z) = 0/0$ when evaluated at the sinusoidal frequencies, i.e., it is undefined. Impulses may then be inserted into the PSD at the locations of the sinusoidal frequencies, yielding the spectrum of Fig. 10.5. Equation (10.27) will be useful in the discussion on the Pisarenko harmonic decomposition technique in Chap. 13.

10.8   APPLICATION TO SUNSPOT NUMBERS

Figure 10.6 illustrates MA and ARMA PSD estimates made from the filtered sunspot numbers using the programs of Appendix 10.A and 10.B. A long AR model of 50 was used to make the ARMA(25,25) estimate. The first two peaks in the ARMA spectrum are located at 0.0938 and 0.2012 cycle per year (10.67 and 4.97 years per cycle). A long AR model of 64 was used to make the MA(32) estimate. The first two peaks in the MA spectrum are located at 0.0923 and 0.181 cycle per year (10.84 and 5.52 years per cycle).

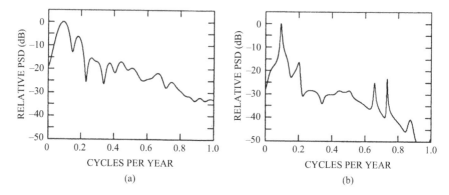

Figure 10.6. Spectral estimates of filtered sunspot numbers. (a) MA(25). (b) ARMA(32,32).

References

Box, G. E. P., and G. M. Jenkins, *Time Series Analysis, Forecasting and Control,* Holden-Day, Inc., San Francisco, 1970.

Bruzzone, S. P., and M. Kaveh, On Some Suboptimum ARMA Spectral Estimators, *IEEE Trans. Acoust. Speech Signal Process.,* vol. ASSP-28, pp. 753–755, December 1980.

Bruzzone, S. P., and M. Kaveh, Information Tradeoffs in Using the Sample Autocorrelation Function in ARMA Parameter Estimation, *IEEE Trans. Acoust. Speech Signal Process.,* vol. ASSP-32, pp. 701–715, August 1984.

Cadzow, J. A., Autoregressive Moving Average Spectral Estimation: A Model Equation Error Procedure, *IEEE Trans. Geosci. Remote Sensing,* vol. GE-19, pp. 24–28, January 1981.

Cadzow, J. A., Spectral Estimation: An Overdetermined Rational Model Equation Approach, *Proc. IEEE,* vol. 70, pp. 907–938, September 1982.

Chow, J. C., On the Estimation of the Order of a Moving Average Process, *IEEE Trans. Autom. Control,* vol. AC-17, pp. 386–387, June 1972a.

Chow, J. C., On Estimating the Orders of an Autoregressive-Moving Average Process with Uncertain Observations, *IEEE Trans. Autom. Control,* vol. AC-17, pp. 707–709, October 1972b.

Cioffi, J. M., and T. Kailath, Fast, Recursive-Least-Squares Transversal Filters for Adaptive Filtering, *IEEE Trans. Acoust. Speech Signal Process.,* vol. ASSP-32, pp. 304–337, April 1984.

Friedlander, B., Recursive Lattice Forms for Spectral Estimation, *IEEE Trans. Acoust. Speech Signal Process.,* vol. ASSP-30, pp. 920–930, December 1982.

Friedlander, B., and K. C. Sharman, Performance Evaluation of the Modified Yule-Walker Estimator, *IEEE Trans. Acoust. Speech Signal Process.,* vol. ASSP-33, pp. 719–725, June 1985.

Graupe, D., D. J. Krause, and J. B. Moore, Identification of Autoregressive Moving Average Parameters of Time Series, *IEEE Trans. Autom. Control,* vol. AC-20, pp. 104–107, February 1975.

Izraelevitz, D., and J. S. Lim, Properties of the Overdetermined Normal Equation Method for Spectral Estimation When Applied to Sinusoids in Noise, *IEEE Trans. Acoust. Speech Signal Process.*, vol. ASSP-33, pp. 406–412, April 1985.

Kaveh, M., High Resolution Spectral Estimation for Noisy Signals, *IEEE Trans. Acoust. Speech Signal Process.*, vol. ASSP-27, pp. 286–287, June 1979.

Kaveh, M., and S. P. Bruzzone, Statistical Efficiency of Correlation-Based Methods for ARMA Spectral Estimation, *IEE Proc.*, Part F, vol. 130, pp. 211–217, April 1983.

Kay, S. M., *Modern Spectral Estimation*, Prentice-Hall, Inc., Englewood Cliffs, N. J., 1987.

Lee, D.T.L., B. Friedlander, and M. Morf, Recursive Ladder Algorithms for ARMA Modeling, *IEEE Trans. Autom. Control*, vol. AC-27, pp. 753–764, August 1982.

Mehra, R. K., On-Line Identification of Linear Dynamic Systems with Applications to Kalman Filtering, *IEEE Trans. Autom. Control*, vol. AC-16, pp. 12–22, February 1971.

Porat, B., and B. Friedlander, Asymptotic Analysis of the Bias of the Modified Yule-Walker Estimator, *IEEE Trans. Autom. Control*, vol. AC-30, pp. 765–767, August 1985.

Sakai, H., and H. Tokumaru, Statistical Analysis of a Spectral Estimator for ARMA Processes, *IEEE Trans. Autom. Control*, vol. AC-25, pp. 122–124, February 1980.

Steiglitz, K., and L. E. McBride, A Technique for the Identification of Linear Systems, *IEEE Trans. Autom. Control*, vol. AC-10, pp. 461–464, October 1965.

Ulrych, T. J., and R. W. Clayton, Time Series Modeling and Maximum Entropy, *Phys. Earth Planet. Inter.*, vol. 12, pp. 188–200, August 1976.

11

PRONY'S METHOD

11.1 INTRODUCTION

Prony's method is a technique for modeling sampled data as a linear combination of exponentials. Although it is not a spectral estimation technique, Prony's method has a close relationship to the least squares linear prediction algorithms used for AR and ARMA parameter estimation. This relationship yields additional insight into spectral estimation methods that use AR and ARMA models. Prony's method seeks to fit a *deterministic* exponential model to the data, in contrast to AR and ARMA methods that seek to fit a *random* model to the second-order data statistics. A spectral interpretation of Prony's method may be obtained by computing the energy spectral density (ESD) of the deterministic exponential model.

Gaspard Riche, Baron de Prony [1795], was led to believe that laws governing expansion of various gases could be represented by sums of damped exponentials. He proposed a method for interpolating data points in his measurements by fitting an exponential model to a few equally spaced measured data points, and then computing the additional values by evaluation of the exponential model at intermediate points. The modern least squares version of the exponential modeling method has evolved significantly from Prony's original procedure. Prony's paper presented a method of *exactly* fitting as many purely *damped* exponentials as needed to fit the N available data points. The modern version of Prony's method generalizes to damped sinusoidal models. It also makes use of least squares analysis to *approximately* fit an exponential model for cases where there are more data points than needed to fit to the assumed number of exponential terms. A modification of the modern Prony method can fit a purely sinusoidal model with no damped components. The original and modified least squares Prony methods are presented in this chapter.

There are three basic steps in the Prony method. Step one determines the linear prediction parameters that fit the available data. In step two, the roots of a polynomial formed from the linear prediction coefficients will yield the estimates of damping and sinusoidal frequencies of each of the exponential terms. Step three involves the solution of a second set of linear equations to yield the estimates of the exponential amplitude and sinusoidal initial phase. The relationship of the linear prediction and autoregressive parameters explored in Chap. 7 leads to an interpretation of steps one and two as solving for the *poles* of an AR process. Thus, any AR or ARMA spectral analysis that involves a study of the pole locations could loosely be termed a Prony procedure.

11.2 SUMMARY

Assuming one has the N complex data samples $x[1], \ldots, x[N]$, the Prony method will estimate $x[n]$ with a p-term complex exponential model

$$\hat{x}[n] = \sum_{k=1}^{p} A_k \exp\big[(\alpha_k + j2\pi f_k)(n-1)T + j\theta_k\big] \qquad (11.1)$$

for $1 \leq n \leq N$, where T is the sample interval in seconds, A_k is the amplitude of the complex exponential, α_k is the damping factor in seconds^{-1}, f_k is the sinusoidal frequency in Hz, and θ_k is the sinusoidal initial phase in radians. The parameters are completely arbitrary. In the case of real data samples, the complex exponentials must occur in complex conjugate pairs of equal amplitude, thus reducing the exponential representation to

$$\hat{x}[n] = \sum_{k=1}^{p/2} 2A_k \exp[\alpha_k(n-1)T]\cos[2\pi f_k(n-1)T + \theta_k] \qquad (11.2)$$

for $1 \leq n \leq N$. If the number of complex exponentials p is even, then there are $p/2$ damped cosines. If p is odd, then there are $(p-1)/2$ damped cosines plus a single purely damped exponential. The periodogram could be interpreted as modeling a time series as a harmonic sinusoidal model. The distinction between the periodogram model and the Prony approach is the method by which frequencies are selected. The periodogram *preselects* a harmonic set of frequencies. The Prony method *estimates* the frequencies from the data.

Figure 11.1 is a flow diagram of the program steps that will yield the Prony amplitude, damping factor, frequency, and initial phase estimates. Two variations

```
• ACQUIRE DATA RECORD
    Total Number of Samples in Record
    Sampling Interval (sec/sample)

• TREND REMOVAL
    Optional (see Chapter 14)

• SELECT EXPONENTIAL FIT PARAMETERS
    Number of Complex Exponentials (choose 2x for real exp.)
    Damped (method=1) or undamped (method=2)

• ESTIMATE COMPLEX EXPONENTIAL PARAMETERS
    Function lstsqs_prony.m
        Produces amplitude, phase, damping factor, frequency

• COMPUTE AND PLOT PRONY ESD (optional)
    Function esd.m, using estimated parameters
```

Figure 11.1. Flowchart of Prony exponential estimation method and Prony ESD.

are possible. If a general damped sinusoidal model is required, set the parameter METHOD = 1. If specialization to the undamped sinusoidal model is desired, set parameter METHOD = 2. The output of Prony estimate results in two complex exponential parameter arrays associated with the assumed p complex exponentials in the signal (see Sec. 11.3). Once the exponential parameters are estimated, they may be used as an exponential signal model for which the spectral density may be calculated based on the model. MATLAB demonstration script spectrum_demo_prony.m, available on the text website, will perform both operations: estimating the exponential parameters from one of three data sources [test1987, doppler_radar, sunspot_numbers] and plotting the resulting spectral density.

Figure 11.2 depicts the Prony energy spectral density (ESD) estimates, based on the 64-point test data sequence test1987, for the two selectable Prony methods cases. For the METHOD = 2 case, the modified Prony method produces a line spectrum because of its sinusoidal model assumption. This results in very good estimates for the four actual sinusoids in the test1987 signal, but is inaccurate in its representation of the colored noise process within the spectrum.

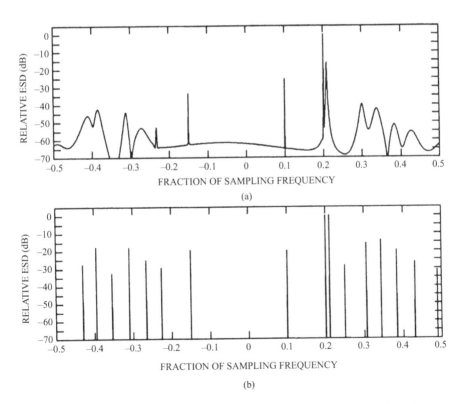

Figure 11.2. Examples of two Prony spectral estimates made from the 64-point test data sequence. (a) Extended Prony method, order 15. (b) Modified Prony method, order 16.

11.3 SIMULTANEOUS EXPONENTIAL PARAMETER ESTIMATION

The p-exponent discrete-time function of Eq. (11.1) may be concisely expressed in the form

$$\hat{x}[n] = \sum_{k=1}^{p} h_k z_k^{n-1}, \tag{11.3}$$

where the complex constants h_k and z_k are defined as

$$h_k = A_k \exp(j\theta_k) \tag{11.4}$$
$$z_k = \exp[(\alpha_k + j2\pi f_k)T]. \tag{11.5}$$

Note that h_k is a complex amplitude that represents a time-*independent* parameter, whereas z_k is a complex exponent that represents a time-*dependent* parameter. Ideally, one would like to minimize the squared error over the N data values

$$\rho = \sum_{n=1}^{N} |\epsilon[n]|^2, \tag{11.6}$$

where

$$\epsilon[n] = x[n] - \hat{x}[n] = x[n] - \sum_{k=1}^{p} h_k z_k^{n-1}, \tag{11.7}$$

with respect to the h_k parameters, the z_k parameters, and the number of exponents p simultaneously. This turns out to be a difficult nonlinear problem, even if the value of p is known. This difficulty can be demonstrated by the single-exponent case. Minimization of the squared error ρ using the damped exponential model

$$\hat{x}[n] = A \exp(\alpha[n-1]T)$$

is obtained by setting to zero the derivatives with respect to A and α

$$\frac{\partial \rho}{\partial A} = c_1 - c_2 A = 0$$
$$\frac{\partial \rho}{\partial \alpha} = c_3 - c_4 A = 0, \tag{11.8}$$

where

$$c_1 = \sum_{n=1}^{N} x[n] \exp(\alpha[n-1]T)$$
$$c_2 = \sum_{n=1}^{N} \exp(\alpha[n-1]T)$$

$$c_3 = \sum_{n=1}^{N} n \, x[n] \exp(\alpha[n-1]T)$$

$$c_4 = \sum_{n=1}^{N} n \exp(\alpha[n-1]T). \tag{11.9}$$

It has been assumed for simplicity that $x[n]$, A, and α are real. From the first equation of (11.8) one obtains $A = c_1/c_2$; substituting this into the second equation of (11.8) yields

$$c_2 c_3 = c_2 c_4. \tag{11.10}$$

This is a highly nonlinear expression in terms of sums involving $\exp(\alpha[n-1]T)$ which must be solved for α. No analytic solution is available.

Iterative algorithms, such as gradient descent procedures or Newton's method, have been devised to minimize Eq. (11.6) simultaneously with respect to all the exponential parameters [McDonough, 1963; McDonough and Huggins, 1968; Evans and Fischl, 1973]. These algorithms are very computationally expensive, sometimes requiring at each step the inversion of matrices of dimension as large as the number of data samples. Gradient descent algorithms for multimodal equations also may not converge to the global minimum. This computational difficulty led to the development of a suboptimum minimization of ρ, known as the least-squares Prony method, that utilizes linear equation solutions. The Prony method embeds the nonlinear aspects of the exponential model into a polynomial factoring, for which reasonably fast solution algorithms are available.

11.4 ORIGINAL PRONY CONCEPT

If as many data samples are used as there are exponential parameters, then an exact exponential fit to the data may be made. Consider the p-exponent discrete-time function

$$x[n] = \sum_{k=1}^{p} h_k z_k^{n-1}. \tag{11.11}$$

Note that $x[n]$, rather than $\hat{x}[n]$, has been used because exactly $2p$ complex samples $x[1], \ldots, x[2p]$ are used to fit an exact exponential model to the $2p$ complex parameters $h_1, \ldots, h_p, z_1, \ldots, z_p$. The p equations of Eq. (11.11) for $1 \leq n \leq p$ may be expressed in matrix form as

$$\begin{pmatrix} z_1^0 & z_2^0 & \cdots & z_p^0 \\ z_1^1 & z_2^1 & \cdots & z_p^1 \\ \vdots & \vdots & & \vdots \\ z_1^{p-1} & z_2^{p-1} & \cdots & z_p^{p-1} \end{pmatrix} \begin{pmatrix} h_1 \\ h_2 \\ \vdots \\ h_p \end{pmatrix} = \begin{pmatrix} x[1] \\ x[2] \\ \vdots \\ x[p] \end{pmatrix}. \tag{11.12}$$

The matrix of time-indexed z elements has a Vandermonde structure. If a method can be found to separately determine the z elements, then Eq. (11.12) represents a set of linear simultaneous equations that can be solved for the unknown vector of complex amplitudes [MATLAB function vandermonde_lineqs in Chap. 3]. Prony's contribution was the discovery of such a method.

The key to the separation is to recognize that Eq. (11.11) is the solution to some homogeneous linear constant-coefficient difference equation. In order to find the form of this difference equation, first define the polynomial $\phi(z)$ that has the z_k exponents as its roots,

$$\phi(z) = \prod_{k=1}^{p} (z - z_k). \tag{11.13}$$

If the products of Eq. (11.13) are expanded into a power series, then the polynomial may be represented as the summation

$$\phi(z) = \sum_{m=0}^{p} a[m] z^{p-m}, \tag{11.14}$$

with complex coefficients $a[m]$ such that $a[0] = 1$. Shifting the index on Eq. (11.11) from n to $n - m$ and multiplying by the parameter $a[m]$ yields

$$a[m]x[n-m] = a[m] \sum_{k=1}^{p} h_k z_k^{n-m-1}. \tag{11.15}$$

Forming similar products $a[0]x[n], \ldots, a[m-1]x[n-m+1]$ and summing produces

$$\sum_{m=0}^{p} a[m]x[n-m] = \sum_{i=0}^{p} h_i \sum_{m=0}^{p} a[m] z_i^{n-m-1}, \tag{11.16}$$

which is valid for $p + 1 \leq n \leq 2p$. Making the substitution $z_i^{n-m-1} = z_i^{n-p} z_i^{p-m-1}$, then

$$\sum_{m=0}^{p} a[m]x[n-m] = \sum_{i=0}^{p} h_i z_i^{n-p} \sum_{m=0}^{p} a[m] z_i^{p-m-1} = 0. \tag{11.17}$$

The right-hand summation in Eq. (11.17) may be recognized as the polynomial defined by Eq. (11.14), evaluated at each of its roots z_i, yielding the zero result indicated. Equation (11.17) is the linear difference equation whose homogeneous solution is given by Eq. (11.11). The polynomial (11.14) is the *characteristic equation* associated with this linear difference equation.

The p equations representing the valid values of $a[n]$ that satisfy Eq. (11.17) may be expressed as the $p \times p$ matrix equation

$$
\begin{pmatrix}
x[p] & x[p-1] & \cdots & x[1] \\
x[p+1] & x[p] & \cdots & x[2] \\
\vdots & \vdots & \ddots & \vdots \\
x[2p-1] & x[2p-2] & \cdots & x[p]
\end{pmatrix}
\begin{pmatrix}
a[1] \\
a[2] \\
\vdots \\
a[p]
\end{pmatrix}
= -
\begin{pmatrix}
x[p+1] \\
x[p+2] \\
\vdots \\
x[2p]
\end{pmatrix}.
\tag{11.18}
$$

Equation (11.18) demonstrates that, with $2p$ complex data samples, it is possible to decouple the h_k and z_k parameters. The complex polynomial coefficients $a[1], \ldots, a[p]$, which are functions of only the time-dependent components z_k of the exponential model, form a linear predictive relationship among the time samples. The matrix of Eq. (11.18) has a Toeplitz structure, so that the solution may be obtained with MATLAB function `toeplitz_lineqs` in Chap. 3.

The Prony procedure to fit p exponentials to $2p$ data samples may now be summarized in three steps. First, solution of Eq. (11.18) for the polynomial coefficients is obtained. Second, the roots of the polynomial defined by Eq. (11.14) are calculated. The damping α_i and sinusoidal frequency f_i may be determined from the root z_i using the relationships

$$
\alpha_i = \ln|z_i|/T \quad \sec^{-1} \tag{11.19}
$$

$$
f_i = \tan^{-1}\left[\mathrm{Im}\{z_i\}/\mathrm{Re}\{z_i\}\right]/2\pi T \quad \text{Hz.} \tag{11.20}
$$

To complete the Prony procedure, the roots computed in the second step are used to construct the matrix elements of Eq. (11.12), which is then solved for the p complex parameters $h[1], \ldots, h[p]$. The amplitude A_i and initial phase θ_i may be determined from each h_i parameter with the relationships

$$
A_i = |h_i| \tag{11.21}
$$

$$
\theta_i = \tan^{-1}\left[\mathrm{Im}\{h_i\}/\mathrm{Re}\{h_i\}\right] \quad \text{radians.} \tag{11.22}
$$

11.5 LEAST SQUARES PRONY METHOD

For practical situations, the number of data points N usually exceeds the minimum number needed to fit a model of p exponentials, i.e., $N > 2p$. In this overdetermined case, the data sequence can only be approximated as an exponential sequence,

$$
\hat{x}[n[= \sum_{k=1}^{p} h_k z_k^{n-1} \tag{11.23}
$$

for $1 \le n \le N$. Denote the approximation error as $\epsilon[n] = x[n] - \hat{x}[n]$. Simultaneously finding the order p and the parameters $\{h_k, z_k\}$ for $k = 1$ to $k = p$ that minimizes the total squared error

$$\rho = \sum_{n=1}^{N} |\epsilon[n]|^2 \tag{11.24}$$

was shown in Sec. 11.3 to be a difficult nonlinear problem. A suboptimum solution that provides satisfactory results may be obtained with a variant of the Prony method presented in Sec. 11.4. Substitution of appropriate linear least squares procedures for the first and third steps of the three-step Prony method yields an exponential modeling procedure that has sometimes been called the *extended Prony method* [Householder, 1950; McDonough, 1963]. This suboptimum approach effectively concentrates the nonlinearity of the exponential fitting problem into a polynomial factoring.

In the overdetermined data case, the linear difference Eq. (11.17) must be modified to

$$\sum_{m=1}^{p} a[m]x[n-m] = e[n] \tag{11.25}$$

for $p + 1 \le n \le N$. The term $e[n]$ represents the *linear prediction* approximation error, in contrast to error $\epsilon[n]$, which represents the *exponential* approximation error. Expression (11.25) is identical to the forward linear prediction error equation, making each $a[m]$ term a linear prediction parameter. In lieu of Eq. (11.17), the $a[m]$ parameters may be selected as those that minimize the linear prediction squared error $\sum_{n=p+1}^{N} |e[n]|^2$, rather than the exponential approximation squared error ρ of Eq. (11.24). This is simply the covariance method of linear prediction [see MATLAB function `covariance_lp` in Chap. 8]. The number of exponentials p (also called the number of poles in deference to AR processing) may be estimated using the same order selection rules established in Chap. 8. Alternatively, singular value decomposition (SVD) analysis similar to that covered in Sec. 11.9 may be used. In either case, the maximum order is limited to $p \le N/2$. The roots of the polynomial constructed from the linear prediction coefficients will yield, using Eqs. (11.19) and (11.20), the estimates of damping and frequency for each exponent.

If $z_1, \ldots, z_p$ have been determined by least squares linear prediction analysis and polynomial factoring, then the exponential approximation $\hat{x}[n]$ of Eq. (11.23) becomes linear in the remaining unknown parameters $h_1, \ldots, h_p$. Minimizing the squared error with respect to each of the h_k parameters yields the complex-valued $p \times p$ matrix normal equation (see Sec. 3.4 for derivation details)

$$\left(\mathbf{Z}^H \mathbf{Z} \right) \mathbf{h} = \left(\mathbf{Z}^H \mathbf{x} \right), \tag{11.26}$$

where the $N \times p$ matrix $\mathbf{Z}$, the $p \times 1$ vector $\mathbf{b}$, and $N \times 1$ data vector $\mathbf{x}$ are defined as

$$\mathbf{Z} = \begin{pmatrix} 1 & 1 & \cdots & 1 \\ z_1 & z_2 & \cdots & z_p \\ \vdots & \vdots & & \vdots \\ z_1^{N-1} & z_2^{N-1} & \cdots & z_p^{N-1} \end{pmatrix}, \quad \mathbf{h} = \begin{pmatrix} h_1 \\ h_2 \\ \vdots \\ h_p \end{pmatrix}, \quad \mathbf{x} = \begin{pmatrix} x[1] \\ x[2] \\ \vdots \\ x[N] \end{pmatrix}. \tag{11.27}$$

The $p \times p$ Hermitian matrix $\mathbf{Z}^H \mathbf{Z}$ has the form

$$\mathbf{Z}^H \mathbf{Z} = \begin{pmatrix} \gamma_{11} & \cdots & \gamma_{1p} \\ \vdots & & \vdots \\ \gamma_{p1} & \cdots & \gamma_{pp} \end{pmatrix}, \tag{11.28}$$

where

$$\gamma_{jk} = \sum_{n=0}^{N-1} (z_j^* z_k)^n = \gamma_{kj}^*. \tag{11.29}$$

A useful relationship that avoids the summation of Eq. (11.29) is

$$\gamma_{jk} = \begin{cases} \dfrac{(z_j^* z_k)^N - 1}{(z_j^* z_k) - 1} & \text{if } z_j^* z_k \neq 1 \\ N & \text{if } z_j^* z_k = 1 \end{cases}. \tag{11.30}$$

A fast computational algorithm has been developed that exploits the least squares structure composed of a product of Vandermonde data matrices. The MATLAB function lstsqsvdm.m is included in the text website library of functions and scripts.

```
function h=lstsqsvdm(x,z)
% Fast computational solution of least squares Vandermonde normal equations
% based on algorithm of Demeure [1989].
%
%        h = lstsqsvdm(x,z)
%
% x -- vector of complex data samples
% z -- vector of complex exponent parameters
% h -- vector of estimated complex amplitude parameters
```

The Prony method will also fit exponentials to any additive noise present in the data because the exponential model does not make a separate estimate of the noise process. An exponential model incorporating additive noise would have the form

$$x[n] = \sum_{k=1}^{p} h_k z_k^{n-1} + \epsilon[n]. \tag{11.31}$$

The function $\epsilon[n]$ has also been used to represent the approximation error of the exponential model. If $x[n] - \epsilon[n]$ is used in place of $x[n]$ in the analysis of Sec. 11.4, then the linear difference equation that describes the process consisting of a sum of exponentials plus white noise becomes

$$x[n] = -\sum_{k=1}^{p} a[m]x[n-m] + \sum_{k=0}^{p} a[m]\epsilon[n-m]. \tag{11.32}$$

This is an ARMA(p,p) with identical AR and MA coefficients and a driving noise process $\epsilon[n]$. The first step of the Prony method uses the linear prediction

$$x[n] = -\sum_{k=1}^{p} a[m]x[n-m] + e[n] \tag{11.33}$$

and attempts to whiten $e[n]$. Comparing Eq. (11.33) with (11.32), the whitened process $e[n]$ does not correspond well to the nonwhite MA process represented by $\sum_{k=0}^{p} a[m]\epsilon[n-m]$. It is for this reason that the Prony method does not often perform well in the presence of significant additive noise; it fails to account for nonwhite noise in the process. When the Prony estimation method is performed in the presence of significant additive noise, the damping terms are significantly misestimated, often being estimated to be much greater than they actually are [McDonough, 1963; Van Blaricum and Mittra, 1975, 1978; Poggio et al., 1978]. Using p larger than the actual number of poles helps to model and account for the noise (see Sec. 11.9).

The MATLAB function lstsqs_prony.m performs all the steps provided in this section (forming complex exponential parameter estimates from a data record):

```
function [h,z]=lstsqs_prony(method,num_exps,x)
% Solves for the exponential model parameters by the least squares Prony method
% using fast QR computational algorithms.
%
%    [h,z] = ls_prony(method,num_exps,x)
%
% method   -- select: 1 -- regular Prony (damped exponentials) , 2 -- modified
%                     Prony (undamped exponentials)
% num_exps -- number of presumed exponentials (must be even for method = 2)
% x        -- vector of data samples
% h        -- vector of complex amplitudes of exponential time series model
% z        -- vector of complex exponents of exponential time series model
```

11.6 MODIFIED LEAST SQUARES PRONY METHOD

The normal least squares Prony method can be modified to approximate a *complex* data sequence by a model consisting of *undamped* ($\alpha = 0$) complex sinusoids [Marple, 1979; Korn and Korn, 1961]. Only the case of an even number of complex exponentials shall be considered here. The $2p$ component model similar to Eq. (11.1) then has the form

$$\hat{x}[n] = \sum_{k=1}^{2p} A_k \exp(j2\pi f_k[n-1]T + j\theta_k) = \sum_{k=1}^{2p} h_k z_k^{n-1} \tag{11.34}$$

for $1 \leq n \leq N$, with $h_k = A_k \exp(j\theta_k)$ and $z_k = \exp(j2\pi f_k T)$. Note that the z_k are of unit modulus, that is, $|z_k| = 1$. If the h_k and z_k occur in complex conjugate pairs and $f_k \neq 0$ or $f_k \neq 1/2T$, then a *real* data sequence may be approximated by a model consisting of p even or odd real undamped sinusoids

$$\hat{x}[n] = \sum_{k=1}^{p} 2A_k \cos(2\pi f_k[n-1]T + \theta_k) = \sum_{k=1}^{p}\left(h_k z_k^{n-1} + h_k^*(z_k^*)^{n-1}\right) \quad (11.35)$$

for $1 \leq n \leq N$. The polynomial constructed with roots that are the z_k of either Eq. (11.34) or (11.35) has the form

$$\phi(z) = \prod_{k=1}^{2p}(z - z_k) = \sum_{k=0}^{2p} a[k] z^{2p-k}, \quad (11.36)$$

where $a[0] = 1$ by definition and the remaining $a[k]$ parameters will be, in general, complex, unless the real sinusoid case of Eq. (11.35) is considered. In the latter situation, the $a[k]$ are all real. It is simple to establish that coefficient $a[2p]$ is given by

$$a[2p] = \prod_{k=1}^{2p} z_k, \quad (11.37)$$

which follows from the definition of $\phi(z)$. Due to the unit modulus property $z_k^{-1} = z_k^*$, it is simple to show that $|a[2p]| = 1$ and that the polynomial

$$a[2p] z^{2p} \phi^*(z) \quad (11.38)$$

has the same roots as $\phi(z)$. Because the polynomial

$$a[2p] z^{2p} \phi^*(z) = \sum_{k=0}^{2p} a[2p] a^*[2p - k] z^{2p-k} \quad (11.39)$$

must yield the same roots, one is led to conclude that the conjugate property $a[k] = a[2p]a^*[2p - k]$ for $k = 0$ to $k = 2p$ must exist between the coefficients. Thus, the homogeneous linear difference equation that has Eq. (11.35) as its solution is

$$a[p]x[n-p] + \sum_{k=1}^{p}\left(a[p-k]x[n-p+k] + a[p+k]x[n-p-k]\right) = 0 \quad (11.40)$$

for $2p + 1 \leq n \leq N$. A more convenient form of Eq. (11.40), obtained by dividing by the value of the center element $a[p]$, yields the conjugate symmetric difference equation

$$x[n - p] + \sum_{k=1}^{p} \left(g_p[k]x[n - p + k] + g_p^*[k]x[n - p - k] \right) = 0, \qquad (11.41)$$

where $g_p[k] = a[p - k]/a[p]$. The conjugate symmetry is due to the property $a[p] = a[2p]a^*[p]$. A similar symmetric difference equation for real sinusoids was presented by Hildebrand [1956].

The modified Prony method replaces in step one the linear prediction error of Eq. (11.33) with the conjugate symmetric linear smoothing error (uses both past and future sample values)

$$e_p^s[n] = x[n] + \sum_{k=1}^{p} \left(g_p[k]x[n + k] + g_p^*[k]x[n - k] \right), \qquad (11.42)$$

defined over the interval $p + 1 \leq n \leq N - p$ (uses only the available data), and minimizes the squared smoothing error

$$\rho_p^s = \sum_{n=p+1}^{N-p} \left| e_p^s[n] \right|^2, \qquad (11.43)$$

rather than the squared linear prediction error given by Eq. (11.24). Setting the complex derivatives of ρ_p^s with respect to $g_p[1]$ through $g_p[p]$ to zero yields the normal equations expressed succinctly by the matrix representation

$$\mathbf{R}_{2p}\mathbf{g}_{2p} = \begin{pmatrix} \mathbf{0}_p \\ 2\rho_p^s \\ \mathbf{0}_p \end{pmatrix}, \qquad (11.44)$$

where centrosymmetric matrix $\mathbf{R}_{2p}$ and conjugate symmetric vector $\mathbf{g}_{2p}$ are

$$\mathbf{R}_{2p} = \begin{pmatrix} r_{2p}[0, 0] & \cdots & r_{2p}[0, 2p] \\ \vdots & & \vdots \\ r_{2p}[2p, 0] & \cdots & r_{2p}[2p, 2p] \end{pmatrix}, \quad \mathbf{g}_{2p} = \begin{pmatrix} g_p[p] \\ \vdots \\ g_p[1] \\ 1 \\ g_p^*[1] \\ \vdots \\ g_p^*[p] \end{pmatrix}. \qquad (11.45)$$

Elements of $\mathbf{R}_{2p}$ are

$$r_{2p}[j, k] = \sum_{n=2p+1}^{N} \left(x^*[n - j]x[n - k] + x[n - p + j]x^*[n - p + k] \right). \qquad (11.46)$$

The reader may have made the observation that $\mathbf{R}_{2p}$ in Eq. (11.45) is identical to the matrix of the modified covariance method, given by Eq. (8.48) in Chap. 8. A fast algorithm was presented in Sec. 8.11 for the solution of the modified covariance normal equations. The fast algorithm that solves Eq. (11.44) can be developed (see Sec. 11.11) by a simple extension of the modified covariance algorithm. The resulting fast algorithm requires $Np + 18p^2$ computations to solve Eq. (11.44). In addition, all lower-order least squares solutions of Eq. (11.44) are solved as part of the fast algorithm solution with no additional computations required. In situations where the number of sinusoids is not known, this provides an ability to check all models from one sinusoid up to p sinusoids. MATLAB function symcovar implements this fast algorithm:

```
function [rho_s,g]=symcovar(p,x)
% Fast algorithm for the solution of the hermitian symmetric covariance least
% squares normal equations.
%
%            [rho_s,g] = symcovar(p,x)
%
% p      -- order of linear smoothing filter (must be even)
% x      -- vector of data samples
% rho_s  -- least squares estimate of linear smoothing variance
% g      -- vector of hermitian symmetric linear smoothing parameters
```

Although complex roots of unit modulus produce polynomials with conjugate symmetric coefficients, the converse is not necessarily true. A conjugate symmetric polynomial only guarantees that if z_i is a root, then its reciprocal z_i^{-1} will also be a root. The root need not be of unit modulus, a requirement if an undamped sinusoidal is to be generated. In practice, the order $2p$ polynomials formed from the vector $\mathbf{g}_{2p}$ of linear smoothing parameters

$$g[p]z^{2p} + \cdots + g[1]z^{p+1} + z^p + g^*[1]z^{p-1} + \cdots + g^*[p] = 0, \qquad (11.47)$$

rarely exhibit reciprocal roots of nonunit modulus. When such roots occur, they often correspond to frequencies of 0 Hz or $1/2T$ Hz. The modified Prony method is completed with the same method as used by the unmodified Prony method to determine the sinusoid amplitudes and initial phases.

11.7 PRONY SPECTRUM

The Prony procedure normally terminates with the computation of the amplitude, damping, frequency, and phase parameter estimates. However, it is possible to carry the processing one further step to yield a Prony "spectrum" [Marple, 1980]. Many possible spectra may be defined, depending on the assumptions made concerning the waveform outside of the observed interval. The Prony spectrum is defined in terms of the exponential approximation $\hat{x}[n]$, rather than in terms of the original time sequence $x[n]$.

One assumption is that the discrete-time exponential sum in Eq. (11.3) is defined over the interval $-\infty < n < \infty$ as the one-sided function

$$\hat{x}_1[n+1] = \begin{cases} \displaystyle\sum_{k=1}^{p} h_k z_k^n & n \geq 0 \\ 0 & n < 0 \end{cases}, \tag{11.48}$$

as depicted in Fig. 11.3(a). If the signal $x[n]$ is real, the exponentials will occur in complex conjugate pairs $\exp(\pm j[2\pi f_k + \theta_k])$ to form a single cosine term $\cos(2\pi f_k + \theta_k)$. The z-transform of Eq. (11.48) is

$$\hat{X}_1(z) = \sum_{k=1}^{p} \left(\frac{h_k}{1 - z_k z^{-1}} \right), \tag{11.49}$$

which converges for $|z_k| < |z|$. It is also assumed that $|z_k| < 1$, that is, all damping parameters are negative, yielding decaying exponentials. If this is so, then

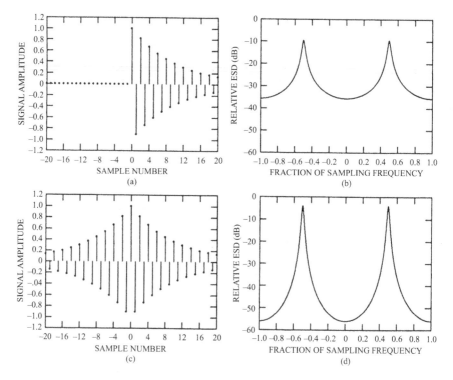

Figure 11.3. Exponential models and associated energy spectral density. (a) One-sided exponential model. (b) ESD of one-sided model. (c) Two-sided models. (d) ESD of two-sided model.

substitution of $z = \exp(j2\pi f T)$ into Eq. (11.49) will yield the discrete-time Fourier transform of the deterministic, finite-energy sequence $\hat{x}_1[n]$. Therefore, the Prony energy spectral density (ESD) for this one-sided exponential model is

$$\hat{S}_1(f) = \left| T \, \hat{X}_1(\exp[j2\pi f T]) \right|^2, \tag{11.50}$$

defined over the frequency interval $-1/2T \leq f \leq 1/2T$. This spectrum is appropriate for transient signals. A typical spectrum is shown in Fig. 11.3(b).

An alternative interpretation of the exponential model is the two-sided function

$$\hat{x}_2[n] = \begin{cases} \displaystyle\sum_{k=1}^{p} h_k z_k^n & n \geq 0 \\[2ex] \displaystyle\sum_{k=1}^{p} h_k (z_k^*)^{-n} & n < 0 \end{cases}, \tag{11.51}$$

where $z_k = \exp(\alpha_k T + j2\pi f_k T)$ and $(z_k^*)^{-1} = \exp(-\alpha_k T + j2\pi f_k T)$. This definition forces the damped portion of the exponential to be symmetrical about the origin, as depicted in Fig. 11.3(c). The z-transform of Eq. (11.51) is

$$\hat{X}_2(z) = \sum_{k=1}^{p} h_k \left(\frac{1}{1 - z_k z^{-1}} - \frac{1}{1 - (z_k^* z)^{-1}} \right)$$

$$= \sum_{k=1}^{p} h_k \left(\frac{(z_k - 1/z_k^*) z^{-1}}{1 - (z_k + 1/z_k^*) z^{-1} + (z_k/z_k^*) z^{-2}} \right), \tag{11.52}$$

which converges for $|z_k| < |z| < |z_k^{-1}|$. If one also assumes that $|z_k| < 1$ and substitutes $z = \exp(j2\pi f T)$ into Eq. (11.52), then the discrete-time Fourier transform of the two-sided exponential model will be

$$\hat{X}_2(f) = T \, \hat{X}_2\big(\exp[j2\pi f T]\big) \tag{11.53}$$

$$= \sum_{k=1}^{p} h_k \left(\frac{T(\exp[\alpha_k T] - \exp[-\alpha_k T]) \exp(j2\pi[f_k - f]T)}{1 - (\exp[\alpha_k T] + \exp[-\alpha_k T]) \exp(j2\pi[f_k - f]T) + \exp(j4\pi[f_k - f]T)} \right)$$

and the resulting Prony ESD is

$$\hat{S}_2(f) = \left| \hat{X}_2(f) \right|^2. \tag{11.54}$$

In general, $\hat{S}_2(f)$ provides a sharper spectral response than $\hat{S}_1(f)$, as depicted in Fig. 11.3(d). The author prefers this second definition because an undamped sinusoid ($\alpha = 0$) is defined over the infinite time interval. The ESD function $\hat{S}_2(f)$ has in this case an infinite response at the sinusoidal frequency, behaving like a discrete impulse function.

Equations (11.50) and (11.54) have been implemented as MATLAB function esd.m, which may be found at the text website:

```
function sd=esd(num_esd,method,T,h,z)
% Computes the one-sided or two-sided energy spectral density
%
%          sd = esd(num_esd,method,T,h,z)
%
% num_esd -- number of ESD values to compute (must be power of two)
% method  -- select ESD model:  1 -- onesided , 2 -- twosided
% T       -- sample interval in seconds
% h       -- vector of complex amplitude parameters
% z       -- vector of complex exponent parameters
% sd      -- vector of energy spectral density values
```

Either of the two possible Prony spectra has the ability to produce both narrowband and wideband spectral shapes. The response widths are a function of the magnitude of the damping factor, as illustrated by Fig. 11.3. The bell-shaped responses have peak ESD magnitudes of $(2A_k/\alpha_k)^2$ and bandwidths (to the 6-dB points) of α/π Hz, so resolution varies as a function of damping. If α is large, the response is broadband. If α is small, then a sharp narrowband response is produced.

Another possible Prony spectrum incorporates all information in the entire data sample sequence. If the exponential approximation residuals

$$\epsilon[n] = x[n] - \sum_{k=1}^{p} h_k z_k^{n-1} \tag{11.55}$$

are computed, then a conventional periodogram analysis may be performed on the residuals and the periodogram spectrum combined with either of the two suggested Prony spectra.

11.8 ACCOUNTING FOR KNOWN EXPONENTIAL COMPONENTS

There are applications in which some of the exponential modes (either the poles or the damping and frequency components) may be known *a priori*, due to the availability of prior measurements or due to the physical or electrical properties of the source generating the signal under analysis. The associated amplitude and phase components of these exponential modes may, however, not be known. The first step of the Prony method can be modified to account for these known poles [Trivett and Robinson, 1981]. Suppose the q exponential components $z_1, \ldots, z_q$ are known. The characteristic polynomial associated with these q known components is

$$\prod_{m=0}^{q} (z - z_k) = \sum_{k=0}^{q} c[k] z^k, \tag{11.56}$$

where $c[q] = 1$. The characteristic polynomial for all p components [Eq. (11.14)] may then be factored as

$$\sum_{m=0}^{p} a[m]z^m = \left(\sum_{k=0}^{q} c[k]z^k\right)\left(\sum_{i=0}^{p-q} \alpha[i]z^i\right), \qquad (11.57)$$

where $\alpha[p-q] = 1$. Equating similar powers of z yields

$$a[m] = \sum_{k=0}^{q} c[k]\alpha[m-k], \qquad (11.58)$$

where $\alpha[i] = 0$ for $i > p-q+1$ or $i < 0$. Substituting Eq. (11.58) for $a[m]$ in Eq. (11.16) yields

$$\sum_{m=1}^{p} a[m]x[n-m] = \sum_{m=1}^{p}\left(\sum_{k=0}^{q} c[k]\alpha[m-k]\right)x[n-m] = 0 \qquad (11.59)$$

defined for $p+1 \le n \le 2p$. Equation (11.59) may be reorganized as

$$\sum_{m=0}^{p-q} \alpha[m]y[n-m] = 0 \qquad (11.60)$$

for $p+1 \le n \le 2p$, where the new sequence $y[n]$ is defined as

$$y[n] = \sum_{k=0}^{q} c[k]x[n-k]. \qquad (11.61)$$

This represents a convolution operation, that is, filtering of the original time sequence $x[n]$ to produce the new sequence $y[n]$.

The Prony method with some components known proceeds as follows. First, the original data is filtered [Eq. (11.61)] using a filter with coefficients determined by the known poles [Eq. (11.58)]. The filtered data samples are then processed as usual by the least squares covariance linear prediction algorithm to yield estimates of the parameters $\alpha[m]$. The roots of the reduced-order polynomial

$$\sum_{i=0}^{p-q} \alpha[i]z^i \qquad (11.62)$$

produce estimates of the unknown poles. These $p-q$ poles and the q known poles are then combined for the least squares operation that will yield the amplitudes and phases of all p components.

11.9 IDENTIFICATION OF EXPONENTIALS IN NOISE

The least squares Prony method has been shown to require the solution of the identical covariance linear prediction normal equations as encountered for AR spectral
analysis in Chap. 8. This common heritage with the AR spectral method suggests use
of the AR order selection criteria, as developed in Sec. 8.10, for the determination
of the number of exponentials present in the data. In situations where the noise level
is low, these criteria are often satisfactory. However, in higher noise levels, the frequency and damping component estimates are usually inaccurate and biased due to
the effect of noise. Distinguishing roots of the Prony characteristic polynomial due
to the weak exponentials from roots due to noise is often difficult. Three methods
may be used to enhance identification of actual exponential signals in the data and
to improve the accuracy of the frequency and damping component estimates. These
methods involve the use of both forward and backward linear prediction polynomial
zeros, high prediction orders, and singular value decomposition (SVD).

In the absence of additive noise, it was shown in Sec. 11.4 that p exponentials
are generated by the forward linear prediction

$$\sum_{m=0}^{p} a[m]x[n-m] = 0, \tag{11.63}$$

in which $a[0] = 1$ and the characteristic polynomial

$$A(z) = \sum_{m=0}^{p} a[m]z^{p-m} \tag{11.64}$$

has roots at $z_k = \exp(s_k)$ for $k = 1$ to $k = p$, in which $s_k = (\alpha_k + j2\pi f_k)T$ contains
the damping factor and frequency of the kth exponential. The same p exponentials
may also be generated in reverse time by the backward linear prediction

$$\sum_{m=0}^{p} b[m]x[n-p+m] = 0, \tag{11.65}$$

in which $b[0] = 1$. The characteristic polynomial

$$B(z) = \sum_{m=0}^{p} b^*[m]z^{p-m} \tag{11.66}$$

formed from the conjugated backward linear prediction coefficients has roots $z_k = \exp(-s_k^*) = \exp([-\alpha_k + j2\pi f_k]T)$ for $k = 1$ to $k = p$. For a decaying damping
factor ($\alpha_k < 0$), the roots of the forward linear prediction characteristic polyno-
mial $A(z)$ fall *inside* the unit z-plane circle, whereas the roots of the backward
linear prediction characteristic polynomial $B(z)$ fall *outside* the unit circle due to

the reciprocal damping factor $\exp(-\alpha_k T)$, which is a growing exponential. These characteristics of the root locations of polynomials $A(z)$ and $B(z)$ are properties of deterministic exponentials.

Consider the process consisting of two complex exponentials in complex additive white Gaussian noise

$$x[n] = A_1 \exp(s_1 n) + A_2 \exp(s_2 n) + w[n], \tag{11.67}$$

in which $s_1 = -0.1 + j2\pi(0.52)$, $s_2 = -0.2 + j2\pi(0.42)$, $A_1 = 1$, $A_2 = 1$, and the noise variance is ρ. With no noise, the second-order characteristic polynomials $A(z)$ and $B(z)$ for the two exponentials will have roots located at the positions indicated in Figs. 11.4(a) and (b). Fifty epochs, each of 25 data points indexed from $n = 0$ to $n = 24$, were generated by using different $w[n]$ sequences at a signal-to-noise ratio of $10\log_{10}(1/\rho) = 20$ dB. Each epoch was processed by the covariance least squares linear prediction algorithm (MATLAB function `covariance_lp`) using order $p = 2$ and the roots (zeros) of $A(z)$ and $B(z)$ were plotted. The backward linear prediction coefficients are obtained simultaneously with the forward linear prediction coefficients by the fast algorithm of Sec. 8.10. Therefore, no additional computations are required in order to get the backward linear prediction coefficients. Figures 11.4(c) and (d), respectively, show the zero locations of the forward and backward characteristic polynomials for all 50 epochs. For comparison, the true zero locations for the two exponentials are also indicated. The additive noise has caused a bias in the estimates of the zeros, which means a bias in the damping and frequency terms. The bias is more severe for the exponential with the larger damping factor.

The bias may be reduced significantly by selecting a linear prediction order much higher than the number of exponentials actually present in the signal, an empirical observation made frequently in the literature (e.g., see Van Blaricum and Mittra [1975] and Kumaresan and Tufts [1982]). Figures 11.4(e) and (f) illustrate the positions of the zeros when the covariance algorithm for order $p = 8$ is selected. The zeros related to the true exponentials in the signal now cluster closer to the correct positions, but the extraneous zeros due to noise fluctuate widely. However, the selection of a higher linear prediction order has a negative effect in that extraneous zeros are introduced, making it difficult to separate the zeros due to the actual exponential signals from the zeros due to noise, at least when examining only the $A(z)$ polynomial roots. The separation is obvious in the $B(z)$ zero locations of Fig. 11.4(f) only due to the clustering effect of 50 independent data sets. The zeros from a single polynomial would not be so obvious. By examining the zeros of *both* $A(z)$ and $B(z)$, one then has a means to distinguish the true signal zeros from the noise zeros. The statistics of a stationary random process do not change when the process is time reversed. Thus, the noise zeros of both $A(z)$ and $B(z)$ tend predominantly to stay within the unit circle, as illustrated by the test ensemble results in Figs. 11.4(e) and (f). The true exponential signal zeros will occur in reciprocal positions along a common radius when examining the zeros of $A(z)$ and $B(z)$ [Kumaresan, 1983].

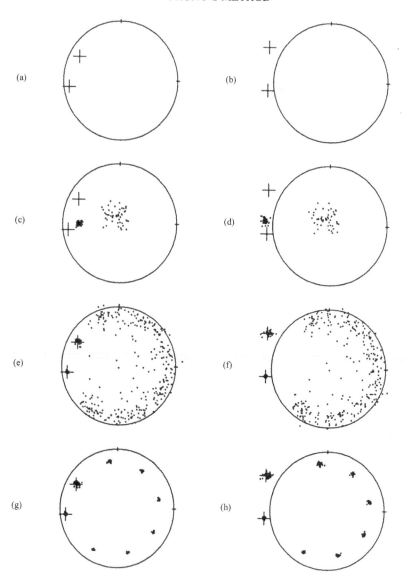

Figure 11.4. Zero locations of forward linear prediction filter polynomial $A(z)$ and backward linear prediction filter polynomial $B(z)$ for 50 independent data realizations of 25 samples each. The unit radius circle of the z-plane is shown for reference. The circled x symbols show the true location of the two exponential signals. For $A(z)$, these correspond to $\exp(s_1)$ and $\exp(s_2)$ inside the unit circle. For $B(z)$, these correspond to $\exp(-s_1^*)$ and $\exp(-s_2^*)$ outside the unit circle. (a) $A(z)$, $p = 2$, no noise. (b) $B(z)$, $p = 2$, no noise. (c) $A(z)$, $p = 2$, noise. (d) $B(z)$, $p = 2$, noise. (e) $A(z)$, $p = 8$, noise. (f) $B(z)$, $p = 8$, noise. (g) $A(z)$, $p = 8$, SVD with two principal eigenvectors. (h) $B(z)$, $p = 8$, SVD with two principal eigenvectors.

The application of singular value decomposition (SVD) can provide further improvement. The forward and backward linear prediction errors of Sec. 8.5.1 may be expressed succinctly as

$$\mathbf{X}_p^f \mathbf{a}_p^f = -\mathbf{x}_p^f + \mathbf{e}_p^f \qquad \mathbf{X}_p^b \mathbf{a}_p^b = -\mathbf{x}_p^b + \mathbf{e}_p^b, \qquad (11.68)$$

in which the Toeplitz data matrices $\mathbf{X}_p^f, \mathbf{X}_p^b$ and data vectors $\mathbf{x}_p^f, \mathbf{x}_p^b$ are defined as

$$\mathbf{X}_p^f = \begin{pmatrix} x[p] & \cdots & x[1] \\ \vdots & & \vdots \\ x[N-1] & \cdots & x[N-p] \end{pmatrix}, \quad \mathbf{x}_p^f = \begin{pmatrix} x[p+1] \\ \vdots \\ x[N] \end{pmatrix},$$

$$\mathbf{X}_p^b = \begin{pmatrix} x[p+1] & \cdots & x[2] \\ \vdots & & \vdots \\ x[N] & \cdots & x[N-p+1] \end{pmatrix}, \quad \mathbf{x}_p^b = \begin{pmatrix} x[1] \\ \vdots \\ x[N-p] \end{pmatrix}, \qquad (11.69)$$

and the forward linear prediction coefficient vector $\mathbf{a}_p^f$, forward linear prediction error vector $\mathbf{e}_p^f$, backward linear prediction coefficient vector $\mathbf{a}_p^b$, and backward linear prediction error vector $\mathbf{e}_p^b$ are defined as

$$\mathbf{a}_p^f = \begin{pmatrix} a^f[1] \\ \vdots \\ a^f[p] \end{pmatrix}, \mathbf{e}_p^f = \begin{pmatrix} e^f[p+1] \\ \vdots \\ e^f[N] \end{pmatrix}, \mathbf{a}_p^b = \begin{pmatrix} a^b[p] \\ \vdots \\ a^b[1] \end{pmatrix}, \mathbf{e}_p^b = \begin{pmatrix} e^b[p+1] \\ \vdots \\ e^b[N] \end{pmatrix}. \qquad (11.70)$$

Based on Eq. (3.96), the data matrices have the following singular value decompositions

$$\mathbf{X}_p^f = \sum_{n=1}^{p} \sigma_n^f \mathbf{u}_n^f (\mathbf{v}_n^f)^H \qquad \mathbf{X}_p^b = \sum_{n=1}^{p} \sigma_n^b \mathbf{u}_n^b (\mathbf{v}_n^b)^H, \qquad (11.71)$$

in which the σ_n^f are positive singular values of $\mathbf{X}_p^f$ and the σ_n^b are positive singular values of $\mathbf{X}_p^b$. The $\mathbf{u}_n$ and $\mathbf{v}_n$ are eigenvectors of the respective data matrix. If a signal consists of m exponentials in additive noise, then the m eigenvectors associated with the m largest singular values primarily span the m exponential components. The $p - m$ eigenvectors of the remaining smaller singular values primarily span the noise components. Chapter 13 will provide more details of this procedure. Assuming the singular values have been ordered by decreasing value, e.g., $\sigma_1^f > \sigma_2^f > \cdots > \sigma_p^f$, then a *reduced rank approximation* to each data matrix may be formed by truncating the SVD relationships of Eq. (11.71) to the m principal singular values

$$\hat{\mathbf{X}}_p^f = \sum_{n=1}^{m} \sigma_n^f \mathbf{u}_n^f (\mathbf{v}_n^f)^H \qquad \hat{\mathbf{X}}_p^b = \sum_{n=1}^{m} \sigma_n^b \mathbf{u}_n^b (\mathbf{v}_n^f)^H. \qquad (11.72)$$

This will reduce the noise contribution to the data matrix, effectively enhancing the SNR [Holt and Antill, 1977; Kumaresan and Tufts, 1982; Tufts and

Kumaresan, 1982]. Minimizing the norms $\|\mathbf{a}_p^f\|_2$ and $\|\mathbf{a}_p^b\|_2$ with respect to the reduced rank data matrices will yield the solutions

$$\mathbf{a}_p^f = -\left(\hat{\mathbf{X}}_p^f\right)^{\#}\mathbf{x}_p^f \qquad \mathbf{a}_p^b = -\left(\hat{\mathbf{X}}_p^b\right)^{\#}\mathbf{x}_p^b \qquad (11.73)$$

in which the pseudoinverse data matrices are defined as

$$\left(\hat{\mathbf{X}}_p^f\right)^{\#} = \sum_{n=1}^{m}(\sigma_n^f)^{-1}\mathbf{v}_n^f(\mathbf{u}_n^f)^H \qquad \left(\hat{\mathbf{X}}_p^b\right)^{\#} = \sum_{n=1}^{m}(\sigma_n^b)^{-1}\mathbf{v}_n^b(\mathbf{u}_n^b)^H. \qquad (11.74)$$

The order p must lie in the range $m \leq p \leq N - m$ in order to keep the rank of $\mathbf{X}_p^f$ amd $\mathbf{X}_p^b$ greater than or equal to m, the assumed number of exponentials. If the number of exponentials is not known, it may be estimated by comparing the relative magnitudes of the singular values. The signal-related singular values tend to be larger than the noise-related singular values. Once the forward and backward linear prediction coefficients have been computed, as indicated by Eq. (11.73), the roots computed as indicated by Eqs. (11.64) and (11.66) will yield the exponential estimates. The data vectors $\mathbf{x}_p^f$ and $\mathbf{x}_p^b$ are not considered for noise effects, even though they are themselves noisy. In order to reduce the effect of noise from both data matrices and data vectors, the total least squares method [Rahman and Yu, 1986] can be applied to give slightly better results than shown here.

Using the SVD approach with a reduced-order approximation produced the results shown in Figs. 11.4(g) and (h) for the 50 data epochs. Selection of $p = 8$ and $m = 2$ was used in this truncated SVD result. The extraneous zeros are much less perturbed and form a uniform pattern around the inside of the unit circle, which is representative of an exponential model approximation to white noise. The zeros due to the actual pair of damped exponentials stand out distinctly in Figs. 11.4(g) and (h). Application of reduced rank SVD in cases of highly damped exponentials or in cases of low SNR may only be of limited success. In both of these situations, the signal singular values are comparable in magnitude to the noise singular values, making it difficult to separate the singular values based strictly on relative magnitude.

11.10 APPLICATION TO SUNSPOT NUMBERS

Figure 11.5 illustrates the energy spectral density estimates of the filtered sunspot numbers when processed by the Prony method. A model order of 36 complex exponentials was fitted. This corresponds to 18 real sinusoids in the modified Prony case. The usual least squares Prony algorithm yielded periods for the four lowest-frequency peaks of 22.71, 10.78, 7.92, and 5.19 years per cycle with an estimated amplitude of 58.78 and an estimated damping of -0.0057 for the 10.78-year component. The modified Prony algorithm yielded periods for the four lowest-frequency peaks of 24.84, 10.96, 8.60, and 5.96 years per cycle with an estimated amplitude of 37.69 for the 10.78-year component, which was the strongest.

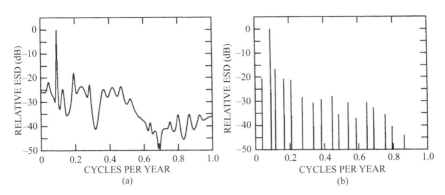

Figure 11.5. Prony ESD estimates of the filtered sunspot numbers. (a) Extended Prony. (b) Modified Prony (undamped sinusoids).

Other applications of Prony's method that the reader may find interesting include bearing estimation [Bucker, 1977] and radar target resonances [Chuang and Moffatt, 1977].

11.11 EXTRA: FAST ALGORITHM TO SOLVE SYMMETRIC COVARIANCE NORMAL EQUATIONS

11.11.1 Introduction

The fast algorithm to solve the symmetric covariance normal equations [Eq. (11.44)] was developed by Marple [1982a, 1982b]. The $2p$th-order complex linear smoothing error may be represented as

$$e_{2p}^s[n] = \mathbf{x}_{2p}^T[n+p]\mathbf{g}_{2p}, \qquad (11.75)$$

where the data vector $\mathbf{x}_{2p}[n+p]$ and symmetric linear smoothing vector $\mathbf{g}_{2p}$ are defined as

$$\mathbf{x}_{2p}[n+p] = \begin{pmatrix} x[n+p] \\ \vdots \\ x[n+1] \\ x[n] \\ x[n-1] \\ \vdots \\ x[n-p] \end{pmatrix}, \quad \mathbf{g}_{2p} = \begin{pmatrix} g_{2p}[p] \\ \vdots \\ g_{2p}[1] \\ 1 \\ g_{2p}^*[1] \\ \vdots \\ g_{2p}^*[p] \end{pmatrix}. \qquad (11.76)$$

Note that $\mathbf{g}_{2p} = \mathbf{J}\mathbf{g}_{2p}^*$, where $\mathbf{J}$ is a $(2p+1) \times (2p+1)$ reflection matrix. Based on measured complex data samples $x[0], \ldots, x[N-1]$, the symmetric covariance method minimizes the sum of squared linear smoothing errors

$$\rho_{2p}^s = \sum_{n=p}^{N-p-1} \left| e_{2p}^s[n] \right|^2, \tag{11.77}$$

resulting in the normal equation

$$\mathbf{R}_{2p}\mathbf{g}_{2p} = \begin{pmatrix} \mathbf{0}_p \\ 2\rho_{2p}^s \\ \mathbf{0}_p \end{pmatrix}, \tag{11.78}$$

where $\mathbf{0}_p$ is a $p \times 1$ all-zero vector. The centrosymmetric matrix $\mathbf{R}_{2p}$ may be expressed as

$$\mathbf{R}_{2p} = \sum_{n=2p}^{N-1} \left[\mathbf{x}_{2p}^*[n]\mathbf{x}_{2p}^T[n] + \mathbf{J}\mathbf{x}_{2p}[n]\mathbf{x}_{2p}^H[n]\mathbf{J} \right]. \tag{11.79}$$

If the smoothing error terms $e_{2p}^s[N - p - 1]$ and $e_{2p}^s[p]$ are not used, and the resulting squared error

$$\rho_{2p}^{s''} = \sum_{n=p+1}^{N-p-2} \left| e_{2p}^s[n] \right|^2 \tag{11.80}$$

is minimized, then the following normal equations are the result

$$\mathbf{R}_{2p}''\mathbf{g}_{2p}'' = \begin{pmatrix} \mathbf{0}_p \\ 2\rho_{2p}^{s''} \\ \mathbf{0}_p \end{pmatrix}, \tag{11.81}$$

where the double prime denotes the solution for this case of omitted error terms. The matrix $\mathbf{R}_{2p}''$ is defined by

$$\mathbf{R}_{2p}'' = \sum_{n=2p+1}^{N-2} \left(\mathbf{x}_{2p}^*[n]\mathbf{x}_{2p}^T[n] + \mathbf{J}\mathbf{x}_{2p}[n]\mathbf{x}_{2p}^H[n]\mathbf{J} \right). \tag{11.82}$$

The following special identities can be obtained by forming the indicated quadratic products:

$$\mathbf{g}_{2p}^H\mathbf{R}_{2p}\mathbf{c}_{2}p = \left(\mathbf{c}_{2p}^H\mathbf{R}_{2p}^H\mathbf{g}_{2p} \right)^* \implies e_{2p}^s[N - p - 1] = 2\rho_{2p}^s c_{2p}[p]$$

$$\mathbf{g}_{2p}^H\mathbf{R}_{2p}\mathbf{d}_{2}p = \left(\mathbf{d}_{2p}^H\mathbf{R}_{2p}^H\mathbf{g}_{2p} \right)^* \implies e_{2p}^s[p] = 2\rho_{2p}^s d_{2p}^*[p]$$

$$\mathbf{g}_{2p}^H\mathbf{R}_{2p}\mathbf{a}_{2}p = \left(\mathbf{a}_{2p}^H\mathbf{R}_{2p}^H\mathbf{g}_{2p} \right)^* \implies g_{2p}[p] = \rho_{2p}^s a_{2p}^*[p]/\rho_{2p}$$

$$\left(0 \quad \mathbf{a}_{2p-1}'^T\mathbf{J} \right) \mathbf{R}_{2p} \begin{pmatrix} 0 \\ \mathbf{g}_{2p-2}^H \\ 0 \end{pmatrix} \implies \mathbf{r}_{2p}^T\mathbf{g}_{2p-2}'^* = -2\rho_{2p-2}^{s''} a_{2p-1}'^*.$$

The fast algorithm developed embeds additional recursive relationships into the modified covariance algorithm of Sec. 8.11. The algorithms run concurrently.

11.11.2 Special Partitions and Auxiliary Parameters

Crucial to the development of a fast algorithm is the following order-index partition of the $(2p+1) \times (2p+1)$ matrix $\mathbf{R}_{2p}$,

$$\mathbf{R}_{2p} = \begin{pmatrix} r_{2p}[0,0] & \mathbf{r}_{2p}^T \mathbf{J} & r_{2p}[0,2p] \\ \mathbf{J}\mathbf{r}_{2p}^* & \mathbf{R}_{2p-2}'' & \mathbf{r}_{2p} \\ r_{2p}[2p,0] & \mathbf{r}_{2p}^H & r_{2p}[2p,2p] \end{pmatrix}, \tag{11.83}$$

where the $(2p-1) \times 1$ column vector $\mathbf{r}_{2p}$ is

$$\mathbf{r}_{2p} = \begin{pmatrix} r_{2p}[1,2p] \\ \vdots \\ r_{2p}[2p-1,2p] \end{pmatrix} \tag{11.84}$$

and $\mathbf{J}$ is a $(2p-1) \times (2p-1)$ reflection matrix. The elements $r_{2p}[i,j]$ are defined by Eq. (8.49). Note also that $r_{2p}[0,0] = r_{2p}[2p,2p]$ and $r_{2p}[0,2p] = r_{2p}^*[2p,0]$. Note that (11.21) is not the same as $\mathbf{r}_{2p}$ of Sec. (8.11). Also crucial is the following time-index partition,

$$\mathbf{R}_{2p} = \mathbf{R}_{2p}'' + \mathbf{x}_{2p}^*[2p]\mathbf{x}_{2p}^T[2p] + \mathbf{J}\mathbf{x}_{2p}[2p]\mathbf{x}_{2p}^H[2p]\mathbf{J}$$
$$+ \mathbf{x}_{2p}^*[N-1]\mathbf{x}_{2p}^T[N-1] + \mathbf{J}\mathbf{x}_{2p}[N-1]\mathbf{x}_{2p}^H[N-1]\mathbf{J}. \tag{11.85}$$

11.11.3 Order Update Recursions

The order update for the linear smoothing coefficient vector $\mathbf{g}_{2p}$ has the form

$$\mathbf{g}_{2p} = \beta_1 \left\{ \begin{pmatrix} 0 \\ \mathbf{g}_{2p-2}'' \\ 0 \end{pmatrix} + \alpha_1 \mathbf{a}_{2p} + \alpha_1^* \mathbf{J}\mathbf{a}_{2p}^* \right\}, \tag{11.86}$$

for which the complex scalar α_1 and real scalar β_1 are defined as

$$\alpha_1 = -\mathbf{r}_{2p}^T \mathbf{g}_{2p-2}''^* / \rho_{2p} = 2\rho_{2p-2}^{s''} a_{2p-1}^{'*} / \rho_{2p} \tag{11.87}$$

$$\beta_1 = \left(1 + 2\mathrm{Re}\{\alpha_1 a_{2p}[p]\}\right)^{-1}. \tag{11.88}$$

The vector $\mathbf{a}_{2p}$ is obtained from the modified covariance fast algorithm. To verify Eq. (11.87), premultiply both sides by $\mathbf{R}_{2p}$ and substitute the order partition (11.84) as appropriate. This will also yield the order update for the linear smoothing squared error

$$\rho_{2p}^s = \beta_1 \rho_{2p-2}^{s''}. \tag{11.89}$$

Observe that the order update $\mathbf{g}_{2p}$ only takes place coincident with the even-order update of the linear prediction coefficient vector $\mathbf{a}_{2p}$, that is, only on alternate order updates of $\mathbf{a}_{2p}$. Also observe that $\beta_1 > 0$ to ensure that ρ_{2p}^s is positive, as it must be because it is a squared error.

11.11.4 Time Update Recursions

The time-index update of the linear smoothing coefficient vector is given by

$$\mathbf{g}''_{2p} = \beta_2 \left[\mathbf{g}_{2p} + \alpha_2 \mathbf{c}_{2p} + \alpha_2^* \mathbf{J}\mathbf{c}_{2p}^* + \alpha_3 \mathbf{d}_{2p} + \alpha_3^* \mathbf{J}\mathbf{d}_{2p}^* \right], \tag{11.90}$$

where the complex scalars α_2 and α_3 are found as the solution of the 4×4 complex matrix

$$\begin{pmatrix} \delta_{2p} & -e_{2p}^{db}[2p] & -e_{2p}^{cf}[2p] & -e_{2p}^{df*}[N-1] \\ -e_{2p}^{db*}[2p] & \delta_{2p} & -e_{2p}^{df}[N-1] & -e_{2p}^{cf*}[2p] \\ -e_{2p}^{cf*}[2p] & -e_{2p}^{df}[N-1] & \gamma_{2p} & -e_{2p}^{cf*}[N-1] \\ -e_{2p}^{df}[N-1] & -e_{2p}^{cf}[2p] & -e_{2p}^{cf}[N-1] & \gamma_{2p} \end{pmatrix} \begin{pmatrix} \alpha_3 \\ \alpha_3^* \\ \alpha_2 \\ \alpha_2^* \end{pmatrix}$$

$$= \begin{pmatrix} e_{2p}^{s}[p] \\ e_{2p}^{s*}[p] \\ e_{2p}^{s*}[N-p-1] \\ e_{2p}^{s}[N-p-1] \end{pmatrix} = 2\rho_{2p}^{s} \begin{pmatrix} d_{2p}^{*}[p] \\ d_{2p}[p] \\ c_{2p}^{*}[p] \\ c_{2p}[p] \end{pmatrix}. \tag{11.91}$$

where the smoothing errors $e_{2p}^{s}[p]$ and $e_{2p}^{s}[N-p-1]$ can use the substitute identities of (11.83). The real scalar β_2 is given simply by

$$\beta_2 = \left[1 + 2\mathrm{Re}\{\alpha_2 c_{2p}[p]\} + 2\mathrm{Re}\{\alpha_3 d_{2p}[p]\} \right]^{-1}. \tag{11.92}$$

The complex gain vectors $\mathbf{c}_{2p},\mathbf{d}_{2p}$, real gain scalars δ_{2p},γ_{2p}, and complex error scalars $e_{2p}^{cf}[2p]$, $e_{2p}^{db}[2p]$, $e_{2p}^{cf}[N-1]$, $e_{2p}^{df}[N-1]$ are available from the modified covariance algorithm of Sec. 8.11. To verify Eq. (11.91), premultiply both sides by $\mathbf{R}''_{2p}$ and substitute the time partition (11.86) as appropriate. This will also yield the time update for the squared linear smoothing error

$$\rho_{2p}^{s''} = \beta_2 \rho_{2p}^{s}. \tag{11.93}$$

11.11.5 Initial Conditions

Only one initial condition is required, namely the adjusted smoothing squared error

$$\rho_0^{s''} = \sum_{n=2}^{N-1} |x[n]|^2. \tag{11.94}$$

11.11.6 Operation Count

The symmetric covariance algorithm requires $3Np + 40p^2$ multiplications, $12p$ real divides, and storage of $N + 10p + 12$ elements. Going from order $2p+1$ to $2p+3$ requires a number of computations proportional to $2N + 36p$ at each iteration.

References

Bucker, H. P., Comparison of FFT and Prony Algorithms for Bearing Estimation of Narrow-Band Signals in Realistic Ocean Environment, *J. Acoust. Soc. Am.*, vol. 61, pp. 756–762, March 1977.

Chuang, C. W., and D. L. Moffatt, Natural Resonances of Radar Targets via Prony's Method and Target Discrimination, *IEEE Trans. Aerosp. Electron. Syst.*, vol. AES-12, pp. 583–589, September 1976.

Evans, A. G., and R. Fischl, Optimal Least-Squares Time-Domain Synthesis of Recursive Digital Filters, *IEEE Trans. Audio Electoacoust.*, vol. AU-21, pp. 61–65, February 1973.

Hildebrand, F. B., *Introduction to Numerical Analysis*, McGraw-Hill Book Company, New York, 1956, Chapter 9.

Holt, J. N., and R. J. Antill, Determining the Number of Terms in a Prony Algorithm Exponential Fit, *Math. Biosci.*, vol. 36, pp. 319–332, 1977.

Householder, A. S., On Prony's Method of Fitting Exponential Decay Curves and Multiple-Hit Survival Curves, *Oak Ridge National Laboratory Report ORNL-455*, Oak Ridge, Tenn., February 1950.

Jenkins, M. A., and J. F. Traub, Algorithm 419: Zeros of a Complex Polynomial, *Commun. ACM*, vol. 15, pp. 97–99, February 1972.

Korn, G. A., and T. M. Korn, *Mathematical Handbook for Scientists and Engineers*, McGraw-Hill Book Company, New York, p. 660, 1961.

Kulp, R. W., An Optimum Sampling Procedure for Use with the Prony Method, *IEEE Trans. Electromagn. Compat.*, vol. EMC-23, pp. 67–71, May 1981.

Kumaresan, R., On the Zeros of the Linear Prediction-Error Filter for Deterministic Signals, *IEEE Trans. Acoust. Speech Signal Process.*, vol. ASSP-31, pp. 217–220, February 1983.

Kumaresan, R., and D. W. Tufts, Estimating the Parameters of Exponentially Damped Sinusoids and Pole-Zero Modeling in Noise, *IEEE Trans. Acoust. Speech Signal Process.*, vol. ASSP-30, pp. 833–840, December 1982.

Marple, S. L., Jr., Spectral Line Analysis by Pisarenko and Prony Methods, *Proceedings of the 1979 IEEE International Conference on Acoustics, Speech, and Signal Processing*, Washington, D.C., pp. 159–161.

Marple, S. L., Jr., Exponential Energy Spectral Density Estimation, *Proceedings of the 1980 IEEE International Conference on Acoustics, Speech, and Signal Processing*, Denver, Colo., pp. 588–591.

Marple, S. L., Jr., Spectral Line Analysis via a Fast Prony Algorithm, *Proceedings of the 1982 IEEE International Conference on Acoustics, Speech, and Signal Processing*, Paris, France, pp. 1375–1378, 1982a.

Marple, S. L., Jr., Fast Algorithms for Linear Prediction and System Identification Filters with Linear Phase, *IEEE Trans. Acoust. Speech Signal Process.*, vol. ASSP-30, pp. 942–953, December 1982b.

McDonough, R. N., Representations and Analysis of Signals. Part XV. Matched Exponents for the Representation of Signals, Ph.D. dissertation, Department of Electrical Engineering, John Hopkins University, Baltimore, Md., April 1963.

McDonough, R. N., and W. H. Huggins, Best Least-Squares Representation of Signals by Exponentials, *IEEE Trans. Autom. Control*, vol. AC-13, pp. 408–412, August 1968.

Poggio, A. J., M. L. Van Blaricum, E. K. Miller, and R. Mittra, Evaluation for a Processing Technique for Transient Data, *IEEE Trans. Antennas Propag.*, vol. AP-26, pp. 165–173, January 1978.

de Prony, Baron (Gaspard Riche), Essai expérimental et analytique: sur les lois de la dilatabilité de fluides élastiques et sur celles de la force expansive de la vapeur de l'eau et de la vapeur de l'alkool, à différentes températures, *J. E. Polytech.*, vol. 1, no. 2, pp. 24–76, 1795.

Rahman, M. A., and Kai-Bor Yu, Improved Frequency Estimation Using Total Least Squares Approach, *Proceedings of the 1986 IEEE International Conference on Acoustics, Speech, and Signal Processing*, Tokyo, Japan, pp. 1397–1400, April 1986.

Trivett, D. H., and A. Z. Robinson, Modified Prony Method Approach to Echo-Reduction Measurements, *J. Acoust. Soc. Am.*, vol. 70, pp. 1166–1175, October 1981.

Tufts, D. W., and R. Kumaresan, Estimation of Frequencies of Multiple Sinusoids: Making Linear Prediction Perform Like Maximum Likelihood, *Proc. IEEE*, vol. 70, pp. 975–989, September 1982.

Van Blaricum, M. L., and R. Mittra, A Technique for Extracting the Poles and Residues of a System Directly from Its Transient Response, *IEEE Trans. Antennas Propag.*, vol. AP-23, pp. 777–781, November 1975.

Van Blaricum, M. L., and R. Mittra, Problems and Solutions Associated with Prony's Method for Processing Transient Data, *IEEE Trans. Antennas Propag.*, vol. AP-26, pp. 174–182, January 1978; correction in vol. AP-28, p. 949, November 1980.

12

MINIMUM VARIANCE
SPECTRAL ESTIMATION

12.1 INTRODUCTION

The minimum variance (MV) spectral estimator was originally introduced by Capon [1969, 1983] for use in multi-dimensional seismic array frequency-wavenumber analysis. Capon promoted the method as a high-resolution spectral analysis technique, although its resolution is actually between that of the AR spectral estimator and that of the classical spectral estimators. Lacoss [1971] reformulated Capon's original space-time analysis for application to one-dimensional time-series analysis and renamed the technique as the *maximum likelihood method* (MLM) of spectral estimation, due to the maximum likelihood and Gaussian process context used by Capon in his 1969 paper. MLM is not an accurate description because it is not a maximum likelihood estimate of a power spectral density (PSD) function. In fact, the MV spectral estimator is not a true PSD function because the area under the MV estimate does not represent the total power in the measured process, which should be a characteristic of a true PSD. The inverse transform of the MV estimator also does not match the autocorrelation sequence that is used to create the MV estimate, unlike AR methods that do match. Thus, MV will be a spectral estimate in that it describes relative component strengths over frequency, but it is not a true PSD estimator. It does have the virtue of yielding peak heights that are linearly proportional to the power of sinusoids present in the process. *Minimum variance* is more descriptive of the origin of the estimator, which is created by minimizing the variance of the output of a narrowband filter that adapts to the spectral content of the input process at each frequency of interest.

12.2 SUMMARY

The MV spectral estimator is given by

$$P_{\mathrm{MV}}(f) = \frac{T}{\mathbf{e}^H(f)\mathbf{R}_p^{-1}\mathbf{e}(f)},$$
(12.1)

in which $\mathbf{R}_p^{-1}$ is the inverse of the known or estimated autocorrelation matrix of dimension $(p+1) \times (p+1)$, $\mathbf{e}(f)$ is the complex sinusoid vector

$$\mathbf{e}(f) = \begin{pmatrix} 1 \\ \exp(j2\pi fT) \\ \vdots \\ \exp(j2\pi fpT) \end{pmatrix}, \tag{12.2}$$

and T is the sample interval. Figure 12.1 is a flowchart of the MV spectral estimator. A MATLAB script `spectrum_demo_minvar.m` is available through the text website for demonstrating this spectral estimation technique. Embedded in the implementation is a fast algorithm suggested by Musicus [1985] which incorporates the autocorrelation estimate implied by the Burg algorithm. The minimum variance spectral estimate of the 64-point test case for order $p = 15$ is shown in Fig. 12.2.

• ACQUIRE DATA RECORD
    Total Number of Samples in Record
    Sampling Interval (sec/sample)

• TREND REMOVAL
    Optional (see Chapter 14)

• SELECT ADAPTIVE FILTER MODEL ORDER

• COMPUTE MINIMUM VARIANCE PSD
    Function minimum_variance_psd.m

• PLOT PSD; PERFORM ORDER CLOSING
    Adjust Filter Order for Variance vs Resolution Trade-Off

Figure 12.1. Flowchart of the minimum variance spectral estimator.

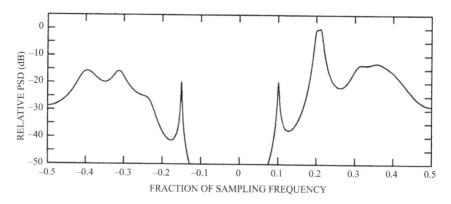

Figure 12.2. Minimum variance spectral estimate of the 64-point test sequence for $p = 15$.

It may be seen that the apparent resolution is poorer than that of the autoregressive spectral estimator of Chap. 8, but better than that of the classical correlogram PSD method of Chap. 5. An order closing procedure can be used to trade-off resolution for spectral variance, as indicated in Fig. 12.1. Cox [1973] has quantified the resolution capability of the MV method.

12.3 DERIVATION OF THE MINIMUM VARIANCE SPECTRAL ESTIMATOR

Consider a transversal filter with $p+1$ coefficients $a[0], a[1], \ldots, a[p]$. The filter output $y[n]$ to the input $x[n]$ is given by the convolution

$$y[n] = \sum_{k=0}^{p} a[k]x[n-k] = \mathbf{x}^T[n]\mathbf{a}, \qquad (12.3)$$

in which the vectors $\mathbf{x}[n]$ and $\mathbf{a}$ of dimension $p+1$ are defined as

$$\mathbf{x}[n] = \begin{pmatrix} x[n] \\ x[n-1] \\ \vdots \\ x[n-p] \end{pmatrix} \quad , \quad \mathbf{a} = \begin{pmatrix} a[0] \\ a[1] \\ \vdots \\ a[p] \end{pmatrix}. \qquad (12.4)$$

The filter output variance is simply

$$\rho = \mathcal{E}\{|y[n]|^2\} = \mathcal{E}\{\mathbf{a}^H\mathbf{x}^*[n]\mathbf{x}^T[n]\mathbf{a}\} = \mathbf{a}^H\mathcal{E}\{\mathbf{x}^*[n]\mathbf{x}^T[n]\}\mathbf{a} = \mathbf{a}^H\mathbf{R}_p\mathbf{a}, \quad (12.5)$$

in which the $(p+1) \times (p+1)$ Toeplitz autocorrelation matrix is

$$\mathbf{R}_p = \begin{pmatrix} r_{xx}[0] & \cdots & r_{xx}^*[p] \\ \vdots & \ddots & \vdots \\ r_{xx}[p] & \cdots & r_{xx}[0] \end{pmatrix}. \qquad (12.6)$$

The output variance of Eq. (12.5) is similar to the linear prediction error filter variance of Eq. (7.4), except that $a[0]$ is arbitrary here rather than constrained to be unity. The coefficients are to be selected so that, at a frequency f_0 under consideration, the frequency response of the filter has unit gain. This constraint can be represented as

$$\sum_{k=0}^{p} a[k]\exp(-j2\pi f_0 pT) = \mathbf{e}^H(f_0)\mathbf{a} = 1, \qquad (12.7)$$

in which the vector $\mathbf{e}(f_0)$ of dimension $p+1$ was defined by Eq. (12.2). A sinusoid of frequency f_0 that is input to the filter would, therefore, be undistorted at the filter output. In order to reject components of the spectrum not near f_0, simply minimize the output variance, Eq. (12.5), subject to the unity frequency response constraint, Eq. (12.7). The constrained minimum variance solution for the filter coefficients can be shown [McDonough, 1983] to satisfy

$$\mathbf{a}_{\text{MV}} = \frac{\mathbf{R}_p^{-1}\mathbf{e}(f_0)}{\mathbf{e}^H(f_0)\mathbf{R}_p^{-1}\mathbf{e}(f_0)}. \tag{12.8}$$

Substitution of Eq. (12.8) into Eq. (12.5) yields the minimum variance

$$\rho_{\text{MV}} = \frac{1}{\mathbf{e}^H(f_0)\mathbf{R}_p^{-1}\mathbf{e}(f_0)}. \tag{12.9}$$

The output variance serves as a good indicator of power in the input spectrum in the vicinity of f_0. A different optimum variance is obtained for each f_0 selected. The filter adapts its shape across the frequency range to the process autocorrelation function.

An appropriate *minimum variance* (MV) *spectral estimator* that may be deduced from Eq. (12.9) is

$$P_{\text{MV}}(f) = T\rho_{\text{MV}} = \frac{T}{\mathbf{e}^H(f)\mathbf{R}_p^{-1}\mathbf{e}(f)} \tag{12.10}$$

defined for $-1/2T \leq f \leq 1/2T$. The sample interval T gives Eq. (12.10) units of power per Hz, appropriate for a power spectral density. The MV spectral estimator is not a true PSD, however. Consider a white noise process of variance ρ_w. It can be shown that the MV spectral estimate of a white noise process is

$$P_{\text{MV}}(f) = \frac{T\rho_w}{p+1}, \tag{12.11}$$

where $p + 1$ is the dimension of the autocorrelation matrix of Eq. (12.6). The true PSD is, of course, $T\rho_w$. Other interpretations and modifications of the MV spectral estimator have been provided by Barnard [1969], Marzetta [1983], Lagunas and Gasull [1984], and Marzetta and Lang [1983, 1984]. Marzetta and Lang [1983] have shown, for example, that the MV relationship (12.10) can be interpreted as providing an upper bound on the weighted power spectral density of an unknown random process, some of whose correlations have been measured.

It can be shown (see Lacoss [1971]) that a complex sinusoid in white noise with autocorrelation

$$r_{xx}[k] = P\exp(j2\pi f_0 kT) + \rho_w\delta[k] \tag{12.12}$$

has an MV spectral estimator value at $f = f_0$ of

$$P_{\text{MV}}(f_0) = \frac{T\rho_w}{P}(1 + pP/\rho_w). \tag{12.13}$$

If the signal-to-noise ratio P/ρ_w is very high, then

$$P_{MV}(f_0) \approx PT, \qquad (12.14)$$

which is the power of the sinusoid scaled by T. Peaks in the MV spectral estimator will be, therefore, approximately *linearly* proportional to power when the signal-to-noise ratio is high and the spectral lines are sufficiently separated, unlike the AR PSD estimator that was proportional to the *square* of the power. However, the area under the MV PSD curve is proportional to the square root of the sinusoid power, an indicator that the MV PSD is not a true PSD. Capon and Goodman [1971] have shown that the MV PSD has a mean and a variance that behaves much like the averaged periodogram (i.e., the variance is reduced linearly as the number of data segments averaged increases).

In practice, an estimate of the autocorrelation matrix, $\hat{\mathbf{R}}_p$, must be used when only data samples are available. A better approach, using the Burg algorithm in lieu of the autocorrelation matrix estimate, is discussed in the next section.

12.4 RELATIONSHIP OF MV AND AR SPECTRAL ESTIMATORS

An explicit relationship exists between the MV and AR spectral estimators when the autocorrelation sequence is known. The inverse of the Toeplitz autocorrelation matrix was shown in Chap. 3 [Eq. (3.166)] to be expressible as

$$\mathbf{R}_p^{-1} = \mathbf{A}_p \mathbf{P}_p^{-1} \mathbf{A}_p^H, \qquad (12.15)$$

in which $\mathbf{A}_p$ is a $(p+1) \times (p+1)$ matrix of the AR parameters from order 0 to order p

$$\mathbf{A}_p = \begin{pmatrix} 1 & 0 & \cdots & 0 \\ a_p[1] & 1 & \cdots & 0 \\ a_p[2] & a_{p-1}[1] & \cdots & 0 \\ \vdots & \vdots & \ddots & \vdots \\ a_p[p] & a_{p-1}[p-1] & \cdots & 1 \end{pmatrix} \qquad (12.16)$$

and $\mathbf{P}_p^{-1}$ is a $(p+1) \times (p+1)$ diagonal matrix of inverted white noise variances for the corresponding orders

$$\mathbf{P}_p^{-1} = \begin{pmatrix} 1/\rho_p & 0 & \cdots & 0 \\ 0 & 1/\rho_{p-1} & \cdots & 0 \\ \vdots & \vdots & \ddots & \vdots \\ 0 & 0 & \cdots & 1/\rho_0 \end{pmatrix}. \qquad (12.17)$$

Substituting Eq. (12.15) into the reciprocal MV spectral function (12.1) yields

$$\frac{1}{P_{MV}(p, f)} = \frac{1}{T}\mathbf{e}^H(f)\mathbf{R}_{xx}^{-1}\mathbf{e}(f) = \frac{1}{T}\mathbf{e}^H(f)\mathbf{A}_p\mathbf{P}_p^{-1}\mathbf{A}_p^H\mathbf{e}(f) \tag{12.18}$$

$$= \frac{1}{T}\sum_{k=0}^{p}\frac{1}{\rho_k}\left|\sum_{m=0}^{k}a_k[m]\exp(-j2\pi fmT)\right|^2 = \sum_{k=0}^{p}\frac{1}{P_{AR}(k, f)},$$

where notation $P(k, f)$ is used to indicate a PSD of order k with frequency f [Burg, 1972]. The assumption $a_k[0] = 1$ is made for all orders $0 \leq k \leq p$. Thus, the reciprocal of the minimum variance spectral estimator is equal to the average of the reciprocals of the autoregressive spectral estimators from order 0 to order p. The lower resolution of the MV spectral estimate relative to the AR estimate observed in practice is, therefore, due to the averaging effect of low-order AR spectra of least resolution mixed with high-order AR spectra of highest resolution. The MV method, though, still has greater resolution than classical PSD estimators in higher signal-to-noise-level conditions. The variance of the MV spectral estimator has been observed empirically for large data records to be less than an AR PSD estimator of identical order p [Lacoss, 1971]. The variance reduction may also be attributed to the averaging effect of Eq. (12.18).

Equation (12.18) represents a *frequency*-domain relationship between the MV and AR spectral estimators. A *time*-domain relationship can also be developed [Musicus, 1985]. The pth order AR PSD estimator may alternatively be expressed as

$$P_{AR}(f) = \frac{T\rho_p}{\left|1 + \sum_{k=1}^{p}a_p[k]\exp(-j2\pi fkT)\right|^2} = \frac{T}{\sum_{k=-p}^{p}\psi_{AR}[k]\exp(-j2\pi fkT)} \tag{12.19}$$

by expanding its denominator. The $\psi_{AR}[k]$ coefficients are calculated from the correlation of the AR parameters

$$\psi_{AR}[k] = \begin{cases} \dfrac{1}{\rho_p}\displaystyle\sum_{i=0}^{p-k}a_p[k+i]a_p^*[i] & \text{for } 0 \leq k \leq p \\ \psi_{AR}^*[-k] & \text{for } -1 \geq k \geq 1 \end{cases} , \tag{12.20}$$

in which $a_p[0] = 1$ is assumed, by definition.

The MV spectral estimator may be expressed in a similar fashion by first recalling the form of the inverse of a Hermitian Toeplitz matrix $\mathbf{R}_p$ is [see Eq. (3.160)]

$$\mathbf{R}_p^{-1} = \frac{1}{\rho_p}\mathbf{T}_p\mathbf{T}_p^H - \frac{1}{\rho_p}\mathbf{S}_p\mathbf{S}_p^H, \tag{12.21}$$

in which the triangular Toeplitz matrix $\mathbf{T}_p$ is

$$
\mathbf{T}_p = \begin{pmatrix}
1 & 0 & \cdots & 0 & 0 \\
a_p[1] & 1 & \cdots & 0 & 0 \\
\vdots & \vdots & \ddots & \vdots & \vdots \\
a_p[p-1] & a_p[p-2] & \cdots & 1 & 0 \\
a_p[p] & a_p[p-1] & \cdots & a_p[1] & 1
\end{pmatrix}
\tag{12.22}
$$

and the triangular Toeplitz matrix $\mathbf{S}_p$ is

$$
\mathbf{S}_p = \begin{pmatrix}
0 & 0 & \cdots & 0 & 0 \\
a_p^*[p] & 0 & \cdots & 0 & 0 \\
\vdots & \vdots & \ddots & \vdots & \vdots \\
a_p^*[2] & a_p^*[3] & \cdots & 0 & 0 \\
a_p^*[1] & a_p^*[2] & \cdots & a_p^*[p] & 0
\end{pmatrix}.
\tag{12.23}
$$

Substituting Eq. (12.21) into Eq. (12.10) for the inverse $\mathbf{R}_p^{-1}$ will yield, after extensive algebra,

$$
P_{\text{MV}}(f) = \frac{T}{\displaystyle\sum_{k=-p}^{p} \psi_{\text{MV}}[k]\exp(-j2\pi f kT)}.
\tag{12.24}
$$

The $\psi_{\text{MV}}[k]$ coefficients are calculated as a linearly weighted (windowed) correlation of AR parameters

$$
\psi_{\text{MV}}[k] = \begin{cases}
\dfrac{1}{\rho_p}\displaystyle\sum_{i=0}^{p-k}(p+1-k-2i)a_p[k+i]a_p^*[i] & \text{for } 0 \le k \le p \\
\psi_{\text{MV}}^*[-k] & \text{for } -1 \ge k \ge -p.
\end{cases}
\tag{12.25}
$$

Computations proportional to p^2 are required to calculate the set of $\psi_{\text{MV}}[k]$ coefficients once. An FFT, such as that in Appendix 2.A, may then be used to evaluate the alternative representation (12.25) of the MV spectral estimator over a range of frequencies. This is more computationally efficient than direct evaluation of the original MV spectral estimator function (12.10), which requires p^3 operations to invert $\mathbf{R}_p$ and p^2 operations for each frequency to be evaluated.

12.5 IMPLEMENTATION OF THE MINIMUM VARIANCE SPECTRAL ESTIMATOR

The most computationally efficient approach to implement the MV spectral estimator uses the four steps

- Compute a sequence of autocorrelation estimates from the data.
- Solve the Yule-Walker equations for the AR parameters using the Levinson algorithm.
- Correlate the AR parameters to yield the $\psi_{MV}[k]$ coefficients; make use of Eq. (12.25).
- Evaluate the MV spectral estimator over frequency using Eq. (12.24) and an FFT algorithm.

Better results can be obtained in practice by replacing the first two steps above with any of the AR parameter estimators of Chap. 8. For example, the first two steps can be replaced with

- Estimate the AR coefficients with the Burg algorithm.

The use of the Burg algorithm means that the implied autocorrelation function is derived from a Toeplitz matrix. MATLAB function `minimum_variance_psd.m` available through the text website incorporates the steps above for computing the MV spectral estimate from a set of complex data samples using the procedure above. It uses the Burg algorithm for the first step.

```
function psd=minimum_variance_psd(num_psd,p,T,x)
% This function computes the minimum variance spectral estimator
%
%          psd = minimum_variance_psd(num_psd,p,T,x)
%
% num_psd -- number of power spectral density values (must be power of two)
% p       -- minimum variance adaptive filter order (number of taps = p + 1)
% T       -- sample interval in seconds
% x       -- vector of data samples
% psd     -- vector of power spectral density values
```

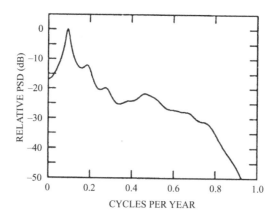

Figure 12.3. Minimum variance spectral estimate of the filtered sunspot numbers, using $p = 32$.

12.6 APPLICATION TO SUNSPOT NUMBERS

Figure 12.3 is a spectral estimate produced from PSD values generated by subroutine MINVAR using the filtered sunspot number sequence as input. An order of $p = 32$ was selected, comparable to the autoregressive spectral estimates of Chap. 8. The two highest peaks are located at 0.0938 cycle per year (10.67 years per cycle) and 0.1870 cycle per year (5.35 years per cycle).

References

Barnard, T. E., Analytical Studies of Techniques for the Computation of High-Resolution Wavenumber Spectra, Special Report No. 9 prepared by *Texas Instruments, Inc.* for the Advanced Research Projects Agency under contract F33657-68-C-0867, May 1969.

Burg, J. P., The Relationship between Maximum Entropy and Maximum Likelihood Spectra, *Geophysics*, vol. 37, pp. 375–376, April 1972.

Capon, J., High-Resolution Frequency-Wavenumber Spectrum Analysis, *Proc. IEEE*, vol. 57, pp. 1408–1418, August 1969.

Capon, J., Maximum-Likelihood Spectral Estimation, Chapter 5 in *Nonlinear Methods of Spectral Analysis*, 2nd ed., S. Haykin, ed., Springer-Verlag, New York, 1983.

Capon, J., and N. R. Goodman, Probability Distribution for Estimators of the Frequency-Wavenumber Spectrum, *Proc. IEEE*, vol. 58, pp. 1785–1786, October 1970; corrections in vol. 59, p. 112, January 1971.

Cox, H., Resolving Power and Sensitivity to Mismatch of Optimum (Line) Array Processors, *J. Acoust. Soc. Am.*, vol. 54, pp. 771–785, 1973.

Lacoss, R. T., Data Adaptive Spectral Analysis Methods, *Geophysics*, vol. 36, pp. 661–675, August 1971.

Lagunas, M. A., and A. Gasull, Measuring True Spectral Density from ML Filters, *Proceedings of the 1984 IEEE International Conference on Acoustics, Speech, and Signal Processing*, San Diego, Calif., pp. 14.10.1–14.10.4, March 1984.

Marzetta, T. L., A New Interpretation for Capon's Maximum Likelihood Method of Frequency-Wavenumber Spectral Estimation, *IEEE Trans. Acoust. Speech Signal Process.*, vol. ASSP-31, pp. 445–449, April 1983.

Marzetta, T. L., and S. W. Lang, New Interpretations for the MLM and DASE Spectral Estimators, *Proceedings of the 1983 IEEE International Conference on Acoustics, Speech, and Signal Processing*, Boston, Mass., pp. 844–846, April 1983.

Marzetta, T. L., and S. W. Lang, Power Spectral Density Bounds, *IEEE Trans. Inform. Theory*, vol. IT-30, pp. 117–122, January 1984.

McDonough, R. N., Application of the Maximum-Likelihood Method and the Maximum-Entropy Method to Array Processing, Chapter 6 in *Nonlinear Methods of Spectral Analysis*, 2nd ed., S. Haykin, ed., Springer-Verlag, New York, 1983.

Musicus, B., Fast MLM Power Spectrum Estimation from Uniformly Spaced Correlations, *IEEE Trans. Acoust. Speech Signal Process.*, vol. ASSP-33, pp. 1333–1335, October 1985.

13

EIGENANALYSIS-BASED FREQUENCY ESTIMATION

13.1 INTRODUCTION

A class of spectral techniques based on an eigenanalysis of an autocorrelation matrix or one of the data matrices of Chap. 8 have been promoted in the research literature as having better resolution and better frequency-estimation characteristics than spectral techniques such as autoregressive or Prony, especially at lower signal-to-noise ratios, where these latter techniques often fail to resolve close sinusoids or other narrowband spectral components. Some support for this claim has previously been demonstrated in Sec. 11.9. Singular value decomposition was used there to provide more accurate estimates of the frequency and damping factors for a process consisting of two damped exponentials in noise. The basis for the improved performance of the eigenanalysis techniques is presented in this chapter. Key to the performance is the division of the information in the autocorrelation matrix or the data matrix into two vector subspaces, one a *signal* subspace and the other a *noise* subspace.

Functions of the vectors in either the signal or noise subspaces can be used to create *frequency estimators* that, when plotted, show sharp peaks at the frequency locations of sinusoids or other narrowband spectral components. These are not true PSD estimators because they do not preserve the measured process power nor can the autocorrelation sequence be recovered by Fourier transforming the frequency estimators. Included in this class of eigenanalysis-based frequency estimators are the Pisarenko harmonic decomposition (PHD) and the multiple signal classification (MUSIC) algorithms.

13.2 SUMMARY

Figure 13.1 is a flowchart of the processing steps needed to generate two prominent eigenanalysis frequency estimators based on noise subspace techniques, the MUSIC and EV algorithms, for which MATLAB functions are included in Sec. 13.6. Figure 13.2 illustrates the eigenvector (EV) algorithm response to the 64-point test case test1987. Figure 13.3 illustrates the MUSIC algorithm response to the same test signal. Both algorithms require as inputs the autocorrelation matrix order p and

• ACQUIRE DATA RECORD
    Total Number of Samples in Record
    Sampling Interval (sec/sample)

• TREND REMOVAL
    Optional (see Chapter 14)

• SELECT EIGENANALYSIS PARAMETERS
    Autocorrelation/Data Matrix Order
    Expected Number of Exponential Signal (less than order)
    Method: MUSIC, EV, or PISARENKO

• COMPUTE NOISE SUBSPACE FREQUENCIES
    Function noise_subspace.m

• PLOT OPTIONAL "PSD"; num_sigs ADJUST?
    Adjust num_sigs for Variance vs Resolution Trade-Off

Figure 13.1. Flowchart of two eigenanalysis-based frequency estimators.

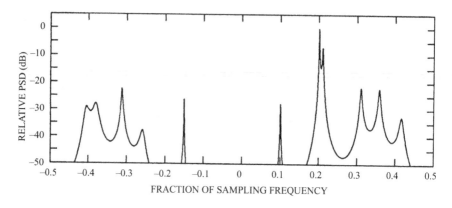

Figure 13.2. Eigenvector method (EV) frequency estimate of the 64-point test sequence using an order $p = 15$ data matrix and $num\_sig = 11$ (4 noise subspace eigenvectors).

an estimate of the number of signal components num_sig. Note that $num\_sig < p$. These figures do not show the ultimate performance of these methods, which work best for processes consisting of narrowband signals in white noise; the colored noise of the test case causes a degradation in the performance because these methods were not designed to handle this case. The Pisarenko harmonic decompostion technique, another prominent eigenanalysis-base frequency estimator, is also summarized in Sec. 13.6. A MATLAB demonstration script, spectrum_demo_eigen.m, is available through the text website to perform any of these three techniques on the three data sets test1987, doppler_radar, sunspot_numbers, and then plot the results.

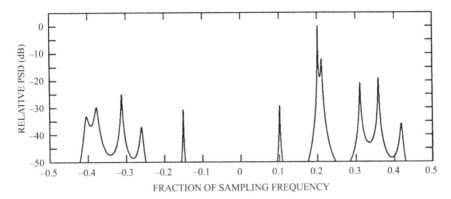

Figure 13.3. MUltiple SIgnal Classification method (MUSIC) frequency estimate of the 64-point test sequence using an order $p = 15$ data matrix and *num_sig* = 11 (4 noise subspace eigenvectors).

13.3 EIGENANALYSIS OF AUTOCORRELATION MATRIX FOR SINUSOIDS IN WHITE NOISE

It was shown in Chap. 4 that the ACS for the wide-sense stationary process consisting of M *complex* sinusoids of random phase in additive complex white noise is

$$r_{xx}[k] = \sum_{i=1}^{M} P_i \exp(j2\pi f_i kT) + \rho_w \delta[k], \qquad (13.1)$$

in which P_i is the power of the ith sinusoid and ρ_w is the noise variance. If the process consists of M *real* sinusoids in additive real white noise, the ACS is

$$r_{xx}[k] = \sum_{i=1}^{M} P_i \cos(2\pi f_i kT) + \rho_w \delta[k] \qquad (13.2)$$

$$= \sum_{i=1}^{M} \frac{P_i}{2} \left[\exp(j2\pi f_i kT) + \exp(-j2\pi f_i kT) \right] + \rho_w \delta[k].$$

The pth-order Toeplitz autocorrelation matrix of dimension $(p+1) \times (p+1)$ is

$$\mathbf{R}_p = \begin{pmatrix} r_{xx}[0] & \cdots & r_{xx}^*[p] \\ \vdots & \ddots & \vdots \\ r_{xx}[p] & \cdots & r_{xx}[0] \end{pmatrix}. \qquad (13.3)$$

For the case of complex sinusoids in white noise, the autocorrelation matrix also has the structure

$$\mathbf{R}_p = \sum_{i=1}^{M} P_i \mathbf{s}_i \mathbf{s}_i^H + \rho_w \mathbf{I}, \tag{13.4}$$

in which $\mathbf{I}$ is a $(p+1) \times (p+1)$ identity matrix and $\mathbf{s}_i$ is a signal vector of dimension $p+1$ carrying the frequency information of the ith sinusoid

$$\mathbf{s}_i = \begin{pmatrix} 1 \\ \exp(j 2\pi f_i T) \\ \vdots \\ \exp(j 2\pi f_i pT) \end{pmatrix}. \tag{13.5}$$

Matrix $\mathbf{R}_p$ can be expressed as the sum of a signal autocorrelation matrix and a noise autocorrelation matrix

$$\mathbf{R}_p = \mathbf{S}_p + \mathbf{W}_p, \tag{13.6}$$

in which

$$\mathbf{S}_p = \sum_{i=1}^{M} P_i \mathbf{s}_i \mathbf{s}_i^H \tag{13.7}$$

$$\mathbf{W}_p = \rho_w \mathbf{I} \tag{13.8}$$

are $(p+1) \times (p+1)$ matrices. If an autocorrelation matrix has an order greater than the number of complex sinusoids ($p > M$), then the signal matrix $\mathbf{S}_p$ will have rank M (because each vector outer product $\mathbf{s}_i \mathbf{s}_i^H$ is a rank 1 matrix). The noise matrix will have the full rank $p+1$.

In an analogous manner, the pth-order autocorrelation matrix for the case of M real sinusoids in white noise has the structure

$$\mathbf{R}_p = \sum_{i=1}^{M} \frac{P_i}{2} \left[\mathbf{s}_i \mathbf{s}_i^H + \mathbf{s}_i^* \mathbf{s}_i^T \right] + \rho_w \mathbf{I}. \tag{13.9}$$

The signal matrix $\mathbf{S}_p$ in this case will have rank $2M$. Discussions from this point will concentrate on the complex case; the real case is generally a simple extension of the complex case with a change in rank from M to $2M$.

The signal matrix will have the eigendecomposition

$$\mathbf{S}_p = \sum_{i=1}^{p+1} \lambda_i \mathbf{v}_i \mathbf{v}_i^H \tag{13.10}$$

in which the eigenvalues have been ordered in decreasing value $\lambda_1 \geq \lambda_2 \geq \ldots \geq \lambda_{p+1}$ and the eigenvectors are orthonormal ($\mathbf{v}_i^H \mathbf{v}_j = 0$ if $i \neq j$ or 1 if $i = j$).

It can be shown [Noble and Daniell, 1977] that a matrix of dimension $p + 1$ with rank $M < p + 1$ will have $p - M + 1$ zero eigenvalues. The eigendecomposition of Eq. (13.10) can then be written as

$$\mathbf{S}_p = \sum_{i=1}^{M} \lambda_i \mathbf{v}_i \mathbf{v}_i^H. \tag{13.11}$$

The eigenvectors $\mathbf{v}_1, \ldots, \mathbf{v}_M$, known as the *principal eigenvectors*, span the same *signal subspace* as the signal vectors $\mathbf{s}_1, \ldots, \mathbf{s}_M$. This means that any principal eigenvector must be expressible as a linear combination of the signal vectors

$$\mathbf{v}_i = \sum_{k=1}^{M} \beta_{ik} \mathbf{s}_k \tag{13.12}$$

for $1 \leq i \leq M$. To be an eigenvector of $\mathbf{S}_p$, then

$$\mathbf{S}_p \mathbf{v}_i = \lambda_i \mathbf{v}_i. \tag{13.13}$$

Substituting Eq. (13.7) into Eq. (13.13) yields

$$\sum_{k=1}^{M} P_k \mathbf{s}_k \mathbf{s}_k^H \mathbf{v}_i = \lambda_i \mathbf{v}_i \tag{13.14}$$

or

$$\mathbf{v}_i = \sum_{k=1}^{M} \left(\frac{P_k}{\lambda_i} \mathbf{s}_k^H \mathbf{v}_i \right) \mathbf{s}_k \tag{13.15}$$

for $1 \leq i \leq M$, which means

$$\beta_{ik} = \frac{P_k}{\lambda_i} \mathbf{s}_k^H \mathbf{v}_i. \tag{13.16}$$

For the case $M = 1$, note that $\lambda_1 = pP$ and $\mathbf{v}_1 = \mathbf{s}_1 / \sqrt{P}$.

An alternative representation of the identity matrix in terms of the orthonormal eigenvectors is

$$\mathbf{I} = \sum_{i=1}^{p+1} \mathbf{v}_i \mathbf{v}_i^H. \tag{13.17}$$

Substitution of Eqs. (13.11) and (13.17) into Eq. (13.6) yields the eigendecomposition of the autocorrelation matrix

$$\mathbf{R}_p = \sum_{i=1}^{M} \lambda_i \mathbf{v}_i \mathbf{v}_i^H + \rho_w \sum_{i=1}^{p+1} \mathbf{v}_i \mathbf{v}_i^H = \sum_{i=1}^{M} (\lambda_i + \rho_w) \mathbf{v}_i \mathbf{v}_i^H + \sum_{i=M+1}^{p+1} \rho_w \mathbf{v}_i \mathbf{v}_i^H. \tag{13.18}$$

Thus, the eigenvectors $\mathbf{v}_{M+1}, \ldots, \mathbf{v}_{p+1}$ span the noise subspace of $\mathbf{R}_p$, all with the identical eigenvalue ρ_w. The principal eigenvectors $\mathbf{v}_1, \ldots, \mathbf{v}_M$ span the signal subspace of both $\mathbf{R}_p$ and $\mathbf{S}_p$, with eigenvalues of $\lambda_1 + \rho_w$, ..., $\lambda_M + \rho_w$. The eigenvalues of the principal eigenvectors, though, are composed of powers of *both* signal and noise, so white noise does contribute to the eigenvalue *weighting* of the noise-free signal subspace eigenvectors.

The eigendecomposition (13.18) of the autocorrelation matrix can be exploited in two ways to generate improved spectral estimators, or more correctly, improved frequency estimators. *Retaining only the information in the signal subspace eigenvectors, that is, forming a lower-rank approximation to $\mathbf{R}_p$, effectively enhances the SNR* because of the omission of the contribution of power in the noise subspace components. This is the basis of *principal component* (signal subspace) *frequency estimators*, discussed in Sec. 13.5. Noting that the eigenvectors are orthogonal and that the principal eigenvectors span the same subspace of the signal vectors, then the *signal vectors are orthogonal to all the vectors in the noise subspace*, including any linear combination

$$\mathbf{s}_i^H \left(\sum_{k=M+1}^{p+1} \alpha_k \mathbf{v}_k \right) = 0 \tag{13.19}$$

for $1 \leq i \leq M$ (or $2M$ in the case of M real sinusoids). This property forms the basis of the *noise subspace frequency estimators*, discussed in Sec. 13.6.

In the special case $p = M$ (or $p = 2M$ in the case of M real sinusoids), the noise subspace will have a single eigenvector, $\mathbf{v}_{p+1}$, with eigenvalue ρ_w. The M signal vectors will be orthogonal to this eigenvector

$$\mathbf{s}_i^H \mathbf{v}_{p+1} = \sum_{k=1}^{M+1} v_{p+1}[k] \exp(-j2\pi f_i kT) = 0 \tag{13.20}$$

for $1 \leq i \leq M$. Thus, the roots of the polynomial

$$\sum_{k=0}^{M} v_{p+1}[k+1] z^{-k} \tag{13.21}$$

will all lie on the unit circle at angles $2\pi f_i T$ for $1 \leq i \leq M$. This is the central concept of the Pisarenko harmonic decomposition (PHD), discussed further in Sec. 13.6.

13.4   EIGENANALYSIS OF DATA MATRIX FOR EXPONENTIALS IN NOISE

The autocorrelation sequence normally is not known, so that the properties of Sec. 13.3 are mostly of theoretical, rather than practical, interest. The concepts, however, can be extended to the covariance and modified covariance data matrices that are a part of the exponential estimation techniques of the Prony method (see Chap. 11). It is shown in this section that the data matrices have eigendecomposition properties similar to the autocorrelation matrix. The principal eigenvectors of the data matrix predominantly span the signal subspace, and the singular values of these principal eigenvectors tend to be larger than the noise subspace singular values. Thus, the singular values determined by an SVD of the data matrix are the basis for the separation of the eigenvectors into a mostly signal subspace and a mostly noise subspace.

The Prony method was introduced in Chap. 11 as a technique for estimation of the parameters of a *damped exponential* model that approximate a given data sequence. Central to this method is the solution of a set of linear equations involving the autocorrelation-like matrix product $T_p^H T_p$, in which T_p is the order p data matrix of the covariance method of linear prediction

$$T_p = \begin{pmatrix} x[p+1] & \cdots & x[1] \\ \vdots & \ddots & \vdots \\ x[N] & \cdots & x[N-p] \end{pmatrix}. \tag{13.22}$$

The modified Prony method, a variation of the Prony method for *undamped sinusoidal* modeling, involves the modified covariance data matrix

$$\begin{pmatrix} T_p \\ T_p^* J \end{pmatrix}. \tag{13.23}$$

The discussion which follows will focus on matrix T_p, but the conclusions will also be valid for the modified covariance data matrix.

Consider the noiseless complex exponential signal sequence

$$x[n] = \sum_{k=1}^{M} h_k z_k^n, \tag{13.24}$$

in which $z_k = \exp([\alpha_k + j2\pi f_k]T)$ and $h_k = A_k \exp(j\phi_k)$. Note that damped exponentials are permitted as valid signals. Matrix T_p formed from the $x[n]$ of Eq. (13.22) will have rank M as long as the selected order p is within the range $M \leq p \leq N - M$ (for the modified covariance matrix, the range is $M \leq p \leq [N-M]/2$). The data matrix T_p can be decomposed [Kumaresan, 1983] as

$$T_p = BC, \tag{13.25}$$

in which B, an $(N-p) \times M$ matrix, and C, an $M \times p$ matrix, are

$$B = \begin{pmatrix} h_1 z_1^p & \cdots & h_M z_M^p \\ h_1 z_1^{p+1} & \cdots & h_M z_M^{p+1} \\ \vdots & & \vdots \\ h_1 z_1^N & \cdots & h_M z_M^N \end{pmatrix}, \quad C = \begin{pmatrix} 1 & z_1^{-1} & \cdots & z_1^{-(p-1)} \\ 1 & z_2^{-1} & \cdots & z_2^{-(p-1)} \\ \vdots & \vdots & & \vdots \\ 1 & z_M^{-1} & \cdots & z_M^{-(p-1)} \end{pmatrix}. \tag{13.26}$$

Using Eq. (13.25), then

$$T_p^H T_p = C^H B^H BC. \tag{13.27}$$

The related $M \times M$ matrix $B^H BCC^H$ is positive definite because both B and C have full rank M. If λ_i for $1 \le i \le M$ designates the eigenvalues and w_i for $1 \le i \le M$ designates the eigenvectors, then

$$(B^H BCC^H)w_i = \lambda_i w_i \tag{13.28}$$

for $1 \le i \le M$. Premultiply Eq. (13.28) by matrix C^H to yield

$$C^H B^H BCC^H w_i = \lambda_i C^H w_i. \tag{13.29}$$

Define the vector

$$v_i = C^H w_i \tag{13.30}$$

and substitute Eq. (13.30) into Eq. (13.29), leading to the result

$$T_p^H T_p v_i = \lambda_i v_i \tag{13.31}$$

for $1 \le i \le M$. The M nonzero eigenvalues of the $p \times p$ matrix $T_p^H T_p$ are, therefore, identical to the eigenvalues of the matrix $B^H BCC^H$. The remaining $p - M$ eigenvalues of $T_p^H T_p$ are zero because $T_p^H T_p$ is of rank M. The corresponding eigenvectors of the nonzero eigenvalues are given by Eq. (13.30). Thus, any principal eigenvector of $T_p^H T_p$ will be a linear combination of the columns of C^H, which is composed of signal vectors as shown in Eq. (13.26). It can also be shown that any principal eigenvector of $T_p T_p^H$ is a linear combination of the columns of B, which is also composed of signal vectors. The matrix T_p will have M nonzero singular values, as these are simply the square roots of the eigenvalues. The eigenvectors of the zero eigenvalues of $T_p^H T_p$ or $T_p T_p^H$ are orthogonal to the M signal subspace, or principal, eigenvectors associated with the nonzero eigenvalues in the signal subspace.

If the data has noise, the properties are not exactly true, but they tend to be approximately true. Thus, the M principal singular values of a T_p matrix composed of noisy samples tend to be larger than the $p - M$ smallest singular values (which were exactly zero in the noiseless case). The M eigenvectors corresponding to the M principal eigenvalues of either $T_p^H T_p$ or $T_p T_p^H$ have fewer noise contributions than the noise subspace eigenvectors corresponding to the $p - M$ smallest singular values. These statements have some justification as a result of the analysis provided in Sec. 13.3.

13.5 SIGNAL SUBSPACE FREQUENCY ESTIMATORS

It has been shown that retention of the signal subspace, or principal, eigenvectors effectively improves the SNR for processes consisting of exponentials in white noise by eliminating much of the noise contribution to an autocorrelation matrix or a data matrix. A class of signal subspace frequency estimators may be formed from any of the PSD estimators of Chaps. 5–12 by simply replacing the autocorrelation matrix or data matrix by its reduced rank approximation in terms of the principal eigenvectors.

Consider the correlogram method PSD estimator, the minimum variance PSD estimator, and the Yule-Walker autoregressive PSD estimator. The correlogram and MV methods reply on the known or estimated autocorrelation matrix $\mathbf{R}_p$ for the definition of the spectral estimate

$$P_{\text{CORR}}(f) = T\mathbf{e}^H(f)\mathbf{R}_p\mathbf{e}(f) \tag{13.32}$$

$$P_{\text{MV}}(f) = T[\mathbf{e}^H(f)\mathbf{R}_p^{-1}\mathbf{e}(f)]^{-1}, \tag{13.33}$$

while the autoregressive (AR) method depends on the autocorrelation matrix to obtain the AR parameters

$$\begin{pmatrix} 1 \\ \mathbf{a}_p \end{pmatrix} = \mathbf{R}_p^{-1} \begin{pmatrix} \rho_p \\ \mathbf{0}_p \end{pmatrix}, \tag{13.34}$$

where the vectors and matrix have the following definitions

$$\mathbf{R}_p = \begin{pmatrix} r_{xx}[0] & \cdots & r_{xx}^*[p] \\ \vdots & \ddots & \vdots \\ r_{xx}[p] & \cdots & r_{xx}[0] \end{pmatrix}$$

$$\mathbf{e}(f) = \begin{pmatrix} 1 \\ \exp(j2\pi fT) \\ \vdots \\ \exp(j2\pi fpT) \end{pmatrix}, \quad \mathbf{a}_p = \begin{pmatrix} a[1] \\ a[2] \\ \vdots \\ a[p] \end{pmatrix}. \tag{13.35}$$

If the orthonormal eigendecomposition of $\mathbf{R}_p$ is

$$\mathbf{R}_p = \sum_{k=1}^{p} \lambda_k \mathbf{v}_k \mathbf{v}_k^H, \tag{13.36}$$

in which the eigenvalues are ranked in decreasing magnitude $\lambda_1 \geq \lambda_2 \geq \ldots \lambda_p$, and there are estimated to be M principal components ($M \leq p$), then the reduced rank principal eigenvector approximations to $\mathbf{R}_p$ and $\mathbf{R}_p^{-1}$,

$$\hat{\mathbf{R}}_p = \sum_{k=1}^{M} \lambda_k \mathbf{v}_k \mathbf{v}_k^H, \quad \hat{\mathbf{R}}_p^{-1} = \sum_{k=1}^{M} \frac{1}{\lambda_k} \mathbf{v}_k \mathbf{v}_k^H, \tag{13.37}$$

may be used in lieu of $\mathbf{R}_p$ and $\mathbf{R}_p^{-1}$ in Eqs. (13.32), (13.33), and (13.34) to create spectral estimators with reduced noise contribution due to the omission of the noise subspace eigenvectors. Owsley [1976, 1985a] originally proposed the signal subspace, or principal components, approach to estimation of the classic correl-ogram method for linear array beamforming, the equivalent of temporal spectral estimation. Johnson [1982] and Owsley [1985b] promoted the signal subspace improvements of the MV method.

The linear prediction algorithms of Chaps. 8 and 11 depend upon autocorrelation-like data matrix products, such as $\mathbf{T}_p^H \mathbf{T}_p$ as described by Eq. (13.22). The eigenvectors and eigenvalues of $\mathbf{T}_p^H \mathbf{T}_p$ are obtained by forming the singu-lar value decomposition of rectangular Toeplitz data matrix $\mathbf{T}_p$, as described by Eq. (3.96). The singular values will be the positive square roots of the eigenvalues of $\mathbf{T}_p^H \mathbf{T}_p$. A reduced rank approximation of $\mathbf{T}_p$ may be formed by omitting the eigenvectors associated with the small singular values, retaining only the principal eigenvectors and their associated singular values. An application of this approach has been demonstrated for the Prony method (Sec. 11.9). Tufts and Kumaresan [1982] have promoted this approach to frequency estimation. They found experi-mentally that the covariance data matrix $\mathbf{T}_p$ was best for damped exponentials in noise, while the modified covariance data matrix of Eq. (13.23) was best for esti-mates of undamped sinusoids in noise [Kumaresan, 1982]. This is not surprising, as the modified covariance matrix was also a central feature of the modified Prony method for sinusoidal estimation. Figure 13.4 illustrates the improvement in effec-tive SNR for a reduced rank modified covariance data matrix based on 25 samples of the two-component complex sinusoidal process

$$x[n] = \exp(j2\pi[0.32]n + \pi/4) + \exp(j2\pi[0.3]n) + w[n] \quad \text{for } n = 0 \text{ to } 24$$
$$(13.38)$$

of 10-dB signal-to-noise ratio. This approach was shown by Tufts and Kumaresan to approach the exact maximum likelihood performance, distinguishing pairs of

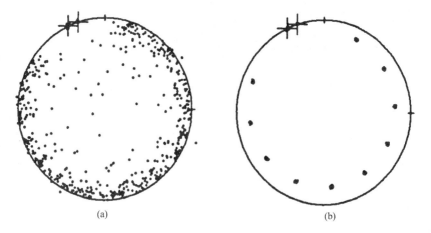

(a)   (b)

Figure 13.4. Location of zeros of modified covariance method for 50 independent data sets. Complex sinusoidals were located at fractional frequencies of 0.32 and 0.3 in white noise of 10-dB SNR. Each realization is 25 samples. (a) Order $p = 12$. (b) SVD of order $p = 12$ data matrix using two principal eigenvectors.

close sinusoidals at low SNRs where linear prediction techniques were unable to distinguish two frequencies accurately.

13.6 NOISE SUBSPACE FREQUENCY ESTIMATORS

All estimators in this class rely on the property that the noise subspace eigenvectors of the autocorrelation matrix or the data matrix are orthogonal to the signal eigenvectors, or any linear combination of the signal eigenvectors. The data matrices most often of interest are the ones associated with the covariance and modified covariance methods of linear prediction, as discussed in Chap. 8.

13.6.1 Pisarenko Harmonic Decomposition (PHD)

The Pisarenko harmonic decomposition (PHD) method was one of the earliest procedures for spectral estimation by eigenanalysis [Pisarenko, 1973]. The PHD assumes that the process consists of M complex sinusoids in additive complex white noise and derives the sinusoidal frequencies, sinusoidal powers, and white noise variance from the *known autocorrelation sequence* $r_{xx}[0]$ to $r_{xx}[M]$. If the process consists of M real sinusoids in additive real white noise, then the ACS must be known for lags 0 to $2M$. It was shown in Sec. 13.3 that the autocorrelation matrix $\mathbf{R}_{M+1}$ formed from the ACS will have, in the complex case, a single noise subspace eigenvector $\mathbf{v}_{M+1}$ with associated eigenvalue ρ_w, corresponding to the white noise variance. This eigenvalue is also the minimum eigenvalue of matrix $\mathbf{R}_{M+1}$. This can be expressed as

$$\mathbf{R}_{M+1}\mathbf{v}_{M+1} = \rho_w \mathbf{v}_{M+1}. \tag{13.39}$$

Given $\mathbf{R}_{M+1}$, MATLAB function `minimum_eigenvalue` in Chap. 3 will solve for the minimum eigenvalue (the noise variance) and its associated eigenvector. This algorithm accounts for the conjugate symmetry of the eigenvector. Once found, the sinusoid frequencies may be determined by factoring the eigenfilter polynomial, Eq. (13.21). The roots are guaranteed to be on the unit circle.

After the frequencies have been determined from the polynomial factoring, the sinusoidal powers can be determined from the autocorrelation sequence $r_{xx}[1], \ldots, r_{xx}[M]$ and Eq. (13.1) [Marple, 1976]. In matrix form, this is

$$\begin{pmatrix} \exp(j2\pi f_1 T) & \exp(j2\pi f_2 T) & \ldots & \exp(j2\pi f_M T) \\ \exp(j2\pi f_1 2T) & \exp(j2\pi f_2 2T) & \ldots & \exp(j2\pi f_M 2T) \\ \vdots & \vdots & & \vdots \\ \exp(j2\pi f_1 MT) & \exp(j2\pi f_2 MT) & \ldots & \exp(j2\pi f_M MT) \end{pmatrix} \begin{pmatrix} P_1 \\ P_2 \\ \vdots \\ P_M \end{pmatrix}$$

$$= \begin{pmatrix} r_{xx}[1] \\ r_{xx}[2] \\ \vdots \\ r_{xx}[M] \end{pmatrix} \tag{13.40}$$

which may be solved for the vector of unknown sinusoid powers $P_1, \ldots, P_M$ using standard linear equation solution routines for complex elements. In practice, one must estimate the ACS from the data samples. The biased autocorrelation estimate must be used to ensure a positive-definite matrix. Determining the order (number of sinusoids) M to use for the PHD method is difficult. A MATLAB function pisarenko.m is available through the text website that implements this noise subspace technique:

```
function [rho,freqs,powers,psd]=pisarenko(num_psd,num_sig,T,x)
% Pisarenko harmonic decomposition frequency/spectral estimator.
%
%      [rho,freqs,powers,psd] = pisarenko(num_psd,num_sig,T,x)
%
% num_psd -- number of spectral points for plot
% num_sig -- number of presumed complex sinusoidals (need 2 for each real sine)
% T       -- sample interval in seconds
% x       -- vector of data samples
% rho     -- white noise power spectral density
% freqs   -- vector of estimated sinusoid frequencies
% powers  -- vector of estimated sinusoid powers
% psd     -- vector of num_psd spectral points
```

Because the noise subspace eigenvalues theoretically repeat when the order is larger than the number of sinusoids [see Eq. (13.18)], a check for repeating eigenvalues of a very large order autocorrelation matrix would seem to be one method of order selection. In practice, however, use of estimated autocorrelation sequences do not usually generate repeated eigenvalues. Thus, the performance of the PHD with short data sequences tends to be poor. Some statistical properties of the PHD have been described by Sakai [1984].

An adaptive variant of Pisarenko's method has been proposed by Thompson [1979] for tracking narrowband signals in broadband noise. Its performance has been studied in depth by Larimore [1983]. It is a time-sequential PHD-type algorithm that, in contrast to the block-data PHD algorithm just presented, updates the estimate of the eigenvector (associated with the minimum eigenvalue) on a sample-by-sample basis. The algorithm is a constrained gradient search version of the LMS algorithm presented in Chap. 9,

$$\mathbf{v}'[k+1] = \mathbf{v}[k] - 2\mu e_k[k+1]\mathbf{x}[k]$$

$$\mathbf{v}[k+1] = \frac{\mathbf{v}'[k+1]}{(\mathbf{v}'[k+1])^H(\mathbf{v}'[k+1])} \quad (13.41)$$

$$e_k[k+1] = x[k+1] + \sum_{n=1}^{M} v[n]x[k+1-n].$$

When converged, $\mathbf{v}$ is a unity norm vector in the order $p = 2M$ subspace associated with the minimum eigenvalue. Experience with this time-recursive algorithm has indicated that convergence may be nonmonotonic and require long windows of data to isolate spectral lines.

13.6.2 Frequency Estimator Functions

Frequency estimator functions can be devised, based on noise subspace principles, to yield spectrum-like plots with sharp peaks. The $p - M$ noise subspace eigenvectors $\mathbf{v}_{M+1}, \cdots, \mathbf{v}_p$ of an autocorrelation matrix, covariance data matrix, or modified covariance data matrix of p total eigenvectors and M principal eigenvectors will theoretically be orthogonal to the sinusoidal signal vectors, so that linear combinations with arbitrary weighting α_k such as

$$\sum_{k=M+1}^{p} \alpha_k \left| \mathbf{e}^H(f)\mathbf{v}_k \right|^2 = \mathbf{e}^H(f) \left(\sum_{k=M+1}^{p} \alpha_k \mathbf{v}_k \mathbf{v}_k^H \right) \mathbf{e}(f), \qquad (13.42)$$

where $\mathbf{e}(f)$ is the complex sinusoid vector

$$\mathbf{e}(f) = \begin{pmatrix} 1 \\ \exp(j2\pi fT) \\ \vdots \\ \exp(j2\pi fMT) \end{pmatrix}, \qquad (13.43)$$

will be zero whenever $\mathbf{e}(f_i) = \mathbf{s}_i$, one of the sinusoidal signal vectors. This means that the frequency estimator function

$$\frac{1}{\displaystyle\sum_{k=M+1}^{p} \alpha_k \left| \mathbf{e}^H(f)\mathbf{v}_k \right|^2} \qquad (13.44)$$

will theoretically have infinite value whenever evaluated at $f = f_i$, one of the sinusoidal signal frequencies. In practice, estimation errors will cause the function of Eq. (13.44) to be finite, but with very sharp peaks at the sinusoidal frequencies. Frequency estimators such as Eq. (13.44) are, of course, not true PSD estimators, but are pseudo spectral estimators useful for estimation of sinusoidal or narrowband spectral component frequencies with a resolution somewhat higher than that of autoregressive spectral estimation techniques.

Two specific frequency estimators have been proposed in the literature. Selecting $\alpha_k = 1$ for all k yields the multiple signal classification (MUSIC) algorithm [Schmidt, 1981, 1986] frequency estimator

$$P_{\text{MUSIC}}(f) = \frac{1}{\mathbf{e}^H(f) \left(\displaystyle\sum_{k=M+1}^{p} \mathbf{v}_k \mathbf{v}_k^H \right) \mathbf{e}(f)}, \qquad (13.45)$$

based strictly on the noise subspace eigenvectors with uniform weighting. Choosing $\alpha_k = 1/\lambda_k$ yields the eigenvector (EV) algorithm [Johnson, 1982] frequency estimator

$$P_{EV}(f) = \cfrac{1}{\mathbf{e}^H(f) \left(\sum_{k=M+1}^{p} \cfrac{1}{\lambda_k} \mathbf{v}_k \mathbf{v}_k^H \right) \mathbf{e}(f)}, \qquad (13.46)$$

which weights each noise subspace eigenvector by the inverse of its associated eigenvalue.

The MATLAB function `noise_subspace.m` will compute either the MUSIC (*method* = 0) or the EV (*method* = 1) noise subspace frequency estimator, based on an SVD of the modified covariance data matrix [see Eq. (13.23)]

$$\begin{pmatrix} x[p] & \cdots & x[1] \\ \vdots & & \vdots \\ x[N-1] & \cdots & x[N-p] \\ x^*[2] & \cdots & x^*[p] \\ \vdots & & \vdots \\ x^*[N-p+1] & \cdots & x^*[N] \end{pmatrix}. \qquad (13.47)$$

```
function psd=noise_subspace(num_psd,method,m,num_sig,x)
% This function computes spectral/frequency estimator by the MUSIC or EV method
% signal subspace analysis from the forward/backward data matrix.
%
%         psd = noise_subspace(num_psd,method,m,num_sig,x)
%
% num_psd -- number of spectral values to evaluate (power of two)
% method  -- select: 1--Min Norm method , 2--MUSIC algorithm , 3--EV algorithm
% m       -- dimension of data matrix ( prediction filter order )
% num_sig -- presumed number of complex sinusoidals in signal subspace
% x       -- vector of data samples
% psd     -- vector of  num_psd  frequency 'spectrum' values
```

Johnson and DeGraaf [1982] have indicated that the EV method produces fewer spurious peaks than MUSIC for a given choice of order p due to the use of inverse eigenvalue weighting. They also claim that the EV method tends to shape the noise spectrum better than MUSIC. The ultimate performance of any frequency estimator, though, will be a function of how well the selection of signal and noise subspaces is made. Examination of the relative magnitudes of the eigenvalues (singular values) of the autocorrelation matrix (data matrix) eigendecomposition (SVD) will be the key tool for deciding the separation of the eigenvectors into two subspaces. Asymptotic performance analyses for the frequency estimator algorithms as a function of SNR, order, and frequency resolution may be found in Kaveh and Barabell [1986], Wang and Kaveh [1986], and Spielman et al. [1986]. The sensitivity of these techniques to colored noise, rather than white noise, has yet to be quantified. However, the fundamental limitation to performance in practice is how well the signal and noise eigenvalues can be separated from the noise-only eigenvalues. Interesting applications of these techniques for spatial processing may be found in Evans et al. [1982] and Gabriel [1986].

13.7 ORDER SELECTION

The simple idea of separating eigenvectors into signal and noise subspaces based upon an examination of either the eigenvalues of the autocorrelation matrix or the singular values of the data matrix does not work well in practice, especially with short sample records. The AIC order-selection criterion first introduced in Chap. 8 has been extended by Wax and Kailath [1985] to handle the subspace separation problem. Assuming $\lambda_1 > \lambda_2 > \cdots > \lambda_p$ are the eigenvalues of the sample autocorrelation matrix $\hat{\mathbf{R}}_p$ and assuming $m < p$, where m is the number of sinusoidal signals actually present in the data, and N data samples, then

$$\text{AIC}[m] = (p - m) \ln \left(\frac{\dfrac{1}{p-m} \displaystyle\sum_{i=m+1}^{p} \lambda_i}{\displaystyle\prod_{i=m+1}^{p} \lambda_i^{-(p-m)}} \right) + m(2p - m). \tag{13.48}$$

The number of sinusoids in the signal subspace is determined by selecting the minimum value of AIC[m]. Wax and Kailath [1986] have reported some preliminary results from use of the criterion of (13.48).

References

Evans, J. E., J. R. Johnson, and D. F. Sun, Applications of Advanced Signal Processing Techniques to Angle of Arrival Estimation of ATC Navigation and Surveillance Systems, *MIT Lincoln Laboratory Technical Report 582*, June 1982.

Gabriel, W. F., Using Spectral Estimation Techniques in Adaptive Processing Antenna Systems, *IEEE Trans. Antennas Propag.*, vol. AP-34, pp. 291–300, March 1986.

Johnson, D. H., The Application of Spectral Estimation Methods to Bearing Estimation Problems, *Proc. IEEE*, vol. 70, pp. 1018–1028, September 1982.

Johnson, D. H., and S. R. DeGraaf, Improving the Resolution of Bearing in Passive Sonar Arrays by Eigenvalue Analysis, *IEEE Trans. Acoust. Speech Signal Process.*, vol. ASSP-30, pp. 638–647, August 1982.

Kaveh, M., and A. J. Barabell, The Statistical Performance of the MUSIC and the Minimum-Norm Algorithms for Resolving Plane Waves in Noise, *IEEE Trans. Acoust. Speech Signal Process.*, vol. ASSP-34, pp. 331–341, April 1986.

Kumaresan, R., Estimating the Parameters of Exponentially Damped or Undamped Sinusoidal Signals in Noise, Ph.D. dissertation, Department of Electrical Engineering, University of Rhode Island, Kingston, R. I., 1982.

Kumaresan, R., On the Zeros of the Linear Prediction Error Filter for Deterministic Signals, *IEEE Trans. Acoust. Speech Signal Process.*, vol. ASSP-31, pp. 217–220, February 1983.

Larimore, M. G., Adaption Convergence of Spectral Estimation Based on Pisarenko Harmonic Retrieval, *IEEE Trans. Acoust. Speech Signal Process.*, vol. ASSP-31, pp. 955–962, August 1983.

Marple, S. L. Jr., Conventional Fourier, Autoregressive, and Special ARMA Methods of Spectrum Analysis, Engineer's dissertation, Department of Electrical Engineering, Stanford University, Stanford, Calif., December 1976.

Noble, B., and J. W. Daniel, Applied Linear Algebra, 2nd ed., Prentice-Hall, Inc., Englewood Cliffs, N. J., 1977.

Owsley, N. L., A Recent Trend in Adaptive Signal Processing for Sensor Arrays: Constrained Adaptation, in *Signal Processing*, J. W. R. Griffiths et al., eds., Academic Press, Inc., New York, 1976.

Owsley, N. L., Sonar Array Processing, Chapter 3 in *Array Signal Processing*, S. Haykin, ed., Prentice-Hall, Inc., Englewood Cliffs, N. J., 1985a.

Owsley, N. L., High-Resolution Spectrum Analysis by Dominant-Mode Enhancement, Chapter 4 in *VLSI and Modern Signal Processing*, S. Y. Kung, H. J. Whitehouse, and T. Kailath, eds., Prentice-Hall, Inc., Englewood Cliffs, N.J., 1985b.

Pisarenko, V. F., The Retrieval of Harmonics from a Covariance Function, *Geophys. J. R. Astron. Soc.*, vol. 33, pp. 347–366, 1973.

Sakai, H., Statistical Analysis of Pisarenko's Method for Sinusoidal Frequency Estimation, *IEEE Trans. Acoust. Speech Signal Process.*, vol. ASSP-32, pp. 95–101, February 1984.

Schmidt, R. O., A Signal Subspace Approach to Multiple Emitter Location and Spectral Estimation, Ph.D. dissertation, Department of Electrical Engineering, Stanford University, Stanford, Calif., November 1981.

Schmidt, R. O., Multiple Emitter Location and Signal Parameter Estimation, *IEEE Trans. Antennas Propag.*, vol. AP-34, pp. 276–280, March 1986.

Spielman, D., A. Paulraj, and T. Kailath, Performance Analysis of the Music Algorithm, *Proceedings of the 1986 IEEE International Conference on Acoustics, Speech, and Signal Processing*, Tokyo, Japan, pp. 1909–1912, April 1986.

Thompson, P. A., An Adaptive Spectral Analysis Technique for Unbiased Frequency Estimation in the Presence of White Noise, *Proceedings of 13th Asilomar Conference on Circuits, Systems, and Computers*, pp. 529–533, November 1979.

Tufts, D. W., and R. Kumaresan, Estimation of Frequencies of Multiple Sinusoids: Making Linear Prediction Perform like Maximum Likelihood, *Proc. IEEE*, vol. 70, pp. 975–989, September 1982.

Wang, H., and M. Kaveh, Performance of Narrowband Signal-Subspace Processing, *Proceedings of the 1986 IEEE International Conference on Acoustics, Speech, and Signal Processing*, Tokyo, Japan, pp. 589–592, April 1986.

Wax, M., and T. Kailath, Detection of Signals by Information Theoretic Criteria, *IEEE Trans. Acoust. Speech Signal Process.*, vol. ASSP-33, pp. 387–392, April 1985.

14

SUMMARY OF SPECTRAL ESTIMATORS

The unifying approach employed in Chaps. 5–13 has been to view each spectral estimation technique in the context of fitting data to an assumed model. The primary
motivation for many of the alternative spectral estimators has been the unsatisfactory performance of classical spectral estimators, especially where the reciprocal of
the data sequence duration is of the same order as the desired resolution. However,
if the data set is sufficiently large and statistically stationary, then classical spectral
estimators as discussed in Chap. 5 will be the most robust. The classical methods
have been the most well characterized estimators of all the methods included in
this text; the only assumption they make is that the data is wide-sense stationary.
The nonclassical methods require additional assumptions. The variations in performance among the various spectral estimation methods may often be attributed
to how well the underlying assumed model matches the process under analysis.
When the model is an accurate representation of the data, spectral estimates can
be obtained with performance exceeding that of the classical spectral estimators
(e.g., the periodogram). If the model is accurate, but a poor spectral estimator of the
parameters is used, poor (increased variance) spectral estimates will result. Unfortunately, few algorithms have been thoroughly analyzed for statistical performance
with finite data records. Most comparisons among competing algorithms have been
based on limited computer simulations. Different models may yield similar results,
but one model may require fewer model parameters and therefore be more efficient
than the other models in its representation of the observed process. Differences in
performance are also found among algorithms that estimate the parameters of a
common model, as was illustrated by the AR PSD estimates of Chap. 8.

The general spectral estimation problem is that of estimation of the power
spectral density or, equivalently, the autocorrelation function of a random process.
Because the PSD is formally determined by an infinite number of autocorrelation
function values, the task of estimating the PSD based on a finite data set is inherently
an ill-defined problem. At best, one could only hope to estimate a finite number
of lags of the autocorrelation function. The classical spectral estimators of Chap. 5
attempt to estimate only the most significant lags from 0 to some maximum M.
The resulting PSD estimator may be severely biased if the true PSD exhibits strong
autocorrelation values for lags greater than M. In an attempt to reduce this spectral
estimator problem to one which is more manageable, various models for the PSD

were proposed in Chaps. 6–14. The class of time-series models that yielded rational PSD functions (AR, MA, ARMA) are typical. The models are parameterized by a finite, and hopefully small, set of coefficients. The incorporation of a model leads to the replacement of the general spectral estimation problem by a parameter estimation problem. Less biased spectral estimators of higher resolution will result to the extent that the model is an accurate representation of the process.

Table 14.1 provides a synopsis of key spectral estimation methods presented throughout Chaps. 5–13. Some PSD estimation methods have more than one parameter estimation algorithm listed. Quadratic matrix forms have been used in this book to concisely describe the structural form of the PSD function. Many of the methods involve a minimization problem on some data matrix to yield a solution for the model parameters; this data matrix is indicated on the table. Some constant scaling terms have been omitted to better highlight the similarities and differences among the various methods. Exact PSD and parameter normal equations may be found in the reference section indicated by Table 14.1. Approximate computational complexity for each method is summarized. Only the highest-order terms of the polynomial representing the computational count are shown. The computational complexities shown assume use of the computer programs listed in the table.

The $p \times 1$ data vector $\mathbf{x}$, a $(p+1) \times 1$ parameter vector $\mathbf{a}$, a $p \times 1$ parameter vector $\mathbf{b}$, and a $(p+1) \times 1$ complex sinusoid vector $\mathbf{e}(f)$ used in Table 14.1 are defined as

$$
\mathbf{x} = \mathbf{x}_p[n] = \begin{pmatrix} x[n] \\ x[n-1] \\ \vdots \\ x[n-p] \end{pmatrix}, \quad \mathbf{a} = \begin{pmatrix} 1 \\ a_p[1] \\ \vdots \\ a_p[p] \end{pmatrix}, \tag{14.1}
$$

$$
\mathbf{b} = \begin{pmatrix} 1 \\ b_p[1] \\ \vdots \\ b_p[p] \end{pmatrix}, \quad \mathbf{e}(f) = \begin{pmatrix} 1 \\ \exp(j2\pi fT) \\ \vdots \\ \exp(j2\pi[p-1]fT) \end{pmatrix}.
$$

The Toeplitz autocorrelation data matrix $\mathbf{X}_1$, Toeplitz covariance data matrix $\mathbf{X}_2$, modified covariance data matrix $\mathbf{X}_3$, and prewindowed data matrix $\mathbf{X}_4$ were defined in Chap. 8 as

$$
\mathbf{X}_1 = \begin{pmatrix} x[1] & \cdots & 0 \\ \vdots & \ddots & \vdots \\ x[p+1] & \cdots & x[1] \\ \vdots & & \vdots \\ x[N] & \cdots & x[N-p] \\ \vdots & \ddots & \vdots \\ 0 & \cdots & x[N] \end{pmatrix}, \quad \mathbf{X}_2 = \begin{pmatrix} x[p+1] & \cdots & x[1] \\ \vdots & \ddots & \vdots \\ x[N-p] & & x[p+1] \\ \vdots & \ddots & \vdots \\ x[N] & \cdots & x[N-p] \end{pmatrix},
$$

$$
\tag{14.2}
$$

Table 14.1. Summary of Spectral Estimation Methods

Class	Model	Method	Key Text Section	Program Appendix [1]	Spectrum Quad. Form [2]	Data Matrix [3]	Relative Complexity [4]	Resol./Var. Control [5]
Classical	None [6]	Periodogram	5.7	—/5.C	$|e^H(f)x|^2$	—	$N\log_2 N$	Window
	None	Correlogram	5.6	—/5.B	$e^H(f)R_A e(f)$	X_1	$N\log_2 N$	Window
Parametric	Autoregressive	Yule-Walker (YW)	8.3	8.A/6.A	$|e^H(f)a|^2$	X_1	N^2	AR order
		Burg	8.4.2	8.B/6.A		—	N^2	
		Covariance	8.5.1	8.C/6.A		X_2	N^2	
		Modified Covar.	8.5.2	8.D/6.A		X_3	N^2	
		Adaptive LMS	9.3	9.B/6.A		—	N [7]	
		Adaptive RLS	9.4.2	9.C/6.A		X_4	N [7]	
	Moving Average	High-Order AR	10.3	10.A/6.A	$|e^H(f)b|^2$	X_1	N^2	MA order
	Autoregressive-Moving Average	Modified YW	10.4	10.B/6.A	$\dfrac{|e^H(f)b|^2}{|e^H(f)a|^2}$	X_1	N^2	AR, MA orders
	Damped Sines	Prony	11.5	11.C/11.E	$\left|DTFT\left(\sum h_k z^k\right)\right|^2$	X_2	N^2	No. Sines
	Undamped Sines	Modified Prony	11.6	11.C/—	Line Spectrum	X_3	N^2	No. Sines
	Sines + Noise	Pisarenko	13.6.1	3.F/—	Line Spectrum	X_1	N^3	No. Sines
Non Parametric	None	Min. Variance	12.4	—/12.A	$[e^H(f)R_A^{-1}e(f)]^{-1}$	X_1	N^2	Dimen. R_A
	None	MUSIC Eigen.	13.6.2	—/13.A	$[e^H(f)R_M e(f)]^{-1}$	X_3	N^3	Sing. Values
	None	EV Eigenanalysis	13.6.2	—/13.A	$[e^H(f)R_E e(f)]^{-1}$	X_3	N^3	Sing. Values

[1] Parameter estimation program/Spectrum evaluation program.

[2] See the text for vector and matrix definitions.

[3] See the text for data matrix definitions.

[4] The relative complexity is rough order-of-magnitude comparisons between spectral techniques. See the appropriate text section for detailed computational complexity analysis.

[5] The key parameter that affects the resolution vs. PSD variance trade-off is given in this column.

[6] The periodogram does have an interpretation as a sinusoidal model with preselected harmonic frequencies (see Chap. 11 Problems).

[7] Computational count per time update.

and

$$\mathbf{X}_3 = \begin{pmatrix} \mathbf{X}_2 \\ \mathbf{X}_2^* \mathbf{J} \end{pmatrix}, \quad \mathbf{X}_4 = \begin{pmatrix} x[1] & \cdots & 0 \\ \vdots & \ddots & \vdots \\ x[p+1] & \cdots & x[1] \\ \vdots & & \vdots \\ x[N] & \cdots & x[N-p] \end{pmatrix},$$

and the $(p+1) \times (p+1)$ Toeplitz autocorrelation matrix $\mathbf{R}_A$ as

$$\mathbf{R}_A = \mathcal{E}\{\mathbf{x}_p^*[n]\mathbf{x}_p^T[n]\} = \begin{pmatrix} r_{xx}[0] & \cdots & r_{xx}^*[p] \\ \vdots & \ddots & \vdots \\ r_{xx}[p] & \cdots & r_{xx}[0] \end{pmatrix}. \tag{14.3}$$

The matrices $\mathbf{R}_M$ and $\mathbf{R}_E$ represent the reduced rank approximation to $\mathbf{R}_A$ for the MUSIC and EV eigenanalysis methods, respectively (see Chap. 13 for exact definitions). Many of the spectral estimation methods in Table 14.1 perform either a constrained minimization or an unconstrained minimization of a quadratic form involving the data matrices $\mathbf{X}_1$ to $\mathbf{X}_4$ or the autocorrelation matrix $\mathbf{R}_A$ to obtain the model parameters. The spectral estimates produced from the 64-sample test sequence of Appendix II by each method listed in Table 14.1 have been illustrated in the Summary section of each of the earlier chapters.

The research literature often attempts to classify the spectral estimation methods of Table 14.1 into data "adaptive" and "nonadaptive" categories. These terms were used to distinguish the classical spectral estimation techniques from the newer techniques, such as autoregressive spectral analysis. However, it has been shown that all spectral estimation methods fit, or "adapt," the model parameters in some manner to the data. The Prony method, for example, makes amplitude, phase, damping, and frequency estimates for an exponential model. A periodogram computed with the DTFS makes only amplitude and phase estimates, because the frequencies of the sinusoidal model have been preselected at harmonic intervals. The *only* appropriate use for the term "adaptive" is for those time-sequential updating algorithms that track time variations in the model parameters, such as the methods in Chap. 9.

In some situations, data must be preprocessed prior to spectral estimation in order to remove significant signal components that could adversely effect the quality of the spectral estimate if not removed. Signal components that are candidates for removal include large means (DC), linear and higher-order polynomial trends, and strong tones (sinusoidal signals). Dominant tones or a high DC level will produce significant sidelobes in periodogram and correlogram method PSD estimates that can mask the presence of weak signal components close in frequency to the strong components. Polynomial trends are nonstationary in character and have been shown to have significant bias effects on the low-frequency end of AR spectral estimates [Kane and Trivedi, 1979]. Removing strong signal components and trends helps to reduce the dynamic range of spectral values, and may therefore be regarded as roughly *prewhitening* the spectrum. Data with bad or missing sample points can use linear prediction techniques of Chap. 8 to first replace these points before spectral estimation is performed [Nuttall, 1976; Bowling and Lai, 1979].

Polynomial trends up to Mth order may be removed with an FIR rejection filter of length $M + 1$

$$y[n] = \sum_{k=0}^{M} h[k]x[n-k]. \tag{14.4}$$

Denoting the trend as

$$x[n] = \sum_{i=0}^{M} b[i](nT)^i, \tag{14.5}$$

where T is the sample interval, then in order to have

$$y[n] = \sum_{i=0}^{M} b[i] \sum_{k=0}^{M} h[k](nT - kT)^i = 0, \tag{14.6}$$

the following equations must be satisfied

$$\sum_{k=0}^{M} (k)^i h[k] = 0 \tag{14.7}$$

for $i = 0$ to $i = M$. This forms a set of $M + 1$ equations for the $M + 1$ unknown filter coefficients. Tones may also be removed by a symmetric FIR filter with linear phase.

Trend and tone rejection filters must be used with care. Not only are the trends and tones removed, but also components of the spectrum that are close to the rejection regions of the filter response [the Fourier transform of Eq. (14.4)]. This tends to be the low-frequency region of the spectrum. Removing the sample mean must also be used cautiously. Because of the zero-mean assumption made in the statistical development of many spectral estimators, one may be tempted to always remove the sample mean

$$\bar{x} = \frac{1}{N} \sum_{i=1}^{N} x[n] \tag{14.8}$$

from the data. To see the impact such action may have, consider the example depicted in Fig. 14.1. A short sample sequence of a portion of sinusoidal cycle produces the sharp peak in the modified covariance AR spectral estimate shown in Fig. 14.1(c). When the sample mean is removed from the data, the modified covariance AR PSD estimate of the modified data, shown in Fig. 14.1(b), now indicates, erroneously, the presence of a DC component. This is a case where mean removal introduces artifacts into the spectrum, rather than improving it. Complete removal of the sample mean is also undesirable when the spectrum may have a random process with finite

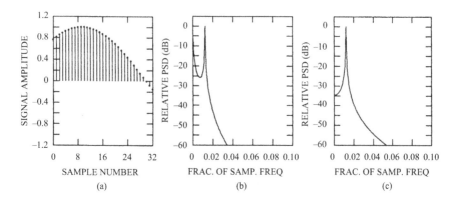

Figure 14.1. Effect of incorrect mean removal. (a) Samples from part of a sinusoidal cycle. (b) AR PSD estimate by modified covariance method with sample mean removed. (b) AR PSD estimate by modified covariance method with sample mean not removed.

power spectral density at zero frequency, such as white noise, even though there is no finite *power* at zero frequency (zero mean). The sample mean should also not be removed from data composed of transient signals. As a rule of thumb, sample mean and trends should only be removed in situations in which they are physically expected or in which they are clearly dominant in the data.

References

Bowling, S. B., and S. Lai, The Use of Linear Prediction for the Interpolation and Extrapolation of Missing Data Prior to Spectral Analysis, *Proceedings of the Rome Air Development Center Workshop on Spectrum Estimation*, pp. 39–50, October 1979.

Kane, R. P., and N. B. Trivedi, Effects of Linear Trend and Mean Value on Maximum Entropy Spectral Analysis, *Institute de Pesquisas Espaciais, Report INPE-1568-RPE/069*, Sao Jose dos Compos, Brasil, 1979 (available as NTIS N79-33949).

Nuttall, A. H., Spectral Analysis of a Univariate Process with Bad Data Points, via Maximum Entropy and Linear Predictive Techniques, *Naval Underwater Systems Center Technical Report TR-5303*, New London, Conn., March 1976.

15

MULTICHANNEL SPECTRAL ESTIMATION

15.1 INTRODUCTION

The discussion to this point in the book has been concerned with spectral estimation of a single stationary sample sequence. The next two chapters will extend these one-dimensional, single-variable spectral estimators to the multichannel case (vector functions of a scalar variable) and the two-dimensional case (scalar functions of a vector variable). This chapter extends the concepts developed earlier in the text to the simultaneous processing of samples from multiple data channels. Such arrays of signals are quite common in phased array radar, sonar, and seismic measurements, to name a few applications. Thus, each section of this chapter will parallel the development presented in Chaps. 5–13. Multichannel linear system theory, multichannel random process theory, multichannel parametric models, multichannel autoregressive PSD, multichannel linear prediction, and multichannel minimum variance spectral estimation are covered. Due to the interest in high-resolution spectral estimation for short data sequences, a significant part of this chapter is devoted to multichannel autoregressive PSD estimation. For additional background material on multichannel time series analysis, consult the text by Robinson [1983].

15.2 SUMMARY

The goal of multichannel spectral analysis of m channels of data is the estimation of the Hermitian power spectral density (PSD) matrix

$$\mathbf{P}(f) = \begin{pmatrix} P_{11}(f) & \ldots & P_{1m}(f) \\ \vdots & \ddots & \vdots \\ P_{m1}(f) & \ldots & P_{mm}(f) \end{pmatrix},$$

which is formally defined by Eq. (15.19). The PSD matrix consists of autospectral terms P_{ii}, for $i = 1$ to m, and cross spectral terms P_{ij}, for $i \neq j$, from all possible channel pairings. Figure 15.1 is a flowchart of five possible multichannel estimation approaches. The MATLAB multichannel demonstration script `spectrum_demo_MC.m` illustrates 2-channel analysis for three 2-channel data sets.

331

```
• ACQUIRE DATA RECORD
  Total Number of Samples in Record
  Number of Channels
  Sampling Interval (sec/sample)

• TREND REMOVAL
  Optional (see Chapter 14)

• SELECT METHOD & ITS PARAMETERS
  Periodogram: segment size, overlap, window
  Correlogram: correlation max lag
  Yule-Walker AR: model order
  Lattice AR: model order
  Minimum Variance Adaptive: filter order

• ESTIMATE DIRECT PSD or PARAMETERS
  periodogoram_psd_MC.m, or [direct PSD]
  correlogram_psd_MC.m, or [direct PSD]
  yule_walker_MC.m, or
  lattice_MC.m, or
  minimum_variance_MC.m [direct PSD]

• COMPUTE AR PSD
  Function ar_psd_MC.m
```

Figure 15.1. Flowchart of five approaches to multichannel PSD estimation.

One set is a 32 sample 2-channel data set created by alternating every other sample in `doppler_radar.dat` into a channel. A second data set is a data record created from the AR(1) test case described in Sec. 15.12. Finally, actual real sunspot and air temperature (`temperature.dat` at the text website) 2-channel data is processed into a 2-channel data record for analysis discussed in Sec. 15.14. Three classes of multichannel spectral estimates are incorporated into the demonstration script. Multichannel PSD analysis by the classical periodogram approach is covered in Sec. 15.5. Multichannel autoregressive (AR) PSD estimation is discussed extensively in Secs. 15.7–15.12. The multichannel minimum variance method is described in Sec. 15.13. The cross spectral terms from these various estimators are most conveniently interpreted in terms of the coherence function [see Eqs. (15.15)–(15.17)], which is produced by the plot output from the demonstration script, supported by MATLAB function `plot_psd_MC.m`.

15.3 MULTICHANNEL LINEAR SYSTEMS THEORY

Let $\mathbf{x}[n]$ denote a vector of samples from an m-channel process at sample index n,

$$\mathbf{x}[n] = \begin{pmatrix} x_1[n] \\ \vdots \\ x_m[n] \end{pmatrix}. \tag{15.1}$$

If $\mathbf{x}[n]$ represents the input $m \times 1$ vector of m channels and $\mathbf{y}[n]$ represents the output $m \times 1$ vector of m channels, then the linear shift-invariant system convolution that relates $\mathbf{x}[n]$ to $\mathbf{y}[n]$ is given by

$$\mathbf{y}[n] = \sum_{k=-\infty}^{\infty} \mathbf{H}[n-k]\mathbf{x}[k] = \mathbf{H}[n] \star \mathbf{x}[n], \qquad (15.2)$$

where the $m \times m$ *multichannel impulse response matrix* is given by

$$\mathbf{H}[k] = \begin{pmatrix} h_{1,1}[k] & \cdots & h_{1,m}[k] \\ h_{2,1}[k] & \cdots & h_{2,m}[k] \\ \vdots & & \vdots \\ h_{m,1}[k] & \cdots & h_{m,m}[k] \end{pmatrix}.$$

An individual output channel $y_i[n]$ would, therefore, have the convolution function

$$y_i[n] = \sum_{j=1}^{m} \sum_{k=-\infty}^{\infty} h_{ij}[n-k]x_j[k].$$

The *multichannel z-transform matrix* of a matrix sequence $\mathbf{H}[k]$ is defined as

$$\mathbf{H}(z) = \sum_{k=-\infty}^{\infty} \mathbf{H}[k]z^{-k} = \mathcal{Z}\{\mathbf{H}[k]\}, \qquad (15.3)$$

where z is a complex scalar. In this case, $\mathbf{H}(z)$ is the *multichannel system response function*. The z-transform of the signal vector $\mathbf{x}[k]$ is defined as

$$\mathbf{X}(z) = \sum_{k=-\infty}^{\infty} \mathbf{x}[k]z^{-k}. \qquad (15.4)$$

The *multichannel discrete-time Fourier transform* (DTFT) *matrix* $\mathbf{H}(f)$ is defined as

$$\mathbf{H}(f) = T\,\mathbf{H}(z)\Big|_{z=\exp(j2\pi fT)} = T \sum_{k=-\infty}^{\infty} \mathbf{H}[k]\exp(-j2\pi fkT) \qquad (15.5)$$

and the multichannel DTFT vector $\mathbf{X}(f)$ is defined as

$$\mathbf{X}(f) = T \sum_{k=-\infty}^{\infty} \mathbf{x}[k]\exp(-j2\pi fkT). \qquad (15.6)$$

The multichannel convolution theorem relates the z-transform of the matrix and vectors in the convolution of Eq. (15.2) as

$$\mathbf{Y}(z) = \mathbf{H}(z)\mathbf{X}(z). \tag{15.7}$$

A causal m-channel linear system described by the constant coefficient matrix difference equation

$$\sum_{k=0}^{p} \mathbf{A}[k]\mathbf{y}[m-k] = \sum_{k=0}^{q} \mathbf{C}[k]\mathbf{x}[n-k], \tag{15.8}$$

in which $\mathbf{A}[k]$ and $\mathbf{C}[k]$ are $m \times m$ coefficient matrices, has the z-transform (assuming zero initial conditions)

$$\mathbf{A}(z)\mathbf{Y}(z) = \mathbf{C}(z)\mathbf{X}(z)$$

or

$$\mathbf{Y}(z) = \mathbf{A}^{-1}(z)\mathbf{C}(z)\mathbf{X}(z). \tag{15.9}$$

15.4 MULTICHANNEL RANDOM PROCESS THEORY

Elementary concepts of two-channel random process theory for wide-sense stationary processes were developed in Chap. 3. The cross correlation between the single-channel processes $x[n]$ and $y[n]$ was defined there as

$$r_{xy}[k] = \mathcal{E}\{x[n+k]y^*[n]\} \tag{15.10}$$

with the property that $r_{xy}[k] = r_{yx}^*[-k]$. The two-channel cross power spectral density was shown to be simply the discrete-time Fourier transform of the cross correlation function

$$P_{xy}(f) = T \sum_{k=-\infty}^{\infty} r_{xy}[k]\exp(-j2\pi fkT). \tag{15.11}$$

This function is, in general, complex and has the complex conjugate symmetry

$$P_{xy}(f) = P_{yx}^*(-f). \tag{15.12}$$

It is straightforward to show that the cross spectrum is less than or equal to the geometric mean of the x and y process autospectra,

$$\left|P_{xy}(f)\right|^2 \leq P_{xx}(f)P_{yy}(f). \tag{15.13}$$

The Hermitian matrix

$$\begin{pmatrix} P_{xx}(f) & P_{xy}(f) \\ P_{yx}(f) & P_{yy}(f) \end{pmatrix} \tag{15.14}$$

is called the *coherence matrix*. Due to property (15.13), this 2×2 matrix must have a nonnegative determinant for all frequencies. The complex dimensionless expression

$$\Phi_{xy}(f) = \frac{P_{xy}(f)}{\sqrt{P_{xx}(f)}\sqrt{P_{yy}(f)}} \tag{15.15}$$

is termed the *coherence function*. Related functions are the *magnitude squared coherence* (MSC)

$$\mathrm{MSC}(f) = \left| \Phi_{xy}(f) \right|^2 = \frac{\left| P_{xy}(f) \right|^2}{P_{xx}(f) P_{yy}(f)} \tag{15.16}$$

and the *coherence phase spectrum* (or simply the "phase" spectrum)

$$\theta(f) = tan^{-1} \left[\mathrm{Im}\{\Phi_{xy}(f)\} / \mathrm{Re}\{\Phi_{xy}(f)\} \right]. \tag{15.17}$$

Note that the MSC must lie between 0 (for frequencies where there is no coherence between channels) and 1 (for frequencies where the channels are perfectly coherent to within some fixed phase relationship). Thus, the MSC may be used to measure, as a function of frequency, the similarity between a pair of signals. Its primary use is the detection of a common signal on two different channels. The coherence phase represents the phase lag or lead of channel x with respect to channel y as a function of frequency.

To generalize the scalar auto/cross correlation and scalar auto/cross power spectral density functions to the multichannel case, define the *multichannel correlation sequence* (MCS) at lag k of the stationary m-channel data vector $\mathbf{x}[n]$ as

$$\mathbf{R}_{xx}[k] = \mathcal{E}\{\mathbf{x}[n+k]\mathbf{x}^H[n]\} = \begin{pmatrix} r_{11}[k] & r_{12}[k] & \ldots & r_{1m}[k] \\ r_{21}[k] & r_{22}[k] & \ldots & r_{2m}[k] \\ \vdots & \vdots & \ddots & \vdots \\ r_{m1}[k] & r_{m2}[k] & \ldots & r_{mm}[k] \end{pmatrix}, \tag{15.18}$$

which forms a correlation matrix whose on-diagonal terms are single-channel autocorrelations and whose off-diagonal terms are two-channel cross correlations. Although $r_{ij}[k] \neq r_{ji}^*[k]$, the property $r_{ij}[k] = r_{ji}^*[-k]$ does hold for $i \neq j$; $\mathbf{R}_{xx}[k]$ will, therefore, *not* have a Hermitian symmetry property, i.e., $\mathbf{R}_{xx}[k] \neq \mathbf{R}_{xx}^H[k]$, but it will have, instead, the property $\mathbf{R}_{xx}[k] = \mathbf{R}_{xx}^H[-k]$. An analogous definition may be provided for the multichannel covariance function, using the vector of channel means. If the multichannel process $\mathbf{x}[n]$ has zero mean in all channels, then the multichannel covariance matrix is equal to the multichannel correlation matrix.

The $m \times m$ discrete-time Fourier transform matrix of the multichannel correlation matrix

$$\mathbf{P_{xx}}(f) = T \sum_{k=-\infty}^{\infty} \mathbf{R_{xx}}[k] \exp(-j2\pi f k T)$$

$$= \begin{pmatrix} P_{11}(f) & P_{12}(f) & \dots & P_{1m}(f) \\ P_{21}(f) & P_{22}(f) & \dots & P_{2m}(f) \\ \vdots & \vdots & \ddots & \vdots \\ P_{m1}(f) & P_{m2}(f) & \dots & P_{mm}(f) \end{pmatrix} \qquad (15.19)$$

is defined to be the *multichannel power spectral density* (PSD) *matrix* of the multichannel random process. It has on-diagonal elements that are the single-channel autospectral densities and off-diagonal elements that are the two-channel cross spectral densities. Note that $\mathbf{P_{xx}}(f)$ is Hermitian, that is, $\mathbf{P_{xx}}(f) = \mathbf{P_{xx}^H}(f)$, and positive semidefinite. This means that every principal minor matrix of $\mathbf{P_{xx}}(f)$ has a nonnegative determinant and, therefore, all coherencies of channel pairs have values bounded between 0 and 1 in magnitude. For example, the two-channel PSD matrix

$$\det \begin{pmatrix} P_{11}(f) & P_{12}(f) \\ P_{21}(f) & P_{22}(f) \end{pmatrix} \geq 0$$

implies

$$0 \leq \frac{|P_{21}(f)|^2}{P_{11}(f)P_{22}(f)} \leq 1.$$

Similar statements may be made for higher-order determinants, implying the existence of multiple coherence functions, such as $\Phi_{ijk}(f)$, with the property

$$0 \leq |\Phi_{ijk}(f)| \leq 1.$$

A zero-mean multichannel random process of special interest is multichannel white noise $\mathbf{w}[n]$, which has a multichannel MCS of

$$\mathbf{R_{ww}}[k] = \begin{cases} \boldsymbol{P}_w & \text{if } k = 0 \\ \mathbf{0} & \text{otherwise} \end{cases}, \qquad (15.20)$$

in which $\boldsymbol{P}_w$ is an $m \times m$ Hermitian matrix. The white noise in each channel is correlated with itself and correlated with the white noise in the other channels only at lag $k = 0$. The multichannel PSD of the multichannel white noise process is, therefore,

$$\mathbf{P_{ww}}(f) = T\,\boldsymbol{P}_w. \qquad (15.21)$$

Using arguments similar to that in Sec. 4.3, it is simple to show that the relationship between the correlation matrix $\mathbf{R_{yy}}[m]$, the output correlation matrix of a multichannel filter with impulse response matrix sequence $\mathbf{H}[m]$, and $\mathbf{R_{xx}}[m]$, the correlation matrix of the multichannel input process, is

$$\mathbf{R_{yy}}[m] = \mathbf{H}[m] \star \mathbf{R_{xx}}[m] \star \mathbf{H}^H[-m]. \tag{15.22}$$

The multichannel z-transform of Eq. (15.22) is then

$$\mathbf{P_{yy}}(z) = \mathbf{H}(z)\mathbf{P_{xx}}(z)\mathbf{H}^H(1/z^*). \tag{15.23}$$

The $m \times m$ matrix PSD between the input and output process of a multichannel linear filter, based on definitions (15.5) and (15.19), is

$$\mathbf{P_{yy}}(f) = \mathbf{H}(f)\mathbf{P_{xx}}(f)\mathbf{H}^H(f), \tag{15.24}$$

in which

$$\mathbf{H}(f) = \sum_{k=-\infty}^{\infty} \mathbf{H}[k]\exp(-j2\pi f k T).$$

Note that $\mathbf{H}(f)$ is not formally a multichannel DTFT because it lacks the scaling factor T.

15.5 MULTICHANNEL CLASSICAL SPECTRAL ESTIMATORS

A discussion of classical methods of autospectral and cross spectral estimation has been presented in Chap. 5 for both the periodogram and the correlogram methods of PSD estimation. The reader may refer to that chapter for algorithms that will compute autospectral estimates of individual channels and cross spectral estimates of channel pairs, which may then be used to fill in the entries of the PSD matrix (15.19).

The multichannel periodogram and correlogram methods for PSD estimation can be expressed more eloquently in multichannel matrix notation. The *multichannel averaged periodogram*, based on averaging K segments, has the form

$$\hat{\mathbf{P}}_{\mathrm{PER}}(f) = \frac{1}{K}\left[\frac{1}{NT}\sum_{k=1}^{K}\mathbf{X}_k(f)\mathbf{X}_k^H(f)\right], \tag{15.25}$$

in which

$$\mathbf{X}_k(f) = T\sum_{n=0}^{N-1}\mathbf{x}_k[n]\exp(-j2\pi f n T)$$

is an $m \times 1$ vector DTFT made from N samples of the m-channel sample vector $\mathbf{x}_k[n]$ taken from the kth segment. Averaging is not only important for purposes of statistical smoothing of the periodogram, it is also important for minimizing the bias in the coherence estimate. Consider the extreme situation of a two-channel case without segment averaging (only a single segment). The auto- and cross-periodogram estimates made from N samples of channel x_1 and channel x_2 will then have the form

$$\hat{P}_{11}(f) = \frac{1}{NT} X_1(f) X_1^*(f)$$

$$\hat{P}_{22}(f) = \frac{1}{NT} X_2(f) X_2^*(f)$$

$$\hat{P}_{12}(f) = \frac{1}{NT} X_1(f) X_2^*(f),$$

where

$$X_1(f) = T \sum_{n=0}^{N-1} x_1[n] \exp(-j2\pi f n T)$$

$$X_2(f) = T \sum_{n=0}^{N-1} x_2[n] \exp(-j2\pi f n T).$$

The magnitude squared coherence estimate will then be unity because

$$\mathrm{MSC}(f) = \frac{\hat{P}_{12}(f)\hat{P}_{12}^*(f)}{\hat{P}_{11}(f)\hat{P}_{22}(f)} = \frac{\left[X_1(f)X_2^*(f)\right]\left[X_1^*(f)X_2(f)\right]}{\left[X_1(f)X_1^*(f)\right]\left[X_2(f)X_2^*(f)\right]} = 1 \quad (15.26)$$

for all f, independent of the values of the data. This case illustrates an extreme situation of the effect of bias that is always present in coherence estimates made from periodogram spectral estimates.

The MATLAB implementation of the multichannel classical periodogram is found in function `periodogram_psd_MC.m` available through the text website:

```
function psd=periodogram_psd_MC(num_psd,window,overlap,seg_size,T,x)
% This function computes the classical multichannel periodogram power spectral
% density matrix over 'num_psd' frequencies.
%
%        psd = periodogram_psd_MC(num_psd,window,overlap,seg_size,T,x)
%
% num_psd  -- number of frequency points in psd (must be a power of two)
% window   -- window selection:  1 -- none , 1 -- Hamming , 2 -- Nuttall
% overlap  -- number of overlap samples between segments
% seg_size -- number of samples per segment (must be even)
% T        -- sample interval in seconds
% x        -- sample data array:  x(sample #,channel #)
% psd      -- block vector of 'num_psd' psd matrices
```

The biased multichannel correlation matrix estimate for lag k can be expressed as

$$\check{\mathbf{R}}_{\mathbf{xx}}[k] = \frac{1}{N} \sum_{n=1}^{N-|k|} \mathbf{x}[n+|k|]\mathbf{x}^H[n], \quad (15.27)$$

in which the multichannel data vector $\mathbf{x}[n]$ was defined in (15.1). The MATLAB implementation of the multichannel correlation (both auto and cross) matrix estimation is found in function `correlation_sequence_MC.m` available through the text website:

```
function r=correlation_sequence_MC(max_lag,bias,x)
% Generates the multichannel matrix correlation sequence from data.
%
%          r = correlation_sequence_MC(max_lag,bias,x)
%
% max_lag -- maximum correlation lag time index
% bias    -- selection: 'unbiased' or 'biased'
% x       -- sample data array:  x(sample #,channel #)
% r       -- block column vector correlation sequence
```

Assuming lags from 0 to $\pm p$ are estimated, the multichannel correlogram method will yield the multichannel PSD estimate

$$\hat{\mathbf{P}}_{\mathrm{CORR}}(f) = T \sum_{k=-p}^{p} \check{\mathbf{R}}_{\mathbf{xx}}[k] \exp(-j2\pi fkT). \tag{15.28}$$

Expression (15.28) can be succinctly expressed as

$$\hat{\mathbf{P}}_{\mathrm{CORR}}(f) = T\underline{\mathbf{e}}_p(f)\underline{\check{\mathbf{R}}}_p\underline{\mathbf{e}}_p^H(f), \tag{15.29}$$

in which the block-Toeplitz correlation matrix of block dimension $(p+1) \times (p+1)$ is defined as

$$\underline{\check{\mathbf{R}}}_p = \begin{pmatrix} \check{\mathbf{R}}_{\mathbf{xx}}[0] & \check{\mathbf{R}}_{\mathbf{xx}}[1] & \dots & \check{\mathbf{R}}_{\mathbf{xx}}[p] \\ \check{\mathbf{R}}_{\mathbf{xx}}[-1] & \check{\mathbf{R}}_{\mathbf{xx}}[0] & \dots & \check{\mathbf{R}}_{\mathbf{xx}}[p-1] \\ \vdots & \vdots & \ddots & \vdots \\ \check{\mathbf{R}}_{\mathbf{xx}}[-p] & \check{\mathbf{R}}_{\mathbf{xx}}[-p+1] & \dots & \check{\mathbf{R}}_{\mathbf{xx}}[0] \end{pmatrix} \tag{15.30}$$

and the complex sinusoid block *row* vector $\underline{\mathbf{e}}_p(f)$ of $p+1$ block elements is defined as

$$\underline{\mathbf{e}}_p(f) = \begin{pmatrix} \mathbf{I}_m & \exp(j2\pi fT)\mathbf{I}_m & \dots & \exp(j2\pi fpT)\mathbf{I}_m \end{pmatrix}. \tag{15.31}$$

Note that each element of block vector $\underline{\mathbf{e}}_p(f)$ is an $m \times m$ identity matrix $\mathbf{I}_m$ scaled by an appropriate complex sinusoidal value.

The MATLAB implementation of the multichannel classical correlogram method of PSD estimation is found in function `correlogram_psd_MC.m` available through the text website:

```
function psd=correlogram_psd_MC(num_psd,window,max_lag,T,x)
% This function produces the classical multichannel Blackman-Tukey power
% spectral density matrix over 'num_psd' frequencies.
%
%          psd = correlogram_psd_MC(num_psd,window,max_lag,T,x)
```

```
%
% num_psd  -- number of frequency points in psd (power of 2)
% window   -- selection:  0 -- none , 1 -- Hamming , 2 -- Nuttall
% max_lag  -- maximum lag for correlation matrix estimate
% T        -- sample interval in seconds
% x        -- matrix of data samples:  x(sample #,channel #)
% psd      -- block vector of 'num_psd' psd matrices
```

Schemes have been devised to utilize classical autospectral estimation algorithms to estimate cross spectral densities. If successful, this would alleviate the need to develop cross spectral estimates and the need for full multichannel algorithms. For example, forming the new complex data processes

$$w[n] = x[n] + jy[n]$$
$$z[n] = x[n] + y[n]$$

from the original data processes $x[n]$ and $y[n]$, and computing the autospectra $P_{xx}(f)$, $P_{yy}(f)$, $P_{ww}(f)$, $P_{zz}(f)$, it may easily be shown that the real and imaginary parts of the cross spectral density between channels x and y are given by

$$\text{Re}\{P_{xy}(f)\} = \frac{1}{2}\big[P_{zz}(f) - P_{xx}(f) - P_{yy}(f)\big]$$
$$\text{Im}\{P_{xy}(f)\} = \frac{1}{2}\big[P_{ww}(f) - P_{xx}(f) - P_{yy}(f)\big]. \qquad (15.32)$$

This method of cross spectral estimation can yield values of the magnitude squared coherence, computed from the $\hat{P}_{xy}$ given by Eq. (15.32), that are *greater* than unity. This is obviously undesirable for detecting coherent components between channels, and therefore this approach should not be considered.

15.6 MULTICHANNEL ARMA, AR, AND MA PROCESSES

In analogy with the single-channel definition, the m-channel ARMA is defined as the vector recursion

$$\mathbf{x}[n] = -\sum_{k=1}^{p} \mathbf{A}[k]\mathbf{x}[n-k] + \sum_{k=0}^{q} \mathbf{C}[k]\mathbf{u}[n-k], \qquad (15.33)$$

in which the $\mathbf{A}[k]$ are the $m \times m$ autoregressive parameter matrices, the $\mathbf{C}[k]$ are the $m \times m$ moving average parameter matrices, and $\mathbf{u}[k]$ is an $m \times 1$ vector representing the input driving noise process, assumed to be a wide-sense stationary multichannel process. The z-transform relating input and output was shown in Eq. (15.9) to be

$$\mathbf{X}(z) = \mathbf{A}^{-1}(z)\mathbf{C}(z)\mathbf{U}(z), \qquad (15.34)$$

for which

$$A(z) = I + \sum_{k=1}^{p} A[k]z^{-k}$$

$$C(z) = I + \sum_{k=1}^{q} C[k]z^{-k}. \qquad (15.35)$$

This means that the matrix transfer function from input to output is

$$H(z) = A^{-1}(z)C(z). \qquad (15.36)$$

The z-transform of the output correlation sequence is obtained by substituting (15.36) into (15.23), yielding

$$P_{xx}(z) = A^{-1}(z)C(z)P_{uu}(z)C^{H}(1/z^*)A^{-H}(1/z^*), \qquad (15.37)$$

where $^{-H}$ denotes the Hermitian transpose of the inverse. If the assumption is made that the multichannel input process is a white noise process with constant covariance matrix P_w, then the output process has a multichannel ARMA PSD function, derived from (15.37), of

$$P_{ARMA}(f) = TA^{-1}(f)C(f)P_w C^{H}(f)A^{-H}(f), \qquad (15.38)$$

defined for $|f| \leq 1/2T$, in which

$$A(f) = A(\exp[j2\pi fT]) = I + \sum_{k=1}^{p} A[k]\exp(-j2\pi fkT)$$

and

$$C(f) = C(\exp[j2\pi fT]) = I + \sum_{k=1}^{q} C[k]\exp(-j2\pi fkT).$$

An alternative expression for the multichannel ARMA PSD function is

$$P_{ARMA}(f) = T\,[\underline{a}_p \underline{e}_p^H(f)]^{-1}[\underline{c}_q \underline{e}_q^H(f)]P_w[\underline{e}_q(f)\underline{c}_q^H][\underline{e}_p(f)\underline{a}_p^H]^{-1}, \qquad (15.39)$$

in which the block *row* vectors $\underline{a}_p$, with $p+1$ block elements, and $\underline{c}_q$, with $q+1$ block elements, are defined as

$$\underline{a}_p = \begin{pmatrix} I & A[1] & \ldots & A[p] \end{pmatrix}$$

$$\underline{c}_q = \begin{pmatrix} I & C[1] & \ldots & C[q] \end{pmatrix}, \qquad (15.40)$$

and $\underline{e}_p(f)$ and $\underline{e}_q(f)$ are complex sinusoid block row vectors, defined by Eq. (15.31).

A multichannel AR(p) process

$$\mathbf{x}[n] = -\sum_{k=1}^{p} \mathbf{A}[k]\mathbf{x}[n-k] + \mathbf{u}[n] \tag{15.41}$$

has the z-transform

$$\mathbf{P}_{AR}(z) = \mathbf{A}^{-1}(z)\boldsymbol{P}_w\mathbf{A}^{-H}(1/z^*) \tag{15.42}$$

and, therefore, has the multichannel PSD function

$$\mathbf{P}_{AR}(f) = T\,[\underline{\mathbf{a}}_p\underline{\mathbf{e}}_p^H(f)]^{-1}\,\boldsymbol{P}_w[\underline{\mathbf{e}}_p(f)\underline{\mathbf{a}}_p^H]^{-1}. \tag{15.43}$$

A multichannel MA(q) process

$$\mathbf{x}[n] = \sum_{k=1}^{q} \mathbf{C}[k]\mathbf{u}[n-k] + \mathbf{u}[n] \tag{15.44}$$

has the z-transform

$$\mathbf{P}_{MA}(z) = \mathbf{C}(z)\boldsymbol{P}_w\mathbf{C}^H(1/z^*) \tag{15.45}$$

and, therefore, has the multichannel PSD function

$$\mathbf{P}_{MA}(f) = T\,[\underline{\mathbf{c}}_q\underline{\mathbf{e}}_q^H(f)]\boldsymbol{P}_w[\underline{\mathbf{e}}_q(f)\underline{\mathbf{c}}_q^H]. \tag{15.46}$$

Although a multichannel process may be a vector autoregression, it does not follow that each channel is individually an autoregressive process [Nuttall, 1976a, 1976b]. Consider the two-channel first-order vector autoregression

$$\mathbf{x}[n] = -\mathbf{A}[1]\mathbf{x}[n-1] + \mathbf{u}[n], \tag{15.47}$$

in which the input process is uncorrelated from channel to channel and the variance for each channel is unity ($\boldsymbol{P}_w = \mathbf{I}$), and

$$\mathbf{A}[1] = \begin{pmatrix} a & b \\ c & d \end{pmatrix},$$

containing complex elements a, b, c, and d. The two-channel AR z-transform for this case is

$$\mathbf{P}_{AR}(z) = \mathbf{A}^{-1}(z)\mathbf{A}^{-H}(1/z^*), \tag{15.48}$$

in which

$$\mathbf{A}(z) = (\mathbf{I} - \mathbf{A}[1]z^{-1}) = \begin{pmatrix} 1 - z^{-1}a & -z^{-1}b \\ -z^{-1}c & 1 - z^{-1}d \end{pmatrix}.$$

Substituting for $\mathbf{A}(z)$ in (15.42) yields the expression

$$\mathbf{P}_{AR}(z) \tag{15.49}$$

$$= \frac{1}{D(z)D(1/z)} \begin{pmatrix} 1 + |b|^2 + |d|^2 + z^{-1}d + zd^* & -(a^*b + c^*d + z^{-1}b + zc^*) \\ -(ab^* + cd^* + z^{-1}c + zb^*) & 1 + |a|^2 + |c|^2 + z^{-1}a + za^* \end{pmatrix}$$

in which the polynomial $D(z)$ is

$$D(z) = \left(1 - z^{-1}(a+d) + z^{-2}(ad - bc)\right). \tag{15.50}$$

Inspection of denominator $D(z)$ reveals that each of the elements of matrix $\mathbf{P}_{AR}(z)$ has *four* poles and *three* zeros in the finite z-plane. Thus, the autospectra of each channel are representative of a *single-channel* ARMA(4,3) process, even though the multichannel problem is a *first-order vector autoregression* with independent white excitation inputs. Generally, an m-channel autoregression of order p with independent white excitation driving each channel will have autospectra and cross spectra with $m^2 p$ poles and $(m^2 - 1)p$ zeros in the finite z-plane (of which p zeros occur at the origin), in contrast to the single-channel autoregression, which only has $2p$ poles. This case illustrates that an autospectrum of a multichannel autoregression has more degrees of freedom than an autospectrum of a single-channel autoregression.

15.7 MULTICHANNEL YULE-WALKER EQUATIONS

The focus in the next five sections will be on the multichannel autoregressive spectral estimate. Multichannel ARMA and MA have received only limited attention in the literature. For those interested, the Yule-Walker equations for a multichannel ARMA is presented as a problem exercise. Recent references to multichannel ARMA include Lee et al. [1982], Friedlander [1982], and Friedlander and Porat [1986].

The stationary m-channel autoregressive process may be expressed as the block vector inner product

$$\mathbf{e}_p^f[n] = \underline{\mathbf{a}}_p \underline{\mathbf{x}}_p[n] \tag{15.51}$$

for which the block *row* vector of multichannel AR matrix coefficients was defined in (15.40) as

$$\underline{\mathbf{a}}_p = \begin{pmatrix} \mathbf{I} & \mathbf{A}_p[1] & \ldots & \mathbf{A}_p[p] \end{pmatrix}, \tag{15.52}$$

$\mathbf{I}$ is an $m \times m$ identity matrix, and the block *column* vector of multichannel data vectors is

$$\underline{\mathbf{x}}_p[n] = \begin{pmatrix} \mathbf{x}[n] \\ \vdots \\ \mathbf{x}[n - p] \end{pmatrix}. \tag{15.53}$$

A superscript f has been added in Eq. (15.51) to distinguish $\mathbf{e}_p^f[n]$ as not only the autoregressive driving sequence (previously written as $\mathbf{u}[n]$), but also as the forward linear prediction error. In order to establish a relationship between the AR matrix coefficients and the matrix correlation sequence for lags 0 to p, postmultiply both sides of (15.51) by $\underline{\mathbf{x}}_p^H[n]$ and take the expectation,

$$\mathcal{E}\{\mathbf{e}^f[n]\underline{\mathbf{x}}_p^H[n]\} = \mathcal{E}\{\underline{\mathbf{a}}_p\underline{\mathbf{x}}_p[n]\underline{\mathbf{x}}_p^H[n]\} = \underline{\mathbf{a}}_p\mathcal{E}\{\underline{\mathbf{x}}_p[n]\underline{\mathbf{x}}_p^H[n]\}. \tag{15.54}$$

Define the block matrix $\underline{\mathbf{R}}_p$ of $(p+1) \times (p+1)$ block entries $\mathbf{R}_{\mathbf{xx}}[k]$ (each of which is dimension $m \times m$) as

$$\begin{aligned}\underline{\mathbf{R}}_p &= \mathcal{E}\{\underline{\mathbf{x}}_p[n]\underline{\mathbf{x}}_p^H[n]\} \\ &= \begin{pmatrix} \mathcal{E}\{\mathbf{x}[n]\mathbf{x}^H[n]\} & \cdots & \mathcal{E}\{\mathbf{x}[n]\mathbf{x}^H[n-p]\} \\ \vdots & & \vdots \\ \mathcal{E}\{\mathbf{x}[n-p]\mathbf{x}^H[n]\} & \cdots & \mathcal{E}\{\mathbf{x}[n-p]\mathbf{x}^H[n-p]\} \end{pmatrix} \\ &= \begin{pmatrix} \mathbf{R}_{\mathbf{xx}}[0] & \mathbf{R}_{\mathbf{xx}}[1] & \cdots & \mathbf{R}_{\mathbf{xx}}[p] \\ \mathbf{R}_{\mathbf{xx}}[-1] & \mathbf{R}_{\mathbf{xx}}[0] & \cdots & \mathbf{R}_{\mathbf{xx}}[p-1] \\ \vdots & \vdots & \ddots & \vdots \\ \mathbf{R}_{\mathbf{xx}}[-p] & \mathbf{R}_{\mathbf{xx}}[-p+1] & \cdots & \mathbf{R}_{\mathbf{xx}}[0] \end{pmatrix}. \end{aligned} \tag{15.55}$$

The matrix $\underline{\mathbf{R}}_p$ has both a *Hermitian* and a *block-Toeplitz* structure. Individual matrix elements of $\underline{\mathbf{R}}_p$ are *not*, in general, Hermitian ($\mathbf{R}_{\mathbf{xx}}[k] \neq \mathbf{R}_{\mathbf{xx}}^H[k]$), although $\mathbf{R}_{\mathbf{xx}}[-k] = \mathbf{R}_{\mathbf{xx}}^H[k]$. The $(p+1) \times 1$ block vector $\mathcal{E}\{\mathbf{e}^f[n]\underline{\mathbf{x}}_p^H[n]\}$ may be simplified to

$$\begin{aligned}\mathcal{E}\{\mathbf{e}^f[n]\underline{\mathbf{x}}_p^H[n]\} &= \mathcal{E}\left\{\mathbf{e}^f[n]\left(\mathbf{e}^{fH}[n] - \sum_{k=1}^{p}\mathbf{x}^H[n-k]\mathbf{A}_p^H[k]\right)\right\} \\ &= \begin{pmatrix} \mathbf{P}_p^f & \mathbf{0} & \cdots & \mathbf{0} \end{pmatrix}, \end{aligned} \tag{15.56}$$

in which $\mathbf{0}$ are $m \times m$ all-zero matrices and $\mathbf{P}_p^f = \mathcal{E}\{\mathbf{e}^f[n]\mathbf{e}^{fH}[n]\} = \mathbf{P}_w$ is the covariance matrix of the AR driving noise. The last equality in Eq. (15.56) follows from the property that the driving noise process is uncorrelated with past values of the AR process, $\mathcal{E}\{\mathbf{e}^f[n]\mathbf{x}^H[n-k]\} = \mathbf{0}$ for $k < n$. Equation (15.54) may be rewritten with this result as

$$\underline{\mathbf{a}}_p\underline{\mathbf{R}}_p = \begin{pmatrix} \mathbf{P}_p^f & \mathbf{0} & \cdots & \mathbf{0} \end{pmatrix}. \tag{15.57}$$

This is the multichannel version of the Yule-Walker normal equations for both a multichannel AR(p) and a multichannel forward linear prediction filter of order p.

The multichannel autoregressive process created by the *backward* filter process at time index $n-1$ is

$$\mathbf{e}_p^b[n-1] = \mathbf{x}[n-p-1] + \sum_{k=1}^{p}\mathbf{B}[k]\mathbf{x}[n-p-1+k] = \underline{\mathbf{b}}_p\underline{\mathbf{x}}_p[n-1], \tag{15.58}$$

for which the block *row* vector of backward multichannel AR matrix coefficients is

$$\underline{\mathbf{b}}_p = \begin{pmatrix} \mathbf{B}_p[p] & \cdots & \mathbf{B}_p[1] & \mathbf{I} \end{pmatrix}. \tag{15.59}$$

The corresponding multichannel Yule-Walker equations for the backward coefficients of a stationary multichannel process are found by constructing

$$\mathcal{E}\left\{\mathbf{e}_p^b[n-1]\underline{\mathbf{x}}_p^H[n-1]\right\} = \mathcal{E}\left\{\underline{\mathbf{b}}_p\underline{\mathbf{x}}_p[n-1]\underline{\mathbf{x}}_p^H[n-1]\right\}$$

to yield

$$\underline{\mathbf{b}}_p\underline{\mathbf{R}}_p = \begin{pmatrix} \mathbf{0} & \cdots & \mathbf{0} & \mathbf{P}_p{}^b \end{pmatrix}, \tag{15.60}$$

for which $\mathbf{P}_p{}^b = \mathcal{E}\left\{\mathbf{e}^b[n-1]\mathbf{e}^{bH}[n-1]\right\}$ is the covariance of the driving noise process for the backward AR process. $\mathbf{P}_p{}^b$ will, in general, be different than $\mathbf{P}_p{}^f$ for multichannel processes. In the single-channel case, they were identical. The Hermitian Toeplitz structure of the correlation matrix in the single-channel case was sufficient to guarantee the conjugate relationship $a_p[k] = b_p^*[k]$ between the forward and backward AR parameters shown in Sec. 3.8.1. In the multichannel case, though, the elements of the block correlation matrix $\underline{\mathbf{R}}_p$ lack the Hermitian property needed to produce a similar property between elements of $\underline{\mathbf{a}}_p$ and $\underline{\mathbf{b}}_p$. The multichannel AR(p) process also has the spectral density z-transform given in terms of the backward filter coefficients

$$\mathsf{P}_{\mathrm{AR}}(z) = \mathsf{B}^{-1}(z)\mathbf{P}_p{}^b\mathsf{B}^{-H}(1/z^*), \tag{15.61}$$

for which $\mathsf{B}(z)$ is the matrix polynomial

$$\mathsf{B}(z) = z^{-p}\mathbf{I} + \sum_{k=1}^p \mathbf{B}[k]z^{-p+k} = z^{-p}\left(\mathbf{I} + \sum_{k=1}^p \mathbf{B}[k]z^k\right). \tag{15.62}$$

The transforms for $\mathsf{P}_{\mathrm{AR}}(z)$ given by Eqs. (15.42) and (15.62) must be equivalent for a stationary multichannel process and, therefore, either expression may be used to obtain the multichannel AR PSD.

15.8 MULTICHANNEL LEVINSON ALGORITHM

A fast algorithm for the simultaneous solution of Yule-Walker equations (15.57) and (15.60) was apparently first developed by Wiggins and Robinson [1965]. It is structured along the same lines as the single-channel Levinson algorithm presented in Sec. 3.8.1. The block-Toeplitz multichannel correlation matrix $\underline{\mathbf{R}}_{p+1}$ may be partitioned in two ways,

$$\underline{\mathbf{R}}_{p+1} = \begin{pmatrix} \mathbf{R}_p & \mathbf{s}_{p+1}^H \\ \underline{\mathbf{s}}_{p+1} & \mathbf{R_{xx}}[0] \end{pmatrix} = \begin{pmatrix} \mathbf{R_{xx}}[0] & \underline{\mathbf{r}}_{p+1} \\ \underline{\mathbf{r}}_{p+1}^H & \underline{\mathbf{R}}_p \end{pmatrix},$$ (15.63)

for which the block *row* vectors $\underline{\mathbf{r}}_{p+1}$ are $\underline{\mathbf{s}}_{p+1}$ are defined as

$$\underline{\mathbf{r}}_{p+1} = \begin{pmatrix} \mathbf{R_{xx}}[1] & \ldots & \mathbf{R_{xx}}[p+1] \end{pmatrix}$$
$$\underline{\mathbf{s}}_{p+1} = \begin{pmatrix} \mathbf{R_{xx}}[-(p+1)] & \ldots & \mathbf{R_{xx}}[-1] \end{pmatrix}.$$ (15.64)

These partitions may then be used to construct the following relationship

$$\begin{pmatrix} \underline{\mathbf{a}}_p & \mathbf{0} \end{pmatrix} \underline{\mathbf{R}}_{p+1} = \begin{pmatrix} P_p{}^f & \mathbf{0} & \ldots & \mathbf{0} & \Delta_{p+1} \end{pmatrix}$$ (15.65)

$$\begin{pmatrix} \mathbf{0} & \underline{\mathbf{b}}_p \end{pmatrix} \underline{\mathbf{R}}_{p+1} = \begin{pmatrix} \nabla_{p+1} & \mathbf{0} & \ldots & \mathbf{0} & P_p{}^b \end{pmatrix},$$ (15.66)

for which the $m \times m$ matrices Δ_{p+1} and ∇_{p+1} are defined as

$$\Delta_{p+1} = \mathbf{R_{xx}}[p+1] + \sum_{k=1}^{p} \mathbf{A}_p[k]\mathbf{R_{xx}}[p+1-k] = \underline{\mathbf{a}}_p \mathbf{s}_{p+1}^H$$ (15.67)

$$\nabla_{p+1} = \mathbf{R_{xx}}[-p-1] + \sum_{k=1}^{p} \mathbf{B}_p[k]\mathbf{R_{xx}}[k-p-1] = \underline{\mathbf{b}}_p \underline{\mathbf{r}}_{p+1}^H.$$ (15.68)

The order update recursion for the forward and backward autoregressive coefficient matrices take the form

$$\underline{\mathbf{a}}_{p+1} = \begin{pmatrix} \underline{\mathbf{a}}_p & \mathbf{0} \end{pmatrix} + \mathbf{A}_{p+1}[p+1]\begin{pmatrix} \mathbf{0} & \underline{\mathbf{b}}_p \end{pmatrix}$$ (15.69)

$$\underline{\mathbf{b}}_{p+1} = \begin{pmatrix} \mathbf{0} & \underline{\mathbf{b}}_p \end{pmatrix} + \mathbf{B}_{p+1}[p+1]\begin{pmatrix} \underline{\mathbf{a}}_p & \mathbf{0} \end{pmatrix}.$$ (15.70)

On a matrix element-by-matrix element basis, the linear prediction filter matrix coefficients then obey the recursions

$$\mathbf{A}_{p+1}[k] = \mathbf{A}_p[k] + \mathbf{A}_{p+1}[p+1]\mathbf{B}_p[p+1-k]$$ (15.71)
$$\mathbf{B}_{p+1}[k] = \mathbf{B}_p[k] + \mathbf{B}_{p+1}[p+1]\mathbf{A}_p[p+1-k]$$ (15.72)

for $k = 1$ to $k = p$. In the multichannel case, there are two distinct reflection coefficient matrices given by

$$\mathbf{A}_{p+1}[p+1] = -\Delta_{p+1}\left(P_p{}^b\right)^{-1}$$ (15.73)

$$\mathbf{B}_{p+1}[p+1] = -\nabla_{p+1}\left(P_p{}^f\right)^{-1}.$$ (15.74)

This is in contrast to the single-channel case, for which there was only a single reflection coefficient, due to the property $b_p[p] = a_p^*[p]$ (Sec. 3.8.1). The validity of Eqs. (15.69) and (15.70) may be checked by postmultiplying each side of these equations by $\underline{\mathbf{R}}_{p+1}$ and using Eqs. (15.65), (15.66), (15.73), and (15.74). A by-product of this check is the order update for the noise prediction error covariance matrices,

$$P_{p+1}^{f} = P_p^{f} + A_{p+1}[p+1]\nabla_{p+1}$$
$$= (I - A_{p+1}[p+1]B_{p+1}[p+1])P_p^{f} \qquad (15.75)$$
$$P_{p+1}^{b} = P_p^{b} + B_{p+1}[p+1]\Delta_{p+1}$$
$$= (I - B_{p+1}[p+1]A_{p+1}[p+1])P_p^{b}. \qquad (15.76)$$

Using the matrix identity

$$(\underline{a}_p \quad 0)\,\underline{R}_{p+1}\begin{pmatrix} 0 \\ \underline{b}_p^{H} \end{pmatrix} = (0 \quad \underline{b}_p)\,\underline{R}_{p+1}\begin{pmatrix} \mathbf{a}_p^{H} \\ 0 \end{pmatrix}, \qquad (15.77)$$

the following lemma may be derived

$$\nabla_{p+1} = \Delta_{p+1}^{H}. \qquad (15.78)$$

The $m \times m$ matrix Δ_{p+1} is simply the cross correlation between the forward and backward linear prediction residuals at one unit of delay (lag),

$$\Delta_{p+1} = \mathcal{E}\{e_p^{f}[n]e_p^{b\,H}[n-1]\} = P_p^{fb}. \qquad (15.79)$$

A definition for a normalized cross correlation, or *normalized partial correlation*, comparable to the single-channel reflection coefficient defined by Eq. (7.22), is

$$\Lambda_{p+1} = (P_p^{f\,1/2})^{-1}(P_p^{fb})(P_p^{b\,1/2})^{-H} \qquad (15.80)$$
$$= (\mathcal{E}\{e_p^{f}[n]e_p^{f\,H}[n]\}^{1/2})^{-1}(\mathcal{E}\{e_p^{f}[n]e_p^{b\,H}[n-1]\})$$
$$\times (\mathcal{E}\{e_p^{b}[n-1]e_p^{b\,H}[n-1]\}^{1/2})^{-H}.$$

In the multichannel case, the partial correlation coefficient matrix and the reflection coefficient matrices are *not* identical, unlike the single-channel case, for which they are the same. The superscript $-H$ denotes the conjugate transpose of the inverse, i.e., the order of application is the inverse first, followed by the conjugate transpose. The superscript $1/2$ denotes the lower triangular matrix obtained by the Cholesky decomposition of the Hermitian matrix to which the operation is applied. Therefore, if M is a Hermitian matrix, then

$$M = (M^{1/2})(M^{1/2})^{H}. \qquad (15.81)$$

The forward and backward reflection coefficients of Eqs. (15.73) and (15.74) may be expressed in terms of Λ_{p+1} as

$$A_{p+1}[p+1] = -(P_p^{f\,1/2})\Lambda_{p+1}(P_p^{b\,1/2})^{-1} \qquad (15.82)$$
$$B_{p+1}[p+1] = -(P_p^{b\,1/2})\Lambda_{p+1}^{H}(P_p^{f\,1/2})^{-1}. \qquad (15.83)$$

The multichannel normalized correlation Λ_p provides a unique parameterization sequence for the multichannel algorithm, just as the reflection coefficient $a_p[p]$

provided a unique parameterization of the single-channel AR process. It has been shown [Morf et al., 1978b], based on the nonnegative definite correlation $\underline{\mathbf{R}}_p$, that the singular values of $\Lambda_{p+1}\Lambda_{p+1}^H$ are within the unit circle. This is the multichannel extension to the requirement that single-channel reflection coefficients have magnitudes less than unity.

Equations (15.67), (15.69), (15.70), (15.73), (15.74), (15.75), and (15.76) constitute the multichannel extension of the Levinson algorithm. Equations (15.80) and (15.83) may be used in lieu of Eqs. (15.73) and (15.74). The recursive-in-order nature of the algorithm requires that $\underline{\mathbf{R}}_q$ be strictly positive definite and nonsingular at any order from $q = 1$ to $q = p$. This will be true as long as $\det|\mathbf{P}_q{}^f| > 0$ and $\det|\mathbf{P}_q{}^b| > 0$ at each order. The initial conditions $\mathbf{P}_0{}^f = \mathbf{P}_0{}^b = \mathbf{R}_{xx}[0]$, required to commence the Levinson recursion, are obtained by inspection of Eqs. (15.65) and (15.66).

The MATLAB implementation of the multichannel Levinson recursion is found in function `levinson_recursion_MC.m` available through the text website:

```
function [rho_a,a,rho_b,b]=levinson_recursion_MC(r)
% Solves for the multichannel AR parameter matrices from the multichannel
% correlation sequence by the multichannel Levinson algorithm.
%
%       [rho_a,a,rho_b,b] = levinson_recursion_MC(r)
%
% r    -- multichannel correlation sequence block column vector
% rho_a -- forward linear prediction error/white noise covariance matrix
% a    -- forward linear prediction/autoregressive (AR) parameter block vector
% rho_b -- backward linear prediction error/white noise covariance matrix
% b    -- backward linear prediction/autoregressive (AR) parameter block vector
```

A useful relationship between the order p multichannel linear prediction errors and the order $p + 1$ linear prediction errors may be obtained by postmultiplying both Eqs. (15.69) and (15.70) by $\underline{\mathbf{x}}_{p+1}[n]$ to yield

$$\mathbf{e}_{p+1}^f[n] = \mathbf{e}_p^f[n] + \mathbf{A}_{p+1}[p+1]\mathbf{e}_p^b[n-1] \qquad (15.84)$$

$$\mathbf{e}_{p+1}^b[n] = \mathbf{e}_p^b[n-1] + \mathbf{B}_{p+1}[p+1]\mathbf{e}_p^f[n]. \qquad (15.85)$$

15.9 MULTICHANNEL BLOCK TOEPLITZ MATRIX INVERSE

The multichannel Levinson algorithm serves as the basis for methods that describe the complete inverse of the Hermitian block-Toeplitz correlation matrix of Eq. (15.55). Akaike [1973] and Watson [1973] were two of the first to publish algorithms for such a block inverse. A form of the Hermitian block-Toeplitz inverse that will be especially useful in this chapter is

$$\underline{\mathbf{R}}_p^{-1} = \underline{\mathbf{A}}_p^H \underline{\mathbf{P}}_p^f \underline{\mathbf{A}}_p - \underline{\mathbf{B}}_p^H \underline{\mathbf{P}}_p^b \underline{\mathbf{B}}_p \qquad (15.86)$$

in which

$$
\underline{\mathbf{A}}_p = \begin{pmatrix} \mathbf{I} & \mathbf{A}_p[1] & \cdots & \mathbf{A}_p[p] \\ \mathbf{0} & \ddots & \ddots & \vdots \\ \vdots & \ddots & \ddots & \mathbf{A}_p[1] \\ \mathbf{0} & \cdots & \mathbf{0} & \mathbf{I} \end{pmatrix}, \quad \underline{\mathbf{P}}_p^f = \begin{pmatrix} (P_p^f)^{-1} & \mathbf{0} & \cdots & \mathbf{0} \\ \mathbf{0} & (P_p^f)^{-1} & \ddots & \vdots \\ \vdots & \ddots & \ddots & \mathbf{0} \\ \mathbf{0} & \cdots & \mathbf{0} & (P_p^f)^{-1} \end{pmatrix}
$$

$$
\underline{\mathbf{B}}_p = \begin{pmatrix} \mathbf{0} & \mathbf{B}_p[p] & \cdots & \mathbf{B}_p[1] \\ \mathbf{0} & \ddots & \ddots & \vdots \\ \vdots & \ddots & \ddots & \mathbf{B}_p[p] \\ \mathbf{0} & \cdots & \mathbf{0} & \mathbf{0} \end{pmatrix}, \quad \underline{\mathbf{P}}_p^b = \begin{pmatrix} (P_p^b)^{-1} & \mathbf{0} & \cdots & \mathbf{0} \\ \mathbf{0} & (P_p^b)^{-1} & \ddots & \vdots \\ \vdots & \ddots & \ddots & \mathbf{0} \\ \mathbf{0} & \cdots & \mathbf{0} & (P_p^b)^{-1} \end{pmatrix}.
$$

This result may be derived using the approach of Sec. 3.8.1 (see also Friedlander et al., 1979].

15.10   MULTICHANNEL AUTOREGRESSIVE SPECTRAL ESTIMATION

Algorithms that directly estimate the parameters of a multichannel AR process from the data are patterned after algorithms that have been developed for the single-channel AR process in Chap. 8. Algorithms based on correlation function estimation, partial correlation estimation, and least squares linear prediction are covered in this section.

15.10.1   Correlation Function Estimation Method

The most obvious approach to AR PSD estimation from the r-channel sample data is to use correlation function estimates in lieu of the unknown correlation function. One estimate based on the N vector samples $\mathbf{x}[1]$ to $\mathbf{x}[N]$ is the biased matrix lag product sum

$$
\hat{\mathbf{R}}_{\mathbf{xx}}[m] = \frac{1}{N} \sum_{m=1}^{N-m} \mathbf{x}[n+m]\mathbf{x}^H[n], \tag{15.87}
$$

defined for $0 \le m \le N - 1$. The autocorrelation terms for negative lags are obtained by the property $\hat{\mathbf{R}}_{\mathbf{xx}}[-m] = \hat{\mathbf{R}}_{\mathbf{xx}}^H[m]$. The use of biased correlation estimates will guarantee that $\underline{\mathbf{R}}_p$ will be positive semidefinite. Combining the multichannel correlation estimates with the multichannel Levinson algorithm will yield the estimates of the AR coefficient matrices, which can then be used in Eq. (15.43) to obtain the AR PSD. This procedure will be termed the *multichannel Yule-Walker algorithm*. The MATLAB implementation of the multichannel Yule-Walker algorithm is found in function yule_walker_MC.m available through the text website:

```
function [rho_a,a,rho_b,b]=yule_walker_MC(p,x)
% Multichannel Yule-Walker autoregressive parameter estimation algorithm.
%
%          [rho_a,a,rho_b,b] = yule_walker_MC(p,x)
%
% p      -- order of multichannel autoregressive (AR) process
% x      -- sample data matrix: x(sample #,channel #)
% rho_a  -- forward linear prediction error/white noise covariance matrix
% a      -- forward linear prediction/autoregressive (AR) parameter block vector
% rho_b  -- backward linear prediction error/white noise covariance matrix
% b      -- backward linear prediction/autoregressive (AR) parameter block vector
```

15.10.2   Partial Correlation Estimation: Vieira-Morf Method

Only the residual cross correlation Δ_p or, equivalently, the partial correlation Λ_p depend on the correlation function $\mathbf{R}_{xx}[m]$. Replacing these with estimates based on the available data would permit the remainder of the Levinson recursion to be incorporated into an AR parameter estimation algorithm. Algorithms that parallel the geometric and harmonic algorithms of Chap. 8 for the estimation of the reflection coefficients have been developed.

The multichannel version of the geometric algorithm was proposed by Vieira and Morf [see Morf et al., 1978a]. They proposed to estimate the partial correlation as

$$\hat{\Lambda}_{p+1} = \left(\hat{P}_p^{f\,1/2}\right)^{-1}\left(\hat{P}_p^{fb}\right)\left(\hat{P}_p^{b1/2}\right)^{-H} \tag{15.88}$$

based on estimates of the residual variance matrices $P_p{}^f$, $P_p{}^b$, and covariance matrix $P_p{}^{fb}$

$$\hat{P}_p^{f} = \frac{1}{N}\sum_{n=p+2}^{N} \mathbf{e}_p^{f}[n]\mathbf{e}_p^{f\,H}[n]$$

$$\hat{P}_p^{b} = \frac{1}{N}\sum_{n=p+2}^{N} \mathbf{e}_p^{b}[n-1]\mathbf{e}_p^{b\,H}[n-1] \tag{15.89}$$

$$\hat{P}_p^{fb} = \frac{1}{N}\sum_{n=p+2}^{N} \mathbf{e}_p^{f}[n]\mathbf{e}_p^{b\,H}[n-1].$$

Note that the factor $1/N$ is canceled when computing $\hat{\Lambda}_{p+1}$.

The residuals $\mathbf{e}_p^{f}[n]$ and $\mathbf{e}_p^{b}[n]$ for $p+1 \le n \le N$ are updated at each order using recursions provided by Eqs. (15.84) and (15.85). With these modifications to estimate $\hat{\Lambda}_{p+1}$, the multichannel Levinson algorithm can proceed as in the known correlation case. This includes the order updates (15.75) and (15.76) for $P_p{}^f$ and $P_p{}^b$; the estimates $\hat{P}_p{}^{f}$ and $\hat{P}_p{}^{b}$ are used *strictly* for estimating $\hat{\Lambda}_{p+1}$.

The Vieira-Morf geometric algorithm is initialized by setting

$$\boldsymbol{P}_0^f = \boldsymbol{P}_0^b = \frac{1}{N} \sum_{n=1}^{N} \mathbf{x}[n]\mathbf{x}^H[n] \qquad (15.90)$$

and

$$\mathbf{e}_0^f[n] = \mathbf{e}_0^b[n] = \mathbf{x}[n] \qquad (15.91)$$

for $1 \leq n \leq N$. This multichannel algorithm reduces to the geometric mean algorithm for a single channel, because $\boldsymbol{P}_p^f = \boldsymbol{P}_p^b = \rho$ in this case.

15.10.3 Partial Correlation Estimation: Nuttall-Strand Method

A multichannel version of the harmonic mean algorithm was independently proposed by Nuttall [1976a,b] and Strand [1977]. They elected to directly estimate the cross correlation coefficient $\boldsymbol{\Lambda}_p$ from the data, rather than to estimate the normalized partial correlation $\boldsymbol{\Delta}_p$. The estimate is given by that matrix $\boldsymbol{\Delta}_p$ which minimizes the trace of a weighted average (arithmetic mean) of the *estimated* forward and backward linear prediction residuals covariance matrices

$$\text{tr}\left\{\mathbf{V}_{p-1}\hat{\boldsymbol{P}}_p^f + \mathbf{W}_{p-1}\hat{\boldsymbol{P}}_p^b\right\} \qquad (15.92)$$

subject to the three constraints

$$\mathbf{e}_p^f[n] = \mathbf{e}_{p-1}^f[n] - \boldsymbol{\Delta}_p(\boldsymbol{P}_p^b)^{-1}\mathbf{e}_{p-1}^b[n-1] \qquad (15.93)$$

$$\mathbf{e}_p^b[n] = \mathbf{e}_p^b[n-1] - \boldsymbol{\Delta}_p^H(\boldsymbol{P}_p^f)^{-1}\mathbf{e}_{p-1}^f[n] \qquad (15.94)$$

$$\mathbf{A}_p[p]\boldsymbol{P}_p^f = \left(\mathbf{B}_p[p]\boldsymbol{P}_p^b\right)^H. \qquad (15.95)$$

The estimates $\hat{\boldsymbol{P}}_p^f$ and $\hat{\boldsymbol{P}}_p^b$ were defined previously by Eq. (15.89). Matrices $\mathbf{V}_p$ and $\mathbf{W}_p$ are arbitrary positive-definite weighting matrices to be selected. Constraints (15.93) and (15.94) follow directly from Eqs. (15.69), (15.70), (15.73), and (15.74). The linear matrix constraint (15.95) follows directly from Eqs. (15.73), (15.74), and (15.78). This last constraint demonstrates that the reflection coefficient matrices cannot be specified independently. These constraints jointly force the solution for $\boldsymbol{\Delta}_p$ to obey as many properties of the multichannel Levinson algorithm as possible. The weighting matrices must be carefully selected so that the resulting solution for $\boldsymbol{\Delta}_p$ will yield a stable correlation sequence (guaranteed if $\boldsymbol{\Lambda}_p$ has singular values less than or equal to unity) and a positive-definite multichannel spectral estimate.

Nuttall [1977] has made an extensive examination of weighting matrices and has determined that the inverse matrices

$$\mathbf{V}_p = \left(\boldsymbol{P}_p^f\right)^{-1} \qquad (15.96)$$

$$\mathbf{W}_p = \left(\boldsymbol{P}_p^b\right)^{-1} \qquad (15.97)$$

will yield the desired stability property. The AR parameter estimates are stable if both $\det[\mathbf{A}_p(z)]$ and $\det[z^{-p}\mathbf{B}_p(1/z)]$ possess all of their zeros within the unit z-plane circle. Stability ensures that the forward and backward correlation extrapolations are damped extensions, i.e., the terms grow smaller rather than increase in magnitude [Nuttall, 1977]. It is desirable not to have the channels with larger errors to receive the most emphasis. The weightings of Eqs. (15.96) and (15.97) seem intuitive in that they weight each channel proportional to its inverse squared error. If these weightings are selected, then a minimization of Eq. (15.92) produces the matrix expression

$$\hat{\boldsymbol{P}}_p{}^f \left(\boldsymbol{P}_p{}^f\right)^{-1} \hat{\mathbf{\Delta}}_{p+1} + \hat{\mathbf{\Delta}}_{p+1}\left(\boldsymbol{P}_p{}^b\right)^{-1} \hat{\boldsymbol{P}}_p{}^b = -2\hat{\boldsymbol{P}}_p{}^{fb}, \qquad (15.98)$$

which contains a mixture of the error covariance matrices $\boldsymbol{P}_p{}^f$ and $\boldsymbol{P}_p{}^b$ from the Levinson recursion and the error covariance matrix $estimates$ $\hat{\boldsymbol{P}}_p{}^f$, $\hat{\boldsymbol{P}}_p{}^b$, and $\hat{\boldsymbol{P}}_p{}^{fb}$. In the single-channel case, each element of Eq. (15.98) reduces to a scalar and the error covariances are identical, i.e., $\boldsymbol{P}_p{}^f = \boldsymbol{P}_p{}^b$. Noting that $\mathbf{A}_{p+1}[p+1] = \hat{\mathbf{\Delta}}_{p+1}\boldsymbol{P}_p{}^f$, the solution for $\mathbf{A}_{p+1}[p+1]$ in the single-channel case reduces to the harmonic mean relationship

$$\mathbf{A}_{p+1}[p+1] = \frac{-2\hat{\boldsymbol{P}}_p{}^{fb}}{\hat{\boldsymbol{P}}_p{}^f + \hat{\boldsymbol{P}}_p{}^b}. \qquad (15.99)$$

Thus, the single-channel case of the Nuttall-Strand multichannel algorithm is identical to the Burg algorithm of Sec. 8.4.

Equation (15.98) is a special case of a bilinear matrix equation

$$\mathbf{AX} + \mathbf{XB} = \mathbf{C}, \qquad (15.100)$$

in which $\mathbf{A}$, $\mathbf{B}$, $\mathbf{C}$, and $\mathbf{X}$ are $m \times m$ matrices. Solutions that obtain a unique matrix $\mathbf{X}$ have been provided by several authors. Nuttall [1976b] selected the solution

$$\mathbf{X} = \mathbf{PQ}^{-1}$$

$$\mathbf{P} = \sum_{k=0}^{m-1}(-1)^k \mathbf{A}_k \mathbf{CB}^{m-1-k}$$

$$\mathbf{Q} = \sum_{k=0}^{m}(-1)^k a_k \mathbf{B}^{m-k} \qquad (15.101)$$

$$a_k = -\frac{1}{k}\mathrm{tr}\left(\mathbf{AD}_{k-1}\right) \quad 1 \le k \le m$$

$$\mathbf{D}_k = \mathbf{AD}_{k-1} + a_k \mathbf{I} \quad 1 \le k \le m,$$

in which $a_0 = 1$ and $\mathbf{A}_0 = \mathbf{I}$. The matrices $\mathbf{P}$ and $\mathbf{Q}$ may be simplified for the two-channel case [Nuttall, 1981] to

$$\mathbf{P} = \mathbf{CB} - \left(\mathbf{A} - \mathrm{tr}(\mathbf{A})\mathbf{I}\right)\mathbf{C} \tag{15.102}$$

$$\mathbf{Q} = \left(\mathrm{tr}\,\mathbf{A} + \mathrm{tr}\,\mathbf{B}\right)\mathbf{B} + \left(\det\mathbf{A} - \det\mathbf{B}\right)\mathbf{I}. \tag{15.103}$$

The Nuttall-Strand algorithm is initialized by setting $P_0{}^f$, $P_0{}^b$, $\mathbf{e}_0^f[n]$, and $\mathbf{e}_0^b[n]$ to the values given by Eqs. (15.90) and (15.91). The MATLAB implementation of the multichannel lattice algorithm by the Nuttall-Strand approach is found in function `lattice_MC.m` available through the text website:

```
function [rho_a,a,rho_b,b]=lattice_MC(method,p,x)
% Computes multichannel autoregressive parameter matrix using either the
% Vieira-Morf algorithm (multi-channel generalization of the single-channel
% geometric lattice algorithm) or the Nuttall-Strand algorithm (multi-channel
% generalization of the single-channel Burg lattice algorithm).
%
%        [rho_a,a,rho_b,b] = lattice_MC(method,p,x)
%
% method -- select: 'VM' -- Vieira-Morf , 'NS' -- Nuttall-Strand
% p      -- order of multichannel AR filter
% x      -- sample data array:  x(sample #,channel #)
% rho_a  -- forward linear prediction error/white noise covariance
%           matrix
% a      -- block vector of forward linear prediction/autoregressive
%           matrix elements
% rho_b  -- backward linear prediction error/white noise covariance
%           matrix
% b      -- block vector of backward linear prediction/autoregressive
%           matrix elements
```

If a one-channel data set is processed by this function (*num_channels=1*), the multichannel function will return numerically identical values as `lattice` of Chap. 8.

15.10.4 Covariance Linear Prediction Method

Algorithms based on multichannel extensions of the single-channel least squares linear prediction algorithms of Sec. 8.5 follow in a straightforward manner. The multichannel forward linear prediction error

$$\mathbf{e}_p^f[n] = \mathbf{x}[n] + \sum_{k=1}^p \mathbf{A}_p[k]\mathbf{x}[n-k] = \underline{\mathbf{a}}_p\underline{\mathbf{x}}_p[n] \tag{15.104}$$

and the backward linear prediction error

$$\mathbf{e}_p^b[n] = \mathbf{x}[n-p] + \sum_{k=1}^p \mathbf{B}_p[k]\mathbf{x}[n-p+k] = \underline{\mathbf{b}}_p\underline{\mathbf{x}}_p[n] \tag{15.105}$$

may be defined for time indices from $n = 1$ to $n = N + p$ if the assumption that $\mathbf{x}[n] = \mathbf{0}$ outside the available data range (assumed to be $n = 1$ to $n = N$) is made. The Hermitian, nonnegative-definite squared forward and backward linear prediction error matrices

$$\boldsymbol{P}_p{}^f = \sum_{n=p+1}^{N} \mathbf{e}_p^f[n]\left(\mathbf{e}_p^f[n]\right)^H = \sum_{n=p+1}^{N} \underline{\mathbf{a}}_p \underline{\mathbf{x}}_p[n]\underline{\mathbf{x}}_p^H[n]\underline{\mathbf{a}}_p^H = \underline{\mathbf{a}}_p\underline{\mathbf{R}}_p\underline{\mathbf{a}}_p^H \quad (15.106)$$

$$\boldsymbol{P}_p{}^b = \sum_{n=p+1}^{N} \mathbf{e}_p^b[n]\left(\mathbf{e}_p^b[n]\right)^H = \underline{\mathbf{b}}_p\underline{\mathbf{R}}_p\underline{\mathbf{b}}_p^H \quad (15.107)$$

and the $(p + 1) \times (p + 1)$ Hermitian block data matrix

$$\underline{\mathbf{R}}_p = \sum_{n=p+1}^{N} \underline{\mathbf{x}}_p[n]\underline{\mathbf{x}}_p^H[n] \quad (15.108)$$

will be formed, however, by summing only over the range from $n = p + 1$ to $n = N$. This uses only the available data and corresponds to the multichannel "covariance" case of linear prediction. It uses only valid errors of nonwindowed multichannel data. The "autocorrelation" and prewindowed cases of linear prediction with index ranges of $1 \le n \le N + p$ and $1 \le n \le N$, respectively, could also be used, but the covariance case yields the best results for short data sequences.

Some multichannel matrix algebra to separately minimize scalars tr $\boldsymbol{P}_p{}^f = \sum\left(\mathbf{e}_p^f\right)^H\mathbf{e}_p^f$ and tr $\boldsymbol{P}_p{}^b = \sum\left(\mathbf{e}_p^b\right)^H\mathbf{e}_p^b$ leads to the following pair of multichannel linear prediction normal equations

$$\underline{\mathbf{a}}_p\underline{\mathbf{R}}_p = \begin{pmatrix} \boldsymbol{P}_p{}^f & \mathbf{0} & \ldots & \mathbf{0} \end{pmatrix} \quad (15.109)$$

$$\underline{\mathbf{b}}_p\underline{\mathbf{R}}_p = \begin{pmatrix} \mathbf{0} & \ldots & \mathbf{0} & \boldsymbol{P}_p{}^b \end{pmatrix}. \quad (15.110)$$

Note that $\underline{\mathbf{R}}_p$ must be positive definite (nonsingular) in order to obtain the solutions. The matrix $\underline{\mathbf{R}}_p$ may be expressed as

$$\underline{\mathbf{R}}_p = \begin{pmatrix} \mathbf{R}_{\mathbf{xx}}[0, 0] & \ldots & \mathbf{R}_{\mathbf{xx}}[0, p] \\ \vdots & & \vdots \\ \mathbf{R}_{\mathbf{xx}}[p, 0] & \ldots & \mathbf{R}_{\mathbf{xx}}[p, p] \end{pmatrix}, \quad (15.111)$$

which has entries

$$\mathbf{R}_{\mathbf{xx}}[i, j] = \sum_{n=p+1}^{N} \mathbf{x}[n - i]\mathbf{x}^H[n - j]. \quad (15.112)$$

Even though $\underline{\mathbf{R}}_p$ is not block Toeplitz, a fast algorithm that solves Eqs. (15.109) and (15.110) has been developed by the author along similar lines as the single-channel covariance algorithm in Chap. 8. As in the single-channel case, the multichannel PSD formed from the forward linear prediction coefficient matrices will yield a

PSD estimate different from that formed from the backward coefficient matrices, due to the separate unconstrained minimizations of the traces of the forward and backward squared errors that lead to $\underline{\mathbf{R}}_p$ not having a block-Toeplitz structure. This is not consistent with the stationary assumption of the underlying model. However, the spectral estimates will still be similar. The stability of the multichannel linear prediction filters is, as with the single-channel case, not guaranteed, i.e., a stable correlation function is not guaranteed.

15.10.5   Multichannel AR Spectral Evaluation

The AR spectral estimate is completed by evaluating the multichannel PSD. The forward autoregressive coefficient spectral density (15.43) or the backward autoregressive coefficient spectral density (15.61) as a function of frequency f

$$\mathbf{P}_{\mathrm{AR}}(f) = T\big[\mathbf{A}(f)\big]^{-1}P_p{}^f\big[\mathbf{A}(f)\big]^{-H} \qquad (15.113)$$

$$\mathbf{P}_{\mathrm{AR}}(f) = T\big[\mathbf{B}(f)\big]^{-1}P_p{}^b\big[\mathbf{B}(f)\big]^{-H}, \qquad (15.114)$$

for which

$$\mathbf{A}(f) = \mathbf{I} + \sum_{k=1}^{p}\mathbf{A}[k]\exp(-j2\pi f kT) \qquad (15.115)$$

$$\mathbf{B}(f) = \exp(-j2\pi f pT)\left[\mathbf{I} + \sum_{k=1}^{p}\mathbf{B}[k]\exp(j2\pi f kT)\right]. \qquad (15.116)$$

The MATLAB implementation of the multichannel autoregressive PSD valuation is found in function `ar_psd_MC.m` available through the text website:

```
function psd=ar_psd_MC(f_or_b,num_psd,Ts,Pa,a,Pb,b)
% This function computes the multi-channel AR power spectral density matrix
% over the specified number of frequency points using the forward or backward
% multi-channel linear linear prediction (LP)/AR parameter block row vectors
%
%       psd = ar_psd_MC(f_or_b,num_psd,Ts,Pa,a,Pb,b)
%
% f_or_b  -- select: 1 -- forward PSD computation, 2 -- backward PSD computation
% num_psd -- number of frequency samples in spectrum
% Ts      -- sample interval in appropriate units (seconds, meters, etc.)
% Pa      -- forward linear prediction/autoregressive white noise variance
% a       -- block vector of forward linear prediction / AR parameters
% Pb      -- backward linear prediction/autoregressive white noise variance
% a       -- block vector of backward linear prediction / AR parameters
% psd     -- block column vector of multi-channel PSD matrices
```

Note that selection of `f_or_b` for either forward or backward will yield the same numerical solution if either `yule_walker_MC` or `lattice_MC` algorithms are used.

15.11 AUTOREGRESSIVE ORDER SELECTION

An aid in the determination of the order of the multichannel AR model is the multichannel version of the Akaike AIC criterion [Jones, 1974; Nuttall, 1976b], which has the form

$$\text{AIC}[p] = N \ln(\det \boldsymbol{P}_p{}^{f}) + 2m^2 p = M \ln(\det \boldsymbol{P}_p{}^{b}) + 2m^2 p. \qquad (15.117)$$

The order to choose is the one for which $\text{AIC}[p]$ is a minimum. A suggested upper bound for the order is $3\sqrt{N}/m$ [Nuttall, 1976b], where m is the number of channels and N is the number of data points per channel. In general, the order selection procedure for multichannel AR analysis has not been researched in depth, so that the $\text{AIC}[p]$ must be used only as a guideline for the approximate order.

15.12 EXPERIMENTAL COMPARISON OF MULTICHANNEL AUTOGRESSIVE PSD ESTIMATORS

Analytical characterization of the performance of the various multichannel AR PSD estimators is mathematically intractable, so one must resort to experimental approaches to yield empirical characterizations. Two types of two-channel data sets are employed here to characterize performance. One is an actual first-order two-channel autoregression. The second is a single-channel process with tones embedded in a colored noise process similar to the single-channel test case.

In the first test, a first-order ($p = 1$) two-channel autoregressive process was used to generate data ensembles. Data were generated according to the recursion

$$\mathbf{x}[n] = -\mathbf{A}[1]\mathbf{x}[n-1] + \mathbf{w}[n], \qquad (15.118)$$

in which the autoregressive coefficient matrix is

$$\mathbf{A}[1] = \begin{pmatrix} -0.85 & +0.75 \\ -0.65 & -0.55 \end{pmatrix} \qquad (15.119)$$

and $\mathbf{w}[n]$ is a white two-channel process, uncorrelated between channels and of unit variance in each channel. Even though this is a first-order autoregressive process, the autospectra of each channel will have four poles and three zeros. Figure 15.2 depicts the true autospectra and coherence for this process. The autospectra for each of the two channels and the MSC will have peaks at a fraction of sampling frequency of 0.12295. The phase of the complex coherence (cross spectrum) is $90°$ at this peak frequency and $0°$ at frequencies 0.0 and 0.5. The peak magnitude of the squared coherence is 0.999013 at this peak frequency. Twenty sequences, each of 100 sample points, were processed by each of the three AR algorithms,

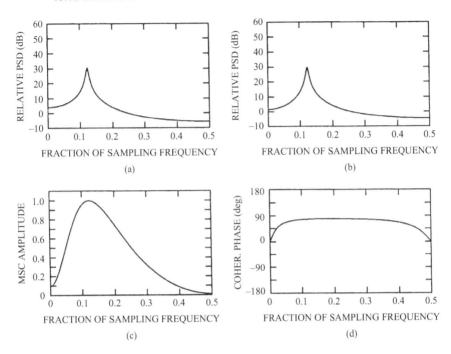

Figure 15.2. True spectra of AR(1) test case. (a) Channel 1 autospectrum. (b) Channel 2 autospectrum. (c) Magnitude squared coherence. (d) Phase of coherence.

with the order p restricted to be unity. The 20 spectral estimates were overlapped, as shown in Figs. 15.3–15.6. The autospectra were plotted with respect to the same reference value of 1500. Of the four sets of plots, the Nuttall-Strand and backward linear prediction algorithms had the least variance in PSD amplitude, while the Vieira-Morf and forward linear prediction algorithms had the least variance of the spectral peak in frequency. Otherwise, all four methods yielded very similar spectral estimates.

The next test consisted of data sequences with three sinusoids in a colored noise process. In channel 1, sinusoids with fractions of the sampling frequency of 0.1, 0.2, and 0.24 were used. The respective amplitudes were 0.1, 1, and 1, while the respective initial phases were $0°$, $90°$, and $235°$. The colored noise process, which had most of its power above the frequencies of the sinusoids, was generated by passing white noise of 0.05 variance through a digital filter with a raised-cosine spectral response between 0.2 and 0.5, identical to the process used for the test sequence test1987. In channel 2, sinusoids of fractional frequencies 0.1, 0.2, 0.4, amplitudes of 0.1, 1, 1, and initial phases of $0°$, $210°$, $25°$ were generated. A colored noise process similar to channel 1, but independently generated, was added to the sinusoids. Note that the sinusoidal components at fractional frequencies 0.1 and 0.2 are common to both channels. A data sequence of 64 points was generated for each channel. Figures 15.7–15.9 depict the autospectra, coherence magnitude,

358   MULTICHANNEL SPECTRAL ESTIMATION

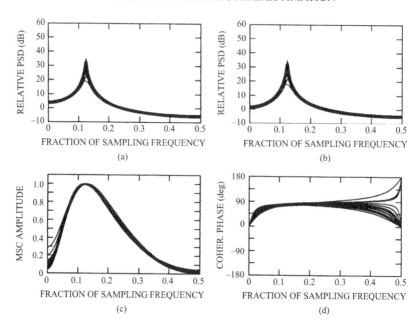

Figure 15.3. Nuttall-Strand algorithm AR(1) estimate. (a) Channel 1 autospectrum. (b) Channel 2 autospectrum. (c) Magnitude squared coherence. (d) Phase of coherence.

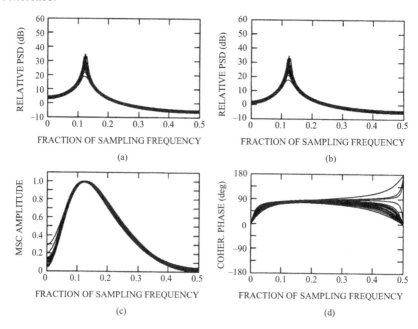

Figure 15.4. Vieira-Morf algorithm AR(1) estimate. (a) Channel 1 autospectrum. (b) Channel 2 autospectrum. (c) Magnitude squared coherence. (d) Phase of coherence.

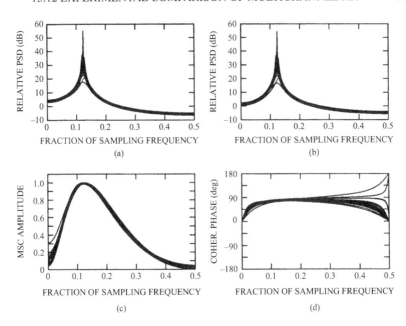

Figure 15.5. Covariance algorithm AR(1) estimate (forward coefficients). (a) Channel 1 autospectrum. (b) Channel 2 autospectrum. (c) Magnitude squared coherence. (d) Phase of coherence.

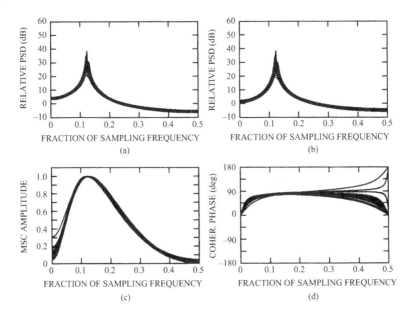

Figure 15.6. Covariance algorithm AR(1) estimate (backward coefficients). (a) Channel 1 autospectrum. (b) Channel 2 autospectrum. (c) Magnitude squared coherence. (d) Phase of coherence.

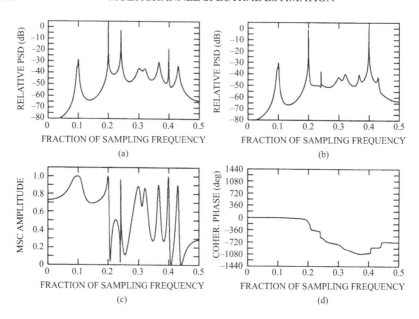

Figure 15.7. Nuttall-Strand algorithm AR(12) estimate for 64 samples. (a) Channel 1 autospectrum. (b) Channel 2 autospectrum. (c) Magnitude squared coherence. (d) Phase of coherence.

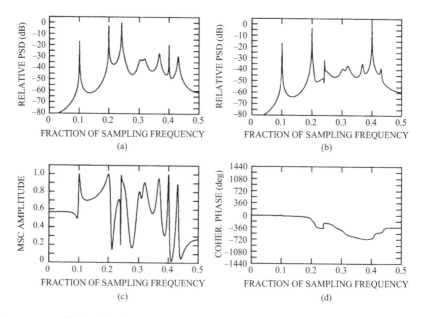

Figure 15.8. Vieira-Morf algorithm AR(12) estimate for 64 samples. (a) Channel 1 autospectrum. (b) Channel 2 autospectrum. (c) Magnitude squared coherence. (d) Phase of coherence.

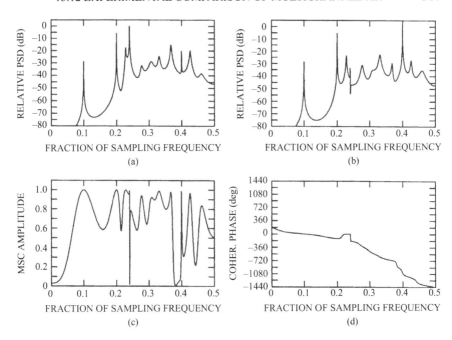

Figure 15.9. Covariance algorithm AR(12) estimate for 64 samples (forward coefficients). (a) Channel 1 autospectrum. (b) Channel 2 autospectrum. (c) Magnitude squared coherence. (d) Phase of coherence.

and coherence phase for order $p = 12$ estimates by the three autoregressive methods. Note that in all plots, some energy of the 0.24 fractional frequency sinusoid, which should be present only in channel 1, has coupled into the autospectral estimate for channel 2 and, conversely, the 0.4 fractional frequency sinusoid, which should be present only in channel 2, has coupled into the autospectral estimate for channel 1. The MSC plots also show spikes at the tone frequencies 0.24 and 0.4, where, ideally, the coherence should be zero. This "feed-across" artifact is an undesirable effect; however, it is common to all the methods presented in this chapter and seems unavoidable whenever processing short data sequences with multichannel AR algorithms [see also Fougere, 1981]. If the data sequences from each channel are processed separately by the single-channel AR methods, such as the Burg algorithm, the autospectra do not give a false tonal indication, as illustrated in Fig. 15.10. By increasing the data sequence length to 500 samples, there is an improvement in the feed-across effect in the multichannel autospectra, as shown in Fig. 15.11, although the coherence still has sharp spikes near unity. Thus it seems that the attractive high-resolution properties of the single-channel linear prediction spectrum analysis when processing short data records do not extend to the multichannel case without problems. The coupling effect is even more pronounced for more than two channels. Although the cross spectral PSD entries are for two channels, the data from all channels contributes to the individual two-channel cross

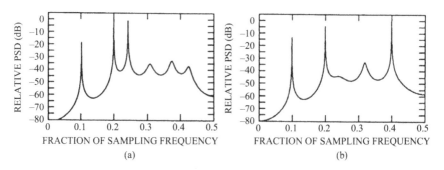

Figure 15.10. Burg single-channel algorithm AR(12) estimate for 64 samples. (a) Channel 1 autospectrum. (b) Channel 2 autospectrum.

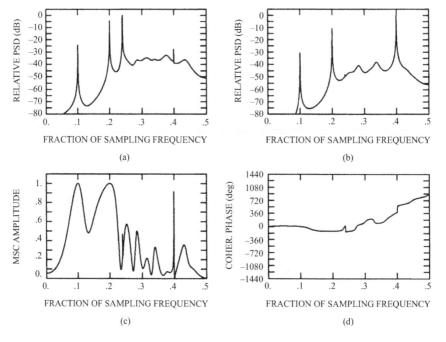

Figure 15.11. Nuttall-Strand algorithm AR(12) estimate for 500 samples. (a) Channel 1 autospectrum. (b) Channel 2 autospectrum. (c) Magnitude squared coherence. (d) Phase of coherence.

spectral estimates. The feed-across effect leads to false indications of narrowband components, especially when processing short data sequences. A detailed analysis [Marple and Nuttall, 1983] of the cause of the feed-across effect has shown it is due to inexact pole-zero cancellation in the multichannel PSD estimate. This example emphasizes the need to confirm details seen in the autospectra generated by a multichannel algorithm with independent single-channel autospectral estimates.

The spectral line-splitting phenomenon observed in single-channel linear prediction spectrum analysis has also been observed under similar tone conditions in some multichannel algorithms, particularly the Nuttall-Strand and Vieira-Morf algorithms. It has not been observed in the multichannel covariance algorithm.

15.13   MULTICHANNEL MINIMUM VARIANCE SPECTRAL ESTIMATION

Similar arguments as those used in Chap. 12 could be used here to develop a multichannel minimum variance spectral estimator, in which the estimator represents the variance at the output of an m-channel finite impulse response filter with filter coefficients constrained to pass a sinusoid at a given frequency without distortion and to attenuate all other spectral components. The resultant multichannel MV PSD estimator will be found to have the form [Lacoss, 1977]

$$\hat{\mathbf{P}}_{MV}(f) = T \left[\underline{\mathbf{e}}_p(f) \underline{\mathbf{R}}_p^{-1} \underline{\mathbf{e}}_p^H(f) \right]^{-1}, \tag{15.120}$$

in which the block complex sinusoid vector $\underline{\mathbf{e}}_p(f)$ is defined in (15.31) and $\underline{\mathbf{R}}_p^{-1}$ is the inverse of the block-Toeplitz matrix defined in (15.55).

One could use an estimate $\hat{\mathbf{R}}_p$ of the block correlation matrix, such as that of (15.87), and then evaluate Eq. (15.120) over a range of frequencies, yielding the MV PSD estimate. However, an alternative approach based on a relation of the correlation matrix inverse to the multichannel autoregressive parameters can greatly reduce the computational effort needed to evaluate multichannel MV PSD function (15.120). It can be shown that an equivalent representation of the multichannel MV PSD is

$$\hat{\mathbf{P}}_{MV}(f) = T \left[\sum_{k=-p}^{p} \Psi[k] \exp(-j2\pi f kT) \right]^{-1}, \tag{15.121}$$

in which the set of $p \times p$ matrices $\Psi[k]$ are calculated as the correlation of the multichannel AR parameter matrices as follows:

$$\Psi[k] = \begin{cases} \sum_{i=0}^{p-k} \left\{ (p+1-k-i)\mathbf{A}_p^H[k+i](\mathbf{P}_p{}^f)^{-1}\mathbf{A}_p[i] \right. \\ \left. -i\,\mathbf{B}_p^H[k+i](\mathbf{P}_p{}^b)^{-1}\mathbf{B}_p[i] \right\} & \text{for } 0 \le k \le p \\ \Psi^H[-k] & \text{for } -1 \le k \le -p. \end{cases} \tag{15.122}$$

The MATLAB implementation of the multichannel minimum variance PSD technique is found in function `minimum_variance_psd_MC.m` available through the text website:

```
function psd=minimum_variance_psd_MC(num_psd,p,T,x)
% This function computes the multichannel minimum variance spectral estimate
```

```
%
%          psd = minimum_variance_psd_MC(num_psd,p,T,x)
%
% num_psd  -- number of frequency samples in PSD (must be power of two)
% p        -- order of minimum variance filter (number of taps = p + 1)
% T        -- sample interval in seconds
% x        -- sample data array:  x(sample #,channel #)
% psd      -- block vector of 'num_psd' multichannel PSD matrix elements
```

15.14 TWO-CHANNEL SPECTRAL ANALYSIS OF SUNSPOT NUMBERS AND AIR TEMPERATURE

Using two-channel spectral analysis, correlative effects have been shown by analysis to exist between sunspot numbers and variations in terrestrial phenomena, such as sea level (tides), surface air temperature, earth rotation interval, air pressure, magnetic declination, and drought or flood cycles [Currie, 1981, 1984]. The possible relationship between sunspot numbers and surface air temperature at one terrestrial location will be investigated here. Appendix III lists the monthly mean temperature in °C of the American city of St. Louis for the years 1845 through 1978. This data was provided courtesy of Dr. R. Currie, who collected and performed an exhaustive analysis of worldwide temperature records. A preliminary periodogram analysis of the 1608 monthly temperature samples is shown in Fig. 15.12(a). The periodogram was computed by averaging the sample periodograms of six segments of 400 samples each with 200 samples of overlap between segments (8 samples were not used). A large DC component and a 1-cycle per year component, due to the seasonal variation of the temperature, are clearly indicated. A weak seasonal harmonic at 2 cycles per year is also just detectable. These dominant components must be eliminated before a two-channel analysis with the filtered sunspot numbers

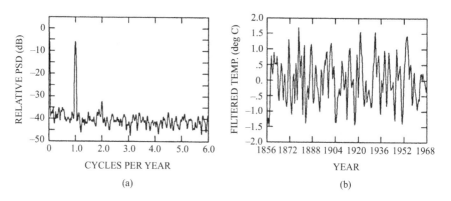

Figure 15.12. (a) Periodogram of unfiltered St. Louis monthly mean temperatures. (b) Temperature data after filtering and decimation by a factor of six.

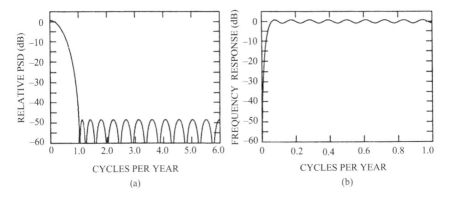

Figure 15.13. Frequency response of temperature data filters. (a) 25-term linear-phase low-pass filter. (b) 37-term linear-phase high-pass filter.

can be performed. Low-pass filtering, sample rate decimation by a factor of 6, and high-pass filtering yielded the filtered temperature data plotted in Fig. 15.12(b). The linear-phase filter responses are shown in Fig. 15.13. Both filters were designed with the Parks-McClellan method. The low-pass filter is 25 samples in duration. It filters out the 1- and 2-cycle per year seasonal periods. The high-pass filter is 37 samples in duration. It attentuates the DC component by 60.45 dB. A total of 227 filtered temperature samples is generated at the rate of two samples per year, from July 1855 to July 1968.

The result of a two-channel periodogram analysis of the 227 samples of filtered temperature and sunspot data is shown in Fig. 15.14. The autospectra are the usual single-channel periodograms. Five Hamming windowed segments of 75 samples each with 38 samples of overlap were used. The magnitude squared coherence and coherence phase were computed as described by Eqs. (15.9) and (15.10), using a cross spectral periodogram of the same five segments. No significant features are apparent in the temperature spectrum of Fig. 15.14(b). The two peaks near the low-frequency end of Fig. 15.14(c) are located at 0.077 and 0.253 cycle per year (12.96 and 3.95 years per cycle) and have coherencies of 0.73 and 0.83, respectively. Thus, it would appear that periodogram analysis does not indicate a strong correlation between sunspot and temperature data.

The result of a two-channel autoregressive analysis by the Nuttall-Strand algorithm is shown in Fig. 15.15. Order 25 was selected. Similar results were obtained with the Vieira-Morf and covariance algorithms. Both autospectra have peaks at the identical frequency of 0.0942 cycle per year (10.61 years per cycle). The MSC indicates peaks at 0.0938 and 0.2471 cycle per year (10.67 and 4.05 years per cycle) with coherences of 0.70 and 0.64, respectively. It seems possible, therefore, that there is a correlation between the dominant sunspot cycle and a component in the temperature data. The coherence phase at 0.0938 cycle per year is -186.5 degrees, which means the modulated component in the temperature data lags the sunspot cycle

366   MULTICHANNEL SPECTRAL ESTIMATION

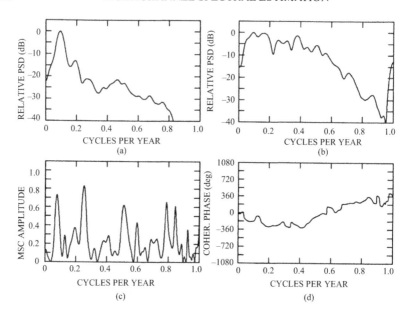

Figure 15.14. Two-channel periodogram analysis of sunspot and temperature data. (a) Autospectrum of filtered sunspot data. (b) Autospectrum of filtered temperature data. (c) Magnitude squared coherence. (d) Phase of coherence (unwrapped).

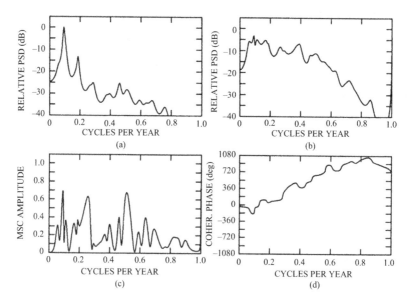

Figure 15.15. Two-channel Nuttall-Strand AR(25) analysis of sunspot and temperature data. (a) Autospectrum of filtered sunspot data. (b) Autospectrum of filtered temperature data. (c) Magnitude squared coherence. (d) Phase of coherence (unwrapped).

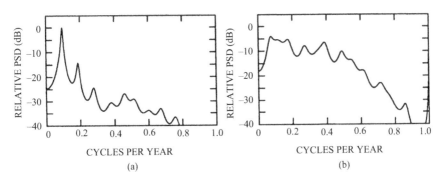

Figure 15.16. Single-channel Burg AR(25) analysis. (a) Autospectrum of filtered sunspot data. (b) Autospectrum of filtered temperature data.

by 5.52 years. A single-channel AR(25) analysis by the Burg algorithm, shown in Fig. 15.16, serves as a check on the two-channel AR analysis. The sharp peak seen in Fig. 15.15(b) does not appear in the single-channel spectrum of Fig. 15.16(b). This suggests that the correlative peaks at 0.0938 cycle per year may be an artifact of the two-channel AR analysis.

References

Akaike, H., Block Toeplitz Matrix Inversion, *SIAM J. Appl. Math.*, vol. 24, pp. 334–341, March 1973.

Currie, R. G., Solar Cycle Signal in Air Temperature in North America: Amplitude, Gradient, Phase, and Distribution, *J. Atmos. Sci.*, vol. 38, pp. 808–818, April 1981.

Currie, R. G., Periodic (18.6-Years) and Cyclic (11-Year) Induced Drought and Flood in Western North America, *J. Geophys. Res.*, vol. 89, pp. 7215–7230, August 1984.

Fougere, P. F., Spontaneous Line Splitting in Multichannel Maximum Entropy Power Spectra, *Proceedings of First ASSP Workshop on Spectral Estimation*, McMaster University, Hamilton, Ontario, Canada, pp. 1.6.1–1.6.3, August 1981.

Friedlander, B., Lattice Methods for Spectral Estimation, *Proc. IEEE*, vol. 70, pp. 990–1017, September 1982.

Friedlander, B., M. Morf, T. Kailath, and L. Ljung, New Inversion Formulas for Matrices Classified in Terms of Their Distance from Toeplitz Matrices, *Linear Algebra and Its Applications*, vol. 27, pp. 31–60, 1979.

Friedlander, B., and B. Porat, Multichannel ARMA Spectral Estimation by the Modified Yule-Walker Method, *Signal Process.*, vol. 10, pp. 49–59, 1986.

Jones, R. H., Identification and Autoregressive Spectrum Estimation, *IEEE Trans. Autom. Control*, vol. AC-19, pp. 894–897, December 1974.

Lacoss, R. T., Autoregressive and Maximum Likelihood Spectral Analysis Techniques, in *Aspects of Signal Processing*, part 2, G. Tucconi, ed., pp. 591–615, Reidel, Boston, Mass., 1977.

Lee, D., B. Friedlander, and M. Morf, Recursive Ladder Forms for ARMA Modeling, *IEEE Trans. Autom. Control*, vol. AC-27, pp. 753–764, August 1982.

Marple, S. L., Jr. and A. H. Nuttall, Experimental Comparison of Three Multichannel Linear Prediction Spectral Estimators, *IEE Proc.*, vol. 130, Part F, pp. 218–229, April 1983.

Morf, M., A. Vieira, D. T. Lee, and T. Kailath, Recursive Multichannel Maximum Entropy Spectral Estimation, *IEEE Trans. Geosci. Electron.*, vol. GE-16, pp. 85–94, April 1978a.

Morf, M., A. Vieira, D. T. Lee, and T. Kailath, Covariance Characterization by Partial Autocorrelation Matrices, *Ann. Stat.*, vol. 6, pp. 643–648, May 1978b.

Nuttall, A. H., FORTRAN Program for Multivariate Linear Predictive Spectral Analysis, Employing Forward and Backward Averaging, *Naval Underwater Systems Center Technical Report 5419*, New London, Conn., May 1976a.

Nuttall, A. H., Multivariate Linear Predictive Spectral Analysis Employing Weighted Forward and Backward Averaging: A Generalization of Burg's Algorithm, *Naval Underwater Systems Center Technical Report 5501*, New London, Conn., October 1976b.

Nuttall, A. H., Positive Definite Spectral Estimate and Stable Correlation Recursion for Multivariate Linear Predictive Spectral Analysis, *Naval Underwater Systmes Center Technical Report 5729*, New London, Conn., November 1977.

Nuttall, A. H., Two-Channel Linear-Predictive Spectral Analysis: Program for the HP 9845 Desk Calculator, *Naval Underwater Systems Center Technical Report 6533*, New London, Conn., October 1981.

Robinson, E. A., *Multichannel Time Series Analysis with Digital Computer Programs*, 2nd ed., Goose Pond Press, Houston, Tex., 1983.

Strand, O. N., Multichannel Complex Maximum Entropy (Autoregressive) Spectral Analysis, *IEEE Trans. Autom. Control*, vol. AC-22, pp. 634–640, August 1977.

Watson, G. A., An Algorithm for the Inversion of Block Matrices of Toeplitz Form, *J. Assoc. Comput. Mach.*, vol. 20, pp. 409–415, July 1973.

Whittle, P., On the Fitting of Multivariate Autoregressions, and the Approximate Canonical Factorization of a Spectral Density Matrix, *Biometrika*, vol. 50, pp. 129–134, 1963.

Wiggins, R. A., and E. A. Robinson, Recursive Solution to the Multichannel Filtering Problem, *J. Geophys. Res.*, vol. 70, pp. 1885–1891, April 1965.

16

TWO-DIMENSIONAL SPECTRAL ESTIMATION

16.1 INTRODUCTION

The high-resolution performance of the one-dimensional (1-D) spectal estimators of Chaps. 5–13 has led many researchers to expect similar performance for two-dimensional (2-D) and higher-dimensional extensions of these spectral estimators, where these extensions exist. Two-dimensional spectral analysis can be used for space-space data arrays (e.g., image processing), space-time data arrays (e.g., sonar, seismic, and synthetic aperture radar processing), or even time-time data arrays (e.g., analysis of radar pulse repetition interval vs. arrival time). Difficulties extending the 1-D techniques to 2-D are often encountered due to differences in the 1-D and 2-D linear systems theory, particularly the inability to factor a 2-D polynomial into polynomials of lower degress. The 1-D concept of isolated poles, zeros, and roots thus cannot be extended, in general, to the 2-D case. Two-dimensional systems have many more degrees of freedom than 1-D systems. The computational burden imposed by some of the advanced 2-D spectral estimation methods have limited their testing and application to small 2-D data sets with simple signal scenarios, such as a few sinusoids in spatially white noise. These two reasons explain why 2-D spectral estimation continues to be an active area of research. Only the 2-D periodogram, 2-D hybrid methods, and 2-D minimum variance method have seen practical application to extensive 2-D data sets, primarily due to the fact that these methods do not depend on 2-D polynomial factorization.

A single chapter can only briefly highlight 2-D spectral estimation techniques. Although the mathematics in this chapter could be expressed for multidimensional spectral analysis, the greatest interest currently is 2-D spectral analysis. It is for this reason that this chapter focuses only on the 2-D case. The issues and mathematical details caused by the significant theoretical differences between 1-D and 2-D spectral techniques require an entire book unto itself. The reader is invited to consult resources, such as the text by Dudgeon and Mersereau [1984] or the tutorial paper by McClellan [1982], for additional details and references on multidimensional digital signal processing and spectral estimation. The developments in this chapter will roughly parallel the topical sequence of Chaps. 2–13 on 1-D techniques.

16.2 SUMMARY

It is assumed that 2-D signals are obtained by sampling over a rectangular grid for both mathematical convenience and ease of presentation. This is the way 2-D discrete signals are usually obtained in practice. Limitations of space permit only two 2-D spectrum analysis programs to be included in this chapter: the 2-D classic periodogram spectral estimator and the 2-D quarter-plane autoregressive spectral estimator. Figure 16.1 is a flowchart of the steps needed to produce spectra from the two programs and the 2-D minimum variance technique. Examples of 2-D spectra are shown in Fig. 16.2. The two methods selected show the potential resolution improvement over classical 2-D spectral estimators.

Following a review of 2-D linear system, Fourier transform, and random process theory in Secs. 16.3 and 16.4, five techniques of 2-D spectral estimation are covered in the remaining sections. Section 16.5 discusses classical 2-D spectral estimators. Section 16.6 demonstrates how the separability of the 2-D exponential may be used to develop a modified classical spectral estimator that is a standard FFT in one dimension and a high-resolution 1-D spectral estimator in the other dimension. This permits all of the 1-D techniques previously developed to be used in a hybrid 2-D spectral method. This approach is especially useful in time-space data, where there is often a wealth of time data and a paucity of space data. Section 16.7 is a lengthy discourse on 2-D autoregressive spectral estimation. Section 16.8 summarizes the 2-D maximum entropy PSD estimator, which is not identical to the 2-D AR spectral estimator. Finally, Sec. 16.9 highlights the 2-D minimum variance spectral estimator.

```
• ACQUIRE DATA RECORD
    Total Number of Samples in Record
    Number of Channels
    Sampling Interval (sec/sample)

• SELECT METHOD & ITS PARAMETERS
    Periodogram: 2-D segment size, 2-D overlap, 2-D window
    Correlogram: 2-D correlation max lag each dimension
    Yule-Walker AR: model order each dimension
    Lattice AR: model order each dimension
    Minimum Variance Adaptive: filter order each dimension

• ESTIMATE DIRECT PSD or PARAMETERS
    peridogoram_psd_2D.m, or [direct PSD]
    correlogram_psd_2D.m, or [direct PSD]
    yule_walker_2D.m, or
    lattice_2D.m, or
    minimum_variance_2D.m [direct PSD]

• COMPUTE 2-D AR PSD
    Function ar_psd_2D.m
```

Figure 16.1. Flowchart of five approaches to 2-D spectral estimation.

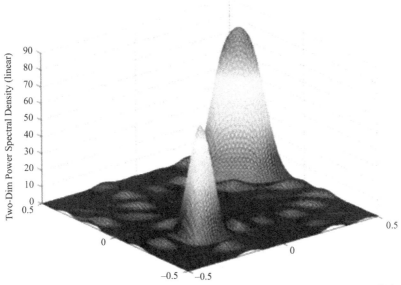

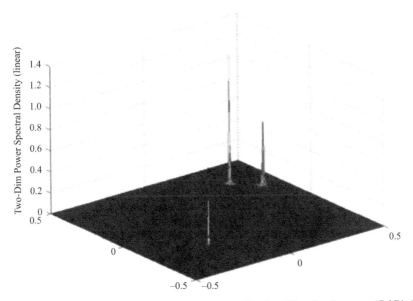

Figure 16.2. Illustration of two 2-D spectral estimates based on known 2-D auto-correlation sequence of three complex 2-D sinusoids in white noise. Top: 2-D correlogram method. Bottom: 2-D autoregressive method.

Only very limited experimental results have been reported in the literature regarding 2-D spectral estimators, so performance comparisons are difficult to make. The high complexity of computations, even with fast algorithms, have limited application of 2-D PSD estimators to small data sets with only a few 2-D parameters. Five 2-D spectral estimation techniques have been implemented in MATLAB software and may be evaluated and plotted using the demonstration script spectrum_demo_2D.m available through the text website. Two simple data sets are incorporated into the script. One is an 8 by 8 data array structured from the 64 samples of the 1-D doppler_radar data record. The second is an 8 by 8 array consisting of three complex sinusoidals in white noise. Two of the sinusoids are close in 2-D frequency as a test of the spectral resolution capability.

16.3   TWO-DIMENSIONAL LINEAR SYSTEMS AND TRANSFORM THEORY

A *discrete 2-D signal* is any function $x[m, n]$ that is a 2-D sequence, or *array*, of real or complex numbers defined for the ordered pair of integers m and n over $-\infty < m, n < \infty$. The discrete 2-D signal may represent a sampled 2-D continuous function $x(t_1, t_2)$, in which t_1 and t_2 range over a continuum of values. Although many different 2-D sampling grids of the continuous function $x(t_1, t_2)$ are possible, only the rectangular grid

$$x[m, n] = x(mT_1, nT_2) \tag{16.1}$$

is considered in this chapter. The sampling intervals along each dimension, denoted as T_1 and T_2, may be in units of time or space. The range of m and n indices over which the 2-D sequence is nonzero is called the *region of support*. A typical finite data record is assumed in this chapter to have a rectangular region of support, $0 \leq m \leq M - 1$ and $0 \leq n \leq N - 1$, as this is the most common situation found in practice.

The 2-D *unit impulse* (sample) *function* has the definition

$$\delta[m, n] = \delta[m]\delta[n] = \begin{cases} 1 & \text{for } m = n = 0 \\ 0 & \text{otherwise} \end{cases}. \tag{16.2}$$

The 2-D *discrete convolution sum* for a 2-D linear shift invariant system is

$$y[m, n] = \sum_{i=-\infty}^{\infty} \sum_{j=-\infty}^{\infty} h[m - i, n - j]x[i, j] = h[m, n] \star x[m, n] \tag{16.3}$$

for input $x[i, j]$ and output $y[m, n]$. The 2-D sequence $h[m, n]$, defined over an infinite range in both dimensions, is the *2-D system impulse response sequence*, obtained as the output sequence when the input is $x[i, j] = \delta[i, j]$. As in the 1-D case, a 2-D linear shift-invariant system is said to be *stable* if its output sequence remains bounded for all bounded input sequences. The stability of 2-D systems is significantly more difficult to test than 1-D stability.

A 1-D linear system was characterized as causal if its output $y[n]$ was based on a convolution sum involving only the current and past inputs $x[k]$ to the system

$$y[n] = \sum_{k=-\infty}^{n} h[n-k]x[k] \qquad (16.4)$$

for n, $n+1$, $n+2$, and so forth. The 1-D discrete impulse response sequence is nonzero only for $h[0], h[1], \ldots, h[\infty]$. This made physical sense when the signals were functions of time. Causality is not inherent, however, for 2-D signals in which one or both dimensions of the signal are functions of space rather than time. One region of support for the 2-D discrete impulse response sequences that has been adopted as a convention for a "causal" 2-D system is the *nonsymmetric half plane* (NSHP), as illustrated in Fig. 16.3(a). This definition of 2-D causality assumes that the output sequence $y[m, n]$ is calculated in a manner analogous to that of a raster on a television: scan line by line top to bottom, moving left to right on each line. If the output $y[m, n]$ is visualized as position n on line m, then the output is generated recursively by first computing $y[m, -\infty]$ to $y[m, \infty]$, followed by incrementing line index m. Note that both input and output sequences may have support over the infinite *full plane* (FP), even though the impulse response has NSHP support.

A special case region of support for causal 2-D systems confines the 2-D impulse response to the *quarter plane* (QP), as illustrated by Fig. 16.3(b). The QP case will be the region of support of most interest for purposes of 2-D spectral analysis. Two other regions of support for the 2-D discrete impulse response are illustrated in Fig. 16.3(c) and (d). The semicausal region of support covers the *symmetric half plane* (SHP). It is causal in one dimension and noncausal in the other dimension. The noncausal region of support is simply the entire *full plane* (FP), as assumed by Eq. (16.3).

A 2-D recursive difference equation relating the input to the output of a 2-D system is

$$\sum_i \sum_j a[i, j]x[m-i, n-j] = \sum_i \sum_j b[i, j]u[m-i, n-j]. \qquad (16.5)$$

The assumption $a[0, 0] = 1$ can be made with no loss in generality, so that Eq. (16.5) can be rewritten as

$$x[m, n] = -\sum_i \sum_j a[i, j]x[m-i, n-j] + \sum_i \sum_j b[i, j]u[m-i, n-j].$$
$$(16.6)$$

The range of the various summations denote the *order* of the 2-D difference equation. The order of array $a[i, j]$, for example, would be $p_1 \times p_2$ if its region of support were the rectangular QP covering $0 \le i \le p_1$ and $0 \le j \le p_2$. Selecting an NSHP

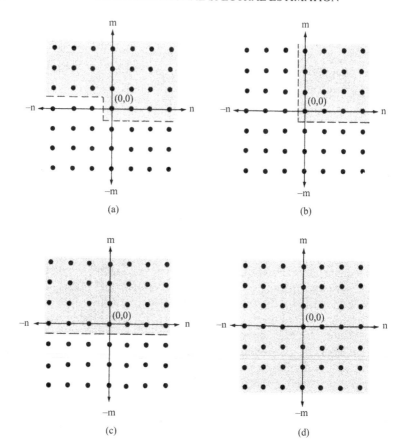

Figure 16.3. Regions of support for discrete system sequence $h[m, n]$. (Nonzero in shaded area.) (a) Causal (NSHP). (b) Causal (QP). (c) Semicausal (SHP). (d) Noncausal (FP).

region of support for both array $a[i, j]$ and array $b[i, j]$ and the appropriate initial conditions for $x[i, j]$ can be shown [Dudgeon and Mersereau, 1984] to create a causal 2-D output sequence, i.e., the discrete impulse response describing Eq. (16.3) also has an NSHP region of support and can, therefore, be computed recursively in the raster fashion previously described.

Exponentials of the form $x[m, n] = z_1^m z_2^n$ are eigenfunctions of 2-D linear shift-invariant systems because the output response $y[m, n]$ to $z_1^m z_2^n$ as an input to the convolution of Eq. (16.3) yields

$$y[m, n] = \mathsf{H}(z_1, z_2) z_1^m z_2^n, \tag{16.7}$$

in which

$$\mathsf{H}(z_1, z_2) = \sum_{i=-\infty}^{\infty} \sum_{j=-\infty}^{\infty} h[i, j] z_1^{-i} z_2^{-j} = \mathcal{Z}\{h[i, j]\} \tag{16.8}$$

is the *discrete 2-D system response function*, or the *discrete 2-D transfer function* of the 2-D discrete impulse response sequence, defined in the region of z_1 and z_2 such that Eq. (16.8) converges. Equation (16.8) is also the general definition of the *2-D z-transform* for any 2-D sequence $h[i, j]$. The 2-D system function of the recursive difference equation (16.5) is easily shown to be the ratio of 2-D polynomials given by

$$H(z_1, z_2) = \frac{B(z_1, z_2)}{A(z_1, z_2)}, \tag{16.9}$$

in which $A(z_1, z_2)$ and $B(z_1, z_2)$ are the z-transforms of the $a[i, j]$ and $b[i, j]$ arrays

$$A(z_1, z_2) = \sum_i \sum_j a[i, j] z_1^{-i} z_2^{-j} \tag{16.10}$$

$$B(z_1, z_2) = \sum_i \sum_j b[i, j] z_1^{-i} z_2^{-j} \tag{16.11}$$

over summation ranges determined by the respective regions of support. A causal 1-D rational system function involved summation ranges of *only* positive index values. The 2-D system function (16.9) for the "causal" NSHP will involve one summation in both (16.10) and (16.11) to use only positive index values and the other summation to have both positive and negative index values. It is said that $H(z_1, z_2)$ has a 2-D "zero" at (z_1, z_2) if $B(z_1, z_2) = 0$ and $A(z_1, z_2) \neq 0$. It is said that $H(z_1, z_2)$ has a 2-D "pole" at (z_1, z_2) if $A(z_1, z_2) = 0$ and $B(z_1, z_2) \neq 0$. Poles and zeros of 2-D rational system functions are quite different from their 1-D counterparts. Roots of 1-D polynomials occur at isolated points in the z-plane, whereas roots of 2-D polynomials are generally continuous contours, rather than isolated points. Consider the simple 2-D polynomial

$$A(z_1, z_2) = 1 - a z_1^{-1} z_2^{-1}, \tag{16.12}$$

which is zero wherever $z_1 z_2 = a$. This is a continuous surface in a 2-D complex z-space. The inability to factor 2-D polynomials into isolated poles and zeros (i.e., first-order polynomials) makes stability testing in 2-D very difficult.

A continuous 2-D signal $x(t_1, t_2)$ has a *2-D continuous-time Fourier transform* (2-D CTFT)

$$X_{CTFT}(f_1, f_2) = \int_{-\infty}^{\infty} \int_{-\infty}^{\infty} x(t_1, t_2) \exp(-j2\pi[f_1 t_1 + f_2 t_2]) \, dt_1 \, dt_2 \tag{16.13}$$

under similar conditions cited for the 1-D CTFT in Chap. 2. If the continuous signal is sampled at intervals of T_1 along the dimension of variable t_1 and T_2 along the dimension of variable t_2, creating the sampled sequence $x[m, n] = x(mT_1, nT_2)$, then the *2-D discrete-time Fourier transform* (2-D DTFT) is given by

$$X_{\text{DTFT}}(f_1, f_2) = T_1 T_2 \, \mathsf{X}(z_1, z_2)\Big|_{\substack{z_1 = \exp(j2\pi f_1 T_1) \\ z_2 = \exp(j2\pi f_2 T_2)}} \tag{16.14}$$

$$= T_1 T_2 \sum_{m=-\infty}^{\infty} \sum_{n=-\infty}^{\infty} x[m, n] \exp(-j2\pi[f_1 m T_1 + f_2 n T_2]). \tag{16.15}$$

If the continuous 2-D signal is temporally/spatially band-limited, such that

$$X_{\text{CTFT}}(f_1, f_2) = 0 \tag{16.16}$$

outside the rectangular region $|f_1| \leq 1/2T_1$ and $|f_2| \leq 1/2T_2$, then

$$X_{\text{DTFT}}(f_1, f_2) = X_{\text{CTFT}}(f_1, f_2) \tag{16.17}$$

only over the region of support $|f_1| \leq 1/2T_1$ and $|f_2| \leq 1/2T_2$. The DTFT is periodic, $X_{\text{DTFT}}(f_1 + \alpha/2T_1, f_2 + \beta/2T_2) = X_{\text{DTFT}}(f_1, f_2)$, for any integers α and β. The original continuous 2-D signal may be recovered from the sampled 2-D sequence in this case by the 2-D interpolation function

$$x(t_1, t_2) = \sum_{m=-\infty}^{\infty} \sum_{n=-\infty}^{\infty} x[m, n] \operatorname{sinc}[(t_1 - m T_1)/T_1] \operatorname{sinc}[(t_2 - n T_2)/T_2], \tag{16.18}$$

a consequence of the 2-D sampling theorem.

The squared magnitude of the 2-D discrete-time Fourier transform may be interpreted as the *2-D energy spectral density*

$$S(f_1, f_2) = \left| X_{\text{DTFT}}(f_1, f_2) \right|^2 \tag{16.19}$$

due to the *2-D energy theorem* relationship

$$\text{Energy} = T_1 T_2 \sum_{m=-\infty}^{\infty} \sum_{n=-\infty}^{\infty} \left| x[m, n] \right|^2$$

$$= \int_{-1/2T_1}^{1/2T_1} \int_{-1/2T_2}^{1/2T_2} \left| X_{\text{DTFT}}(f_1, f_2) \right|^2 df_1 \, df_2. \tag{16.20}$$

The subscript DTFT will be omitted for convenience in later references to this transform in this chapter.

Assume that a discrete 2-D sequence is periodic with period M in one dimension and period N in the other dimension, that is,

$$x[m + \alpha M, n + \beta N] = x[m, n] \tag{16.21}$$

for any integers α and β, so that the sequence is completely specified by the MN values in the QP region $0 \leq m \leq M - 1$ and $0 \leq n \leq N - 1$. The *2-D discrete-time*

Fourier series

$$X_{\mathrm{DTFS}}[k,l] = T_1 T_2 \sum_{m=0}^{M-1} \sum_{n=0}^{N-1} x[m,n] \exp(-j2\pi[mk/M + nl/N]) \qquad (16.22)$$

may be defined for this periodic sequence. The DTFS is itself periodic with periods of M and N in the respective dimensions. The 2-D DTFS would require $M^2 N^2$ complex operations (multiplies and adds) in order to compute all MN transform values. A one-dimensional FFT may be used to reduce this computational burden to approximately $(MN\log_2 MN)/2$ complex operations. This is achieved by taking advantage of the separability of the exponential argument in Eq. (16.22) to first compute the intermediate array, which has been transformed in one dimension,

$$X_{\mathrm{INT}}[m,l] = T_2 \sum_{n=0}^{N-1} x[m,n] \exp(-j2\pi nl/N), \qquad (16.23)$$

followed by the completion of the computation of the DTFS from the intermediate array by transforming along the other dimension

$$X_{\mathrm{DTFS}}[k,l] = T_1 \sum_{m=0}^{M-1} X_{\mathrm{INT}}[m,l] \exp(-j2\pi mk/M). \qquad (16.24)$$

This forms a row-column decomposition of the 2-D DTFS into two sets of 1-D DTFS operations, which may be evaluated as a series of N one-dimensional FFTs of the *columns* of the original data array, followed by a series of M FFTs of the *rows* of the intermediate array.

16.4 TWO-DIMENSIONAL RANDOM PROCESS THEORY

The concept of a 2-D power spectral density (PSD) requires, as in the 1-D case, that the 2-D array $x[m,n]$ of samples from a 2-D random process be wide-sense stationary. Such 2-D random processes are said to form *homogeneous* sequences because the 2-D mean of the process is independent of the location on the 2-D plane and the 2-D autocorrelation sequence (ACS) is a function only of the differential distance between two points in the 2-D plane. The discrete 2-D mean, autocorrelation sequence, autocovariance sequence, and cross correlation sequence for homogeneous random processes are, respectively,

$$\bar{x} = \mathcal{E}\{x[m,n]\} = \text{a constant}$$
$$r_{xx}[k,l] = \mathcal{E}\{x[m+k,n+l]x^*[m,n]\}$$
$$c_{xx}[k,l] = \mathcal{E}\{(x[m+k,n+l] - \bar{x})(x^*[m,n] - \bar{x}^*)\} = r_{xx}[k,l] - |\bar{x}|^2$$
$$r_{yx}[k,l] = \mathcal{E}\{y[m+k,n+l]x^*[m,n]\}. \qquad (16.25)$$

It will generally be assumed that the mean is zero, so that the 2-D autocorrelation and autocovariance sequences are equivalent. The autocorrelation $r_{xx}[0, 0]$ is real and positive. The 2-D ACS may be shown [Dudgeon and Mersereau, 1984] to form a positive-semidefinite sequence, that is,

$$\sum_i \sum_j \sum_k \sum_l a[i, j]a^*[k, l]r_{xx}[i - k, j - l] \geq 0$$

for all $a[i, j]$ sequences. The positive-definite property will assure that the 2-D DTFT of the 2-D ACS will be nonnegative and that correlation matrices formed from the 2-D ACS are invertible.

The *2-D power spectral density* (PSD) $P_{xx}(f_1, f_2)$ is defined, in a similar manner as that of the 1-D PSD definition (4.33), as the 2-D discrete-time Fourier transform of the 2-D autocorrelation sequence

$$P_{xx}(f_1, f_2) = T_1 T_2 \sum_{m=-\infty}^{\infty} \sum_{n=-\infty}^{\infty} r_{xx}[m, n] \exp(-j2\pi[f_1 m T_1 + f_2 n T_2])$$

(16.26)

over the interval $|f_1| \leq 1/2T_1$ and $|f_2| \leq 1/2T_2$. Equation (16.26) is the 2-D version of the Wiener-Khintchine theorem. It may be shown that the 2-D PSD is real and positive.

The assumption of ergodicity in the 2-D autocorrelation sequence justifies the use of the time average

$$r_{xx}[m, n] = \lim_{\substack{N \to \infty \\ M \to \infty}} \frac{1}{2M + 1} \frac{1}{2N + 1} \sum_{i=-M}^{M} \sum_{j=-N}^{N} x[i + m, j + n]x^*[i, j] \quad (16.27)$$

of a single realization of the 2-D random process to yield the 2-D ACS. Using a similar approach as that used in Chap. 4, it can be shown that an equivalent representation of the 2-D PSD function is

$$P_{xx}(f_1, f_2) \quad (16.28)$$

$$= \lim_{\substack{N \to \infty \\ M \to \infty}} \mathcal{E} \left\{ \frac{T_1}{2M + 1} \frac{T_2}{2N + 1} \left| \sum_{m=-M}^{M} \sum_{n=-N}^{N} x[m, n] \exp(-j2\pi[f_1 m T_1 + f_2 n T_2]) \right|^2 \right\}.$$

An important 2-D random process is the 2-D white noise sequence $w[m, n]$, which is defined to have the 2-D ACS

$$r_{ww}[k, l] = \rho_w \delta[k, l], \quad (16.29)$$

that is, a process that is uncorrelated with itself everywhere except at lag $[0, 0]$. The PSD for the 2-D white noise process is simply $P_{ww}(f_1, f_2) = T_1 T_2 \rho_w$.

If $y[m, n]$ is the 2-D random process response resulting from passing stationary 2-D random process $x[m, n]$ through the linear system of (16.3), then the output ACS $r_{yy}[k, l]$ and input ACS $r_{xx}[k, l]$ are related by

$$r_{yy}[k, l] = h[k, l] \star h^*[-k, -l] \star r_{xx}[k, l]. \tag{16.30}$$

If the 2-D z-transform of the 2-D ACS $r_{xx}[k, l]$ is

$$\mathsf{P}_{xx}(z_1, z_2) = \sum_{k=-\infty}^{\infty} \sum_{l=-\infty}^{\infty} r_{xx}[k, l] z_1^{-k} z_2^{-l},$$

and if $\mathsf{P}_{yy}(z_1, z_2)$ is the similar z-transform of $r_{yy}[k, l]$, then the z-transform of (16.30) will relate $\mathsf{P}_{xx}(z_1, z_2)$ and $\mathsf{P}_{yy}(z_1, z_2)$ as follows:

$$\mathsf{P}_{yy}(z_1, z_2) = \mathsf{H}(z_1, z_2)\mathsf{H}^*(1/z_1^*, 1/z_2^*)\mathsf{P}_{xx}(z_1, z_2), \tag{16.31}$$

in which $\mathsf{H}(z_1, z_2)$ is the z-transform (16.8) of the system response function. The 2-D power spectral density relationship between input and output processes is, therefore,

$$P_{yy}(f_1, f_2) = T_1 T_2 \mathsf{P}_{yy}(z_1, z_2)\Big|_{\substack{z_1=\exp(j2\pi f_1 T_1) \\ z_2=\exp(j2\pi f_2 T_2)}} = \left| H(f_1, f_2) \right|^2 P_{xx}(f_1, f_2). \tag{16.32}$$

If the input process is white with ACS given by Eq. (16.29) and the filter has a 2-D transfer function given by the rational function (16.9) of 2-D polynomials, then the output process is that of a *2-D autoregressive-moving average process* (2-D ARMA) with PSD of

$$P_{\text{ARMA}}(f_1, f_2) = T_1 T_2 \rho_w \left| \frac{B(f_1, f_2)}{A(f_1, f_2)} \right|^2, \tag{16.33}$$

in which

$$A(f_1, f_2) = \sum_m \sum_n a[m, n] \exp(-j2\pi[f_1 m T_1 + f_2 n T_2])$$

$$B(f_1, f_2) = \sum_m \sum_n b[m, n] \exp(-j2\pi[f_1 m T_1 + f_2 n T_2])$$

are defined over some arbitrary region of support. If $B(f_1, f_2) = 1$, then the PSD represents a *2-D autoregressive process* (2-D AR). If $A(f_1, f_2) = 1$, then the PSD represents a *2-D moving average process* (2-D MA). Two-dimensional processes also permit different process representations in each dimension. If the numerator, for example, is strictly a funtion of f_1, i.e., $B(f_1, f_2) = B(f_1)$, then the 2-D process is ARMA in the dimension corresponding to f_1 and AR in the dimension corresponding to f_2.

16.5 CLASSICAL 2-D SPECTRAL ESTIMATION

The two-dimensional versions of the classical spectral estimators follow simply from the one-dimensional forms of the periodogram and correlogram methods. It will be convenient for presentation purposes to assume that the data has a quarter-plane region of support, that is, MN data samples $x[m, n]$ are available over the rectangular region $0 \le m \le M - 1$ and $0 \le n \le N - 1$. Other nonrectangular regions of support are possible, but are not considered here because they are not often encountered in practice.

The unbiased 2-D autocorrelation estimate at lag $[k, l]$ is given by

$$\hat{r}_{xx}[k, l] \qquad\qquad (16.34)$$

$$= \begin{cases} \dfrac{1}{(M - k)(N - l)} \displaystyle\sum_{m=0}^{M-1-k} \sum_{n=0}^{N-1-l} x[m + k, n + l]x^*[m, n] & \text{for } k \ge 0, l \ge 0 \\[3ex] \dfrac{1}{(M - k)(N - l)} \displaystyle\sum_{m=0}^{M-1-k} \sum_{n=-l}^{N-1} x[m + k, n + l]x^*[m, n] & \text{for } k \ge 0, l < 0 \\[3ex] \hat{r}_{xx}^*[-k, -l] & \text{for } k \le 0, \text{ any } l \end{cases}$$

over a lag range of $|k| \le p_1$ and $|l| \le p_2$. The maximum lag values are limited to $p_1 \le M - 1$ and $p_2 \le N - 1$. The biased 2-D ACS $\breve{r}_{xx}[k, l]$ may by computed by simply substituting MN for the divisor $(M - k)(N - l)$ in Eq. (16.34).

The MATLAB implementation of the two-dimensional autocorrelation estimate is found in function `correlation_sequence_2D.m` available through the text website:

```
function r=correlation_sequence_2D(max_lag1,max_lag2,x)
% Estimate the two-dimensional (2-D) autocorrelation sequence (ACS) based on
% the biased estimator of equation (16.34). The computation is most efficient
% if the row and column dimensions of the data array are powers of 2.  A 2-D
% ACS array of dimension (max_lag1+1) x (2*max_lag2+1) is returned.  The (0,0)
% time lag index falls in the middle of the top row of array at element
% (1,max_lag2+1).
%
%    r = correlation_sequence_2D(max_lag1,max_lag2,x)
%
% max_lag1 -- maximum 2-D autocorrelation sequence (ACS) lag index along rows
% max_lag2 -- maximum 2-D autocorrelation sequence lag index along columns
% x        -- two-dimensional data array:  x(row sample #,column sample #)
% r        -- two-dimensional ACS array of dimension  (max_lag1+1) rows x
%             (2*max_lag2+1) columns
%
% Elements of r are stored as follows:
%
%        ACS Lag Index (row,col)              Actual Array Index (row,col)
%    (0,-max_lag2) . . . . (0,max_lag2)       (1,1). . . . (1,2*max_lag2+1)
%        .                  .         -->      .                 .
%        .                  .                  .                 .
%(max_lag1,-max_lag2).(max_lag1,max_lag2)(max_lag1+1,1).(max_lag1+1,2*max_lag2+1)
```

The correlogram method PSD estimator for 2-D data samples is

$$\hat{P}_{\text{CORR}}(f_1, f_2) = T_1 T_2 \sum_{k=-p_1}^{p_1} \sum_{l=-p_2}^{p_2} w[k, l]\hat{r}[k, l]\exp(-j2\pi[f_1 kT_1 + f_2 lT_2]),$$
(16.35)

where $w[k, l]$ is a lag window sequence used, as in the 1-D case, to trade off resolution, sidelobe bias, and estimator variance. There is not much in the literature on 2-D window functions for spectral estimation. Two-dimensional windows are often constructed as the product of 1-D windows, such as the 2-D Hamming window

$$w[k, l] = [0.54 + 0.46\cos(\pi k/p_1)][0.54 + 0.46\cos(\pi l/p_2)]$$

for $|k| \leq p_1$ and $|l| \leq p_2$. Huang [1972] has studied 2-D windows with circular, rather than rectangular, symmetry. If the biased ACS $\check{r}_{xx}[k, l]$ is used (i.e., set $w[k, l]$ to a triangular window), then an alternative matrix-vector product expression for the correlogram method PSD is

$$\hat{P}_{\text{CORR}}(f_1, f_2) = T_1 T_2 \underline{\mathbf{e}}^H(f_1, f_2)\underline{\check{\mathbf{R}}}\,\underline{\mathbf{e}}(f_1, f_2),$$
(16.36)

in which the block-Toeplitz autocorrelation matrix of block dimension $(p_1 + 1) \times (p_1 + 1)$

$$\underline{\check{\mathbf{R}}} = \begin{pmatrix} \check{\mathbf{R}}[0] & \check{\mathbf{R}}[1] & \ldots & \check{\mathbf{R}}[p_1] \\ \check{\mathbf{R}}[-1] & \check{\mathbf{R}}[0] & \ldots & \check{\mathbf{R}}[p_1 - 1] \\ \vdots & \vdots & \ddots & \vdots \\ \check{\mathbf{R}}[-p_1] & \check{\mathbf{R}}[-p_1 + 1] & \ldots & \check{\mathbf{R}}[0] \end{pmatrix}$$
(16.37)

has Toeplitz matrix elements of dimension $(p_2 + 1) \times (p_2 + 1)$

$$\check{\mathbf{R}}[k] = \begin{pmatrix} \check{r}_{xx}[k, 0] & \check{r}_{xx}[k, 1] & \ldots & \check{r}_{xx}[k, p_2] \\ \check{r}_{xx}[k, -1] & \check{r}_{xx}[k, 0] & \ldots & \check{r}_{xx}[k, p_2 - 1] \\ \vdots & \vdots & \ddots & \vdots \\ \check{r}_{xx}[k, -p_2] & \check{r}_{xx}[k, -p_2 + 1] & \ldots & \check{r}_{xx}[k, 0] \end{pmatrix}.$$
(16.38)

The 2-D complex sinusoid block vector $\underline{\mathbf{e}}(f_1, f_2)$ is defined as

$$\underline{\mathbf{e}}(f_1, f_2) = \begin{pmatrix} \mathbf{e}_0(f_1, f_2) \\ \mathbf{e}_1(f_1, f_2) \\ \vdots \\ \mathbf{e}_{p_1}(f_1, f_2) \end{pmatrix}, \mathbf{e}_k(f_1, f_2) = \exp(j2\pi f_1 kT_1) \begin{pmatrix} 1 \\ \exp(j2\pi f_2 T_2) \\ \vdots \\ \exp(j2\pi f_2 p_2 T_2) \end{pmatrix}.$$
(16.39)

Note that vectors $\mathbf{e}_k(f_1, f_2)$ each have $(p_2 + 1)$ elements.

The MATLAB implementation of the two-dimensional correlogram-based PSD estimator by the Blackman-Tukey technique is found in function `correlogram_psd_2D.m` available through the text website:

```
function psd=correlogram_psd_2D(num_psd1,num_psd2,max_lag1,max_lag2,T1,T2,x)
% Two-dimensional(2-D) classical Blackman-Tukey classical autospectral estimator
% based on the 2-D Fourier transform of the 2-D autocorrelation array estimate.
%
%     psd = correlogram_psd_2D(num_psd1,num_psd2,max_lag1,max_lag2,T1,T2,x)
%
% num_psd1 -- number of row PSD values to evaluate (must be a power of two)
% num_psd2 -- number of column PSD values to evaluate (must be a power of two)
% max_lag1 -- maximum row lag index of 2-D autocorrelation sequence
% max_lag2 -- maximum column lag index of 2-D autocorrelaiton sequence
% T1       -- sample interval along rows in seconds
% T2       -- sample interval along columns in seconds
% x        -- two dimensional sample data array
% psd      -- two dimensional power spectral density (PSD) array
```

The 2-D periodogram follows the form of the 1-D periodogram,

$$\hat{P}_{\mathrm{PER}}(f_1, f_2) = \frac{T_1 T_2}{MN} \left| \sum_{m=0}^{M-1} \sum_{n=0}^{N-1} w[m, n] x[m, n] \exp(-j2\pi[f_1 m T_1 + f_2 n T_2]) \right|^2,$$

(16.40)

in which $w[m, n]$ is an appropriate data window. The variance of the 2-D periodogram estimator may be reduced by sectioning the data into smaller 2-D arrays, which may be overlapped, taking the 2-D periodogram of each subarray, and averaging the subarray periodograms. Data sequences from space-time arrays are often so limited along the spatial domain that subarray averaging can only be performed along the time dimension; the loss in spectral resolution by averaging subarrays along the spatial dimension usually cannot be tolerated. If the individual subarrays are statistically independent, the variance will be reduced by the number of subarrays averaged. If they are not independent, then the variance reduction will be less. Appendix 16.A contains subroutine TDPERIOD, which computes the 2-D periodogram using the 1-D FFT algorithm.

The MATLAB implementation of the two-dimensional periodogram PSD estimator is found in function `periodogram_psd_2D.m` available through the text website:

```
function psd=periodogram_psd_2D(num_psd1,num_psd2,window,overlap1,overlap2,...
                                seg_size1,seg_size2,T1,T2,x)
% Classical two-dimensional periodogram power spectral density estimate by
% subarray averaging procedure of equation (16.40).
%
%     psd = periodogram_psd_2D(num_psd1,num_psd2,window,overlap1,overlap2,...
%                              seg_size1,seg_size2,T1,T2,x)
%
% num_psd1 -- number of row PSD values to evaluate (must be power of two)
% num_psd2 -- number of column PSD values to evaluate (must be power of two)
% window   -- window selection:  none in this version
% overlap1 -- subarray overlap in row dimension (# samples)
% overlap2 -- subarray overlap in column dimension (# samples)
% seg_size1 -- row dimension of data subarray segment (# samples)
```

```
% seg_size2 -- column dimension of data subarray segment (# samples)
% T1        -- sample interval along rows in seconds
% T2        -- sample interval along columns in seconds
% x         -- two-dimensional sample data array
% psd       -- two dimensional power spectral density array
```

The statistical performance of the 2-D classical estimators appears to be similar to that of the 1-D classical spectral estimators, although analytical results are not readily available. The resolution in each dimension depends upon the data extent in each dimension and will be different, in general, if $MT_1 \neq NT_2$. The 1-D resolution analyzed in Chap. 5 basically holds in the 2-D case for spectral components aligned along one of the two frequency dimensions.

16.6 MODIFIED CLASSICAL 2-D SPECTRAL ESTIMATORS

The 2-D discrete-time Fourier transform (DTFT) used to compute the periodogram was shown in Sec. 16.3 to be computable as a sequence of 1-D DTFTs of the columns followed by a sequence of 1-D DTFTs of the rows. A *hybrid 2-D spectral estimator* may be devised by substituting any of the 1-D high-resolution spectral techniques presented in Chaps. 6–13 for the row DTFTs, as illustrated in Fig. 16.4. The intermediate array generated by the column DTFTs will be composed of complex-valued transform values, so the complex form of the high-resolution techniques must be used. The intent of the hybrid scheme is to improve the resolution in the dimension where the data record may be limited as often occurs with space-time data arrays, where the spatial resolution can be inadequate due to a small number of spatial sensors relative to the larger number of time samples available from each sensor. Mathematically, the two steps of the hybrid approach to 2-D spectral estimation are

$$X_{\text{INT}}[m, f_2] = T_2 \sum_{n=0}^{N-1} x[m, n] \exp(-j2\pi f_2 n T_2) \tag{16.41}$$

$$\hat{P}_{\text{HYBRID}}(f_1, f_2) = \text{High-resolution 1-D PSD of } X_{\text{INT}}[m, f_2] \tag{16.42}$$

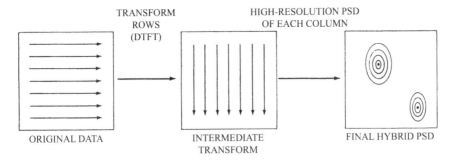

TRANSFORM   HIGH-RESOLUTION PSD
ROWS   OF EACH COLUMN
(DTFT)

ORIGINAL DATA   INTERMEDIATE   FINAL HYBRID PSD
                     TRANSFORM

Figure 16.4. Hybrid 2-D spectral estimation.

at each frequency f_2 computed in Eq. (16.41). Joyce [1979] describes one such hybrid 2-D spectral estimator based on the 1-D FFT and the 1-D complex Burg algorithm. Hybrid schemes tend to work best when the signal-to-noise ratio is fairly high in the limited-record dimension.

Another approach that seeks to improve the resolution of the classical 2-D spectral estimators is the use of linear prediction techniques to extrapolate the original 2-D data array to an enlarged data array. The enlarged data array is then processed in the usual fashion by either of the two classical spectral estimators. Frost [1980], Frey [1982], Pendrel [1979], and Ulrych and Walker [1981] have examined 2-D quarter-plane linear prediction procedures for simultaneous extension of the data array in both dimensions. The data extrapolation procedure appears to provide limited resolution improvement of perhaps up to a factor of 2 in high signal-to-noise scenarios.

16.7 TWO-DIMENSIONAL AUTOREGRESSIVE SPECTRAL ESTIMATION

Parametric models of 2-D random processes are used, as in the 1-D case, with the expectation of improved spectral estimation performance over that achievable with the classical 2-D periodogram and correlogram methods. The 2-D autoregressive (AR) model has received the greatest attention in the literature and will, therefore, be the only parametric 2-D model considered in this chapter. Two-dimensional moving average spectral estimation has not, to this author's knowledge, been studied in the literature. One method of 2-D autoregressive-moving average spectral estimation has been proposed by Cadzow and Ogino [1981].

16.7.1 The 2-D Autoregressive Process

A 2-D autoregressive sequence $x[m, n]$ is generated by driving a 2-D linear shift-invariant filter with a 2-D white noise sequence $w[m, n]$,

$$x[i, j] = -\sum_m \sum_n a[m, n]x[i - m, j - n] + w[i, j], \qquad (16.43)$$

in which the range of the indices m and n depends on the assumed region of support for the AR parameter array. The corresponding 2-D AR power spectral density was shown in Sec. 16.4 to have the form

$$P_{AR}(f_1, f_2) = \frac{T_1 T_2 \rho_w}{\left| 1 + \sum_m \sum_n a[m, n] \exp(-j2\pi [f_1 m T_1 + f_2 n T_2]) \right|^2} \qquad (16.44)$$

in which ρ_w is the 2-D white noise variance.

A primary issue with 2-D AR modeling is the choice of a region of support [Marzetta, 1980]. Only causal 2-D AR processes will be examined here, although the constraint of a stable and causal autoregressive process does not necessarily produce a better 2-D spectral estimate than would be produced by a semicausal or noncausal autoregressive process. Even with the assumption of causality, the 2-D AR process is not unique. Both NSHP and QP support regions may be defined. An upper-half-plane NSHP autoregressive parameter array $a_U[m, n]$ and a lower-half-plane NSHP autoregressive parameter array $a_L[m, n]$ may be defined as

$$a[m, n] = \begin{cases} a_U[m, n] & -p_1 \leq m \leq p_1(1 \leq n \leq p_2), 1 \leq m \leq p_1(n = 0) \\ a_L[m, n] & -p_1 \leq m \leq p_1(-p_2 \leq n \leq -1), -p_1 \leq m \leq -1(n = 0) \end{cases}$$
(16.45)

for positive integers p_1 and p_2. The total number of NSHP AR parameters is, therefore, $2p_1p_2 + p_1 + p_2$. Figure 16.5 illustrates the region of support for these two NSHP autoregressive arrays. First-, second-, third-, and fourth-quadrant QP autoregressive parameter arrays $a_1[m, n]$, $a_2[m, n]$, $a_3[m, n]$, and $a_4[m, n]$ may be defined as

$$a[m, n] = \begin{cases} a_1[m, n] & 0 \leq m \leq p_1(1 \leq n \leq p_2), 1 \leq m \leq p_1(n = 0) \\ a_2[m, n] & -p_1 \leq m \leq 0(1 \leq n \leq p_2), -p_1 \leq m \leq -1(n = 0) \\ a_3[m, n] & -p_1 \leq m \leq 0(-p_2 \leq n \leq -1), -p_1 \leq m \leq -1(n = 0) \\ a_4[m, n] & 0 \leq m \leq p_1(-p_2 \leq n \leq -1), 1 \leq m \leq p_1(n = 0) \end{cases} .$$
(16.46)

The total number of QP AR parameters in any of the four quadrant cases is $p_1p_2 + p_1 + p_2$. Figure 16.6 illustrates the region of support for these four QP autoregressive arrays.

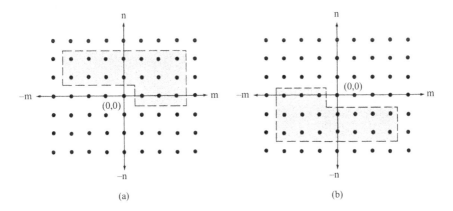

Figure 16.5. Typical NSHP regions of support for the 2-D autoregressive parameter array for $p_1 = 3$ and $p_2 = 2$. (a) Upper half NSHP. (b) Lower half NSHP.

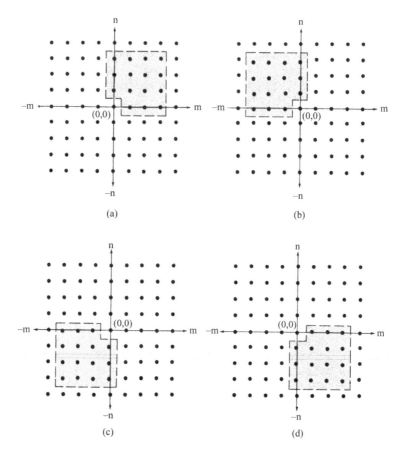

Figure 16.6. Typical QP regions of support for the 2-D autoregressive parameter arrays for $p_1 = p_2 = 3$. (a) First quadrant. (b) Second quadrant. (c) Third quadrant. (d) Fourth quadrant.

16.7.2   Two-Dimensional Linear Prediction

A 2-D linear prediction estimate of the array sample $x[m, n]$ will take the form

$$\hat{x}[m, n] = -\sum_i \sum_j \alpha[i, j] x[m - i, n - j], \qquad (16.47)$$

in which $\alpha[i, j]$ is a 2-D linear prediction coefficient. If the causal region of support for the linear prediction coefficient array is selected, then the 2-D linear prediction coefficients that minimize the variance of the error $x[m, n] - \hat{x}[m, n]$,

$$\rho_{\text{LP}} = \mathcal{E}\big\{|x[m, n] - \hat{x}[m, n]|^2\big\},$$

will yield a linear prediction error that is a 2-D white process. This is analogous to the whitening filter interpretation of the 1-D linear prediction error filter examined in Chap. 7. The normal equations that yield the 2-D linear prediction coefficients from the 2-D ACS are identical to the 2-D Yule-Walker equations for a causal 2-D AR process, to be described in Sec. 16.7.3. Reflection coefficients in 2-D, not covered here, are discussed by Marzetta [1980].

The use of 2-D linear prediction filters with semicausal or noncausal regions of support have also been studied for use in 2-D spectral estimation by Pendrel [1979], Jain [1981], Jain and Ranganath [1981], and Nikias et al. [1984]. The semicausal 2-D linear prediction spectral estimator seems to have slightly better peak location accuracy than either the NSHP or QP causal 2-D linear prediction spectral estimators [Jain, 1981]. It can be shown [Jain, 1981] that the semicausal linear prediction error variance has a spectral density that is white in the causal dimension and moving average in the noncausal dimension, yielding a spectral density expressed as a z-transform of the form

$$P_{SCLP}(z_1, z_2) = \frac{\rho_{LP}\left[1 + \sum \alpha[i, 0]z_1^{-i}\right]}{A(z_1, z_2)A^*(1/z_1^*, 1/z_2^*)}, \tag{16.48}$$

where $A(z_1, z_2)$ is defined as $A(z_1, z_2) = 1 + \sum \sum a[i, j]z_1^{-i}z_2^{-j}$. This is no longer a 2-D AR but represents a 2-D ARMA. Similarly, it can be shown that the noncausal linear prediction error variance has a spectral density that is a moving average process in both dimensions, yielding a spectral density, also expressed as a z-transform, of the form

$$P_{NCLP}(z_1, z_2) = \frac{\rho_{LP}}{A(z_1, z_2)}, \tag{16.49}$$

in which the $A^*(1/z_1^*, 1/z_2^*)$ term appears in both the numerator and denominator, thereby canceling. The semicausal and noncausal linear prediction spectral estimators do not guarantee a nonnegative power spectral density when the model parameters are estimated from finite length data records, whereas the causal linear prediction spectral estimators are guaranteed to always be nonnegative.

16.7.3 Two-Dimensional Yule–Walker Equations

The 2-D Yule-Walker equations for a causal 2-D AR process are obtained by multiplying Eq. (16.39) by $x^*[m - k, n - l]$ and taking the expectation, yielding

$$\sum_i \sum_j a[i, j]r_{xx}[k - i, l - j] = \begin{cases} \rho_w & \text{for } [k, l] = [0, 0] \\ 0 & \text{otherwise} \end{cases}.$$

The summation ranges can be selected to be any one of the six support regions of Eq. (16.45) or (16.46). The relationship between the 2-D AR parameters and the 2-D ACS is, therefore, through a set of linear equations.

The 2-D AR process with QP region of support has received, by far, the most attention in the literature. A key reason is the close relationship between the 2-D AR process and the multichannel AR process described in Chap. 15. This close relationship will provide a means of devising a 2-D Levinson-like algorithm, based on the multichannel Levinson algorithm, to solve the 2-D Yule-Walker equations for the 2-D AR parameter array.

The 2-D Yule-Walker equations for the QP support regions can be arranged, or ordered, into at least two convenient block-matrix forms by ordering the 2-D AR parameters either by rows or by columns. The row-ordered 2-D AR Yule-Walker equations for the first-quadrant region of support can be succinctly expressed as

$$\underline{\mathbf{R}}^r \underline{\mathbf{a}}_1^r = \underline{\boldsymbol{\rho}}_1^r, \tag{16.50}$$

in which the block matrix

$$\underline{\mathbf{R}}^r = \begin{pmatrix} \mathbf{R}^r[0] & \mathbf{R}^r[1] & \cdots & \mathbf{R}^r[p_1] \\ \mathbf{R}^r[-1] & \mathbf{R}^r[0] & \cdots & \mathbf{R}^r[p_1-1] \\ \vdots & \vdots & \ddots & \vdots \\ \mathbf{R}^r[-p_1] & \mathbf{R}^r[-p_1+1] & \cdots & \mathbf{R}^r[0] \end{pmatrix} \tag{16.51}$$

is defined in terms of the matrix elements

$$\mathbf{R}^r[i] = \begin{pmatrix} r_{xx}[i,0] & r_{xx}[i,1] & \cdots & r_{xx}[i,p_2] \\ r_{xx}[i,-1] & r_{xx}[i,0] & \cdots & r_{xx}[i,p_2-1] \\ \vdots & \vdots & \ddots & \vdots \\ r_{xx}[i,-p_2] & r_{xx}[i,-p_2+1] & \cdots & r_{xx}[i,0] \end{pmatrix}, \tag{16.52}$$

and the block vectors

$$\underline{\mathbf{a}}_1^r = \begin{pmatrix} \mathbf{a}_1^r[0] \\ \mathbf{a}_1^r[1] \\ \vdots \\ \mathbf{a}_1^r[p_1] \end{pmatrix}, \quad \underline{\boldsymbol{\rho}}_1^r = \begin{pmatrix} \boldsymbol{\rho}_1^r \\ \mathbf{0} \\ \vdots \\ \mathbf{0} \end{pmatrix}$$

are defined in terms of the vector elements

$$\mathbf{a}_1^r[i] = \begin{pmatrix} a_1[i,0] \\ a_1[i,1] \\ \vdots \\ a_1[i,p_2] \end{pmatrix}, \quad \boldsymbol{\rho}_1^r = \begin{pmatrix} \rho_w \\ 0 \\ \vdots \\ 0 \end{pmatrix}. \tag{16.53}$$

Each bold $\mathbf{0}$ is a column vector of $p_2 + 1$ zeros and, by definition, $a_1[0,0] = 1$. The block Toeplitz matrix $\underline{\mathbf{R}}^r$ is composed of $(p_1 + 1) \times (p_1 + 1)$ autocorrelation matrices $\mathbf{R}^r[i]$, which are themselves Toeplitz. Matrix $\underline{\mathbf{R}}^r$ is, therefore, said to be *doubly Toeplitz* or *Toeplitz-block Toeplitz*. A superscript r is used to remind the reader that row ordering has been used. Each matrix $\mathbf{R}^r[i]$ has dimension $(p_2 + 1) \times (p_2 + 1)$. The block vector $\underline{\mathbf{a}}_1^r$ of $p_2 + 1$ autoregressive parameters, which has a subscript 1 to designate this as a set of first-quadrant AR parameters, is composed of $p_1 + 1$ vectors $\mathbf{a}_1^r[i]$, each of dimension $p_2 + 1$. The block vector $\underline{\rho}_1^r$ has all zero entries, except for the top entry, which is the noise variance ρ_w.

The column-ordered 2-D AR Yule-Walker equations for the first-quadrant region of support has the form

$$\underline{\mathbf{R}}^c \underline{\mathbf{a}}_1^c = \underline{\rho}_1^c, \tag{16.54}$$

in which the block matrix

$$\underline{\mathbf{R}}^c = \begin{pmatrix} \mathbf{R}^c[0] & \mathbf{R}^c[1] & \cdots & \mathbf{R}^c[p_2] \\ \mathbf{R}^c[-1] & \mathbf{R}^c[0] & \cdots & \mathbf{R}^c[p_2 - 1] \\ \vdots & \vdots & \ddots & \vdots \\ \mathbf{R}^c[-p_2] & \mathbf{R}^c[-p_2 + 1] & \cdots & \mathbf{R}^c[0] \end{pmatrix} \tag{16.55}$$

is defined in terms of the matrix elements

$$\mathbf{R}^c[j] = \begin{pmatrix} r_{xx}[0, j] & r_{xx}[1, j] & \cdots & r_{xx}[p_1, j] \\ r_{xx}[-1, j] & r_{xx}[0, j] & \cdots & r_{xx}[p_1 - 1, j] \\ \vdots & \vdots & \ddots & \vdots \\ r_{xx}[-p_1, j] & r_{xx}[-p_1 + 1, j] & \cdots & r_{xx}[0, j] \end{pmatrix}, \tag{16.56}$$

and the block vectors

$$\underline{\mathbf{a}}_1^c = \begin{pmatrix} \mathbf{a}_1^c[0] \\ \mathbf{a}_1^c[1] \\ \cdots \\ \mathbf{a}_1^c[p_2] \end{pmatrix}, \quad \underline{\rho}_1^c = \begin{pmatrix} \rho_1^c \\ \mathbf{0} \\ \vdots \\ \mathbf{0} \end{pmatrix}$$

are defined in terms of the vector elements

$$\mathbf{a}_1^c[j] = \begin{pmatrix} a_1[0, j] \\ a_1[1, j] \\ \vdots \\ a_1[p_1, j] \end{pmatrix}, \quad \rho_1^c = \begin{pmatrix} \rho_w \\ 0 \\ \vdots \\ 0 \end{pmatrix}. \tag{16.57}$$

In this case, $\underline{\mathbf{R}}^c$ has block matrix dimension $(p_2 + 1) \times (p_2 + 1)$, both $\underline{\mathbf{a}}_1^c$ and $\underline{\rho}_1^c$ have block vector dimension $(p_2 + 1)$, $\mathbf{R}^c[j]$ has matrix dimension $(p_1 + 1) \times (p_1 + 1)$, both $\mathbf{a}_1^c[j]$ and ρ_1^c have vector dimension $(p_1 + 1)$, and $\mathbf{0}$ is a column vector of $p_1 + 1$ zeros.

The 2-D Yule-Walker equations for the second-, third-, and fourth-quadrant support regions may also be succinctly expressed in terms of either the row-ordered block matrix $\underline{\mathbf{R}}^r$ or the column-ordered block matrix $\underline{\mathbf{R}}^c$. The row-ordered forms, for example, are

$$\underline{\mathbf{R}}^r \underline{\mathbf{a}}_2^r = \underline{\rho}_2^r \tag{16.58}$$

$$\underline{\mathbf{R}}^r \underline{\mathbf{a}}_3^r = \underline{\rho}_3^r \tag{16.59}$$

$$\underline{\mathbf{R}}^r \underline{\mathbf{a}}_4^r = \underline{\rho}_4^r, \tag{16.60}$$

in which

$$\underline{\mathbf{a}}_2^r = \begin{pmatrix} \mathbf{a}_2^r[0] \\ \mathbf{a}_2^r[1] \\ \vdots \\ \mathbf{a}_2^r[p_1] \end{pmatrix}, \quad \mathbf{a}_2^r[i] = \begin{pmatrix} a_2[i, p_2] \\ \vdots \\ a_2[i, 1] \\ a_2[i, 0] \end{pmatrix}, \quad \underline{\rho}_2^r = \begin{pmatrix} \rho_2^r \\ 0 \\ \vdots \\ 0 \end{pmatrix}, \quad \rho_2^r = \begin{pmatrix} 0 \\ \vdots \\ 0 \\ \rho_w \end{pmatrix}$$

and

$$\underline{\mathbf{a}}_3^r = \begin{pmatrix} \mathbf{a}_3^r[p_1] \\ \vdots \\ \mathbf{a}_3^r[1] \\ \mathbf{a}_3^r[0] \end{pmatrix}, \quad \mathbf{a}_3^r[i] = \begin{pmatrix} a_3[i, p_2] \\ \vdots \\ a_3[i, 1] \\ a_3[i, 0] \end{pmatrix}, \quad \underline{\rho}_3^r = \begin{pmatrix} 0 \\ \vdots \\ 0 \\ \rho_3^r \end{pmatrix}, \quad \rho_3^r = \begin{pmatrix} 0 \\ \vdots \\ 0 \\ \rho_w \end{pmatrix}$$

and

$$\underline{\mathbf{a}}_4^r = \begin{pmatrix} \mathbf{a}_4^r[p_1] \\ \vdots \\ \mathbf{a}_4^r[1] \\ \mathbf{a}_4^r[0] \end{pmatrix}, \quad \mathbf{a}_4^r[i] = \begin{pmatrix} a_4[i, 0] \\ a_4[i, 1] \\ \vdots \\ a_4[i, p_2] \end{pmatrix}, \quad \underline{\rho}_4^r = \begin{pmatrix} 0 \\ \vdots \\ 0 \\ \rho_4^r \end{pmatrix}, \quad \rho_4^r = \begin{pmatrix} \rho_w \\ 0 \\ \vdots \\ 0 \end{pmatrix}.$$

The autocorrelation matrices $\mathbf{R}^r[i]$ are not Hermitian, but they do satisfy the relationship $\mathbf{R}^r[-i] = (\mathbf{R}^r[i])^H$. This is sufficient, however, to make the block matrix $\underline{\mathbf{R}}^r$ Hermitian, that is, $\underline{\mathbf{R}}^r = (\underline{\mathbf{R}}^r)^H$. Matrix $\underline{\mathbf{R}}^r$ also has the property

$$\mathbf{J}\underline{\mathbf{R}}^r \mathbf{J} = (\underline{\mathbf{R}}^r)^*, \tag{16.61}$$

in which $\mathbf{J}$ is a $(p_1 + 1)(p_2 + 1) \times (p_1 + 1)(p_2 + 1)$ reflection matrix. This property follows from the Hermitian and Toeplitz properties of $\underline{\mathbf{R}}^r$. Noting that $\mathbf{JJ}=\mathbf{I}$, where $\mathbf{I}$ is the identity matrix, then Eqs. (16.59) and (16.60) may be expressed as

$$\mathbf{J}(\underline{\mathbf{R}}^r)^* \mathbf{J} \mathbf{J} \underline{\mathbf{a}}_3^r = \mathbf{J}\underline{\rho}_3^r \tag{16.62}$$

$$\mathbf{J}(\underline{\mathbf{R}}^r)^* \mathbf{J} \mathbf{J} \underline{\mathbf{a}}_4^r = \mathbf{J}\underline{\rho}_4^r. \tag{16.63}$$

Observing that $\mathbf{J}\underline{\rho}_3^r = \underline{\rho}_1^r$ and $\mathbf{J}\underline{\rho}_4^r = \underline{\rho}_2^r$ and using property (16.61), then Eqs. (16.62) and (16.63) become

$$\underline{\mathbf{R}}^r(\mathbf{J}\underline{\mathbf{a}}_3^r)^* = (\underline{\rho}_1^r)^* = \underline{\rho}_1^r \qquad (16.64)$$

$$\underline{\mathbf{R}}^r(\mathbf{J}\underline{\mathbf{a}}_4^r)^* = (\underline{\rho}_2^r)^* = \underline{\rho}_2^r, \qquad (16.65)$$

which uses the fact that $\rho_w = \rho_w^*$ because the variance ρ_w is real. Comparing Eq. (16.64) with (16.58) and Eq. (16.65) with (16.59), it can be concluded that

$$\underline{\mathbf{a}}_1^r = (\mathbf{J}\underline{\mathbf{a}}_3^r)^* \qquad (16.66)$$

$$\underline{\mathbf{a}}_2^r = (\mathbf{J}\underline{\mathbf{a}}_4^r)^*, \qquad (16.67)$$

that is, the third quadrant AR parameters are the complex conjugates of the first-quadrant AR parameters,

$$a_3[m, n] = a_1^*[m, n], \qquad (16.68)$$

and the fourth-quadrant AR parameters are the complex conjugates of the second-quadrant AR parameters,

$$a_4[m, n] = a_2^*[m, n]. \qquad (16.69)$$

Equations (16.68) and (16.69) are similar to the conjugate relationship of the forward and backward linear prediction coefficients found in 1-D AR analysis.

A unique relationship was shown in Chap. 7 to exist between a sequence of $p + 1$ ACS lags and the sequence of p 1-D autoregressive parameters and the noise variance ρ_w. Given one sequence, the other sequence could be uniquely determined. This one-to-one mapping is often described as the *autocorrelation matching property* of 1-D autoregressive analysis. This autocorrelation matching property is lost for 2-D AR process. An examination of the normal equations (16.54) and (16.58)–(16.60) will reveal that $2p_1 p_2 - p_1 - p_2$ unique 2-D ACS terms are required to solve for the $p_1 p_2 + p_1 + p_2$ 2-D AR parameters and the variance ρ_w. There are, therefore, insufficient AR parameters to uniquely determine the ACS terms. This implies that no one-to-one relationship can exist between these two alternative representations of a 2-D AR process. Inverse transforming the 2-D power spectral density function, Eq. (16.44), will generate a 2-D autocorrelation sequence that generally will not match the ACS values used to compute the 2-D AR parameters from the 2-D Yule-Walker equations. They will, however, tend to *approximate* the original ACS values. It should also be pointed out that a positive-definite $\mathbf{R}^r$ matrix will guarantee a unique solution of the 2-D Yule-Walker equations, but it will not necessarily guarantee a stable AR model. The autocorrelation matching issue will be central to the 2-D maximum entropy method of Sec. 16.8.

16.7.4   Solving the 2-D Quarter-Plane Yule-Walker Equations

A fast algorithm for solving the 2-D QP Yule-Walker equations can be derived by developing a relationship with the multichannel Yule-Walker equations of Chap. 15, which have the multichannel Levinson algorithm for their solution [Therrien, 1981].

The procedure permits computation of the AR parameters for all four quarter planes simultaneously. The algorithm provided here is a row-ordered 2-D Levinson algorithm in which column order p_2 is fixed and the solution for row order p_1 is obtained recursively, starting from order 1. A companion column-ordered 2-D Levinson algorithm can also be developed by simply exchanging the roles of p_1 and p_2. There can be computational advantages to selecting one ordering over another, especially if $p_1 \gg p_2$ or $p_2 \gg p_1$, even though the orderings will lead to the same solution. Other 2-D Levinson algorithms are also possible [Justice, 1977], such as one that operates recursively on both dimensions simultaneously or one that operates alternately on each dimension.

The multichannel Yule-Walker equations for a $(p_2 + 1)$-channel AR process of order p_1, in which the channels are numbered from 0 to p_2, was shown [Eq. (15.57)] to have the form

$$
\begin{pmatrix}
\mathbf{R_{xx}}[0] & \mathbf{R_{xx}}[1] & \cdots & \mathbf{R_{xx}}[p_1] \\
\mathbf{R_{xx}}[-1] & \mathbf{R_{xx}}[0] & \cdots & \mathbf{R_{xx}}[p_1-1] \\
\vdots & \vdots & \ddots & \vdots \\
\mathbf{R_{xx}}[-p_1] & \mathbf{R_{xx}}[-p_1+1] & \cdots & \mathbf{R_{xx}}[0]
\end{pmatrix}
\begin{pmatrix}
\mathbf{I} \\
\mathbf{A}^H_{p_1}[1] \\
\vdots \\
\mathbf{A}^H_{p_1}[p_1]
\end{pmatrix}
=
\begin{pmatrix}
\mathbf{P}^f_{p_1} \\
\mathbf{0} \\
\vdots \\
\mathbf{0}
\end{pmatrix}
\tag{16.70}
$$

for the forward multichannel AR parameters, and

$$
\begin{pmatrix}
\mathbf{R_{xx}}[0] & \cdots & \mathbf{R_{xx}}[p_1-1] & \mathbf{R_{xx}}[p_1] \\
\vdots & \vdots & \ddots & \vdots \\
\mathbf{R_{xx}}[-p_1+1] & \cdots & \mathbf{R_{xx}}[0] & \mathbf{R_{xx}}[1] \\
\mathbf{R_{xx}}[-p_1] & \cdots & \mathbf{R_{xx}}[-1] & \mathbf{R_{xx}}[0]
\end{pmatrix}
\begin{pmatrix}
\mathbf{B}^H_{p_1}[p_1] \\
\vdots \\
\mathbf{B}^H_{p_1}[1] \\
\mathbf{I}
\end{pmatrix}
=
\begin{pmatrix}
\mathbf{0} \\
\vdots \\
\mathbf{0} \\
\mathbf{P}^b_{p_1}
\end{pmatrix}
\tag{16.71}
$$

for the backward multichannel AR parameters, in which

$$
\mathbf{R_{xx}}[k] =
\begin{pmatrix}
r_{00}[k] & \cdots & r_{0p_2}[k] \\
\vdots & & \vdots \\
r_{p_20}[k] & \cdots & r_{p_2p_2}[k]
\end{pmatrix}
= \mathbf{R}^H_{xx}[-k]
\tag{16.72}
$$

is the multichannel correlation matrix at lag k. The correlation matrix displayed in Eqs. (16.70) and (16.71) is Hermitian and has a block Toeplitz structure. These properties were also characteristic of the 2-D QP Yule-Walker equations, which suggests that the solutions of the 2-D Yule-Walker equations and the multichannel Yule-Walker equations are related. Comparing Eq. (16.54) with Eq. (16.70), a one-to-one correspondence between $\underline{\mathbf{R}}_{p_1}$ and $\underline{\mathbf{R}}^r$ can be made by associating the multichannel crosscorrelation $r_{ij}[k]$ with the 2-D autocorrelation $r_{xx}[k, l]$ as follows

$$
r_{ij}[k] = r_{xx}[k, i - j]
\tag{16.73}
$$

for $0 \le i, j \le p_2$. If one now compares Eq. (16.54) with Eq. (16.70), then it is clear that if

$$
\rho^r_1 = \mathbf{R}^f_{p_1} \mathbf{a}^r_1[0],
\tag{16.74}
$$

393 16.7 2-D AR SPECTRAL ESTIMATION

then it follows that

$$\mathbf{a}_1^r[k] = \mathbf{A}_{p_1}^H[k]\mathbf{a}_1^r[0] \tag{16.75}$$

for $1 \leq k \leq p_1$. Thus, *the multichannel Levinson algorithm can be used to compute a solution of the 2-D Yule-Walker equations* by first solving for matrices $\mathbf{P}_{p_1}^f$ and $\mathbf{A}_{p_1}[k]$ for $k = 1$ to p_1, and then using Eqs. (16.74) and (16.75) to yield the 2-D autoregressive parameters and variance.

The second-quadrant 2-D AR parameters may similarly be computed from the multichannel solution using

$$\rho_2^r = \mathbf{P}_{p_1}^f\mathbf{a}_2^r[0] \tag{16.76}$$

and

$$\mathbf{a}_2^r[k] = \mathbf{A}_{p_1}^H[k]\mathbf{a}_2^r[0] \tag{16.77}$$

for $1 \leq k \leq p_1$.

The Toeplitz structure of the 2-D autocorrelation matrices $\mathbf{R}_{\mathbf{xx}}[k]$ can be exploited to further reduce the computational complexity of the multichannel Levinson algorithm. It can be shown that the doubly Toeplitz structure will force the relationship

$$\mathbf{B}_{p_1}[k] = \mathbf{A}_{p_1}^H[k] \tag{16.78}$$

for $1 \leq k \leq p_1$ and the relationship

$$\mathbf{P}_{p_1}^f = \mathbf{P}_{p_1}^b \tag{16.79}$$

to be true, although these do not hold in the general multichannel case. These relationships will simplify the multichannel Levinson algorithm when applied to the 2-D case as follows:

$$\mathbf{\Delta}_{p_1+1} = \begin{pmatrix} \mathbf{I} & \mathbf{A}_{p_1}[1] & \cdots & \mathbf{A}_{p_1}[p_1] \end{pmatrix} \begin{pmatrix} \mathbf{R}_{\mathbf{xx}}[p_1+1] \\ \mathbf{R}_{\mathbf{xx}}[p_1] \\ \vdots \\ \mathbf{R}_{\mathbf{xx}}[1] \end{pmatrix} \tag{16.80}$$

$$\mathbf{A}_{p_1+1}[p_1+1] = -\mathbf{\Delta}_{p_1+1}\mathbf{P}_{p_1}^f)^{-1} \tag{16.81}$$

$$\mathbf{A}_{p_1+1}[k] = \mathbf{A}_{p_1}[k] + \mathbf{A}_{p_1+1}[p_1+1]\mathbf{A}_{p_1}^H[p_1-k] \quad \text{for } 1 \leq k \leq p_1 \tag{16.82}$$

$$\mathbf{P}_{p_1+1}^f = (\mathbf{I} - \mathbf{A}_{p_1+1}[p_1+1]\mathbf{A}_{p_1+1}^H[p_1+1])\mathbf{P}_{p_1}^f \tag{16.83}$$

with initial condition $\mathbf{P}_0^f = \mathbf{R}_{\mathbf{xx}}[0]$. This reduces the computational burden of the general multichannel Levinson by approximately one-half [Kalouptsidis et al., 1984; Wax and Kailath, 1983]. Appendix 16.B contains subroutine TDAR that solves for the 2-D first- and second-quadrant AR parameters, given the 2-D ACS samples.

The MATLAB implementation of the two-dimensional quarter-plane recursive Levinson algorithm is found in function `levinson_recursion_2D.m` available through the text website:

```
function [p_Q1,a_Q1,p_Q4,a_Q4]=levinson_recursion_2D(r)
% Compute 2-D quarter plane AR parameter arrays given 2-D known or estimated
% ACS using two-dimensional Levinson algorithm that exploits the double
% Toeplitz structure. Only Q1 and Q4 quadrant AR parameter arrays are computed;
% Q3 and Q2 quarter arrays are simply the complex conjugates, respectively,
% of the Q1 and Q4 arrays.  Most computationally efficient when  p2 > p1 ,
% else switch orders.
%
%    [p_Q1,a_Q1,p_Q4,a_Q4] = levinson_recursion_2D(r)
%
% r    -- two-dimensional autocorrelation sequence (ACS) array
% p_Q1 -- Q1-quarter plane white noise variance
% a_Q1 -- two-dimensional Q1-quarter plane AR parameter matrix
% p_Q4 -- Q4-quarter plane white noise variance
% a_Q4 -- two-dimensional Q4-quarter plane AR parameter matrix
```

16.7.5 A Modified Quarter-Plane AR Spectral Estimator

The 2-D first-quadrant autoregressive spectral estimator

$$
\hat{P}_{\text{AR1}}(f_1, f_2) = \frac{T_1 T_2 \hat{\rho}_w}{\left| \displaystyle\sum_{m=0}^{p_1} \sum_{n=0}^{p_2} \hat{a}_1[m, n] \exp(-j2\pi[f_1 m T_1 + f_2 n T_2]) \right|^2}
$$

$$
= \frac{T_1 T_2 \hat{\rho}_w}{\left| \mathbf{e}(f_1, f_2)\hat{\mathbf{a}}_1^r \right|^2}, \tag{16.84}
$$

for which $\underline{e}(f_1, f_2)$ is the 2-D complex sinusoid block vector of Eq. (16.39) and $\underline{\mathbf{a}}_1^r$ is the first-quadrant autoregressive parameter vector estimate of (16.53) (note that $\hat{a}_1[0, 0] = 1$ by definition), and the 2-D second-quadrant autoregressive spectral estimator

$$
\hat{P}_{\text{AR2}}(f_1, f_2) = \frac{T_1 T_2 \hat{\rho}_w}{\left| \displaystyle\sum_{m=-p_1}^{0} \sum_{n=0}^{p_2} \hat{a}_2[m, n] \exp(-j2\pi[f_1 m T_1 + f_2 n T_2]) \right|^2}, \tag{16.85}
$$

in which $\hat{a}_2[0, 0] = 1$, will yield asymmetrically biased spectral estimates, as illustrated in Fig. 16.7 for the case of a known 2-D ACS for a single sinusoid in white noise. The spectral responses are seen to have opposing elliptical skews. Jackson and Chien [1979] suggested that a combined QP AR estimator $\hat{P}_{\text{ARC}}(f_1, f_2)$ be formed as

$$
\frac{1}{\hat{P}_{\text{ARC}}(f_1, f_2)} = \frac{1}{\hat{P}_{\text{AR1}}(f_1, f_2)} + \frac{1}{\hat{P}_{\text{AR2}}(f_1, f_2)}. \tag{16.86}
$$

The combined QP estimator will yield the circular response shown in Fig. 16.7(c). Jackson and Chien also found in practice that less spurious peaks were likely to be

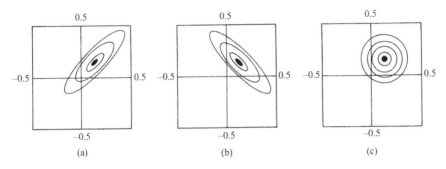

Figure 16.7. Illustration of the bias in the QP AR spectral estimator for a single sinusoid in white noise. (a) First-quadrant 2-D AR PSD estimate. (b) Second-quadrant 2-D AR PSD estimate. (c) Combined 2-D AR PSD estimate.

introduced into the spectral estimate when using the combined estimator because the first- and second-quadrant estimators are not likely to be zero at the same place unless there is actually a signal component existing at that location in the 2-D frequency plane. The MATLAB implementation of the two-dimensional quarter-plane autoregressive PSD is found in function ar_psd_2D.m available through the text website:

```
function psd=ar_psd_2D(psd_type,num_psd1,num_psd2,T1,T2,p_Q1,a_Q1,p_Q4,a_Q4)
% Generate the first, fourth, or combined quarter-plane two-dimensional AR
% power spectral density.
%
%     psd = ar_psd_2D(psd_type,num_psd1,num_psd2,T1,T2,p_Q1,a_Q1,p_Q4,a_Q4)
%
% psd_type -- select: 1--Q1 plane PSD, 2--Q4 plane PSD, 3--combined Q1 & Q4
% num_psd1 -- number of row PSD values to evaluate (power of two)
% num_psd2 -- number of column PSD values to evaluate (power of two)
% T1       -- sample interval along rows in seconds
% T2       -- sample interval along columns in seconds
% p_Q1     -- first-quadrant AR white noise variance
% a_Q1     -- two-dimensional Q1-plane AR parameter array
% p_Q4     -- fourth-quadrant AR white noise variance
% a_Q4     -- two-dimensional Q4-plane AR parameter array
% psd      -- two-dimensional quarter-plane AR power spectral density matrix
```

Note that the FFT-based fast algorithm of ar_psd_2D simultaneously provides the AR parameter vector solution for both the quadrants needed to evaluate Eq. (16.86).

16.7.6 Estimating 2-D Autoregressive Parameters from Data

The algorithmic techniques employed in Chap. 8 for 1-D AR parameter estimation may be extended to the 2-D case. The data can be used to estimate a set of 2-D ACS values over a finite support region, and the 2-D Yule-Walker equations then solved for the desired QP or NSHP autoregressive parameter array. Least squares

2-D linear prediction algorithms based on the covariance technique of Chap. 8 may also be used [Nikias et al., 1984; Nikias and Raghuveer, 1985; Ulrych and Walker, 1981]. Assuming data samples $x[m, n]$ over a rectangular region of support $1 \le m \le M$ and $1 \le n \le N$, then the total squared error E_1 of the first-quadrant linear predictor is

$$E_1 = \frac{1}{2NM} \sum_{m=p_1+1}^{M} \sum_{n=p_2+1}^{N} \left| \sum_{i=0}^{p_1} \sum_{j=0}^{p_2} a_1[i, j] x[m - i, n - j] \right|^2 , \qquad (16.87)$$

in which $a_1[0, 0] = 1$ is assumed. This may be minimized to yield a set of 2-D covariance linear prediction equations in terms of the linear prediction parameters $a_1[i, j]$. A modified covariance linear prediction normal equation set may also be formulated by minimizing the sum of the first- and third-quadrant squared errors, $E_1 + E_3$, subject to the complex conjugate relationship (16.68) that relates the first- and third-quadrant 2-D AR parameters.

The MATLAB implementation of the two-dimensional AR estimator based on the 2-D extension of the Yule-Walker technique is found in function yule_walker_2D.m available through the text website:

```
function [rho_Q1,a_Q1,rho_Q4,a_Q4]=yule_walker_2D(p1,p2,x)
% Two-dimensional Yule-Walker autoregressive parameter estimation algorithm.
%
%      [rho_Q1,a_Q1,rho_Q4,a_Q4] = yule_walker_2D(p1,p2,x)
%
% p1     -- row order of 2-D autoregressive (AR) process
% p2     -- column order of 2-D autoregressive (AR) process
% x      -- sample data array: x(row sample #,column sample #)
% rho_Q1 -- first quadrant (Q1) white noise variance
% a_Q1   -- first quadrant (Q1) 2-D autoregressive parameter array
% rho_Q4 -- fourth quadrant (Q4) white noise variance
% a_Q4   -- fourth quadrant (Q4) 2-D autoregressive parameter array
```

The MATLAB implementation of the two-dimensional AR (linear prediction) estimator based on the 2-D extension of the 1-D lattice linear prediction technique is found in function lattice_2D.m available through the text website:

```
function [p_Q1,a_Q1,p_Q4,a_Q4,P,A]=lattice_2D(p1,p2,x)
% Computes 2-D quarter-plane support AR parameter arrays given 2-D data set
% x  using Marple algorithm to exploit the doubly Toeplitz structure. Only
% Q1 and Q4 quadrant AR parameter arrays are computed; Q2 and Q3 quarter
% plane arrays are simply the complex conjugates, respectively, of the Q4
% and Q1 arrays, ie, a_Q3=conj(a_Q1), p_Q3=p_Q1, a_Q2=conj(a_Q4), p_Q2=p_Q4.
% Most computationally efficient when p2 >= p1, where p1 is the fixed order
% and p2 is the variable order; else switch roles. Reduces to 1-D Burg lattice
% algorithm results if (N1=1,p1=0). Reduces to modified covariance algorithm
% results if (N2=1,p2=0). Must have p1 < N1 and p2 < N2.
%
%      [p_Q1,a_Q1,p_Q4,a_Q4] = lattice_2D(p1,p2,x)
%
```

```
% p1    -- row order of 2-D autoregressive (AR) quarter-plane filter
% p2    -- column order of 2-D autoregressive quarter-plane filter
% x     -- 2-D sample data array:  x(row sample#,column sample #)  N1 x N2
% p_Q1  -- Q1-quarter-plane white noise variance estimate
% a_Q1  -- two-dimensional Q1-quarter-plane AR parameter matrix estimates
% p_Q4  -- Q4-quarter-plane white noise variance estimate
% a_Q4  -- two-dimensional Q4-quarter-plane AR parameter matrix estimates
```

16.8   TWO-DIMENSIONAL MAXIMUM ENTROPY SPECTRAL ESTIMATION

The 2-D maximum entropy method PSD estimator $P_{\text{MEM}}(f_1, f_2)$ maximizes the 2-D entropy rate

$$h = \int_{-1/2T_1}^{1/2T_1} \int_{-1/2T_2}^{1/2T_2} \ln P_{\text{MEM}}(f_1, f_2)\, df_1\, df_2, \qquad (16.88)$$

subject to the constraints that it match the known 2-D ACS values

$$r_{xx}[k, l] = \int_{-1/2T_1}^{1/2T_1} \int_{-1/2T_2}^{1/2T_2} P_{\text{MEM}}(f_1, f_2) \exp(j2\pi[f_1 k T_1 + f_2 l T_2])\, df_1\, df_2. \qquad (16.89)$$

Unlike the parametric 2-D spectral estimators such as the QP AR estimator, the MEM PSD does not require that the known lags of the 2-D ACS have a uniform support grid; arbitrary lags can be used as the constraints. The optimum maximum entropy spectrum in 1-D was shown in Chap. 7 to be the inverse of a positive polynomial that could be factored as the squared magnitude of a finite-order polynomial. The lack of a factorization theorem for 2-D polynomials means that the inverse polynomial of the 2-D maximum entropy PSD cannot be factored, requiring nonlinear means of solution to find the 2-D maximum entropy PSD, which is not always guaranteed to exist given any arbitrary set of 2-D ACS values. It has been shown in Sec. 16.7.3 that the 2-D AR process will not possess the ACS matching property that existed with 1-D AR processes. Thus, *2-D MEM and 2-D AR do not yield identical spectral estimators*, in contrast to the 1-D case for which they were identical.

If the known ACS lags fall on a uniform quarter-plane support grid, then the MEM solution can be shown [Jain and Ranganath, 1981] to have the form

$$P_{\text{MEM}}(f_1, f_2) = \frac{1}{\displaystyle\sum_{k=-p_1}^{p_1} \sum_{l=-p_2}^{p_2} \alpha[k, l] \exp(-j2\pi[f_1 k T_1 + f_2 l T_2])}, \qquad (16.90)$$

whenever it exists. This is the same form as the noncausal 2-D linear prediction spectral estimator. It is not generally possible to factor the denominator into two conjugate polynomials similar to Eq. (16.48). No closed-form solution for the $\alpha[k, l]$ has been found, but techniques have been devised that compute $P_{\mathrm{MEM}}(f_1, f_2)$ directly from the known 2-D ACS values without the need to solve for the $\alpha[k, l]$ parameters. These techniques require either iterative algorithms which have trouble converging, or nonlinear optimizations which are guaranteed to converge but which are computationally intensive [Lim and Malik, 1981; Malik and Lim, 1982; Wernecke, 1977; Lang, 1981; Lang and McClellan, 1982; Lang and Marzetta, 1984; Sharma and Chellappa, 1985].

Equation (16.90) implies an extension of the 2-D ACS. The AR parameters in 1-D were used to extrapolate the 1-D ACS function while maintaining the positive-definite property of the extended autocorrelation sequence. Given a finite, positive-definite 1-D ACS, it is always possible to extrapolate to a valid 1-D ACS. In 2-D, however, a finite 2-D positive-definite 2-D ACS cannot always be extrapolated to a valid 2-D ACS (i.e., it may not be *extendible*) [Dickinson, 1980]. It is true, though, that as long as $P_{\mathrm{MEM}}(f_1, f_2) > 0$ for all f_1 and f_2, then the inverse DTFT of Eq. (16.90) will always generate a valid positive-definite 2-D ACS [Lang and McClellan, 1983; Nikias and Venetsamopoulos, 1985]. Whether a better spectral estimator than 2-D AR can be obtained by constraining the estimator to match the available ACS values has not yet been satisfactorily confirmed by the limited experiments performed to date, especially when using estimated ACS values rather than exactly known ACS values.

16.9 TWO-DIMENSIONAL MINIMUM VARIANCE SPECTRAL ESTIMATION

The 2-D minimum variance spectral (MV) estimator was originally developed as a frequency-wavenumber spectral estimator by Capon [1969]. The extension of the 1-D minimum variance spectral estimator to 2-D requires a mathematical development that very closely parallels the development of Chap. 12 and Sec. 15.13, based on the idea of using a 2-D finite impulse response filter to pass a 2-D complex sinusoid of a certain frequency while simultaneously minimizing the power out of the filter. Unlike the parametric methods of 2-D PSD analysis, the 2-D MV PSD is not constrained to use uniformly-sampled ACS values. If the first QP grid of ACS lags is known, however, then the 2-D minimum variance spectral estimator will have the form

$$P_{\mathrm{MV}}(f_1, f_2) = \frac{T_1 T_2}{\underline{e}^H(f_1, f_2)(\underline{\mathbf{R}}^r)^{-1}\underline{e}(f_1, f_2)}, \tag{16.91}$$

in which $\underline{e}(f_1, f_2)$ is the complex sinusoid block vector of Eq. (16.39) and $(\underline{\mathbf{R}}^r)^{-1}$ is the inverse of the doubly-Toeplitz block autocorrelation matrix of Eq. (16.51). Other variations of Eq. (16.91) will depend on the selected region of support. Dowla and Lim [1984] have shown that the 2-D first-quadrant MV PSD and the 2-D first-quadrant AR PSD have a comparable relationship as that of Eq. (12.18). This means that the first-quadrant MV PSD suffers from the same bias as that of the 2-D first-quadrant AR PSD. Defining a combined MV PSD that uses both the first- and second-quadrant MV PSD estimators

$$\frac{1}{\hat{P}_{MVC}(f_1, f_2)} = \frac{1}{\hat{P}_{MV1}(f_1, f_2)} + \frac{1}{\hat{P}_{MV2}(f_1, f_2)} \tag{16.92}$$

will alleviate this bias.

The inverse of $\underline{R}^r$ may be obtained simply by using the inversion formula (15.86) with a further reduction in computation possible by using properties (16.78) and (16.79), due to the doubly-Toeplitz property of $\underline{R}^r$. Comparable expressions to Eqs. (12.24) and (12.25) may then be developed for the 2-D MV PSD.

The MATLAB implementation of the two-dimensional minimum variance PSD technique is found in function minimum_variance_psd_2D.m available through the text website:

```
function psd=minimum_variance_psd_2D(num_psd1,num_psd2,p1,p2,T1,T2,x)
% This functioncomptues the 2-D minimum variance spectral estimate
%
%    psd = minimum_variance_psd_2D(num_psd1,num_psd2,p1,p2,T1,T2,x)
%
%
% num_psd1 -- number of row frequency samples in  psd (must be power of 2)
% num_psd2 -- number of column frequency samples in psd (must be power of 2)
% p1  -- row order of minimum variance filter (number of row taps = p1+1)
% p2  -- column order of minimum variance filter (number of column taps = p2+1)
% T1  -- row sample interval (in seconds)
% T2  -- column sample interval (in seconds)
% x   -- sample data array of dimension N1 x N2
% psd -- first quadrant 2-D minimum variance spectrum (num_psd1 x num_psd2)
```

Dudgeon and Mersereau [1984] have reported some problems with the 2-D minimum variance spectral estimator when the data is insufficient to yield good autocorrelation estimates to be used in the matrix inversion of $\underline{R}^r$. Nikias et al. [1984] have reported that better results in processing ECG medical sensor data were obtained with 2-D MV than with either the 2-D periodogram or the 2-D AR.

References

Cadzow, J. A., and K. Ogino, Two-Dimensional Spectral Estimation, *IEEE Trans. Acoust. Speech Signal Process.*, vol. ASSP-29, pp. 396–401, June 1981.

Capon, J., High Resolution Frequency-Wavenumber Spectrum Analysis, *Proc. IEEE*, vol. 57, pp. 1408–1418, August 1969.

Dickinson, B. W., Two-Dimensional Markov Spectrum Estimates Need Not Exist, *IEEE Trans. Inf. Theory*, vol. IT-26, pp. 120–121, January 1980.

Dowla, F. U., and J. S. Lim, Relationship between Maximum Likelihood Method and Autoregressive Modeling in Multidimensional Power Spectrum Estimation, *IEEE Trans. Acoust. Speech Signal Process.*, vol. ASSP-32, pp. 1083–1087, October 1984.

Dudgeon, D. E., and R. M. Mersereau, *Multidimensional Digital Signal Processing*, Prentice-Hall, Inc., Englewood Cliffs, N. J., 1984.

Frey, E. J., Two-Dimensional Spectral Estimation: Comparison of Current Techniques, M. S. thesis, University of Colorado, Boulder, Colo., April 1982.

Frost, O. L. III, High-Resolution 2-D Spectral Analysis at Low SNR, *Proceedings of the 1980 IEEE International Conference on Acoustics, Speech, and Signal Processing*, Denver, Colo., pp. 580–583, April 1980.

Huang, T. S., Two-Dimensional Windows, *IEEE Trans. Audio Electroacoust.*, vol. AU-20, pp. 88–90, March 1972.

Jackson, L. B., and H. C. Chien, Frequency and Bearing Estimation by Two-Dimensional Linear Prediction, *Proceedings of the 1979 IEEE International Conference on Acoustics, Speech, and Signal Processing*, Washington, D.C., pp. 665–668, April 1979.

Jain, A. K., Advances in Mathematical Models for Image Processing, *Proc. IEEE*, vol. 69, pp. 502–528, May 1981.

Jain, A. K., and S. Ranganath, High Resolution Spectrum Estimation in Two Dimensions, *Proceedings of the First ASSP Workshop on Spectrum Estimation*, Hamilton, Ontario, Canada, pp. 3.4.1–3.4.5, August 1981.

Joyce, L. S., A Separable 2-D Autoregressive Spectral Estimation Algorithm, *Proceedings of the 1979 IEEE International Conference on Acoustics, Speech, and Signal Processing*, Washington, D.C., pp. 677–680, April 1979.

Justice, J. H., A Levinson-Type Algorithm for Two-Dimensional Wiener Filtering Using Bivariate Szego Polynomials, *Proc. IEEE*, vol. 65, pp. 882–886, June 1977.

Kalouptsidis, N., G. Carayannis, and D. Manolakis, Fast Algorithms for Block Toeplitz Matrices with Toeplitz Entries, *Signal Process.*, vol. 6, pp. 77–81, January 1984.

Lang, S. W., and T. L. Marzetta, Image Spectral Estimation, Chapter 4 in *Digital Image Processing Techniques*, M. P. Ekstrom, ed., Academic Press, Inc., Orlando, Fla., 1984.

Lang, S. W., and J. H. McClellan, Multidimensional MEM Spectral Estimation, *IEEE Trans. Acoust. Speech Signal Process.*, vol. ASSP-30, pp. 880–887, December 1982.

Lang, S. W., and J. H. McClellan, Spectral Estimation for Sensor Arrays, *IEEE Trans. Acoust. Speech Signal Process.*, vol. ASSP-31, pp. 349–358, April 1983.

Lim, J. S., and N. A. Malik, A New Algorithm for Two-Dimensional Maximum Entropy Power Spectrum Estimation, *IEEE Trans. Acoust. Speech Signal Process.*, vol. ASSP-29, pp. 401–413, June 1981.

Malik, N. A., and J. S. Lim, Properties of Two-Dimensional Maximum Entropy Power Spectrum Estimates, *IEEE Trans. Acoust. Speech Signal Process.*, vol. ASSP-30, pp. 788–797, October 1982.

Marzetta, T. L., Two-Dimensional Linear Prediction: Autocorrelation Arrays, Minimum-Phase Prediction Error Filters, and Reflection Coefficient Arrays, *IEEE Trans. Acoust. Speech Signal Process.*, vol. ASSP-28, pp. 725–733, December 1980.

McClellan, J. H., Multidimensional Spectral Estimation, *Proc. IEEE*, vol. 70, pp. 1029–1039, September 1982.

Nikias, C. L., and M. R. Raghuveer, Multi-Dimensional Parametric Spectral Estimation, *Signal Process.*, vol. 9, pp. 191–205, October 1985.

Nikias, C. L., P. D. Scott, and J. H. Siegel, Computer Based Two-Dimensional Spectral Estimation for the Detection of Prearrhythmic States after Hypothermic Myocardial Preservation, *Comput. Biol. Med.*, vol. 14, pp. 159–178, 1984.

Nikias, C. L., and A. N. Venetsanopoulos, Sufficient Condition for Extendibility and Two-Dimensional Power Spectrum Estimation, *Proceedings of the 1985 IEEE International Conference on Acoustics, Speech, and Signal Processing*, Tampa, Fla., pp. 792–795, March 1985.

Pendrel, J. V., The Maximum Entropy Principle in Two-Dimensional Spectral Analysis, Ph.D. dissertation, York University, Ontario, Canada, November 1979.

Sharma, G., and R. Chellappa, A Model-Based Approach for Estimation of Two-Dimensional Maximum Entropy Power Sectra, *IEEE Trans. Inform. Theory*, vol. IT-31, pp. 90–99, January 1985.

Therrien, C. W., Relations between 2-D and Multichannel Linear Prediction, *IEEE Trans. Acoust. Speech Signal Process.*, vol. ASSP-29, pp. 454–456, June 1981.

Ulrych, T. J., and C. J. Walker, High Resolution Two-Dimensional Power Spectral Estimation, in *Applied Time Series II*, D. F. Findley, ed., Academic Press, Inc., New York, pp. 71–99, 1981.

Wax, M., and T. Kailath, Efficient Inversion of Toeplitz-Block Toeplitz Matrix, *IEEE Trans. Acoust. Speech Signal Process.*, vol. ASSP-31, pp. 1218–1221, October 1983.

Wernecke, S. J., Two-Dimensional Maximum Entropy Reconstruction of Radio Brightness, *Radio Sci.*, vol. 12, pp. 831–844, September–October 1977.

INDEX